Solid Mechanics and Its Applications

Volume 282

The fundamental questions arising in mechanics are: Why?, How?, and How much? The aim of this series is to provide lucid accounts written by authoritative researchers giving vision and insight in answering these questions on the subject of mechanics as it relates to solids. The scope of the series covers the entire spectrum of solid mechanics. Thus it includes the foundation of mechanics; variational formulations; computational mechanics; statics, kinematics and dynamics of rigid and elastic bodies; vibrations of solids and structures; dynamical systems and chaos; the theories of elasticity, plasticity and viscoelasticity; composite materials; rods, beams, shells and membranes; structural control and stability; soils, rocks and geomechanics; fracture; tribology; experimental mechanics; biomechanics and machine design. The median level of presentation is the first year graduate student. Some texts are monographs defining the current state of the field; others are accessible to final year undergraduates; but essentially the emphasis is on readability and clarity.

Springer and Professors Barber and Klarbring welcome book ideas from authors. Potential authors who wish to submit a book proposal should contact Dr. Mayra Castro, Senior Editor, Springer Heidelberg, Germany, email: mayra.castro@springer.com

Indexed by SCOPUS, Ei Compendex, EBSCO Discovery Service, OCLC, ProQuest Summon, Google Scholar and SpringerLink.

Sergey Kanaun

Self-Consistent Methods for Composites

Volume 1: Static Problems

Second Edition

Springer

Sergey Kanaun
Atizapán de Zaragoza, Mexico

ISSN 0925-0042 ISSN 2214-7764 (electronic)
Solid Mechanics and Its Applications
ISBN 978-3-031-96592-0 ISBN 978-3-031-96739-9 (eBook)
https://doi.org/10.1007/978-3-031-96739-9

This Springer imprint is published by the registered company Springer Nature Switzerland AG
The registered company address is: Gewerbestrasse 11, 6330 Cham, Switzerland

Preface to the Second Edition

From the time of publishing the book in 2008, the role of self-consistent methods in the theory of heterogeneous materials has been changed. Exponential growth of computer capacities and appearance of available efficient software make numerical methods to be an efficient tool for simulations of physical and mechanical properties of the materials with complex microstructures. The commercial software based on finite element and boundary element methods allow calculating effective property of the composites that consist of arbitrary complex components. Nevertheless, straightforward numerical simulations require exceedingly large volume of computational work. In many cases, e.g., for rock materials and geological structures, the precise values of physical properties are not known with sufficient accuracy, and numerical simulations must embrace a wide region of possible values of such parameters, not to speak about uncertainty in definition of statistical properties of microstructures of actual composites and geological materials. In many cases, reliable information on the heterogeneous martial behavior can be obtained by long-time and cumbersome computational work. Self-consistent methods simplify substantially the solution of the homogenization problems for heterogeneous media. These methods involve a small number of structural parameters but they allow assessing bounds for the effective properties that are narrower than the energy (Hashin-Strikman) bounds. This is why self-consistent methods, e.g., Mori-Tanaka method, are widely used for solution of the homogenization problems for conductive and elastic composites at present time.

Combination of self-consistent and numerical methods opens another possibility for efficient solution of the homogenization problems for composites with complex micro structures. By such a combination, the one-particle problems for structural elements of arbitrary shapes and properties can be solved numerically, and then, these solutions are used in the framework of self-consistent methods for accounting inclusion interactions. This strategy reduces substantially the volume of computational work in comparison with numerical homogenization on the basis of representative volume elements of the composites. Even in the case of ellipsoidal inclusions, when the solution of the one-particle problem is known with precision to several integrals, numerical calculation of these integrals combined with self-consistent methods result in fast and efficient solution of the homogenization problem for anisotropic elastic media containing random sets of anisotropic inclusions.

New Chaps. 2 and 3 of the first volume of the book are devoted to analysis of the general structure of electrostatic magneto-static and elastic fields in homogenous anisotropic media. Attention is focused on partition of these fields into divergence-free and rotor-free parts. The projectors that realize such a partition play the crucial role in integral presentations of electromagnetic and elastic fields in heterogeneous media. These presentations are an important step in application of self-consistent methods for solution of the homogenization problems.

In new Chap. 8, the one-particle problem for spherical radially transverse isotropic inclusions is considered in detail. For the multilayered inclusions, conventional numerical methods are not efficient, and the method proposed in this chapter provides a fast and accurate algorithm of solution for inclusions with an arbitrary number of the layers.

The conductivity and elasticity problems for multilayered radially transverse isotropic inclusions relate to the problem of neutral inclusions in homogeneous host media. Such inclusions do not disturb external fields applied to the medium, and therefore, they are "invisible" for external observers. This problem is considered in Chap. 9 of the new book edition.

New Chap. 12 is devoted to application of the effective medium method for solution of the homogenization problem for polycrystalline materials.

In the new edition, more attention is focused on the difference in predictions of various self-consistent methods, and reliability of the methods predictions for composites of specific microstructures.

Atizapán de Zaragoza, Mexico Sergey Kanaun
April 2025

Preface to the First Edition

The theory of heterogeneous materials has been intensively developed during the last few decades. The principal reason for the interest to this part of the mechanics and physics of solids is a wide region of application of heterogeneous materials in modern material engineering. Self-consistent methods form a well-known branch of the theory of heterogeneous materials. In most books on the mechanics and physics of heterogeneous media, the self-consistent methods are presented and discussed. But there are no books covering the entire spectrum of self-consistent methods in application to the calculation of static and dynamic effective properties of heterogeneous materials. This book has the aim of covering this gap.

In the book, engineers can find the equations for the effective properties of composites reinforced with various types of fillers. The principal advantage of self-consistent methods is that they provide relatively simple equations for the effective properties of composites. Such equations for static and dynamic effective properties of matrix composites reinforced with various types of inclusions are widely used in engineering practice. These solutions as well as new self-consistent solutions for various composite structures are presented in the book.

The book is written also for scholars who develop the theory of heterogenous media. In the book, they find the basic ideas and algorithms for the construction of self-consistent solutions and discussions of the ways for improving such solutions.

The book is divided into two volumes. The first volume is devoted to the static conductivity and elasticity problems of heterogeneous materials. Part I of the first volume is devoted to solution of one-particle problems, e.g., physical fields in a homogeneous medium containing an isolated inclusion. The inclusions that are important fillers for actual composite materials are considered: ellipsoidal inclusions, stiff flakes and ribbons, short and long stiff fibers, cracks and crack-like inclusions, multilayered spherical inclusions. Effective physical properties of composites containing random sets of isolated inclusions are considered in the second part of this volume. Various versions of self-consistent methods are applied to derive the expressions for effective elastic and conductive properties of the composites. Predictions of the methods are compared with experimental data and numerical solutions existing in the literature.

The regions of application of the methods are indicated, and the character of possible errors is discussed.

The second volume is devoted to the problems of wave propagation in heterogeneous materials. Self-consistent methods are used for calculation of phase velocities and attenuation factors of electromagnetic and elastic waves propagating in composites and polycrystals.

The two volumes are almost independent.

The history of self-consistent methods is long, and the number of works where these methods were used (explicitly or implicitly) for solution of various problems of the mechanics and physics of heterogeneous media is overwhelming. It is impossible to mention all the authors who made contributions to the subject. In the reference list, only the works that, in our opinion, contain either important new ideas or notable results are mentioned. It is natural that the book reflects the authors' vision of self-consistent methods and it is based mainly on the authors' works.

Atizapán de Zaragoza, Mexico Sergey Kanaun

Contents

Notations

$u_\alpha,\ \varepsilon_{\alpha\beta},\ C_{\alpha\beta\lambda\mu},\ ....$	Lower case Greek indices are tensorial
$\sigma_{\alpha\beta} = C_{\alpha\beta\gamma\delta}\varepsilon_{\gamma\delta}$	Summation with respect to repeating indices is implied
$T_{(\alpha\beta)kl} = \frac{1}{2}\left(T_{\alpha\beta\gamma\delta} + T_{\beta\alpha\gamma\delta}\right)$	Index symmetrization operation
$a,\ b,\ T$ or $\mathbf{a},\mathbf{b},\mathbf{T},...$	Non-index notations for tensors
$x,\mathbf{x}$	Point and vector of point in 3D(2D)-space
$\mathbf{a}\otimes\mathbf{b} \Rightarrow a_\alpha b_\beta$	Tensor product of vectors and tensors
$\mathbf{a}\cdot\mathbf{b} = a_\alpha b_\alpha$	Scalar product of vectors and tensors
$\mathbf{a}\times\mathbf{b} \Rightarrow \epsilon_{\alpha\beta\gamma}a_\beta b_\gamma$	Vector product of vectors and tensors
$\delta_{\alpha\beta}$	Kronecker symbol
$\epsilon_{\alpha\beta\gamma}$	Levi-Civita symbol
$E^1,\ E^2,\dots,\ E^6$	4 rank tensor E-basis (Appendix A.1)
$P^1,\ P^2,\dots,\ P^6$	4 rank tensor P-basis (Appendix A.2)
$T^1,\ T^2,\dots,\ T^{12}$	4 rank tensor T-basis (Appendix A.3)
$\nabla_\alpha = \frac{\partial}{\partial x_\alpha}$	Nabla (gradient) operator
$\mathrm{def}\mathbf{u} \Rightarrow \nabla_{(\alpha}u_{\beta)}$	Deformation operator
$\mathrm{div}\mathbf{T} \Rightarrow \nabla_\alpha T_{\alpha\beta\gamma...}$	Divergence of tensor field T
$\mathrm{rot}\mathbf{T} = \nabla\times\mathbf{T} \Rightarrow$ $\Rightarrow \epsilon_{\alpha\beta\gamma}\nabla_\beta T_{\gamma\delta...}$	Rotor of a tensor field T
$\nabla = \frac{\partial^2}{\partial x_1^2} + \frac{\partial^2}{\partial x_2^2} + \frac{\partial^2}{\partial x_3^2}$	Laplace operator
$\widetilde{f}(k) =$ $= \int f(x)\exp(i\mathbf{k}\cdot\mathbf{x})dx$	Fourier transform of function $f(x)$
$k,\ \mathbf{k}$	Point and vector of a point in Fourier transform space
$pv\int ...dx =$ $= \lim_{\delta\to 0}\int\limits_{\vert x\vert>\delta}\dots dx$	Cauchy principal value of the integral
$\delta(x)$	Dirac's delta-function
Ω	Surface in 3D-space
$\Omega(x)$	Delta-function concentrated on Ω
V	Region in 3D(2D)-space

$V(x) = 1 \text{ if } x \in V$ $V(x) = 0 \text{ if } x \notin V$	Characteristic function of region V
$\langle f(x) \rangle$	Mean value of a random function $f(x)$
$\langle f(x) \vert x \rangle$	Mean value of $f(x)$ under the condition that $x \in V$
$C^0, C,$ $C^1 = C - C^0$	Property tensors of the host medium and the inclusions
C^*	Proprty tensor of the effective medium
$\lambda_0, \mu_0;\ \lambda, \mu;\ \kappa_0 = \frac{\lambda_0+\mu_0}{\lambda_0+2_0}$ λ_*, μ_*	Lame parameters of the host medium, inclusions, and the effective medium
K_0, K, K_*	Bulk moduli of the matrix, inclusions, and the effective medium
E_0, E, E_*	The Young moduli of the matrix, inclusions, and the effective medium
ν_0, ν, ν_*	The Poisson ratio of the matrix, inclusions and the effective medium
$K_1 = K - K_0$ $\mu_1 = \mu - \mu_0$ $\lambda_1 = \lambda - \lambda_0$	Deviation of the elastic coefficients in inclusions

In the book, the following index notation concept is accepted: index notations are used in the initial equations only. If the index notations can be easily restored from the previous equations, the indexes are omitted.

Chapter 1
Introduction

Abstract The homogenization problem for heterogeneous materials is formulated. The place of self-consistent methods among other methods of solution of the homogenization problems is indicated. Two principal self-consistent methods are pointed out. The stricture and the contests of the book are discussed.

Keywords Heterogeneous materials · Homogenization problem · Self-consistent methods · Effective medium method · Effective field method

The intensive development of the theory of heterogeneous media in the past decades is related to the constantly increasing area of application of composite materials in modern industry. Nowadays, composites successfully compete with traditional homogeneous constructive materials: metals, homopolymers, alloys, ceramics, etc. Modern nanotechnologies open the possibilities for dictating the microstructure of composite materials and creating materials with prescribed physical properties. The direct synthesis of composite materials requires comprehensive knowledge of the influence of details of their microstructure on macroproperties.

In this book, important classes of heterogeneous materials—the so-called matrix composites and polycrystals—are considered. The microstructure of matrix composites consists of a homogeneous matrix phase and a set of isolated inclusions homogeneously distributed in the matrix. In applications, there are composites with polymer, metal, ceramic, and other types of matrices. Particles of quartz, sand, glass balls, caoutchouc globules, thin metallic flakes and ribbons, glass fibers, hard carbide whiskers, etc., are used as fillers in matrix composites. Traditional reinforcement elements are long glass and carbon fibers. Other important classes of heterogeneous materials that are within the scope of this book are polycrystals, porous media, and homogeneous media with cracks and crack-like defects.

As a rule, the microstructure of the composite materials is stochastic, and the properties and sizes of the reinforcing particles are random as well as their spatial distributions. As a result, physical fields inside the composites are also random. An important problem of the theory of composites is determination of the mean values of these fields by non-random external loadings. This is the content of the so-called homogenization problem. The solution of this problem allows us to calculate effective properties of composite materials and, for strain-stress analysis of composite

S. Kanaun, *Self-Consistent Methods for Composites*, Solid Mechanics and Its Applications 282, https://doi.org/10.1007/978-3-031-96739-9_1

constructions, to change the actual composite for a homogeneous material with the same reaction on the external loading. Information about higher statistical moments of physical fields in the composites (multipoint correlation functions) is necessary for the description of structurally sensitive processes, like dispersed damage and plastic flow.

Many important problems of the mechanics and physics of heterogeneous media are reduced to the solution of the linear operator equation

$$u - Kc_1 u = u_0. \tag{1.1}$$

Here, $u(x)$ is the physical field in the medium, $c_1(x)$ is a known random function that describes the medium physical properties, $u_0(x)$ is a known non-random external field applied to the medium, x is point of the medium. In this equation, K is a non-random operator that can be presented as an integral operator

$$(Kf)(x) = \int K(x - x')f(x')dx', \tag{1.2}$$

where the kernel $K(x)$ is a known generalized function. The solution $u(x)$ of Eq. (1.1) is a random function statistically dependent on $c_1(x)$. Presence of the product $c_1 u$ in Eq. (1.1) and dependence of u on c_1 make this problem statistically non-linear.

If the function $c_1(x)$ is small, the solution of Eq. (1.1) can be presented in the form of Neumann's series

$$u = u_0 + Kc_1 u_0 + Kc_1 Kc_1 u_0 + ... = \sum_{n=0}^{\infty} (Kc_1)^n u_0. \tag{1.3}$$

The mean value $\langle u(x) \rangle$ of the field $u(x)$ can be obtained by averaging the both parts of this equation over the ensemble realizations of the random function $c_1(x)$. As the result, the mean field $\langle u(x) \rangle$ is the following sum

$$\langle u \rangle = u_0 + K \langle c_1 \rangle u_0 + K \langle c_1 K c_1 \rangle u_0 + K \langle c_1 K c_1 K c_1 \rangle u_0.... \tag{1.4}$$

The right-hand side of this equation contains an infinite number of multipoint correlation functions of the random function $c_1(x)$ that describes statistical properties of the heterogeneous medium

$$\langle c_1(x) \rangle, \quad \langle c_1(x) c_1(x_1) \rangle, \quad \langle c_1(x) c_1(x_1) c_1(x_2) \rangle, ... \tag{1.5}$$

Construction of the multipoint correlations functions of the property tensors for actual heterogeneous materials is not a trivial task that involves cumbersome numerical calculations.

Most actual composite materials consist of contrasting components, and the corresponding function $c_1(x)$ in Eq. (1.1) is not small. Therefore, the number of the terms in the series (1.4) should be sufficiently large in order to obtain a reliable estimation of the average field in the random composite material. Methods of approximate summations of series (1.4) were discussed in a number of works [1–3]. By some assumptions about the statistical properties of the random function $c_1(x)$ that simplify the structure of the correlation functions (1.5) (perfectly disordered materials), the series in Eq. (1.4) can be summed, and finite equations for the mean $\langle u(x)\rangle$ and therefore, for the effective properties of the composites can be obtained. The materials with such properties are called unstructured blends, but the region of application of this approximation remains unclear.

Another branch of the theory of composite materials was opened by the well-known work of Hashin and Strikman [4], where variational principles were used for the calculation of bounds for the effective properties of composites. This branch was extensively developed during the past few decades, and its results are presented, e.g., in the books of Milton [5], Torquato [6], and series of surveys mentioned in these books. For composites with contrast properties of the components, the variational bounds turn out to be rather wide. But the variational formulation is a solid basis for the construction of approximate solutions. Such solutions can be obtained by an appropriate choice of the so-called trial fields that approximate the actual fields in composites with precision to a finite number of constants. For composites with ellipsoidal inclusions, the simplest trial field can be taken as constant inside inclusions. Starting with the works of Walpole [7, 8], and then in a series of works by Willis and coauthors [9–13], this approach was successfully used for the solution of the homogenization problem in statics and dynamics (long wave propagation) of the composites.

A specific object of studies was composites with periodic distributions of the components. The exact asymptotic theory of homogenization of periodic composites was developed in the 1970s and 1980s, and the results of this theory are presented in the books by Sanches-Palensia [14], Bakhvalov and Panasenko [15], Pobedrya [16], Zhikov, Koslov, and Oleinik [17]. In the framework of this theory, the homogenization problem is reduced to a problem for an elementary cell of the composite, and numerical or analytical solutions of this problem allow calculating the composite effective properties. In the 2D-case, many exact results for the effective elastic properties of composites with periodic microstructures were presented in the book by Grigoliuk and Filshtinsky [18] before the appearance of the asymptotic theory of homogenization.

Extensive growth of computer capacities during the last decades has opened the possibility of direct numerical simulation of the properties of composite materials. The finite element and boundary element numerical methods become efficient tools for solution of the homogenization problems for linear and non-linear composite materials, [19]. Application of Fast Fourier transform algorithms for numerical solution of the integral equations of heterogeneous media accelerates substantially solution of the homogenization problems [20, 21]. But numerical simulations require powerful computers and large volume of calculations. Such simulations become extremely difficult in the case of wave propagations when the length of the propagating wave is comparable to or shorter than the sizes of the inclusions.

Self-consistent methods allow reducing the difficulties in solution of the homogenization problems. These methods have been used for solution of the multiparticle problems in classical and quantum physics for decades. For random structures, the homogenization problems have no exact solutions, and the principal idea of self-consistent methods is to reduce the multiparticle problems to the problem for one-particle. For this purpose, simplifying hypotheses on the field structures in random ensembles of interacting particles are accepted. Self-consistent methods are efficient if the field that acts on each particle in the ensemble does not depend strongly on the specific configuration of the particle set, but it is a joint field of all the interacting particles. In the pioneering works of Clausius, Mossoti, Lorenz and Lorentz, Maxwell, and Rayleigh in the nineteenth century, self-consistent methods were used for calculation of the effective dielectric permittivity of composites with spherical particles. The obtained equations were simple, and they satisfactorily described experimental data in quasistatic (long wave) regions. (The historical survey of these works can be found in [22]). Later on, in the quantum theory of the nucleus (Hartree-Fock method), the theory of phase-transition (Weiss method, Landau theory) self-consistent approaches were successfully used to obtain the approximate solutions that turned out to be in good agreement with experimental data. A survey of the application of self-consistent methods to the analysis of the physical properties of disordered crystalline solids is presented in [23].

The first self-consistent solutions in the mechanics of heterogeneous media were obtained in the middle of the last century in the works of Hershey [24], Kerner [25], Kröner [26], Budiansky [27], Hill [28]. In these works, self-consistent methods were used for calculation of effective elastic properties of polycrystals and matrix composite materials. Nowadays, numerous modifications of self-consistent methods are used to simulate static and dynamic behavior of heterogeneous media. The self-consistent methods can be divided in two groups: the effective field and effective medium methods. These two methods are based on different hypotheses and results in different outcomes when applied to the same composite materials. The methods are based on two types of hypotheses: the first one reduces the multiparticle problem to the problem for one-particle, and the second hypothesis is the condition of self-consistency. It is necessary to emphasize that the hypotheses of the methods are introduced in a heuristic manner, and the region of their validity can hardly be evaluated ad hoc. Only the comparisons of the predictions of the methods with experimental data and exact solutions (when the latter exist) allow us to estimate the reliable regions of their applications.

Let a composite material consists of a homogeneous host medium (matrix) and a random set of spherical inclusions with volume fraction p. The dielectric permittivity of the matrix phase is c_0, and c that for the inclusions. The effective medium method (EMM) is based on two hypotheses. The first one is that each inclusion can be considered as an isolated one on the medium with the effective properties of the composite. The second hypothesis is that the field that acts on each inclusion is the average field over the heterogeneous medium (the condition of self-consistency). These hypotheses result in the equation for the effective dielectric permittivity c_* of the composite in the form

$$c_* = c_0 + p\frac{3(c - c_0)c_*}{2c_* + c}. \tag{1.6}$$

This algebraic equation for the effective dielectric permittivity c_* known as Bruggeman's formula, was obtained in 1935 in [29]. Its analogies for conductive, dielectric, and elastic media with ellipsoidal inclusion and for polycrystals with ellipsoidal grains are presented in volume I (for statics) and volume II (for dynamics) of this book.

The effective field method (EFM) is based on two hypotheses that are different from the hypotheses of the EMM. The first hypothesis is that each inclusion in the composite can be considered as an isolated one in the host homogeneous medium with known properties. The local exciting field that acts on this inclusion is the sum of the external field applied to the medium and the field perturbations due to surrounding inclusions. In the simplest version of the method (the quasicrystalline approximation) this field (the effective field) is assumed to be constant and the same for all the inclusions. This is the second hypothesis of the method (the condition of self-consistency). This method results in another equation for the effective dielectric permittivity of the composite with spherical inclusions

$$c_* = c_0 + \frac{3p(c - c_0)c_0}{3c_0 + (1 - p)(c - c_0)}. \tag{1.7}$$

This is the so-called Maxwell-Garnet equation that was derived in 1904 in [30]. The difference of the EMM and the EFM is clear seen from Eqs. (1.6), (1.7). The comparison of these solutions, and the regions of their applications are discussed in this book.

For application of self-consistent methods, the heterogeneous media should have a specific structure: a typical particle of the structure should be identified, and the composite is a homogeneous in space random ensemble of such particles. The particles can be inclusions in the matrix composites, pores in a porous medium, cracks in damaged solids, crystal grain in the polycrystalline aggregate, etc. The one-particle problem is the determination of the physical field in an infinite homogeneous medium with an isolated particle subjected to external fields of a simple structure: homogeneous fields for static problems or plane harmonic waves for stationary dynamic problems. Because of an important role of this problem for the self-consistent methods, Part I of volume I of this book is devoted to the one-particle problems for inclusions of various types.

In Chaps. 2 and 3 of Part I, general structures of electrostatic, magnetostatic and elastic fields in homogeneous anisotropic media are considered. It is shown that the projection operators on the divergence-free and rotor-free functional subspaces play a crucial role for integral presentation of the fields in homogeneous media subjected to external and internal field sources. Similar projection operators arise by integral presentations of the fields in heterogeneous media by external loading.

The integral equations for electric and elastic fields in homogeneous media with an isolated inclusion are considered in Chap. 4. The attention is focused on ellipsoidal inclusions when the one-particle problem is reduced to calculation of a finite number of 2D-regular integrals on account of a remarkable polynomial conservation property: a polynomial external field induces a polynomial field of the same degree inside an ellipsoidal inclusion. Solutions of the one-particle problem for a homogeneous ellipsoidal inclusion and its particular forms: oblate and prolate spheroids, elliptical cylindrical inclusions and elliptical disks are presented in this chapter.

The integral equations of the crack problems of elasticity and conductivity are considered in Chap. 5. It is shown that the one-particle problem for an elliptical crack in an anisotropic medium is reduced to calculation of 2D improper integrals.

In Chap. 6, thin inclusions which materials are much stiffer or much softer that the material of the host medium are considered. The one-particle problems for such inclusions are reduced to solution of 2D-integral equations on the middle surfaces of the inclusions. Explicit solutions of these equations for thin ellipsoidal inclusions are presented in this chapter.

A stiff axisymmetric fiber in a homogeneous host medium is considered in Chap. 7. The one-particle problem of elasticity for the fiber is reduced to 1D-integral equation on the fiber axis. It is shown that the principal term of the stress field in the fiber satisfies the ordinary differential equation of the second order. Explicit solutions of these equations are found for cylindrical, ellipsoidal, and double-conical fiber forms.

Chapter 8 is devoted to the one-particle problem for spherical inclusions with radially transverse isotropic elastic stiffness tensors. Mellin transform method is used for solution of the problem. The method allows revealing the tensor structure of the solution with precision to three scalar functions of radial coordinate. Ordinary differential equations for these functions arise in the process of realization of the method. For multilayered inclusions with constant elastic properties inside layers, an efficient numerical algorithm of solution is proposed.

In Chap. 9, the neutral inclusion problem for a homogeneous medium subjected to constant external fields is considered. A neutral inclusion does not disturb external fields applied to the medium, and therefore, such an inclusion is "invisible" for external observers. An isotropic homogeneous inclusion cannot be neutral for arbitrary external fields, but a radially transverse isotropic inclusion can be neutral by a specific choice of elastic (conductive) coefficients of the inclusion material. The conductivity and elasticity coefficients of a neutral radially transverse isotropic inclusion in the given isotropic medium are revealed. Neutral inclusions with isotropic cores and radially transverse isotropic external layers are also considered in this chapter.

Part II of volume I is devoted to self-consistent methods in application to the solution of the homogenization problem and calculation of the effective static properties of heterogeneous materials. In Chap. 10, the homogenization problem is formulated, and specific features of this problem in the cases of matrix composite materials and polycrystals are considered.

The effective medium method (EMM) is applied to solution of the homogenization problem for matrix composite materials in Chap. 11. Various versions of the method are discussed, and the predictions of the methods for the effective elastic stiffness and conductivity tensors of particulate composites are compared with experimental data and numerical simulations.

In Chap. 12, the EMM is used for calculation of the effective elastic and thermoelastic properties of polycrystalline materials. The case of the polycrystals with spheroidal transverse isotropic grains is considered in detail.

Chapters 13–16 are devoted to the effective field method (EFM). In this method, the principal unknown is the local exciting field $\sigma^*(x)$ that acts on each inclusion in the composite. In the simplest version of the method (the quasicrystalline approximation), the field σ^* is assumed to be constant and the same for all the inclusions. This version results in explicit equations for the effective conductive and elastic stiffness tensors of matrix composites with ellipsoidal inclusions, thin soft and stiff inclusions, stiff short and long fibers. These equations contain a specific two-point correlation function of the random set of inclusions, and details of the spatial distributions of the inclusions affect the form of this function. The method is applied to calculation of the effective elastic and conductive properties of composites with random and periodic sets of inclusions.

In Chap. 17, the composites with several different families of inclusions in a homogeneous host medium (hybrid composites) are considered. In this case, the quasicrystalline approximation results in the effective conductivity and elastic stiffness tensors that are not symmetric with respect to permutation of indices. This defect of the EFM can be corrected if the effective field σ^* is assumed to be different for the inclusions of different families. Examples of application of this version of the EFM for the composites containing spherical inclusion and circular cracks, cylindrical fibers, and spheroidal inclusions are presented in this chapter.

Chapter 18 is devoted to an advanced version of the EFM. In the quasicrystalline approximation, the actual local exiting field $\sigma^*(x)$ acting on the inclusions in the composite is changed with a constant effective field σ^* that is the mean $\langle\sigma^*(x)|x\rangle$ of $\sigma^*(x)$ by the condition that point x belongs to the region of inclusions. In the new version of the EFM, the mean $\langle\sigma^*(x_1)|x_1, x_2\rangle$ under the condition that points x_1 and x_2 are in different inclusions is introduced. The approximate equation for this mean is obtained, and the effective property tensor is expressed in term of the two-point mean $\langle\sigma^*(x_1)|x_1, x_2\rangle$. Predictions of this version of the EFM correspond better to experimental data and numerical simulations in the region of large volume fractions of inclusions.

For rapidly changing (in space and time) external fields, the effective properties of the composites turns out to be non-local, and the constitutive relations for averaged fields contain integral operators. It is shown in Chap. 19 that self-consistent methods allow us to estimate the non-local effects in composites and derive approximate expressions for the kernels of the integral operators in the constitutive equations for the averaged fields.

References

1. Beran M (1968) Statistical continuum theories. Wiley-Interscience, New York, p 424
2. Kröner E (1971) Statistical continuum mechanics. Springer, Udine, p 157

3. Shermergor T (1977) The theory of elasticity of microheterogeneous media. Nauka, Moskow, p 399
4. Hashin Z, Strikman S (1963) A variational approach to the theory of the elastic behavior of multiphase materials. J Mech Phys Solids 11:127–140
5. Milton G (2002) The theory of composites. Cambridge University Press, Oxford, p 780
6. Torquato S (2002) Random heterogeneous materials: microstructure and macroscopic properties. Springer, New York, p 728
7. Walpole L (1966) On bounds for the overall elastic modules of inhomogeneous systems. I, II J Mech Phys Solids 14:151–301
8. Walpole L (1969) On the overall elastic modules of composite materials. J Mech Phys Solids 17:235–251
9. Willis J (1980) A polarization approach to the scattering of elastic waves. I. Scattering by a single inclusion, II. Multiple scattering from inclusions. J Mech Phys Solids 28:287–327
10. Talbot D, Willis J (1982) Variational estimates for dispersion and attenuation of waves in random composites. I General theory. Int J Solids Struct 18:673–683
11. Talbot D, Willis J (1982) Variational estimates for dispersion and attenuation of waves in random composites. II Isotropic composites. Int J Solids Struct 18:685–698
12. Willis J (1983) Variational characterization of waves in fiber-reinforced materials. In: Hashin Z, Herakovich C (eds) Mechanics of composite materials: recent advances. Pergamon, Oxford, pp 191–205
13. Talbot D, Willis J (1983) Variational estimates for dispersion and attenuation of waves in random composites-III, Fiber-reinforced materials. Int J Solids Struct 19:793–811
14. Sanches-Palencia E (1984) Heterogeneous media and oscillation theory. Mir, Moscow, p 472
15. Bakhvalov N, Panasenko G (1984) Homogenization of periodical media. Nauka, Moscow, p 352
16. Pobedrya B (1984) Mechanics of composite materials. Moscow University, Moscow, p 336
17. Zhikov V, Koslov S, Oleinik O (1993) Homogenization of differential operators. Nauka, Moscow, p 398
18. Grigoliuk E, Filshtinsky L (1970) Perforated plates and shells. Nauka, Moscow, p 556
19. Segurado J, Llorca J (2002) A numerical approximation to the elastic properties of sphere-reinforced composites. J Mech Phys Solids 50:2107–2121
20. Moulinec H, Suquet P (1994) A fast numerical method for computing the linear and nonlinear mechanical properties of composites. Compt Rend A Paris 318:1417–1423
21. Kanaun S (2021) An ellipsoidal inclusion and an elliptical crack in the elastic media with ellipsoidal anisotropy. Int J Fract 227:133–136
22. Landauer R (1978) Electrical conductivity in inhomogeneous media. In: Proceedings of conference on electrical transport and optical properties of inhomogeneous media (ETOPIM 1), vol. 2 (AIP Conf Proc), pp 2–46
23. Elliott R, Krumhansl J, Leath P (1974) The theory and properties of randomly disordered crystals and related physical system. Rev Mod Phys 46:465–543
24. Hershey A (1954) The elasticity of an isotropic aggregate of anisotropic cubic crystals. J Appl Mech 21:236–240
25. Kerner E (1956) The elastic and thermoelastic properties of composite media. Proc Phys Soc 69(B):808–813
26. Kröner E (1958) Berechnung der elastischen Konstanten des Vielkristalls aus den Konstanten des Einkristalls. Z Phys 151:504–518
27. Budiansky B (1965) On the elastic moduli of some heterogeneous materials. J Mech Phys Solids 13:223–234
28. Hill R (1965) Self-consistent mechanics of composite materials. J Mech Phys Solids 13:213–222
29. Bruggeman D (1935) Berechnung verschiedener physikalischer Konstanten von heterogenen Substanzen. Ann Phys 24:636–679
30. Garnet J (1904) Colors in metal glasses and metallic films. Phil Trans Royal Soc Lond 203:385–420

Chapter 2
Electro- and Magnetostatic Fields in Homogenous Anisotropic Media

Abstract Electrostatic and magnetostatic fields in homogeneous anisotropic media are considered. Using Green's function of the system of the governing equations for these fields, the electrostatic fields caused by the electric charge distributions are presented in the integral forms. A similar presentation is obtained for the magnetostatic fields generated by distributions of stationary currents. Projection operators for partition of arbitrary vector fields in diverge-free and rotor-free parts are introduced.

Keywords Electrostatic fields · Magnetostatic fields · Anisotropic medium · Green's function · Divergence-free and rotor-free subspaces · Projection operators

2.1 Governing Equations for the Electrostatic Fields in Homogeneous Anisotropic Media

We consider a homogeneous anisotropic dielectric medium with the permittivity tensor C that occupies the entire 3D-space. Second-rank tensor C is symmetric and definitely positive. For linear dielectric media, this tensor relates vectors of electric field $e_\alpha(x)$ and electric displacement $d_\alpha(x)$ by the equation [1]

$$d_\alpha(x) = C_{\alpha\beta} e_\beta(x). \tag{2.1}$$

Here, $x = x(x_1, x_2, x_3)$ is point of the medium, x can be interpreted also as a vector of point x with Cartesian coordinates x_1, x_2, x_3. The components of tensors are considered in Cartesian bases, summation with respect to repeating Greek indices is implied.

The sources of electric fields in dielectric media can be electric charges and conditions at infinity (boundary conditions). Further, it is assumed that electric charges of volume density $q(x)$ are distributed in a finite region of 3D-space, and the field $e(x)$ vanishes at infinity ("zero" boundary conditions at infinity). The vector fields $d(x)$ and $e(x)$ satisfy the system of differential equations [1]

$$\nabla_\alpha d_\alpha(x) = q(x), \; \epsilon_{\alpha\gamma\beta} \nabla_\gamma e_\beta(x) = 0, \; \nabla_\alpha = \partial/\partial x_\alpha. \tag{2.2}$$

S. Kanaun, *Self-Consistent Methods for Composites*, Solid Mechanics and Its Applications 282, https://doi.org/10.1007/978-3-031-96739-9_2

Here, $\epsilon_{\alpha\beta\gamma}$ is the antisymmetric Levi-Civita tensor. The system of governing equations (2.1), (2.2) can be presented in the symbolic form

$$\mathrm{div} d(x) = q(x), d(x) = Ce(x), \mathrm{rot} e(x) = 0. \tag{2.3}$$

Here, div is divergence operator ($\mathrm{div} d = \nabla_\alpha d_\alpha$), and rot is the rotor (curl) operator ($\mathrm{rot}_{\alpha\beta} = \epsilon_{\alpha\gamma\beta}\nabla_\gamma$). It follows from the third equation of the system (2.3) that the field $e(x)$ is irrotational (rotor-free). It is known that each irrotational field can be presented as the gradient of a scalar fields $\varphi(x)$ that is called the potential of the electric field (electric potential)

$$e(x) = -\nabla\varphi(x). \tag{2.4}$$

The equation for $\varphi(x)$ follows from the system (2.3)

$$\nabla_\alpha C_{\alpha\beta}\nabla_\beta\varphi(x) = -q(x). \tag{2.5}$$

Green's function $g(x)$ of the differential operator in the left-hand side of Eq. (2.5) is a vanishing at infinity solution of the equation

$$\nabla C\nabla g(x) = -\delta(x). \tag{2.6}$$

Here, $\delta(x)$ is 3D-Dirac's delta-function. If $g(x)$ is known, the solution of Eq. (2.5) can be presented in the integral form as follows

$$\varphi(x) = \int g(x - x')q(x')dx'. \tag{2.7}$$

To prove this equation we apply the operator $\nabla C\nabla$ to its both parts and using the properties of the delta-function, we obtain

$$\nabla C\nabla\varphi(x) = \int \nabla C\nabla g(x - x')q(x')dx' =$$
$$= -\int \delta(x - x')q(x')dx' = -q(x). \tag{2.8}$$

2.2 Green's Function $g(x)$ for an Anisotropic Medium

First, we consider an isotropic medium when the permittivity tensor

$$C_{\alpha\beta} = \delta_{\alpha\beta}, \tag{2.9}$$

where $\delta_{\alpha\beta}$ is Kronecker's symbol. In this case, Eq. (2.6) for $g(x)$ takes the form

$$\Delta g(x) = -\delta(x), \tag{2.10}$$

where $\Delta = \nabla_\alpha \nabla_\alpha$ is Laplace operator. Application of Fourier transform to this equation results in the equation for Fourier transform $\widetilde{g}(k)$ of Green's function

$$k^2 \widetilde{g}(k) = 1 \text{ or } \widetilde{g}(k) = \frac{1}{k^2},\ k^2 = k_\alpha k_\alpha, \tag{2.11}$$

$$\widetilde{g}(k) = \int g(x) \exp(\mathrm{i}k \cdot x) dx,\ k \cdot x = k_\alpha x_\alpha,\ \mathrm{i} = \sqrt{-1}. \tag{2.12}$$

Here, integration is spread to entire 3D-space. Inverse Fourier transform of $\widetilde{g}(k)$ (Appendix B.2) yields the x-presentation of the function $g(x)$

$$g(x) = \frac{1}{(2\pi)^3} \int \widetilde{g}(k) \exp(-\mathrm{i}k \cdot x) dk = \frac{1}{4\pi r}, \tag{2.13}$$

$$r = |x| = \sqrt{x_\alpha x_\alpha}. \tag{2.14}$$

For solution of Eq. (2.6) in the case of an anisotropic medium, we introduce new coordinate system (x_1', x_2', x_3')

$$x_\alpha' = A_{\alpha\beta} x_\beta,\ x_\alpha = A^{-1}_{\alpha\beta} x_\beta', \tag{2.15}$$

where $A_{\alpha\beta}$ is a constant nondegenerate tensor. In the new variables, Eq. (2.6) for $g(x)$ takes the form

$$\partial_\alpha' A_{\alpha\beta} C_{\beta\gamma} A_{\gamma\eta} \partial_\eta' g(x') = -\delta(A^{-1}x'). \tag{2.16}$$

If the tensor A is such as

$$A_{\alpha\eta} C_{\eta\gamma} A_{\gamma\beta} = \delta_{\alpha\beta} \text{ or } C^{-1}_{\alpha\beta} = A_{\alpha\gamma} A_{\gamma\beta} = A^2_{\alpha\beta}. \tag{2.17}$$

Equation (2.6) in the new variables takes the form

$$\partial_\alpha' \partial_\alpha' g(x') = -(\det A)\, \delta(x'),\quad \det A = \frac{1}{\sqrt{\det C}}. \tag{2.18}$$

Here, $\det A$ is the determinant of a matrix $A_{\alpha\beta}$. Equation (2.18) coincides with the equation for Green's function in Eq. (2.13) in the case of an isotropic medium. Thus from Eqs. (2.10) and (2.13), we yield

$$g(x') = \frac{1}{4\pi \overline{r}(x')},\quad \overline{r}(x') = \sqrt{(\det C)\, x_\alpha' x_\alpha'}. \tag{2.19}$$

Returning to the original coordinate system, we obtain Green's function of the operator $\nabla C \nabla$ for a homogenous anisotropic medium

$$g(x) = \frac{1}{4\pi \bar{r}(x)}, \quad \bar{r}(x) = \sqrt{(\det C)\,(x_\alpha C_{\alpha\beta} x_\beta)}. \tag{2.20}$$

Using this function $g(x)$ and Eq. (2.7), the electric field $e(x)$ and the electric displacement $d(x)$ in the medium with charge density $q(x)$ are presented in the forms

$$e_\alpha(x) = -\int \nabla_\alpha g(x - x')q(x')dx', \tag{2.21}$$

$$d_\alpha(x) = -C_{\alpha\beta} \int \nabla_\beta g(x - x')q(x')dx'. \tag{2.22}$$

These are the integral presentations of the vanishing at infinity solutions of the system (2.3) for electrostatic field and electric displacement in a homogeneous anisotropic medium.

2.3 Magnetostatic Fields in Homogenous Anisotropic Media

A steady magnetic field in homogeneous anisotropic media is described by the magnetic flux vector $b(x)$ and the magnetic field vector $h(x)$. For linear homogeneous magnetic media, these vectors relate by the equation [1]

$$b_\alpha(x) = \mu_{\alpha\beta} h_\beta(x), \tag{2.23}$$

where μ is the tensor of magnetic permeability. It is known that in contrast with the electric charges, magnetic charges do not exist in the nature. In absence of external magnetic fields, magnetic fields in the medium can be caused by a steady electric current $j(x)$ only. The system of equations of magnetostatics for vectors b and h has the form [1]

$$\mathrm{div} b(x) = 0, b(x) = \mu h(x), \ \mathrm{rot} h(x) = j(x). \tag{2.24}$$

It follows from the third equation of this system that $\mathrm{div} j = 0$, and therefore, j is a divergence-free vector field. Such a field can be presented as rotor of another vector field $m(x)$

$$j_\alpha(x) = -\mathrm{rot}_{\alpha\beta} m_\beta(x), \tag{2.25}$$

and $m(x)$ is called the magnetization vector.

Equations (2.3) and (2.24) show that determination of electrostatic and magnetostatic fields in homogeneous anisotropic media is associated with two mathematical problems.

Problem 1. Find a vanishing at infinity vector field $v(x)$ that satisfies the equations

$$\mathrm{div}v = q,\ \mathrm{rot}C^{-1}v = 0. \tag{2.26}$$

Problem 2. Find a vanishing at infinity vector field $v(x)$ that satisfies the equations

$$\mathrm{div}v = 0,\ \mathrm{rot}C^{-1}v = j. \tag{2.27}$$

In these equations, the field sources q and j can be considered as generalized functions with finite supports, C is a second rank definitely positive tensor. In the previous sections, the solution of Problem 1 was presented in the integral form (2.22). Our next aim is derivation of a similar integral presentation of the solution of Problem 2.

First, we consider the case of an isotropic medium with the property tensor $C_{\alpha\beta} = \delta_{\alpha\beta}$. For such a tensor C, the systems (2.26) and (2.27) take forms

$$\mathrm{div}v = q,\ \mathrm{rot}v = 0; \tag{2.28}$$

$$\mathrm{div}v = 0,\ \mathrm{rot}v = j. \tag{2.29}$$

Then, we consider the well-known identity of vector calculus for an arbitrary vector field v

$$\Delta v = \nabla \mathrm{div}v - \mathrm{rot}\left(\mathrm{rot}v\right). \tag{2.30}$$

Application of the integral operator which the kernel $g(x) = 1/(4\pi|x|)$ (Green's function of Laplace operator Δ) to this equation yields

$$\int g(x-x')\Delta' v(x')dx' = \int \Delta g(x-x')v(x')dx' =$$
$$= -\int \delta(x-x')v(x')dx' = -v(x). \tag{2.31}$$

Here, Gauss' theorem (Appendix C.1) and the condition of vanishing $v(x)$ at infinity are used. Therefore, vector field $v(x)$ can be presented in the following integral form

$$v(x) = \int g(x-x')\mathrm{rot}'\mathrm{rot}'v(x')dx' - \int g(x-x')\nabla'\mathrm{div}'v(x')dx' =$$
$$= \int \mathrm{rot}g(x-x')\mathrm{rot}'v(x')dx' - \int \nabla g(x-x')\mathrm{div}'v(x')dx'. \tag{2.32}$$

It follows from this equation that the solution of Problem 1 ($\mathrm{rot}v = 0$, $\mathrm{div}v = q$) is presented in the integral form similar to (2.21)

$$v(x) = -\int \nabla g(x - x')q(x')dx', \tag{2.33}$$

and the integral presentation of the solution of Problem 2 ($\text{div}v = 0, \text{rot}v = j$) is

$$v(x) = \int \text{rot}g(x - x')j(x')dx'. \tag{2.34}$$

For deriving similar results in the case of anisotropic media, we use the analogy of identity (2.30) in the anisotropic case. First, we consider the well-known formula that relates Levi-Civita and Kronecker symbols

$$\epsilon_{\lambda\alpha\gamma}\epsilon_{\gamma\nu\mu} = \delta_{\alpha\mu}\delta_{\lambda\nu} - \delta_{\alpha\nu}\delta_{\lambda\mu}. \tag{2.35}$$

Then, after convolution of this equation over three indices λ, μ, ν with $C_{\beta\nu}C_{\mu\rho}\nabla_\rho\nabla_\lambda$, we obtain the operator identity

$$C_{\alpha\beta}\overline{\Delta} = \overline{\nabla}_\alpha\overline{\nabla}_\beta - C_{\alpha\nu}\overline{\text{rot}}_{\nu\rho}\text{rot}_{\rho\beta}. \tag{2.36}$$

Here,

$$\overline{\Delta} = C_{\alpha\beta}\nabla_\alpha\nabla_\beta,\ \overline{\nabla}_\alpha = C_{\alpha\beta}\nabla_\beta,\ \overline{\text{rot}}_{\alpha\beta} = \epsilon_{\alpha\beta\gamma}\ C_{\gamma\nu}\nabla_\nu. \tag{2.37}$$

Convolutions of Eq. (2.36) with tensor C from the left and tensor $C^{-1}v$ from the right result in the equation

$$\overline{\Delta}v_\alpha = \overline{\nabla}_\alpha\nabla_\beta v_\beta - C_{\alpha\nu}\overline{\text{rot}}_{\nu\rho}\text{rot}_{\rho\mu}C^{-1}_{\mu\delta}v_\delta. \tag{2.38}$$

After application to this equation the integral operator which kernel $g(x)$ is Green's function of the operator $\overline{\Delta}$, we yield the integral presentation of the field $v(x)$ in the form

$$v_\alpha(x) = C_{\alpha\beta}\int \overline{\text{rot}}_{\beta\lambda}g(x - x')\text{rot}'_{\lambda\mu}C^{-1}_{\mu\delta}v_\delta(x')dx' -$$
$$-C_{\alpha\beta}\int \nabla_\beta g(x - x')\text{div}'v(x')dx'. \tag{2.39}$$

Here, it is taken into account that $\nabla'_\alpha g(x - x') = -\nabla_\alpha g(x - x')$. This equation coincides with Eq. (2.32) when $C_{\alpha\beta} = \delta_{\alpha\beta}$. It follows from this equation that the solution of Problem 1 ($\text{div}v = q$, $\text{rot}C^{-1}v = 0$) for anisotropic media is presented in the integral form as follows

$$v_\alpha(x) = -C_{\alpha\beta}\int \nabla_\beta g(x - x')q(x')dx', \tag{2.40}$$

and the integral presentation of the solution of Problem 2 ($\text{div}v = 0, \text{rot}C^{-1}v = j$) is

$$v_\alpha(x) = C_{\alpha\beta} \int \overline{\text{rot}}_{\beta\lambda} g(x - x') j_\lambda(x') dx'. \tag{2.41}$$

Note that the function

$$G_\alpha(x) = C_{\alpha\beta} \nabla_\beta g(x) \tag{2.42}$$

in Eq. (2.40) satisfies the equations

$$\nabla_\alpha G_\alpha(x) = -\delta(x), \text{rot}_{\alpha\beta} C^{-1}_{\beta\gamma} G_\gamma(x) = 0. \tag{2.43}$$

Thus, this function can be interpreted as Green's function of Problem 1. Meanwhile the function

$$Z_{\alpha\beta}(x) = C_{\alpha\nu} \epsilon_{\nu\beta\mu} C_{\mu\delta} \nabla_\delta g(x) \tag{2.44}$$

can be interpreted as Green's function of Problem 2. The function $Z_{\alpha\beta}(x)$ satisfies the equations

$$\text{div}_\alpha Z_{\alpha\beta}(x) = 0, \ \text{rot}_{\alpha\gamma} C^{-1}_{\gamma\delta} Z_{\delta_\beta}(x) = S_{\alpha\beta}(x), \tag{2.45}$$

$$S_{\alpha\beta}(x) = -\delta_{\alpha\beta}\delta(x) - C_{\alpha\gamma} \nabla_\gamma \nabla_\beta g(x). \tag{2.46}$$

It will be shown in the next section that the integral operator with the kernel $S_{\alpha\beta}(x)$ is the projector on the subspace of divergence-free vector functions. It means that for any function $v(x)$ which divergence is equal to zero ($\text{div} v = 0$), the following equation holds

$$\int S_{\alpha\beta}(x - x') v_\beta(x') dx' = v_\alpha(x). \tag{2.47}$$

Therefore, the function $S_{\alpha\beta}(x)$ in Eq. (2.45) is an analogy of the delta-function ($\delta(x)\delta_{\alpha\beta}$) in the subspace of the divergence-free functions.

2.4 Projection Operators

It is known that each vanishing at infinity vector field $v(x)$ can be presented as a sum of rotor-free and divergence-free vector fields

$$v(x) = v^{(1)}(x) + v^{(2)}(x), \tag{2.48}$$

$$\text{rot} v^{(1)}(x) = 0, \ \text{div} v^{(2)}(x) = 0. \tag{2.49}$$

It follows from these equations that Fourier transforms of the functions $v^{(1)}(x)$ and $v^{(2)}(x)$ satisfy the equations

$$k \times \widetilde{v}^{(1)}(k) = 0, \; k \cdot \widetilde{v}^{(2)}(k) = 0, \tag{2.50}$$

where $\times$ means vector product and $\cdot$ is scalar product of two vectors. The problem of partition of a vanishing at infinity vector field $v(x)$ in the rotor-free and divergence-free parts can be solved on the basis of the integral presentation (2.32) of this field. Using Gauss' theorem and overthrowing the derivatives on the kernel in Eq. (2.32), we obtain

$$v_\alpha(x) = -\int \nabla_\alpha \nabla_\beta g(x - x') v_\beta(x') dx' + \\ + \int \mathrm{rot}_{\alpha\beta} \mathrm{rot}_{\beta\gamma} g(x - x') v_\gamma(x') dx'. \tag{2.51}$$

This equation needs explanations. The kernels of the integral operators in Eq. (2.32) are proportional to the first derivative $\nabla g(x)$ of Green's function $g(x)$ that has an integrable singularity of the order of $-2 : \nabla g(x) = O(|x|^{-2})$. The second derivatives of $g(x)$ in Eq. (2.51) have the singularities of the order of -3 at $x = 0$. As a result, the integrals in Eq. (2.51) formally diverge. The regularization formulas for generalized functions that are derivatives of Green's functions are obtained in [2]. In particular, the first integral in Eq. (2.51) should be understood in the following sense

$$-\int \nabla_\alpha \nabla_\beta g(x - x') v_\beta(x') dx' = \\ = A_{\alpha\beta} v_\beta(x) - pv \int \nabla_\alpha \nabla_\beta g(x - x') v_\beta(x') dx', \; A_{\alpha\beta} = \frac{1}{3} \delta_{\alpha\beta}, \tag{2.52}$$

$$pv \int \nabla_\alpha \nabla_\beta g(x - x') v_\beta(x') dx' = \lim_{\varepsilon \to 0} \int_{R^3 / \mathfrak{v}_\varepsilon(x)} \nabla_\alpha \nabla_\beta g(x - x') v_\beta(x') dx'. \tag{2.53}$$

Here, $\mathfrak{v}_\varepsilon(x)$ is the region of a ball of radius ε with the center at point x, $R^3/\mathfrak{v}_\varepsilon(x)$ is the complement of $\mathfrak{v}_\varepsilon$ to the entire space, symbol pv means Cauchy principal value of the integral. The first term on the right-hand side of Eq. (2.52) is called the singular part of the integral, and the second term is its regular part. Note that partition of the integrals in Eq. (2.52) in singular and regular parts is not unique (see Appendix B.2).

It is easy to verify that the functions

$$v_\alpha^{(1)}(x) = \int K_{\alpha\beta}(x - x') v_\beta(x') dx', \tag{2.54}$$

$$v_\alpha^{(2)}(x) = \int S_{\alpha\beta}(x - x') v_\beta(x') dx', \tag{2.55}$$

$$K_{\alpha\beta}(x) = -\nabla_\alpha \nabla_\beta g(x), \; S_{\alpha\beta}(x) = -\mathrm{rot}_{\alpha\gamma} \mathrm{rot}_{\gamma\beta} g(x) \tag{2.56}$$

satisfy the conditions (2.49). For the integrals in Eqs. (2.54), (2.55) are convolutions, their Fourier transforms are products of Fourier transforms of convoluting functions. For derivative in the x-space corresponds to product with the vector $(-ik)$ in Fourier space, for $\widetilde{v}(k)$, we have the presentation

$$\widetilde{v}_\alpha(k) = \widetilde{K}_{\alpha\beta}(k)\widetilde{v}_\beta(k) + \widetilde{S}_{\alpha\beta}(k)\widetilde{v}_\beta(k), \tag{2.57}$$

$$\widetilde{K}_{\alpha\beta}(k) = \frac{1}{k^2}k_\alpha k_\beta, \ \widetilde{S}_{\alpha\beta}(k)\widetilde{v}_\beta(k) = -\frac{1}{k^2}\epsilon_{\alpha\beta\gamma}\epsilon_{\gamma\delta\lambda}k_\beta k_\delta \widetilde{v}_\lambda(k). \tag{2.58}$$

Here, $\widetilde{K}(k)$ and $\widetilde{S}(k)$ are Fourier transforms of the kernels $K(x)$ and $S(x)$. It follows from these equations that

$$\widetilde{S}_{\alpha\beta}(k) + \widetilde{K}_{\alpha\beta}(k) = \delta_{\alpha\beta}, \tag{2.59}$$

$$\widetilde{K}_{\alpha\beta}(k)\widetilde{K}_{\beta\gamma}(k) = \widetilde{K}_{\alpha\gamma}(k), \ \widetilde{S}_{\alpha\beta}(k)\widetilde{S}_{\beta\gamma}(k) = \widetilde{S}_{\alpha\gamma}(k), \tag{2.60}$$

$$\widetilde{K}_{\alpha\beta}(k)\widetilde{S}_{\beta\gamma}(k) = 0, \epsilon_{\alpha\beta\gamma}k_\beta \widetilde{K}_{\gamma\lambda}(k) = 0, \ k_\alpha \widetilde{S}_{\alpha\beta}(k) = 0. \tag{2.61}$$

These equations show that the operator K and S are orthogonal projectors on the subspaces of the rotor-free and divergence-free vector functions.

Now we consider a space of vanishing at infinity vector functions $v(x)$ and perform partition of this space into two subspaces associated with Problems 1 and 2 for homogeneous anisotropic media. The objective is presentation of an arbitrary vector filed $v(x)$ as the sum of div-free and rotC^{-1}-free vector fields

$$v(x) = v^{(1)}(x) + v^{(2)}(x), \tag{2.62}$$

$$\text{rot}C^{-1}v^{(1)}(x) = 0, \text{div}v^{(2)}(x) = 0. \tag{2.63}$$

For this aim, we present Eq. (2.51) in the form

$$v(x) = \int \overline{K}(x-x')v(x')dx' + \int \overline{S}(x-x')v(x')dx'. \tag{2.64}$$

where Fourier transforms of the kernels $\overline{K}(x)$ and $\overline{S}(x)$ in this equation are

$$\widetilde{\overline{K}}_{\alpha\beta}(k) = \frac{1}{(kCk)}C_{\alpha\gamma}k_\gamma k_\beta, \ \widetilde{\overline{S}}_{\alpha\beta}(k) = \delta_{\alpha\beta} - \widetilde{\overline{K}}_{\alpha\beta}(k). \tag{2.65}$$

The tensors $\widetilde{K}(k)$ and $\widetilde{S}(k)$ satisfy the identities similar to (2.59)–(2.61)

$$\widetilde{\overline{K}}(k) \cdot \widetilde{\overline{K}}(k) = \widetilde{K}(k), \ \widetilde{\overline{S}}(k) \cdot \widetilde{\overline{S}}(k) = \widetilde{\overline{S}}(k), \tag{2.66}$$

$$\widetilde{\overline{K}}(k) \cdot \widetilde{\overline{S}}(k) = 0, \ k \times C^{-1}\widetilde{\overline{K}}(k) = 0, \ k \cdot \widetilde{\overline{S}}(k) = 0. \tag{2.67}$$

These equations show that the integral operators in Eq. (2.64) are orthogonal projectors: $\overline{K}$ is the projector on the subspace of rotC^{-1}-free vector functions $v(x)$ such as rot$C^{-1}v(x) = 0$, $\overline{S}$ is the projector on the subspace of divergence-free vector functions with div$v(x) = 0$.

2.5 Conclusion

It is shown in this Chapter that electrostatic and magnetostatic fields in infinite homogeneous anisotropic media have specific structures. The vanishing at infinity fields induced by the field sources with finite supports can be presented as sums of divergence-free and rotC^{-1}-free fields. Such a partition of a stationary electromagnetic field can be performed by the projectors which explicit forms are obtained in this chapter. Similar projection operators appear in the integral presentations of the electro- and magnetostatic fields in heterogeneous media (Chap. 4).

The systems of equations considered in this chapter are similar to the systems of equations for physical fields of other nature in homogeneous anisotropic media. For example, the steady electric current j and electric field e in homogeneous conductive media satisfy the system of equations

$$\mathrm{div}j(x) = 0, \ \ j(x) = Ce(x), \mathrm{rot}e(x) = 0. \tag{2.68}$$

Here, C is second rank tensor of electro-conductivity, and equation $j = Ce$ is Ohm's law. Because there are no steady electric charges in conductive media, these fields can be caused by conditions at infinity (boundary conditions) only.

The system of equations of electrostatic is mathematically equivalent to the system of equations of thermo-conductivity. In this case, the field potential is temperature $T(x)$, and the system of equation for the heat flux Q and temperature gradient $\Upsilon = -\nabla T$ has the form

$$\mathrm{div}Q(x) = q(x), Q(x) = C\Upsilon(x), \mathrm{rot}\Upsilon(x) = 0. \tag{2.69}$$

Here, C is the tensor of thermo-conductivity, and q is the density of heat sources.

References

1. Cheng D (1992) Field and wave electromagnetics. Addison-Wesley Publishing Co., Massachusetts
2. Gelfand I, Shilov G (1964) Generalized functions. Academic, New York, p 470

Chapter 3
Homogenous Elastic Media with Sources of External and Internal Stresses

Abstract Homogeneous elastic media subjected to two types of field sources: body forces (external sources) and dislocations (internal sources) are considered. The strain and stress fields caused by these sources are presented in the integral forms using Green's function of the system of the governing equations of elasticity. Relations of the integral operators in these presentations with the projectors on the divergence-free and rot-free tensor subspaces are revealed. Discontinuities of the elastic fields in the medium subjected to the stress sources are studied. For presentations of elastic fields far from the sources, the concept of multipole expansions of the source densities is introduced.

Keywords Anisotropic elastic medium · Body forces · Dislocations · Green's function for external and internal sources · Projectors on div-free and rot-free subspaces · Multipole expansion

3.1 Homogeneous Elastic Media with the External Stress Sources

We consider a homogeneous anisotropic elastic medium with the stiffness tensor $C_{\alpha\beta\gamma\delta}$. The tensor C is definitely positive and symmetric with respect to the first and second pairs of indices as well as about the index pair permutations. The medium occupies the entire 3D-space, and it is free from stresses at infinity. The stress-strain state of the medium is described by the second rank symmetric strain $\varepsilon_{\alpha\beta}(x)$ and stress $\sigma_{\alpha\beta}(x)$ tensors. If the medium is subjected to body forces of the density $q(x)$, the stress and strain fields are called external, and $q(x)$ is the source of the external stresses. The system of differential equations for external stresses has the form

$$\nabla_\beta \sigma_{\beta\alpha}(x) = -q_\alpha(x), \sigma_{\alpha\beta}(x) = C_{\alpha\beta\gamma\lambda}\varepsilon_{\gamma\lambda}(x), \mathrm{Rot}_{\alpha\beta\gamma\lambda}\varepsilon_{\gamma\lambda}(x) = 0,$$

$$\mathrm{Rot}_{\alpha\beta\lambda\mu}\varepsilon_{\lambda\mu}(x) = \epsilon_{\alpha\nu\lambda}\epsilon_{\beta\rho\mu}\nabla_\nu\nabla_\rho\varepsilon_{\lambda\mu}(x)\,. \tag{3.1}$$

S. Kanaun, *Self-Consistent Methods for Composites*, Solid Mechanics and Its Applications 282, https://doi.org/10.1007/978-3-031-96739-9_3

Here, $\epsilon_{\alpha\nu\lambda}$ is Levi-Civita tensor. The first equation of this system follows from equilibrium of infinitesimal elements of the medium, the second equation is Hooke's law. Symbol Rot in Eq. (3.1) is the compatibility operator. In the symbolic form, this operator can be presented in terms of the double rot-operator of vector calculus

$$\text{Rot}\varepsilon = \text{rot}(\text{rot}\varepsilon)^T = \nabla \times \varepsilon \times \nabla. \tag{3.2}$$

Here, $(..)^T$ is the transposition operator. Equation $\text{Rot}\varepsilon = 0$ is the compatibility equation, and it is equivalent to the statement that tensor ε is symmetrized gradient of a vector potential $u(x)$

$$\varepsilon_{\alpha\beta}(x) = (\text{def}u)_{\alpha\beta}(x), \tag{3.3}$$

$$(\text{def}u)_{\alpha\beta}(x) = \nabla_\alpha u_\beta(x)|_{(\alpha\beta)} = \frac{1}{2}\left(\nabla_\alpha u_\beta(x) + \nabla_\beta u_\alpha(x)\right). \tag{3.4}$$

Here and further, the parentheses in indices mean symmetrization. Vector $u(x)$ is the displacement of point x caused by body forces. It follows from Eq. (3.1) that $u(x)$ satisfies the differential equations

$$(L_{\alpha\beta}u_\beta)(x) = q_\alpha(x), \quad L_{\alpha\beta} = -\nabla_\lambda C_{\lambda\alpha\beta\mu}\nabla_\mu. \tag{3.5}$$

The solution of this equation can be presented in the integral form

$$u_\alpha(x) = \int G_{\alpha\beta}(x - x')q_\beta(x')dx', \tag{3.6}$$

where $G_{\alpha\beta}(x)$ is Green's function of static elasticity for displacements. This function is a vanishing at infinity solution of the differential equation

$$(L_{\alpha\lambda}G_{\lambda\beta})(x) = \delta(x)\delta_{\alpha\beta}. \tag{3.7}$$

Application of the Fourier transform operator to this equation results in the expression for Fourier transform $\widetilde{G}(k)$ of Green's function

$$\widetilde{G}(k) = L^{-1}(k), \quad L_{\alpha\beta}(k) = k_\gamma C_{\gamma\alpha\beta\lambda}k_\lambda. \tag{3.8}$$

It follows from this equation that $G(x)$ is an even homogeneous function of the order of -1 that can be presented in the form

$$G_{\alpha\beta}(x) = \frac{1}{(2\pi)^3}\int \widetilde{G}_{\alpha\beta}(k)\exp(-\mathrm{i}\,(k\cdot x))dk =$$

$$= \frac{1}{|x|}G_{\alpha\beta}(n), k\cdot x = k_\alpha x_\alpha, \quad n_\alpha = \frac{x_\alpha}{|x|}. \tag{3.9}$$

The explicit expressions for $G(x)$ can be obtained for isotropic and transverse isotropic media [1]. In the case of an isotropic medium with Lame's constants λ and μ, the functions $\widetilde{G}(k)$ and $G(x)$ are (Appendix B)

$$\widetilde{G}_{\alpha\beta}(k) = \frac{1}{\mu k^2}\left(\delta_{\alpha\beta} - \kappa\frac{k_\alpha k_\beta}{k^2}\right), k^2 = k_\alpha k_\alpha, \tag{3.10}$$

$$G_{\alpha\beta}(x) = \frac{1}{8\pi\mu|x|}\left((2-\kappa)\delta_{\alpha\beta} + \kappa n_\alpha n_\beta\right), \quad \kappa = \frac{\lambda+\mu}{\lambda+2\mu}. \tag{3.11}$$

Equation (3.1) can be presented in the form of the system for external stresses $\sigma(x)$

$$\operatorname{div}\sigma = -q, \ \operatorname{Rot}C^{-1}\sigma = 0. \tag{3.12}$$

For Eq. (3.6) holds the solution of this system is

$$\sigma_{\alpha\beta}(x) = C_{\alpha\beta\lambda\mu}\int \nabla_\lambda G_{\mu\rho}(x-x')q_\rho(x')dx'. \tag{3.13}$$

Substituting in this equation $q(x)$ from the system (3.12) and using Gauss' theorem, we obtain the identity

$$\sigma_{\alpha\beta}(x) = -C_{\alpha\beta\lambda\mu}\int \nabla_\lambda\nabla_\gamma G_{\mu\rho}(x-x')\sigma_{\gamma\rho}(x')dx' =$$

$$= C_{\alpha\beta\gamma\delta}\int K_{\gamma\delta\mu\nu}(x-x')\sigma_{\mu\nu}(x')dx', \tag{3.14}$$

where the kernel $K(x)$ and its Fourier transform $\widetilde{K}(k)$ are

$$K_{\alpha\beta\lambda\mu}(x) = -\nabla_\alpha\nabla_\lambda G_{\beta\mu}(x)|_{(\alpha\beta)(\lambda\mu)}, \tag{3.15}$$

$$\widetilde{K}_{\alpha\beta\lambda\mu}(k) = k_\alpha k_\lambda \widetilde{G}_{\beta\mu}(k)|_{(\alpha\beta)(\lambda\mu)}. \tag{3.16}$$

It follows from Eq. (3.8) that the tensor

$$\widetilde{\overline{K}}_{\alpha\beta\lambda\mu}(k) = C_{\alpha\beta\gamma\delta}\widetilde{K}_{\gamma\delta\lambda\mu}(k) \tag{3.17}$$

satisfies the equation

$$\widetilde{\overline{K}}_{\alpha\beta\gamma\delta}(k)\widetilde{\overline{K}}_{\gamma\delta\lambda\mu}(k) = \widetilde{\overline{K}}_{\alpha\beta\lambda\mu}(k). \tag{3.18}$$

Therefore, the integral operator with the kernel $\overline{K}(x) = CK(x)$ is a projector. It follows from definition (3.1) of the operator Rot that $\overline{K}(x)$ satisfies the equation

$$\mathrm{Rot}C^{-1}\overline{K}(x) = 0. \tag{3.19}$$

Therefore, $\overline{K}$ is the projector on the subspace of tensor functions $\sigma(x)$ that satisfy the equation $\mathrm{Rot}C^{-1}\sigma(x) = 0$. In this subspace, operator $\overline{K}$ is the identical operator as it follows from Eq. (3.14).

Substituting in Eq. (3.14) $\sigma(x) = C\varepsilon(x)$, and multiplying the result with tensor C^{-1}, we obtain the identity

$$\varepsilon_{\alpha\beta}(x) = \int K_{\alpha\beta\lambda\mu}(x - x')C_{\lambda\mu\gamma\delta}\varepsilon_{\gamma\delta}(x')dx'\,. \tag{3.20}$$

The operator with the kernel $K'(x) = K(x)C$ in this equation is also projector because the equation $\widetilde{K}'(k)\widetilde{K}'(k) = \widetilde{K}'(k)$ holds. For the property

$$\mathrm{Rot}K'(x) = \mathrm{Rot}K(x)C = -\mathrm{Rot}\nabla\nabla G(x)C \equiv 0\,, \tag{3.21}$$

the integral operator K' with the kernel $K'(x)$ is the projector on the subspace of Rot-free tensor fields. If the field $\varepsilon(x)$ belongs to this subspace, K' is the identical operator.

Note that the function $G(x)$ has singularity of the order of $|x|^{-1}$ when $x \to 0$, and $K(x)$ (second derivatives of $G(x)$) is a non-integrable function with the singularity of the order of $|x|^{-3}$. It is shown in Appendix B.2 that the integral in Eqs. (3.14) and (3.20) should be understood in the following sense

$$\int K(x - x')q(x')dx' = A^0 q(x) + pv\int K(x - x')q(x')dx\,. \tag{3.22}$$

$$A^0 = \frac{1}{4\pi}\int\limits_{\Omega_1}\widetilde{K}(k)d\Omega\,. \tag{3.23}$$

Here, $\widetilde{K}(k)$ is Fourier transform of the function $K(x)$, Ω_1 is the surface of the unit sphere in the k-space of Fourier transforms, pv is Cauchy principal value of the integral. This integral exists because for the function $K(x)$ in Eq. (3.15), the following equation holds [2]

$$\int\limits_{\Omega_1} K(n)d\Omega_n = 0.$$

Here, Ω_1 is unit sphere in the x-space, n is a unit vector on this sphere.

3.2 External Stress Sources Concentrated at Points, Lines and Surfaces

1. *Concentrated force.*
Let the density $q_\alpha(x)$ in (3.1) and (3.6) has the form

$$q_\alpha(x) = Q_\alpha \delta(x)\,, \tag{3.24}$$

where Q_α is a constant vector. Such a density corresponds to a concentrated force Q_α applied at the origin ($x = 0$) (Fig. 3.1a). It follows from Eq. (3.6) that the vector of displacements due to this source is

$$u_\alpha(x) = G_{\alpha\beta}(x) Q_\beta\,. \tag{3.25}$$

2. *Force dipole.*
Let a concentrated force Qe_α act at point $x^0 = 0$, in the direction of a unit vector e_α and the force $-Qe_\alpha$ act at the point with the coordinates $x^h_\alpha = he_\alpha$ (Fig. 3.1b). According to Eqs. (3.6), (3.25), the displacement vector due to action of these two forces is

$$u_\alpha(x) = \left(G_{\alpha\beta}(x) - G_{\alpha\beta}(x - x^h)\right) Qe_\beta\,. \tag{3.26}$$

Suppose that $Q = h^{-1}Q^0$ and let $h \to 0$ in Eq. (3.26). As a result, we obtain the source that is called an elementary force dipole of intensity Q^0 applied at the origin. The displacement field $u(x)$ from this source is

$$u_\alpha(x) = Q^0 \nabla_\lambda G_{\alpha\beta}(x) e_\lambda e_\beta\,. \tag{3.27}$$

It follows from Eq. (3.6) and the property of the derivative of the generalized function $\delta(x)$ [3] that the body force

$$q_\alpha(x) = Q^0 e_\alpha e_\beta \nabla_\beta \delta(x) \tag{3.28}$$

corresponds to the force dipole. For such a body force, the displacement field $u(x)$ in Eq. (3.6) coincides with the field in Eq. (3.27).

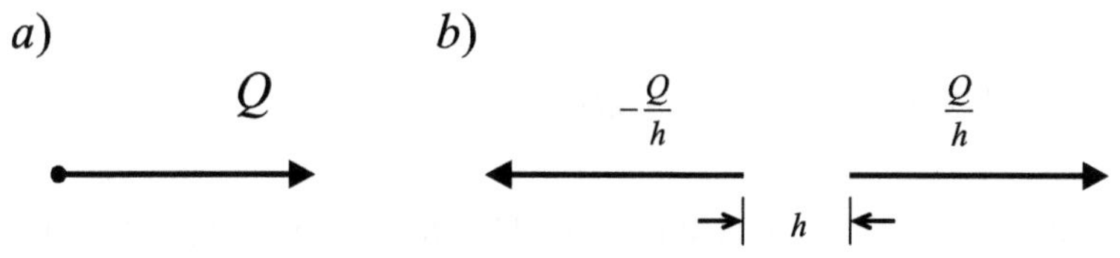

Fig. 3.1 Concentrated force (**a**), force dipole (**b**)

Let the density of the external forces be defined by the equation

$$q_\alpha(x) = Q^{(1)}_{\alpha\beta}\nabla_\beta\delta(x)\,, \tag{3.29}$$

where $Q^{(1)}$ is a second rank tensor. Note that the total force $\overline{Q}_\alpha$ of such a force distribution is equal to zero

$$\overline{Q}_\alpha = \int q_\alpha(x)dx = Q^{(1)}_{\alpha\beta}\int\nabla_\beta\delta(x)dx = 0\,. \tag{3.30}$$

If $Q^{(1)}$ is a symmetric tensor, there exists a Cartesian basis $e^{(i)}$ ($i = 1, 2, 3$) in which $Q^{(1)}$ is represented in the form

$$Q^{(1)}_{\alpha\beta} = Q_1 e^{(1)}_\alpha e^{(1)}_\beta + Q_2 e^{(2)}_\alpha e^{(2)}_\beta + Q_3 e^{(3)}_\alpha e^{(3)}_\beta\,. \tag{3.31}$$

Here, Q_i ($i = 1, 2, 3$) are the eigenvalues of the tensor $Q^{(1)}$. Therefore, the source (3.31) is the sum of three elementary dipoles with the intensities Q_1, Q_2, Q_3 that act along the principal axes of the tensor Q^2.

The total moment M_α of the forces in Eq. (3.29) is

$$M_\alpha = \int \epsilon_{\alpha\beta\lambda}x_\beta Q^{(1)}_{\lambda\mu}\nabla_\mu\delta(x)dx = -\epsilon_{\alpha\beta\lambda}Q^{(1)}_{[\lambda\beta]}\,, \tag{3.32}$$

$$Q^{(1)}_{[\alpha\beta]} = \frac{1}{2}(Q^{(1)}_{\alpha\beta} - Q^{(1)}_{\beta\alpha})\,. \tag{3.33}$$

Thus, the antisymmetric part $Q^{(1)}_{[\alpha\beta]}$ of the tensor $Q^{(1)}$ determines the moment of the generalized source in Eq. (3.29). In what follows, the tensor $Q^{(1)}$ is called the generalized force moment. Such a stress source causes the following displacement field in the medium

$$u_\alpha(x) = \nabla_\lambda G_{\alpha\beta}(x)Q^{(1)}_{\beta\lambda}\,. \tag{3.34}$$

Let $q_{\alpha\beta}(x)$ be the density of generalized force moments distributed in a finite region of 3D-space. The displacement field from such a source is defined by the equation

$$u_\alpha(x) = \int \nabla_\lambda G_{\alpha\beta}(x - x')q_{\beta\lambda}(x')dx'. \tag{3.35}$$

This field coincides with the field in Eq. (3.34) when $q_{\alpha\beta}(x) = Q^{(1)}_{\alpha\beta}\delta(x)$. It follows from Gauss' theorem (Appendix C.1) that the field in Eq. (3.35) coincides with the displacements from the body force density

$$\overline{q}_\alpha(x) = \nabla_\beta q_{\alpha\beta}(x). \tag{3.36}$$

Thus, the density $\overline{q}_\alpha(x)$ in this equation is equivalent to the density of the force moments $q_{\alpha\beta}(x)$ because they cause the same displacement field in the medium. *Vice versa*, any body force distribution $q_\alpha(x)$ is equivalent to a distribution of some force moments. It follows from Eq. (3.36) that the density $q_{\alpha\beta}(x)$ of such moments is not determined uniquely but with precision to the field $q^0_{\alpha\beta}(x)$ that satisfies the equation $\nabla_\beta q^0_{\alpha\beta}(x) = 0$.

Taking two closely placed dipoles with the densities $Q^{(1)}/h$ and $-Q^{(1)}/h$, where h is the distance between the dipoles, and considering the limit $h \to 0$ we obtain the source of a more complex structure (quadrupole). Repeating this procedure, we obtain a sequence of the sources that are called multipoles. The density of the body forces that corresponds to the multipole of the nth order is

$$q_\alpha(x) = Q^{(n)}_{\alpha\beta_1\beta_2\ldots\beta_n} \nabla_{\beta_1}\nabla_{\beta_2}\ldots\nabla_{\beta_n}\delta(x). \tag{3.37}$$

3. *The external stress sources distributed on surfaces and lines.*
Let Ω be a smooth Lyapunov surface in 3D-space, and $\Omega(x)$ is the delta-function concentrated on Ω. The function $\Omega(x)$ acts on an arbitrary smooth function $\varphi(x)$ by the rule [3, 4]:

$$\int \varphi(x)\Omega(x)dx = \int_\Omega \varphi(x)d\Omega, \tag{3.38}$$

where the integral in the left-hand side is spread to entire 3D-space. The distribution of the body forces with density $Q(x)$ on the surface Ω is described by the generalized external stress source (Fig. 3.2)

$$q_\alpha(x) = Q_\alpha(x)\Omega(x). \tag{3.39}$$

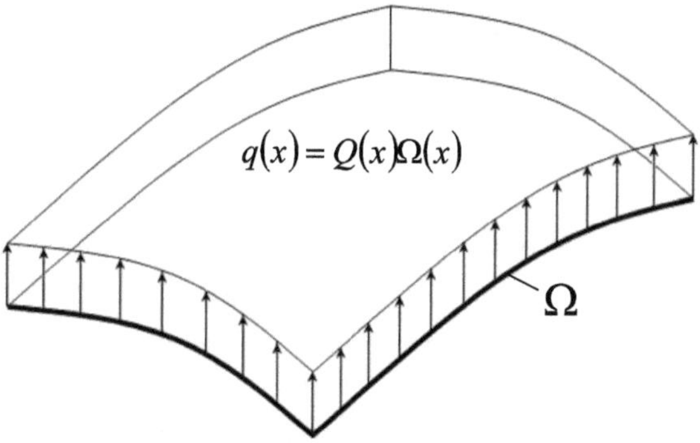

Fig. 3.2 Sources of external stresses distributed on surface Ω

Such a body force distribution is called a simple layer concentrated on surface Ω. The displacement field caused by a simple layer is

$$u_\alpha(x) = \int G_{\alpha\beta}(x-x')q_\beta(x')dx' = \int_\Omega G_{\alpha\beta}(x-x')Q_\beta(x')d\Omega', \tag{3.40}$$

The function $u(x)$ in this equation is called the potential of the simple layer of static elasticity.

Let the force moments of density $Q_{\alpha\beta}(x)$ be distributed on the surface Ω. The corresponding force moment density is

$$q_{\alpha\beta}(x) = Q_{\alpha\beta}(x)\Omega(x). \tag{3.41}$$

The displacement field $u(x)$ caused by this source follows from Eq. (3.40) in the form

$$u_\alpha(x) = \int_\Omega \nabla_\lambda G_{\alpha\beta}(x-x')Q_{\lambda\beta}(x')d\Omega'. \tag{3.42}$$

For a particular density

$$Q_{\alpha\beta}(x) = -C^0_{\alpha\beta\lambda\mu}n_\lambda(x)b_\mu(x), \tag{3.43}$$

potential (3.42) is called the potential of a double layer of static elasticity. In Eq. (3.43), $n(x)$ is the normal to the surface Ω, $b(x)$ is an arbitrary vector field on Ω.

Similar to $\Omega(x)$, a generalized function $l(x)$ concentrated on the line l in 3D-space can be introduced

$$\int \varphi(x)l(x)dx = \int_l \varphi(x)dl. \tag{3.44}$$

Distributions of the body forces and force moments on l are described by the generalized functions

$$q_\alpha(x) = Q_\alpha(x)l(x), \quad q_{\alpha\beta}(x) = Q_{\alpha\beta}(x)l(x), \tag{3.45}$$

The displacement fields due to such sources have the forms

$$u_\alpha(x) = \int_l G_{\alpha\beta}(x-x')Q_\beta(x')dl', \tag{3.46}$$

$$u_\alpha(x) = \int_l \nabla_\lambda G_{\alpha\beta}(x-x')Q_{\beta\lambda}(x')dl'. \tag{3.47}$$

3.3 Homogeneous Elastic Medium with the Internal Stresses Sources

Stresses in elastic media can exist in the absence of external (body) forces. Such stresses are called internal, and they can be caused by temperature fields, plastic deformations in a finite region of the medium, phase transitions accompanied by changing the parameters of the crystalline lattices, etc.

Let plastic deformation $m(x)$ occurs in a finite region V of a homogeneous elastic medium. Because the region V is constrained by the surrounding material, internal stresses σ and elastic strains $\varepsilon^e = C^{-1}\sigma$ associated with these stresses appear in the medium. The total strain tensor ε is the sum of the elastic ε^e and "inelastic" m components ($\varepsilon = \varepsilon^e + m$), and this sum should satisfy the compatibility Eq. (3.1)

$$\mathrm{Rot}_{\alpha\beta\lambda\mu}\varepsilon_{\lambda\mu} = \mathrm{Rot}_{\alpha\beta\lambda\mu}(\varepsilon^e_{\lambda\mu} + m_{\lambda\mu}) = 0 \, . \tag{3.48}$$

In the absence of external forces, stress tensor $\sigma(x)$ satisfies the equilibrium equation ($\mathrm{div}\sigma = 0$). Thus, the system of equations for the internal stresses can be written in the form

$$\nabla_\beta \sigma_{\beta\alpha} = 0 \, , \;\; \sigma_{\alpha\beta} = C_{\alpha\beta\lambda\mu}\varepsilon^e_{\lambda\mu}, \;\; \mathrm{Rot}_{\alpha\beta\lambda\mu}\varepsilon^e_{\lambda\mu} = -\eta_{\alpha\beta} \, , \tag{3.49}$$

$$\eta_{\alpha\beta} = \mathrm{Rot}_{\alpha\beta\lambda\mu} m_{\lambda\mu} \, . \tag{3.50}$$

The second rank tensor $\eta_{\alpha\beta}(x)$ in this equation is called the dislocation density tensor, and $m_{\alpha\beta}(x)$ is the tensor of the dislocation moments [4]. These tensors can be interpreted as the sources of internal stresses. If $\eta_{\alpha\beta}(x)$ is a finite integrable function, the solution of the system (3.49) can be presented in the integral form

$$\sigma_{\alpha\beta}(x) = \int Z_{\alpha\beta\lambda\mu}(x - x')\eta_{\lambda\mu}(x')dx' \, . \tag{3.51}$$

Here, $Z_{\alpha\beta\lambda\mu}(x)$ is Green's tensor for internal stresses. The explicit expression for this tensor for anisotropic media is presented in [4]. Taking into account the relation $\eta_{\alpha\beta}(x) = \mathrm{Rot}_{\alpha\beta\lambda\mu} m_{\lambda\mu}(x)$ and using Gauss' theorem, we obtain

$$\sigma_{\alpha\beta}(x) = \int S_{\alpha\beta\lambda\mu}(x - x')m_{\lambda\mu}(x')dx', \tag{3.52}$$

$$S_{\alpha\beta\lambda\mu}(x) = \mathrm{Rot}_{\alpha\beta\tau\rho} Z_{\tau\rho\lambda\mu}(x) \, . \tag{3.53}$$

The function $S(x)$ in this equation can be expressed in terms of Green's function $G(x)$ of the host medium. In order to derive this expression, we consider a homogeneous anisotropic elastic medium subjected to the sources of external $q(x)$ and internal $m(x)$ stresses. It follows from Eqs. (3.13) and (3.52) that the stress field $\sigma(x)$ in the medium with such sources is the solution of the system

$$\nabla_\beta \sigma_{\beta\alpha}(x) = -q_\alpha(x), \quad \mathrm{Rot}_{\alpha\beta\lambda\mu} C^{-1}_{\lambda\mu\nu\rho} \sigma_{\nu\rho} = -\eta_{\alpha\beta} \,. \tag{3.54}$$

Using Green's functions $G(x)$ for displacements and $Z(x)$ for internal stresses, the stress tensor $\sigma(x)$ is presented in the form

$$\sigma_{\alpha\beta}(x) = \int C_{\alpha\beta\lambda\mu} \nabla_\lambda G_{\mu\rho}(x - x') q_\rho(x') dx' + \\ + \int Z_{\alpha\beta\lambda\mu}(x - x') \eta_{\lambda\mu}(x') dx'. \tag{3.55}$$

Substituting in this equation the left-hand sides of Eqs. (3.54) for $q_\alpha(x)$ and $\eta_{\alpha\beta}(x)$ and using Gauss' theorem, we obtain

$$\sigma_{\alpha\beta}(x) = \int I_{\alpha\beta\rho\tau} \delta(x - x') \sigma_{\rho\tau}(x') dx' = \\ = -C_{\alpha\beta\lambda\mu} \int \nabla_\lambda G_{\mu\rho}(x - x') \nabla'_\tau \sigma_{\tau\rho}(x') dx' - \\ - \int Z_{\alpha\beta\lambda\mu}(x - x') \mathrm{Rot}_{\lambda\mu\rho\tau} C^{-1}_{\rho\tau\nu\delta} \sigma_{\nu\delta}(x') dx' = \\ = \int \left(C_{\alpha\beta\lambda\mu} K_{\lambda\mu\rho\tau}(x - x') - S_{\alpha\beta\lambda\mu}\left(x - x'\right) C^{-1}_{\lambda\mu\rho\tau} \right) \sigma_{\rho\tau}\left(x'\right) dx', \tag{3.56}$$

where $I_{\alpha\beta\rho\tau} = \delta_{\alpha\rho}\delta_{\tau\beta}\big|_{(\alpha\beta)}$ is unit fourth rank tensor, the function $K(x)$ is defined in Eq. (3.15). Comparing the left and right sides of this equation, we yield

$$I_{\alpha\beta\rho\tau} \delta(x) = C_{\alpha\beta\lambda\mu} K_{\lambda\mu\tau\rho}(x) - S_{\alpha\beta\lambda\mu}(x) C^{-1}_{\lambda\mu\rho\tau} \,. \tag{3.57}$$

Thus, the equation for the function $S_{\alpha\beta\lambda\mu}(x)$ takes the form

$$S_{\alpha\beta\lambda\mu}(x) = C_{\alpha\beta\nu\rho} K_{\nu\rho\tau\delta}(x) C_{\tau\delta\lambda\mu} - C_{\alpha\beta\lambda\mu} \delta(x) \,. \tag{3.58}$$

The function $S(x)$ is a generalized homogeneous functions of the order of -3. The integral operator with the kernel $S(x)$ acts on an arbitrary finite function $m(x)$ by the rule that follows from Eq. (3.22)

$$\int S_{\alpha\beta\lambda\mu}(x - x') m_{\lambda\mu}(x') dx' = \\ = D_{\alpha\beta\lambda\mu} m_{\lambda\mu}(x) + pv \int S_{\alpha\beta\lambda\mu}(x - x') m_{\lambda\mu}(x') dx' \,, \tag{3.59}$$

$$D_{\alpha\beta\lambda\mu} = \frac{1}{4\pi} \int_{\Omega_1} \widetilde{S}_{\alpha\beta\lambda\mu}(k) d\Omega \,, \tag{3.60}$$

$$\widetilde{S}_{\alpha\beta\lambda\mu}(k) = C_{\alpha\beta\rho\tau}\widetilde{K}_{\rho\tau\delta\nu}(k)C_{\delta\nu\lambda\mu} - C_{\alpha\beta\lambda\mu} \,. \tag{3.61}$$

Here, $\widetilde{S}(k)$ is Fourier transform of the function $S(x)$ in Eq. (3.58), Ω_1 is the surface of a unit sphere in the k-space.

Note that the stress field $\sigma(x)$ in Eq. (3.52) satisfies automatically the equilibrium equation (div $\sigma = 0$) for any integrable function $m(x)$. This fact is a consequence of the definition of the operator Rot in Eq. (3.1)

$$\nabla_\beta S_{\beta\alpha\lambda\mu}(x) = \nabla_\beta \text{Rot}_{\beta\alpha\rho\tau} Z_{\rho\tau\lambda\mu} \equiv 0 \,. \tag{3.62}$$

It follows from Eq. (3.61) that the functions $\widetilde{\overline{S}}(k) = \widetilde{S}(k)C^{-1}$ and $\widetilde{\overline{K}}(k) = C\widetilde{K}(k)$ satisfy the equations

$$\widetilde{\overline{S}}(k)\widetilde{\overline{S}}(k) = \widetilde{\overline{S}}(k), \; \widetilde{\overline{S}}(k)\widetilde{\overline{K}}(k) = 0, \tag{3.63}$$

$$\widetilde{\overline{K}}(k) + \widetilde{\overline{S}}(k) = I. \tag{3.64}$$

Therefore, the integral operators with the kernels $\overline{K}(x) = CK(x)$ and $\overline{S}(x) = S(x)C^{-1}$ are orthogonal projectors. As a result, any vanishing at infinity tensor field $\sigma(x)$ can be presented as the sum

$$\sigma(x) = \sigma^1(x) + \sigma^2(x), \tag{3.65}$$

where

$$\sigma^1(x) = \int \overline{K}(x - x')\sigma(x')dx', \sigma^2(x) = \int \overline{S}(x - x')\sigma(x')dx'. \tag{3.66}$$

The field σ^1 satisfies the equation Rot$C^{-1}\sigma^1 = 0$, and $\sigma^1(x)$ can be interpreted as the stress field caused by the body forces of the density $q(x) = -\text{div}\,\sigma^1(x)$. The field σ^2 satisfies the equation div $\sigma^2 = 0$, and $\sigma^2(x)$ can be interpreted as the stress field caused by the dislocation density $\eta(x) = -\text{Rot}C^{-1}\sigma^2(x)$.

Note that the integral operator with the kernels

$$\overline{K}'_{\alpha\beta\lambda\mu}(x) = -\nabla_\alpha\nabla_\gamma G_{\beta\delta}(x)C_{\gamma\delta\lambda\mu}|_{(\alpha\beta)}, \tag{3.67}$$

$$\overline{S}'_{\alpha\beta\lambda\mu}(x) = I_{\alpha\beta\lambda\mu}\delta(x) - \overline{K}'_{\alpha\beta\lambda\mu}(x) \tag{3.68}$$

are orthogonal projectors also. Operator $\overline{K}'$ is the projector on the subspace of the Rot-free functions, and operator $\overline{S}'$ is the projector on the subspace of div C-free functions. These operators allow partitioning an arbitrary tensor field $\varepsilon(x)$ in the Rot-free and div C-free fields

$$\varepsilon(x) = \varepsilon^{(1)}(x) + \varepsilon^{(2)}(x), \tag{3.69}$$

$$\varepsilon^{(1)}(x) = \int \overline{K}'(x - x')\varepsilon(x')dx', \varepsilon^{(2)}(x) = \int \overline{S}'(x - x')\varepsilon(x')dx', \tag{3.70}$$

where

$$\mathrm{Rot}\varepsilon^{(1)}(x) = 0, \mathrm{div}\, C\varepsilon^{(2)}(x) = 0. \tag{3.71}$$

3.4 Stress Fields in a Homogeneous Elastic Medium for Particular Distributions of Dislocation Moments

1. Dislocation moments of a constant density in the entire space.
If $m(x) = m^0$ is a constant tensor of dislocation moments in the entire space, expression (3.52) for the stress tensor $\sigma(x)$ in the medium take the form

$$\sigma(x) = \int S(x - x')m^0 dx'. \tag{3.72}$$

Because $S(x)$ is a homogeneous function of the order of -3,

$$S(x) = \frac{1}{|x|^3} S(n), n = \frac{x}{|x|}, \tag{3.73}$$

the integral in Eq. (3.52) formally diverges at $x = x'$ and at infinity. In order to define the value of this integral, we consider the following model problem. Let internal stresses be generated by temperature field $T(x)$ and at $T = 0$, the stresses absent in the medium. The density of the dislocation moments $m_{\alpha\beta}(x)$ associated with the temperature field $T(x)$ is

$$m_{\alpha\beta}(x) = \alpha_{\alpha\beta} T(x)\,, \tag{3.74}$$

where $\alpha_{\alpha\beta}$ is the tensor of temperature expansion coefficients of the medium. For a constant temperature field $T = \text{const}$, the tensor $m_{\alpha\beta}$ is also constant. It is obvious that in this case, the stress state of the medium depends on the condition at the infinity. If the medium is not constrained at infinity, i.e., it can expand (contract) freely, the temperature change does not result in appearance of internal stresses. Thus, the value of the integral in Eq. (3.52) should be equal to zero:

$$\int S(x - x')m^0 dx' = 0\,. \tag{3.75}$$

But if the medium is totally constrained at infinity, the total strain tensor is equal to zero: $\varepsilon = \varepsilon^e + m^0 = 0$. In this case, there appears internal stresses in the medium that are determined by the elastic part of the total strain tensor

$$\sigma_{\alpha\beta} = C_{\alpha\beta\lambda\mu}\varepsilon^e_{\lambda\mu} = -C_{\alpha\beta\lambda\mu}m^0_{\lambda\mu} = -C_{\alpha\beta\lambda\mu}\alpha_{\lambda\mu}T\,. \tag{3.76}$$

From this equation and Eq. (3.52) follows that for a constrained medium, the following equation holds

$$\int S(x-x')m^0dx' = -Cm^0\,. \tag{3.77}$$

Equations (3.75) and (3.76) determine the regularizations of the divergent integral (3.72) for $m^0 = \text{const}$. Note that a natural regularization of this integral for $m = \text{const}$ does not exist, and its value depends on the physical sense of this integral in the specific problems of elasticity.

A consequence of Eqs. (3.75), (3.77) and (3.58) is the definition of action of the integral operator with the kernel $K(x)$ (3.21) on a constant q^0

$$\int K(x-x')q^0dx' = \begin{cases} 0, & \text{if the medium is constrained} \\ C^{-1}q^0, & \text{if the medium is not constrained} \end{cases}. \tag{3.78}$$

Note that this integral can be interpreted as the strain tensor in the medium subjected to generalized force moments of a constant density q^0.

Absence of a unique regularization of the integral operators with the kernels $S(x)$ and $K(x)$ on constants becomes clear if the integral Eq. (3.52) is calculated using Fourier transforms of the integrand functions

$$\int S(x-x')m^0dx' = \int \widetilde{S}(k)m^0\delta(k)\exp(-ik\cdot x)dk = \widetilde{S}(0)m^0. \tag{3.79}$$

Here, the equality $\int \exp(ik\cdot x)dx = (2\pi)^3\delta(k)$ is taken into account [5]. It follows from Eqs. (3.16) and (3.61) that Fourier transform $\widetilde{S}(k)$ of the kernel $S(x)$ is a homogeneous function of the order of 0. Hence,

$$\widetilde{S}(k) = \widetilde{S}(m), \quad m_\alpha = \frac{k_\alpha}{|k|}. \tag{3.80}$$

Therefore, the value of $\widetilde{S}(k)$ at $k = 0$ depends on the direction in which vector k tends to the origin ($k = 0$). The consequence of Eqs. (3.75), (3.77) is the equation

$$\widetilde{S}(0) = \begin{cases} -C, & \text{if the medium is constrained} \\ 0, & \text{if the medium is not constrained} \end{cases}. \tag{3.81}$$

Because the function $\widetilde{K}(k)$ is also a homogeneous function of the order of 0, from Eq. (3.78) we have

$$\widetilde{K}(0) = \begin{cases} 0, & \text{if the medium is constrained} \\ C^{-1}, & \text{if the medium is not constrained} \end{cases} . \tag{3.82}$$

2. Homogeneous distribution of dislocation moments in the half-space
Let the dislocation moments of the density m^0 occupy the half-space $x_3 > 0$

$$m_{\alpha\beta}(x_1, x_2, x_3) = m^0_{\alpha\beta} H(x_3) \quad H(x_3) = \begin{cases} 1, & x_3 > 0 \\ 0, & x_3 < 0 \end{cases} , \tag{3.83}$$

where x_1, x_2, x_3 are Cartesian coordinates, and $H(x_3)$ is Heaviside's function. Using Fourier transforms of the integrand functions, we can present the convolution integral (3.52) as follows

$$\sigma(x) = \int S(x - x')m(x')dx' =$$
$$= (2\pi)^{-3} \int \widetilde{S}(k)\, m^0 \widetilde{H}(k) \exp(-ik \cdot x)\, dk\,. \tag{3.84}$$

Here, Fourier transform $\widetilde{H}(k)$ of the function $H(x_3)$ has the form [5].

$$\widetilde{H}(k_1, k_2, k_3) = (2\pi)^2 \delta(k_1)\delta(k_2) \left(\pi\delta(k_3) + ik_3^{-1}\right) . \tag{3.85}$$

Then, accounting for the integral in Eq. (3.84), we obtain

$$\sigma(x) = \int S(x - x')m(x')dx' = \frac{1}{2}\left(\widetilde{S}(0) + \widetilde{S}(n)\,\mathrm{sign}\,(-n \cdot x)\right) m^0. \tag{3.86}$$

Here, it is taken into account that $\widetilde{S}(0, 0, k_3) = \widetilde{S}(n)$, where n is the external normal to the region $x_3 > 0$ occupied by the dislocations. The value of the tensor $\widetilde{S}(0)$ depends on the conditions at infinity (see Eq. (3.81)). Note that the jump of the stress tensor on the boundary of the region containing dislocation moments does not depend on these conditions and it is determined by the equation that follows from Eqs. (3.52) and (3.86)

$$[\sigma] = \sigma^+ - \sigma^- = -\widetilde{S}(n)m^0\,. \tag{3.87}$$

Here, σ^+ is the limit value of $\sigma(x)$ when point x tends to the plane $x_3 = 0$ from the side of the normal n, and σ^- is the value of $\sigma(x)$ when x tends to this plane from the opposed side. The equation

$$n_\alpha \widetilde{S}_{\alpha\beta\lambda\mu}(n) = n_\alpha C_{\alpha\beta\lambda\mu} n_\rho \widetilde{G}_{\nu\delta}(n) n_\tau C_{\tau\delta\lambda\mu} - n_\alpha C_{\alpha\beta\lambda\mu} = 0\,, \tag{3.88}$$

follows from the definition of the function $\widetilde{G}(n)$

$$\widetilde{G}(n) = L(n)^{-1},\ L_{\alpha\beta}(n) = n_\lambda C_{\lambda\alpha\beta\mu} n_\mu\,. \tag{3.89}$$

Therefore, the stress vector $t_\alpha(x) = \sigma_{\alpha\beta}(x) n_\beta$ is continuous at the boundary of the region containing dislocation moments of a constant density

$$\left[n_\alpha \sigma_{\alpha\beta}\right] = n_\alpha \widetilde{S}_{\alpha\beta\lambda\mu}(n) m^0_{\lambda\mu} \equiv 0\,. \tag{3.90}$$

The total strain tensor $\varepsilon(x)$ in the medium with the dislocation moment of the density $m(x)$ in Eq. (3.83) is presented in the form

$$\begin{aligned} \varepsilon(x) &= C^{-1}\sigma(x) + m(x) = \int K(x - x') C m^0 H(-n \cdot x') dx' = \\ &= \frac{1}{2}\left(\widetilde{K}(0) + \operatorname{sign}(-n \cdot x)\widetilde{K}(n)\right) C m^0. \end{aligned} \tag{3.91}$$

It follows from this equation that the jump of the strain tensor $\varepsilon(x)$ on the boundary of the half-space ($x_3 = 0$) is defined by the equation

$$[\varepsilon] = \varepsilon^+ - \varepsilon^- = -\widetilde{K}(n) C m^0\,. \tag{3.92}$$

3. Constant plastic deformation in a spherical region.
Let a spherical region V of the radius a in an isotropic elastic medium be subjected to a constant plastic deformation ε^p. In this case $m(x) = \varepsilon^p V(x)$, and the tensor of internal stresses in the medium is defined by the equation

$$\sigma_{\alpha\beta}(x) = \int S_{\alpha\beta\gamma\delta}(x - x') V(x') dx' \varepsilon^p_{\gamma\delta}. \tag{3.93}$$

Here, $V(x)$ is the characteristic function of the region V

$$V(x) = 1, x \in V;\ V(x) = 0, x \notin V. \tag{3.94}$$

Using the property of the integral convolution, the integral in this equation can be calculated by transferring to Fourier transforms of the integrand functions (Parceval formula)

$$D(x) = \int S(x - x') V(x') dx' = \frac{1}{(2\pi)^3} \int \widetilde{S}(k) \widetilde{V}(k) \exp(-ik \cdot x) dk, \tag{3.95}$$

where the functions $\widetilde{S}(k)$ and $\widetilde{V}(k)$ are (Appendix B.1)

$$\widetilde{S}(k) = -2\mu \left(P^1(m) + (2\kappa - 1)P^2(m)\right), m_\alpha = \frac{k_\alpha}{|k|}, \tag{3.96}$$

$$\widetilde{V}(k) = a^3 f(a|k|), \;\; f(z) = 4\pi \frac{\sin(z) - z\cos(z)}{z^3}, \kappa = \frac{\lambda + \mu}{\lambda + 2\mu}. \tag{3.97}$$

Here, λ, μ are Lame's parameters of the medium, $P^1(m)$, $P^2(m)$ are the elements of the tensor P-basis (Appendix A.2)

$$P^1_{\alpha\beta\gamma\delta} = \theta_{\alpha\gamma}\theta_{\beta\delta}|_{(\alpha\beta)(\gamma\delta)}, \;\; P^2_{\alpha\beta\gamma\delta} = \theta_{\alpha\beta}\theta_{\gamma\delta}, \;\; \theta_{\alpha\beta} = \delta_{\alpha\beta} - m_\alpha m_\beta. \tag{3.98}$$

Calculating the integral in Eq. (3.95) first over unit sphere Ω_1 in the k-space and then, over $|k|$ from zero to infinity, we obtain

$$D(x) = d_1(r)E^1 + d_2(r)E^2 + d_3(r)\left(E^3(n) + E^4(n)\right) +$$

$$+ d_5(r)E^5(n) + d_6(r)E^6(n), \; r = |x|, \;\; n_\alpha = x_\alpha/r. \tag{3.99}$$

Here, E^i are the element of the tensor E-basis (Appendix A.1)

$$E^1_{\alpha\beta\lambda\mu} = \delta_{\alpha\lambda}\delta_{\beta\mu}|_{(\alpha\beta)}, \;\; E^2_{\alpha\beta\lambda\mu} = \delta_{\alpha\beta}\delta_{\lambda\mu},$$

$$E^3_{\alpha\beta\lambda\mu} = \delta_{\alpha\beta}n_\lambda n_\mu, \; E^4_{\alpha\beta\lambda\mu} = n_\alpha n_\beta \delta_{\lambda\mu}, \tag{3.100}$$

$$E^5_{\alpha\beta\lambda\mu} = n_\alpha \delta_{\beta\mu} n_\lambda|_{(\alpha\beta)(\lambda\mu)}, \;\; E^6_{\alpha\beta\lambda\mu} = n_\alpha n_\beta n_\lambda n_\mu.$$

The coefficients $d_i(r)$ are defined by the equations

$$d_1(r) = \mu_0 \begin{cases} -\frac{1}{15}(5 + 4\kappa_0), & r < a \\ \frac{2}{3}(1 - \kappa_0)\left(\frac{a}{r}\right)^3 + \frac{2}{5}\kappa_0\left(\frac{a}{r}\right)^5, & r > a \end{cases}, \tag{3.101}$$

$$d_2(r) = \mu_0 \begin{cases} \frac{1}{15}(5 - 12\kappa_0), & r < a \\ \frac{1}{3}(-2 + 3\kappa_0)\left(\frac{a}{r}\right)^3 + \frac{1}{5}\kappa_0\left(\frac{a}{r}\right)^5, & r > a \end{cases}, \tag{3.102}$$

$$d_3(r) = d_5(r) = d_6(r) = 0, \;\; r < a, \tag{3.103}$$

$$d_3(r) = \mu_0\left((1 - \kappa_0)\left(\frac{a}{r}\right)^3 - \kappa_0\left(\frac{a}{r}\right)^5\right), r > a, \tag{3.104}$$

$$d_5(r) = 2\mu_0\left((2\kappa_0 - 1)\left(\frac{a}{r}\right)^3 - 2\kappa_0\left(\frac{a}{r}\right)^5\right), r > a, \tag{3.105}$$

$$d_6(r) = \mu_0 \kappa_0 \left(-5 \left(\frac{a}{r}\right)^3 + 7 \left(\frac{a}{r}\right)^5 \right), r > a. \tag{3.106}$$

It follows from these equations the tensor D (and the stress tensor σ) are constant inside the ball V. The principal terms of the stress field $\sigma(x)$ outside of V vanish as r^{-3} at infinity.

4. Somigliana dislocation

Let V_h be in a thin region in 3D-space such as one of its characteristic sizes be substantially smaller than two other sizes, and Ω be the middle surface of V_h. We denote as $h(x)$ the transversal size of V_h along the normal $n(x)$ to Ω. Then, we cut the region V_h from the medium and subject it to the plastic strains $\varepsilon^p(x)$ in V_h. Suppose that $\varepsilon^p(x)$ is constant along the normal to the surface Ω and it is determined by the equation

$$\varepsilon^p_{\alpha\beta}(x) = \frac{1}{h(x)} b_{(\alpha}(x) n_{\beta)}(x) , \quad x \in \Omega . \tag{3.107}$$

Here, $b(x)$ is a vector function defined at the points of the middle surface Ω. Freezing the plastic strains, we subject the region V_h to the surface forces that cause only elastic deformations and return the region V_h to its previous form. Keeping the applied forces, we put the material back into the medium and perform agglutination along the border of the region V_h. After that we remove the external forces (Fig. 3.3) In result, the internal stresses appear in the medium. Let us find the limit distribution of these stresses when $h \to 0$. The density of the dislocation moments $m(x)$ corresponding to such a stress source is determined by the equation

$$m_{\alpha\beta}(x) = \lim_{h\to 0} \frac{1}{h(x)} b_{(\alpha}(x) n_{\beta)} V_h(x) = b_{(\alpha}(x) n_{\beta)}(x) \Omega(x) , \tag{3.108}$$

where $V_h(x)$ is the characteristic function of the region V_h, $\Omega(x)$ is the generalized function concentrated on the middle surface of region V_h (Eq. (3.38)).

The density of dislocation moments in Eq. (3.108) is called Somigliana dislocation. There is another simple interpretation of this source of internal stresses. Let us make a cut along a smooth non-closed surface Ω in a homogeneous elastic medium, separate the edges of the cut by external forces in such a way that the points located on the opposed sides of the cut and coincided in the initial state drift apart on a vector $b(x)$. The obtained void is filled by the material of the medium and stuck to the void border. After removing the external forces, internal stresses appear in the medium. The corresponding density of the dislocation moments has the form (3.108). The stresses in the medium are defined by the equation that follows from Eq. (3.52)

$$\sigma_{\alpha\beta}(x) = \int_{\Omega} S_{\alpha\beta\lambda\mu}(x - x') n_\lambda(x') b_\mu(x') d\Omega' . \tag{3.109}$$

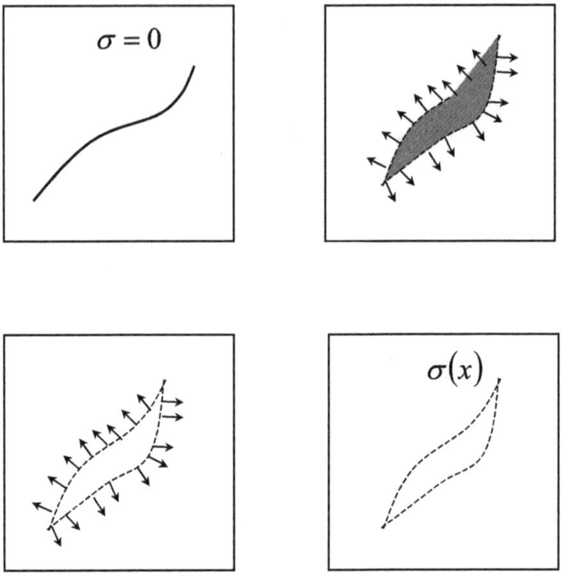

Fig. 3.3 Somigliana's dislocation

Somigliana dislocation relates to the elasticity problem for a crack in a homogeneous elastic medium. Let the medium be cut along a smooth non-closed surface Ω and subjected to some external forces such as the sides of the crack separate on a vector $b(x)$. We denote by $\sigma^0(x)$ the stress field that would have existed in the medium without the crack by action of the same external forces. The stress field in the medium with the crack is the sum of the field $\sigma^0(x)$ and the field in Eq. (3.109):

$$\sigma_{\alpha\beta}(x) = \sigma^0_{\alpha\beta}(x) + \int_\Omega S_{\alpha\beta\lambda\mu}(x - x')n_\lambda(x')b_\mu(x')d\Omega' . \tag{3.110}$$

Because of Eq. (3.62), the field $\sigma(x)$ satisfies the equilibrium equation div$\sigma = -q$ in the entire medium, and using the conditions on the crack surface we obtain the equation for the vector $b(x)$. If the crack surfaces are free from stresses, this condition is

$$n_\alpha(x)\sigma_{\alpha\beta}(x) = 0, \quad x \in \Omega. \tag{3.111}$$

From this equation and Eq. (3.110) follows the equation for the vector $b(x)$ in the form

$$\int_\Omega T_{\alpha\beta}(x, x')b_\beta(x')d\Omega' = n_\beta(x)\sigma^0_{\beta\alpha}(x) , \quad x \in \Omega , \tag{3.112}$$

$$T_{\alpha\beta}(x, x') = -n_\lambda(x)S_{\lambda\alpha\beta\mu}(x - x')n_\mu(x') . \tag{3.113}$$

This equation is considered in details in Chap. 5.

For reduction of the crack problem to an integral equation, the double layer potential of static elasticity can be also used. This potential is determined by Eqs. (3.42), (3.43), and the displacement field in the medium with a crack can be presented in the form

$$u_\alpha(x) = u^0_\alpha(x) - \\ - \int_\Omega \nabla_\lambda G_{\alpha\beta}(x-x')C_{\lambda\beta\mu\nu}n_\mu(x')b_\lambda(x')d\Omega', \tag{3.114}$$

where $u^0(x)$ is the displacement vector in the medium without the crack and the same external forces. It is known that the potential of the double layer is discontinuous, and vector $u(x)$ in Eq. (3.114) has a jump on Ω that is equal to the potential density $b(x)$ (Appendix C.2). For the displacement field in Eq. (3.114), the stress tensor has the form

$$\sigma_{\alpha\beta}(x) = \sigma^0_{\alpha\beta}(x) - \\ - C_{\alpha\beta\lambda\mu}\int_\Omega \nabla_\lambda\nabla_\nu G_{\mu\rho}(x-x')C_{\nu\rho\tau\delta}n_\tau(x')b_\delta(x')d\Omega'. \tag{3.115}$$

Comparison of this equation with Eq. (3.110) shows that their right-hand sides differ on the singular term $Cn(x)b(x)\Omega(x)$ concentrated on the crack surface Ω. Outside of Ω the stress tensors in Eqs. (3.110) and (3.115) coincide. This difference has the following background. If the solution of the crack problem is taken in the form of the potential of a double layer (3.114), the crack is simulated by dipole moments of the density $Q(x) = -Cn(x)b(x)\Omega(x)$ distributed on Ω. As a result, the stress field in the medium contains a singular component proportional to $\Omega(x)$, and it does not satisfies the equation of equilibrium ($\mathrm{div}\sigma = 0$) in the entire space. Indeed, from Eq. (3.115) follows the equations

$$\nabla_\beta\sigma_{\beta\alpha} = -q^s_\alpha, \quad q^s_\alpha(x) = \nabla_\lambda C_{\lambda\alpha\beta\mu}n_\beta(x)b_\mu(x)\Omega(x), \tag{3.116}$$

where $q^s(x)$ is the generalized function concentrated on Ω. That is why modeling a crack by the surface distribution of the dislocation moments (3.108) is a more correct procedure. The stress field in Eq. (3.110) that corresponds to the crack dislocation model satisfies the equilibrium equation in the entire space including the surface Ω, and it does not contain singular components (delta-functions). The corresponding stress vector $t_\alpha(x) = n_\beta\sigma_{\beta\alpha}(x)$ is continuous on the surface Ω. Because in the limit $x \to \Omega$, the values of the stress tensor in Eq. (3.110) and in Eq. (3.115) coincide, Eq. (3.110) for $b(x)$ that follows from the boundary condition (3.111) remains the same if the crack is simulated by a distribution of the force dipoles on Ω.

3.5 Field Discontinuities of in Homogeneous Media with External and Internal Stress Sources

Let the elastic field sources be distributed in a region V which boundary Ω is a closed Lyapunov surface. For a continuous inside V body force density $q_\alpha(x)$, the displacement and strain fields in the medium are presented in the form of integrals with weak (integrable) singularities

$$u_\alpha(x) = \int_V G_{\alpha\beta}(x - x')q_\beta(x')dx', \tag{3.117}$$

$$\varepsilon_{\alpha\beta}(x) = \int_V \nabla_{(\alpha} G_{\beta)\mu}(x)q_\mu(x')dx'. \tag{3.118}$$

These integrals are continuous functions in the entire space.

If the force moments of a continuous density $q_{\alpha\beta}(x)$ are distributed in the region V, the displacements and strains fields are presented in the integral forms in Eqs. (3.35) and (3.36)

$$u_\alpha(x) = \int_V \nabla_\lambda G_{\alpha\beta}(x - x')q_{\lambda\beta}(x')dx', \tag{3.119}$$

$$\varepsilon_{\alpha\beta}(x) = -\int_V K_{\alpha\beta\lambda\mu}(x - x')q_{\lambda\mu}(x')dx'. \tag{3.120}$$

Because the integrand function in Eq. (3.119) has a weak singularity, $u(x)$ is a continuous function in the entire space, but $\varepsilon(x)$ can be discontinuous on the surface Ω. For determination of the jump of $\varepsilon(x)$ on Ω, we present the integral in Eq. (3.120) in the form of the sum

$$\varepsilon(x) = -\int_V K(x - x')\left(q(x') - q(x)\right)dx' - \int_V K(x - x')dx'q(x). \tag{3.121}$$

The function $q(x)$ outside V is defined by an arbitrary smooth continuation. For a continuous bounded function $q(x)$, the first integral in the right-hand side of Eq. (3.121) has a weak singularity and therefore, it is continuous on Ω. In order to find the limit of the second integral when x tends to the point $x_0 \in \Omega$ from the outside and inside of the region V, we introduce the local Cartesian coordinate system (y_1, y_2, y_3) with the origin at the point x_0 and the y_3-axis directed along the external normal $n(x_0)$ to Ω (Fig. 3.4). Then, we consider the limiting value of the integral

$$J(y) = \int_V K(y - y')dy' \tag{3.122}$$

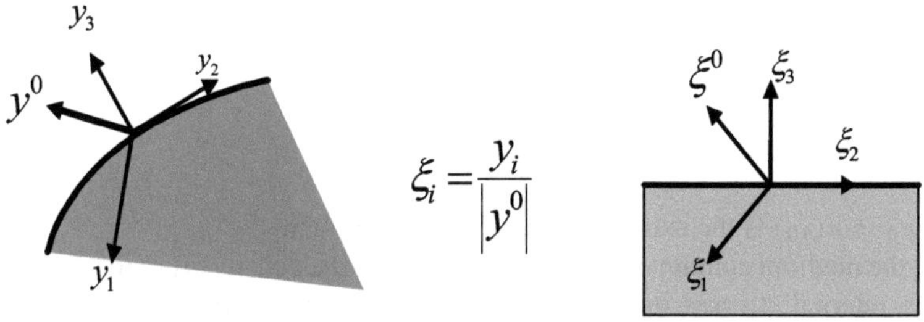

Fig. 3.4 Local coordinate systems on the border of the region containing stress sources, $|y^0| \to 0$, $|\xi^0| = 1$

when y tends to the origin $y = 0$ from the side of the normal $n(x_0)$.

Let $y = y_0$ be a fixed point outside Ω and $\xi_i = y_i/|y_0|$, $(i = 1, 2, 3)$. Because $K(y)$ is a homogeneous function of the order of -3, we have

$$J(y) = J(\xi|y_0|) = \int K(\xi - \xi')V(\xi')d\xi', \tag{3.123}$$

where $V(\xi)$ is the characteristic function of the region V in the variables ξ_i. If y_0 tends to zero, $\xi_0 = y_0/|y_0|$ is the unit vector of the direction along which point y_0 tends to the origin. In the limit $y_0 \to 0$, the region $V(\xi)$ is transformed (in the coordinates ξ_i) into the half-space $\xi_3 < 0$, i.e., $V(\xi_1, \xi_2, \xi_3) \to H_1(\xi_1, \xi_2, \xi_3) = 1 - H(\xi_3)$, where $H(\xi_3)$ is Heaviside's function. Thus, we yield the following equation

$$\lim_{y_0 \to 0} J(y_0) = \int \widetilde{K}(\xi_0 - \xi')H_1(\xi')d\xi' =$$
$$= (2\pi)^{-3} \int \widetilde{K}(k)\widetilde{H}_1(k)\exp(-ik \cdot \xi_0)dk\,, \tag{3.124}$$

Taking into account equation Eq. (3.85) for $\widetilde{H}(k)$ we obtain

$$J^+(0) = \lim_{y_0 \to 0} J(y_0) = \frac{1}{2}\left(\widetilde{K}(0) - \widetilde{K}(n)\right), \quad y_0 \notin V\,. \tag{3.125}$$

The same procedure gives us the limit of $J(y_0)$ if $y_0 \to 0$ from inside of V

$$J^-(0) = \lim_{y_0 \to 0} J(y_0) = \frac{1}{2}\left(\widetilde{K}(0) + \widetilde{K}(n)\right), \quad y_0 \in V\,. \tag{3.126}$$

Hence the jump of the integral $J(y)$ on the boundary Ω of the region V is

$$[J(0)] = J^+(0) - J^-(0) = -\widetilde{K}(n)\,. \tag{3.127}$$

From Eq. (3.121) follows the equation for the jump of the potential $\varepsilon(x)$ in Eq. (3.120) on the surface Ω

$$[\varepsilon(x_0)] = \varepsilon^+(x_0) - \varepsilon^-(x_0) = \widetilde{K}(n_0)q(x_0)\,, \tag{3.128}$$

Here, $n_0 = n(x_0)$ is the external normal to Ω at point x_0.

Let the medium contain dislocation moments of the density $m(x)$ in a finite region V. The internal stresses in the medium is determined by Eq. (3.52). If $m(x)$ is a continuous bounded function, the tensor $\sigma(x)$ can be presented as the sum

$$\sigma(x) = \int\limits_V S(x-x')\left(m(x') - m(x)\right)dx' + \int\limits_V S(x-x')dx'm(x), \tag{3.129}$$

where the first term is continuous on the boundary Ω of V as the integral with a weak singularity. For determination of the jumps of the second integral in this equation, the same method as for determination of the jump of the integral in Eq. (3.122) can be used. As a result, the jump of the stress tensor $\sigma(x)$ in Eq. (3.119) at point $x_0 \in \Omega$ is defined by the equation

$$[\sigma(x_0)] = \sigma^+(x_0) - \sigma^-(x_0) = -\widetilde{S}(n_0)m(x_0)\,, \tag{3.130}$$

where $\sigma^+(x_0)$ is the limit value of $\sigma(x)$ when $x \to x_0$ from the side of the external normal $n_0 = n(x_0)$, $\sigma^-(x_0)$ is the same limit when x tends to x_0 from the opposed side. Function $\widetilde{S}(k)$ is defined in (3.61).

3.6 Elastic Fields Far from the Stress Sources

Let body forces with density $q(x)$ be distributed inside a bounded region V in a homogeneous elastic medium. If $q(x)$ is an integrable function inside V, the displacement field $u(x)$ outside V is an infinitely differentiable function in Eq. (3.6). Expansion of Green's function $G(x - x')$ in the Taylor series in the vicinity of the point $x - x_0$ ($x_0 \in V$, $x \notin V$) has the form

$$G_{\alpha\beta}(x-x') = \sum_{k=0}^{\infty} \frac{1}{k!}\left(\nabla_{\lambda_1}\nabla_{\lambda_2}...\nabla_{\lambda_k} G_{\alpha\beta}(x-x_0)\right) \times$$
$$\times \left(x_0 - x'\right)_{\lambda_1}\left(x_0 - x'\right)_{\lambda_2} \cdots \left(x_0 - x'\right)_{\lambda_k}\,. \tag{3.131}$$

Substituting this expansion into the right-hand side of Eq. (3.6), we yield

$$u_\alpha(x) = \sum_{k=0}^{\infty} \nabla_{\lambda_1}\nabla_{\lambda_2}...\nabla_{\lambda_k} G_{\alpha\beta}(x-x_0)\,Q^k_{\beta\lambda_1\lambda_2...\lambda_k}(x_0)\,, \tag{3.132}$$

$$Q^k_{\beta\lambda_1\lambda_2\ldots\lambda_k}(x_0) = \\ = \frac{(-1)^k}{k!}\int\limits_V (x' - x_0)_{\lambda_1}(x' - x_0)_{\lambda_2}\ldots(x' - x_0)_{\lambda_k} q_\beta(x')dx'. \tag{3.133}$$

The same result can be obtained if the function $q(x)$ in (3.6) is presented in the form of the following series of generalized functions

$$q_\beta(x) = \sum_{m=0}^{\infty} Q^{(m)}_{\beta\lambda_1\lambda_2\ldots\lambda_m}(x_0)\nabla_{\lambda_1}\nabla_{\lambda_2}\ldots\nabla_{\lambda_k}\delta(x - x_0)\,. \tag{3.134}$$

Note that the first term of this expansion

$$Q^{(0)}_\beta(x_0)\delta(x - x_0) = \int\limits_V q_\beta(x')dx'\delta(x - x_0) \tag{3.135}$$

can be interpreted as the total force of the distribution $q_\alpha(x)$ applied to point x_0. The second term in series (3.134) is the force dipole with the total moment

$$Q^{(1)}_{\beta\lambda_1} = -\int\limits_V (x' - x_0)_{\lambda_1} q_\beta(x')dx', \tag{3.136}$$

applied to the point x_0. Other terms in Eq. (3.134) can be interpreted as the concentrated multipoles of higher orders. Series (3.134) is called expansion of the force density $q(x)$ in the series of multipoles concentrated at the point $x_0 \in V$ (Fig. 3.5). The series (3.134) should be understood in the following sense. After convolution of the both parts of Eq. (3.134) with an arbitrary analytical function $\varphi(x)$, the right and left-hand sides of the resulting equation coincide.

The kth term of series (3.132) decays as $|x|^{-(k+1)}$ when $|x| \to \infty$. Therefore, the larger is the distance from the point x to the region V the better is convergence of this series. In order to calculate the elastic fields far from the region V we can take only several first terms of the multipole expansion (3.134). Keeping three terms only, we obtain for the displacement field the equation

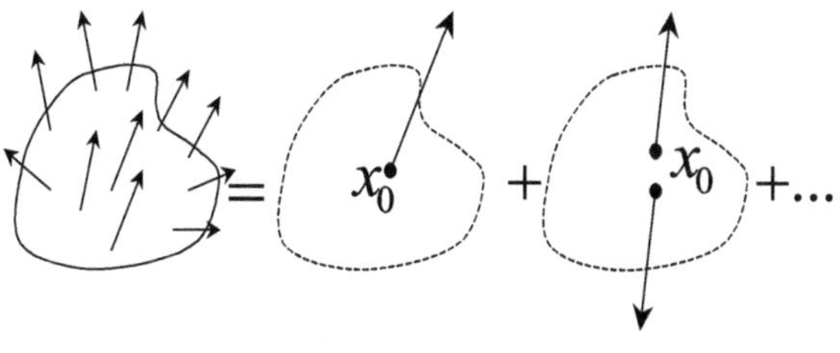

Fig. 3.5 Expansion of the distribution of body forces inside a region in the multipole series

$$u_\alpha(x) = G_{\alpha\beta}(x - x_0)Q_\beta^{(0)} + \nabla_\lambda G_{\alpha\beta}(x - x_0)Q_{\beta\lambda}^{(1)} +$$

$$+ \nabla_\lambda \nabla_\mu G_{\alpha\beta}(x - x_0)Q_{\beta\lambda\mu}^{(2)} . \tag{3.137}$$

Here, $G_{\alpha\beta}(x - x_0)Q_\beta^0$ is the principal term of the field $u(x)$ when $|x| \to \infty$.

The density of the force dipoles or dislocation moments distributed in a finite region V can be also expanded in the multipole series similar to (3.134). In this case, the strain $\varepsilon(x)$ and stress $\sigma(x)$ tensors far from the region V can be approximated by first terms of the series

$$\varepsilon_{\alpha\beta}(x) = \sum_{m=0}^{\infty} \nabla_{\lambda_1}...\nabla_{\lambda_m} K_{\alpha\beta\mu\nu}(x - x_0)C_{\lambda\mu\tau\rho}M_{\tau\rho\lambda_1...\lambda_m}(x_0) , \tag{3.138}$$

$$\sigma_{\alpha\beta}(x) = \sum_{m=0}^{\infty} \nabla_{\lambda_1}...\nabla_{\lambda_m} S_{\alpha\beta\mu\nu}(x - x_0)M_{\mu\nu\lambda_1...\lambda_m}(x_0) , \tag{3.139}$$

$$M_{\alpha\beta\lambda_1...\lambda_m}(x_0) = \frac{(-1)^m}{m!} \int_V (x' - x_0)_{\lambda_1} ... (x' - x_0)_{\lambda_m} m_{\alpha\beta}(x')\, dx'. \tag{3.140}$$

3.7 Notes

The projection operators on the div-free and rotC^{-1}-free subspaces in application to electrostatics, magnetostatics, and elasticity were introduced in [6, 7].

The general structure of Green's function for anisotropic homogeneous elastic media was revealed in [1]. Regularizations of the generalized functions $K(x)$ and $S(x)$ (Green's function of elasticity for strains and internal stresses) follows from the algorithm of regularization of this type of generalized functions presented in [3]. The method of regularization of the operators K and S on constants was proposed in [8].

Green's tensor $Z(x)$ for internal stresses in the case of anisotropic homogeneous elastic media was obtained in [4]. Fundamental results on the internal geometry of elastic media with dislocation sources of various types are obtained in [9–11].

The equations for the jumps of elastic fields on the borders of the regions containing the sources of external and internal stresses are derived in [12].

References

1. Lifshitz I, Rosenzweig L (1947) Green's tensor for an anisotropic infinite elastic medium. J Exper Teor Phys 17:83–791
2. Mikhlin S (1962) Multi-dimensional Singular Integrals and Integral Equations. Phys Math Lit, Moscow, 254

3. Gelfand I, Shilov G (1964) Generalized functions. Academic, New York, p 470
4. Kunin I (1964) Internal stresses in an anisotropic elastic medium. Appl Math Mech (PMM) 28:612–622
5. Brychkov Y, Prudnikov A (1989) Integral transforms of generalized functions. CRC Press, p 344
6. Kunin I (1965) Methods of tensor analysis in the theory of dislocations. US Department of Commerce, Clearing House for Fed Sci Techn Inform. Springfield, VA, p 22151
7. Kunin I (1981) An algebra of tensor operators and its application to elasticity. Int J Eng Sci 19:1551–1561
8. Kanaun S (1975) The method of self-consistent field in the problem of effective elastic properties of composites. J Appl Mech Techn Phys 4:194–203
9. Kröner E (1958) Kontinuumstheorie der Versetzungen and Eigenspannungen. Springer, Berlin Göttingen Heidelberg, p 179
10. Eshelby (1961) Elastic inclusions and inhomogeneities. In: Sneddon I, Hill R (eds) Progress in solid mechanics, vol 2, pp 88–140
11. Kosevich A (1978) Dislocations in the theory of elasticity. Naukova Dumka, Kiev, p 220
12. Kanaun S, Levin V (1994) Effective field method on the mechanics of matrix composite materials. In: Markov K (ed) Advances in mathematical modeling of composite materials. Series on advances in mathematics for applied sciences, vol 15, World Scientific, pp 1–58

Chapter 4
Homogeneous Media with an Isolated Inclusion

Abstract Static problems of conductivity and elasticity for a homogeneous medium with an isolated inclusion are considered. The problem are reduced to volume integral equations for the field in the regions occupied by the inclusion. For homogeneous ellipsoidal inclusions, the property of polynomial conservation of the solution is indicated. The solution for ellipsoidal inclusions subjected to homogeneous external fields is reduced to calculation of regular 2D integrals.

Keywords Isolated inclusion · Volume integral equations · Ellipsoidal inclusions · Polynomial conservation · Solution for homogeneous external fields

In this chapter, we consider static problems of conductivity and elasticity for a homogeneous medium with an isolated inclusion (the one-particle problem). This problem is crucial for self-consistent methods in the theory of heterogeneous media. The principal objectives in these problems are the fields inside the inclusions. For homogeneous ellipsoidal inclusions, the solution has a remarkable property: a polynomial external field induces a polynomial field of the same power inside the inclusions. In this case, the one-particle problem consists in calculation of regular integrals and inversion of forth rank tensors of specific symmetry.

4.1 Integral Equations for the Strain and Stress Tensors in an Elastic Medium with an Isolated Heterogeneous Inclusion

We consider an infinite homogeneous host medium with the elastic stiffness tensor C^0 containing an isolated inclusion in a finite region V. It is assumed that the elastic stiffness tensor $C(x)$ of the inclusion is a piece-wise smooth tensor function. The stiffness tensor of the medium with the inclusion is presented in the form

$$C(x) = C^0 + C^1(x)V(x), \quad C^1(x) = C(x) - C^0, \tag{4.1}$$

where $V(x)$ is the characteristic function of the region V

S. Kanaun, *Self-Consistent Methods for Composites*, Solid Mechanics and Its Applications 282, https://doi.org/10.1007/978-3-031-96739-9_4

$$V(x) = 1 \text{ if } x \in V, \quad V(x) = 0 \text{ if } x \notin V. \tag{4.2}$$

If the medium is subjected to external body forces $q(x)$, the displacement vector $u(x)$ is the solution of the partial differential equation

$$L_{\alpha\beta} u_\beta(x) = q_\alpha(x)\,, \quad L_{\alpha\beta} = L^0_{\alpha\beta} + L^1_{\alpha\beta}\,, \tag{4.3}$$

$$L^0_{\alpha\beta} = -\nabla_\lambda C^0_{\lambda\alpha\beta\mu} \nabla_\mu\,, \quad L^1_{\alpha\beta} = -\nabla_\lambda C^1_{\lambda\alpha\beta\mu}(x) V(x) \nabla_\mu\,. \tag{4.4}$$

Let $u^0(x)$ be the displacement field in the homogeneous host medium subjected to the body forces $q(x)$ and prescribed conditions at infinity. The field $u^0(x)$ satisfies the equation

$$(L^0_{\alpha\beta} u^0_\beta)(x) = q_\alpha(x)\,. \tag{4.5}$$

In what follows, the field $u^0(x)$ and the corresponding strain $\varepsilon^0(x) = \mathrm{def}\, u^0(x)$ and stress $\sigma^0(x) = C^0 \varepsilon^0(x)$ fields are called external fields. The solution $u(x)$ of Eq. (4.3) can be presented in the form

$$u(x) = u^0(x) + u^1(x)\,, \tag{4.6}$$

where $u^1(x)$ is the disturbance of the external field $u^0(x)$ due to the presence of the inclusion, $u^1(x) \to 0$ when $|x| \to \infty$. Substituting Eq. (4.6) into Eq. (4.3), we yield the equation for the field $u^1(x)$

$$(L^0 + L^1) u^1 = -L^1 u^0\,. \tag{4.7}$$

Application of the operator $G = (L^0)^{-1}$ and then, the operator def to Eq. (4.7) results in the equation

$$\varepsilon^1 + \mathrm{def}\, G L^1 u^1 = -\mathrm{def}\, G L^1 u^0, \quad \varepsilon^1 = \mathrm{def}\, u^1. \tag{4.8}$$

Taking into account Eq. (4.4) for the operator L^1, we can rewrite this equation in the form

$$\varepsilon^1(x) + (K C^1 \varepsilon)(x) = 0\,, \tag{4.9}$$

$$K = -\mathrm{def}\, G\, \mathrm{def}, \quad \varepsilon = \mathrm{def}(u^0 + u^1)\,. \tag{4.10}$$

Adding to the both sides of Eq. (4.10) the external strain field ε^0 results in the final equation for the strain tensor $\varepsilon(x)$ in the medium with the inclusion

$$\varepsilon(x) + (K C^1 \varepsilon)(x) = \varepsilon^0(x). \tag{4.11}$$

In the index notations, this equation is

$$\varepsilon_{\alpha\beta}(x) + \int K_{\alpha\beta\lambda\mu}(x - x')C^1_{\lambda\mu\nu\rho}(x')\varepsilon_{\nu\rho}(x')V(x')dx' = \varepsilon^0_{\alpha\beta}(x)\,, \tag{4.12}$$

$$K_{\alpha\beta\lambda\mu}(x) = -\nabla_\alpha \nabla_\lambda G_{\beta\mu}(x)|_{(\alpha\beta)(\lambda\mu)}\,. \tag{4.13}$$

Note that the function $K(x)$ coincides with the kernel of the operator K in Eq. (3.120) for the strain field in a homogeneous elastic medium subjected to generalized force moments. Product $K(x)C^0$ is the kernel of the projection operator on the Rot-free subspace of the second rank tensor function (see Eq. (3.67)). The kernel $K(x)$ is a homogeneous function of the order of -3. Action of such a function on a continuous finite function is defined in Eq. (3.22). The existence and uniqueness of the solution of Eq. (4.12) follows from the equivalence of this equation to the system of differential equations of elasticity (4.7).

Let us rewrite Eq. (4.11) in terms of the stress tensor $\sigma(x)$ in the medium with an inclusion. Using Hook's law, we have

$$\varepsilon(x) = B(x)\sigma(x), \;\; \varepsilon^0(x) = B^0\sigma^0(x), \tag{4.14}$$

$$C^1(x)\varepsilon(x) = C^1(x)B(x)\sigma(x) = -C^0B^1(x)\sigma(x), \tag{4.15}$$

$$B(x) = C(x)^{-1}, \;\; B^0 = \left(C^0\right)^{-1}, \;\; B^1(x) = B(x) - B^0. \tag{4.16}$$

Then, the equation for $\sigma(x)$ follows from Eq. (4.11) in the form

$$B^0\sigma(x) + B^1(x)\sigma(x) - KC^0B^1(x)\sigma(x) = B^0\sigma^0(x). \tag{4.17}$$

After multiplication this equation with the tensor C^0, we yield

$$\sigma(x) - SB^1(x)\sigma(x) = \sigma^0(x), \tag{4.18}$$

where the operator S is

$$S = C^0KC^0 - C^0I. \tag{4.19}$$

The detailed form of Eq. (4.18) is

$$\sigma_{\alpha\beta}(x) - \int S_{\alpha\beta\lambda\mu}(x - x')B^1_{\lambda\mu\nu\rho}(x')\sigma_{\nu\rho}(x')V(x')dx' = \sigma^0_{\alpha\beta}(x)\,, \tag{4.20}$$

$$S_{\alpha\beta\lambda\mu}(x) = C^0_{\alpha\beta\gamma\delta}K_{\gamma\delta\tau\upsilon}(x)C^0_{\tau\upsilon\lambda\mu} - C^0_{\alpha\beta\lambda\mu}\delta(x). \tag{4.21}$$

Note that the function $S(x)$ coincides with Green's function for internal stresses in Eq. (3.58), and operator $S\left(C^0\right)^{-1}$ is the projector on the subspace of div-free second rank tensors.

4.2 Thermo-elasticity Problem for a Homogeneous Medium with an Isolated Inclusion

Let the host medium and the inclusion have different elastic properties and thermo-expansion coefficients. Thus, the stiffness tensor $C(x)$ and the second rank tensor of thermo-expansion coefficients $\alpha(x)$ of the medium can be presented in the forms

$$C(x) = C^0 + C^1(x)V(x), \quad \alpha(x) = \alpha^0 + \alpha^1(x)V(x), \tag{4.22}$$

where α^0 is the tensor of thermo-expansion coefficients of the host medium, $\alpha^1(x) = \alpha(x) - \alpha^0$. If the medium is subjected to external body forces $q(x)$ and temperature field $T(x)$, the total strain tensor $\varepsilon(x)$ is the sum of elastic ε^e and temperature $\alpha(x)T(x)$ strains

$$\varepsilon(x) = \varepsilon^e(x) + \alpha(x)T(x). \tag{4.23}$$

The stress tensor $\sigma(x)$ relates to the elastic strain tensor $\varepsilon^e(x)$ by Hook's law $\sigma(x) = C(x)\varepsilon^e(x)$, and the complete tensor $\varepsilon(x)$ satisfies to the compatibility Eq. (3.1). Thus, the system of differential equation of thermo-elasticity has the form

$$\text{div}\sigma(x) = -q(x) \quad \sigma(x) = C(x)\varepsilon^e(x), \tag{4.24}$$

$$\text{Rot}\varepsilon^e(x) = -\text{Rot}\left(\alpha(x)T(x)\right). \tag{4.25}$$

This system can be written in the form of the governing equations for the homogeneous host medium with the stiffness tensor C^0 subjected to fictitious sources of external and internal stresses as follows

$$\text{div}\sigma(x) = -q(x), \quad \sigma(x) = C^0\varepsilon'(x), \quad \text{Rot}\varepsilon'(x) = -\text{Rot}\, m(x), \tag{4.26}$$

where the auxiliary strain $\varepsilon'(x)$ and dislocation moment $m(x)$ tensors are

$$\varepsilon'(x) = \varepsilon^e(x) - B^1(x)\sigma(x), \quad m(x) = B^1(x)\sigma(x) + \alpha(x)T(x), \tag{4.27}$$

$$B^1(x) = B(x) - B^0, \quad B(x) = C^{-1}(x), \quad B^0 = (C^0)^{-1}. \tag{4.28}$$

Let $\sigma^0(x)$ be the external stress field that satisfies the equations

$$\text{div}\sigma^0(x) = -q(x), \quad \sigma^0(x) = C^0\varepsilon^0(x), \quad \text{Rot}\varepsilon^0(x) = 0 \tag{4.29}$$

and given conditions at infinity. Using Eq. (3.52) for the tensor $\sigma(x)$ in the medium with the sources of internal stresses, the solution of the system (4.29) can be presented in the integral form

$$\sigma(x) = \sigma^0(x) + \int S(x - x')m(x')dx' . \tag{4.30}$$

Here, the function $S(x)$ is defined in Eq. (3.58)

$$S_{\alpha\beta\lambda\mu}(x) = C^0_{\alpha\beta\nu\rho} K_{\nu\rho\tau\delta}(x) C^0_{\tau\delta\lambda\mu} - C^0_{\alpha\beta\lambda\mu}\delta(x) . \tag{4.31}$$

Substituting in Eq. (4.30) the function $m(x)$ from Eq. (4.27), we yield the equation for the stress tensor in the medium with a heterogeneous inclusion

$$\sigma_{\alpha\beta}(x) - \int S_{\alpha\beta\lambda\mu}(x - x') B^1_{\lambda\mu\tau\rho}(x')\sigma_{\tau\rho}(x')V(x')dx' =$$
$$= \sigma^0_{\alpha\beta}(x) + \int S_{\alpha\beta\lambda\mu}(x - x')\alpha_{\lambda\mu}(x')T(x')dx', \tag{4.32}$$

When $T = 0$, the integral term in the right-hand side of this equation vanishes, and we obtain integral equation (4.20) for the stress tensor in the medium with an isolated inclusion. If the external field $\sigma^0(x)$ is equal to zero, Eq. (4.32) is the integral equation for temperature stresses in the medium with a heterogeneous inclusion subjected to temperature field $T(x)$ only

$$\sigma(x) - \int S(x - x')B^1(x')\sigma(x')V(x')dx' =$$
$$= \int S_{\alpha\beta\lambda\mu}(x - x')\alpha_{\lambda\mu}(x')T(x')dx' . \tag{4.33}$$

We consider important properties of the integral equations (4.12) and (4.20).

1. The strain tensor $\varepsilon(x)$ in the medium with a heterogeneous inclusion can be presented in the form

$$\varepsilon(x) = \varepsilon^0(x) + \varepsilon^1(x) , \tag{4.34}$$

where $\varepsilon^1(x)$ is the perturbation of the strain field due to the inclusion. The function $\varepsilon^1(x)$ satisfies the equation that follows from Eq. (4.12)

$$\varepsilon^1(x) + \int K(x - x')C^1(x')\varepsilon^1(x')V(x')dx' =$$
$$= -\int K(x - x')C^1(x')\varepsilon^0(x')V(x')dx'. \tag{4.35}$$

It follows from this equation that the field $\varepsilon^1(x)$ depends on the value of the external field $\varepsilon^0(x)$ inside the region V occupied by the inclusion only.

2. If strain $\varepsilon^+(x) = \varepsilon(x)V(x)$ and stress $\sigma^+(x) = \sigma(x)V(x)$ fields inside the region V are known, the fields outside V are reconstructed from Eqs. (4.12) and (4.20):

$$\varepsilon(x) = \varepsilon^0(x) - \int K(x - x')C^1(x')\varepsilon^+(x')dx', \tag{4.36}$$

$$\sigma(x) = \sigma^0(x) + \int S(x - x')B^1(x')\sigma^+(x')dx'. \tag{4.37}$$

The integral equations for the fields $\varepsilon^+(x)$ and $\sigma^+(x)$ inside the inclusion follow from Eqs. (4.12) and (4.20)

$$\varepsilon^+(x) + \int_V K(x, x')C^1(x')\varepsilon^+(x')dx' = \varepsilon^0(x), \quad x \in V \tag{4.38}$$

$$\sigma^+(x) - \int_V S(x, x')B^1(x')\sigma^+(x')dx' = \sigma^0(x), \quad x \in V. \tag{4.39}$$

3. Let $C^1(x) \to -C^0$, $C(x) = C^0 + C^1(x) \to 0$, $x \in V$ (a cavity in the region V). In this case, Eq. (4.38) for the strain tensor $\varepsilon^+(x)$ takes the form

$$\varepsilon^+(x) - \int_V K(x - x')C^0\varepsilon^+(x')dx' = \varepsilon^0(x), \quad x \in V. \tag{4.40}$$

Because in the limit $C \to 0$,

$$B^1\sigma^+ = \left(C^{-1} - (C^0)^{-1}\right)C\varepsilon^+ \to \varepsilon^+, \tag{4.41}$$

Equation (4.37) for the stress tensor $\sigma(x)$ is

$$\sigma(x) = \sigma^0(x) + \int_V S(x - x')\varepsilon^+(x')dx'. \tag{4.42}$$

Hence after constructing tensor $\varepsilon^+(x)$ from Eq. (4.40), we can find the stress field outside the cavity from Eq. (4.42).

Taking into account Eq. (4.31) for the function $S(x)$, the kernel of the integral operator in Eq. (4.40) can be presented as follows

$$K(x)C^0 = (C^0)^{-1}S(x) + I\delta(x). \tag{4.43}$$

As a result, Eq. (4.40) is transformed into the equation

$$\int_V S(x - x')\varepsilon^+(x')dx' = C^0\varepsilon^0(x). \tag{4.44}$$

It follows from Eq. (3.53) for $S(x)$ and Gauss' theorem that homogeneous equation (4.44) (for $\varepsilon^0 = 0$) is equivalent to the following one:

$$\int_V S(x-x')\varepsilon^+(x')dx' = \int_V Z(x-x')\mathrm{Rot}\,\varepsilon^+(x')dx' = 0\,. \tag{4.45}$$

Hence each function $\varepsilon^+(x)$ that satisfies the compatibility equation $\mathrm{Rot}\varepsilon^+ = 0$ is a solution of the homogeneous Eq. (4.44). Therefore, the solution of Eq. (4.44) is not unique. Further, for the strain field inside a cavity, we take the limiting value of the solution of Eq. (4.12) when $C^1 \to -C^0$.

4. Let $B^1(x) \to -B^0$, or $B \to 0$ inside the region V (an absolutely rigid inclusion). In this case, Eq. (4.20) for the stress tensor in V takes the form

$$\sigma^+(x) - \int_V S(x-x')B^0\sigma^+(x')dx' = \sigma^0(x)\,, \quad x \in V\,. \tag{4.46}$$

Because in the limit $B \to 0$,

$$C^1\varepsilon^+ = \left(B^{-1} - (B^0)^{-1}\right)B\sigma^+ \to \sigma^+, \tag{4.47}$$

Equation (4.36) for the strain tensor in the medium with absolutely rigid inclusion takes the form

$$\varepsilon(x) = \varepsilon^0(x) - \int_V K(x-x')\sigma^+(x')dx'. \tag{4.48}$$

This equation determines the strain field $\varepsilon(x)$ outside of a rigid inclusion if the stress field $\sigma^+(x)$ inside the inclusion is known.

Taking into account Eq. (4.31) for $S(x)$, we obtain the kernel of the integral operator in Eq. (4.46)

$$S(x)B^0 = C^0K(x) - I\delta(x)\,. \tag{4.49}$$

It follows from this equation, Eq. (4.13) for $K(x)$ and Gauss' theorem that the homogeneous equation (4.46) ($\sigma^0 = 0$) is equivalent to the following equation

$$C^0\int_V K(x-x')\sigma^+(x')dx' = -C^0\int_V \nabla G(x-x')\mathrm{div}'\sigma^+(x')dx' = 0\,. \tag{4.50}$$

Thus, an arbitrary stress field $\sigma^+(x)$ that satisfies the equilibrium equation $\mathrm{div}\sigma^+ = 0$ is the solution of the homogeneous equation (4.20), and therefore, the solution of Eq. (4.20) is not unique. Further, the limiting solution of Eq. (4.20) when $B^1 \to -B_0$ inside the inclusion is taken for the stress field σ^+ in the absolutely rigid inclusion.

4.3 Conditions on the Interface of Two Elastic Media

Equation (4.12) for the strain tensor $\varepsilon(x)$ in a homogeneous medium with an isolated inclusion can be presented as the following sum

$$\varepsilon(x) = \varepsilon^0(x) + \varepsilon^1(x)\,, \quad \varepsilon^1(x) = -\int K(x - x')C^1(x')\varepsilon(x')V(x')dx' \tag{4.51}$$

Let V be a finite region with a smooth boundary Ω and $C^1(x)$ be a continuous function inside V. If the external field $\varepsilon^0(x)$ is a continuous function, the function $\varepsilon(x)$ is continuous everywhere except the boundary Ω of V. The jump of $\varepsilon(x)$ coincides with the jump of the potential $\varepsilon^1(x)$ in Eq. (4.51). Comparison of Eq. (4.51) with Eq. (3.120) shows that the potential $\varepsilon^1(x)$ can be interpreted as the strain field in the homogeneous elastic medium with the stiffness tensor C^0 subjected to distributed force moments of the density $q_{\alpha\beta}(x) = C^1_{\alpha\beta\gamma\delta}(x)\varepsilon_{\gamma\delta}(x)V(x)$.

The jumps of the strain field on the boundary of the region containing force moments of density $q(x)$ are considered in Sect. 3.5. Let $n(x_0)$ be the external normal to Ω at point $x_0 \in \Omega$. If $\varepsilon^+(x_0)$ is the limiting value of the strain tensor $\varepsilon(x)$ inside the inclusion when $x \to x_0$, and $\varepsilon^-(x)$ is the similar limit when x is outside V. Then, it follows from Eqs. (3.128) and (4.51) that the jump of the strain tensor on the surface Ω is

$$[\varepsilon(x_0)] = \varepsilon^-(x_0) - \varepsilon^+(x_0) = \widetilde{K}(n_0)C^{1+}(x_0)\varepsilon^+(x_0)\,. \tag{4.52}$$

Here, $n_0 = n(x_0)$, the tensor $\widetilde{K}(k)$ is determined in (3.16), $C^{1+}(x_0)$ is the limit of $C^1(x)$ when $x \to x_0$ inside V (Fig. 4.1). Equation (4.52) implies that the limiting values of the strain tensor in the medium $\varepsilon^-(x_0)$ and in the inclusion $\varepsilon^+(x_0)$ at a point x_0 of the inclusion interface Ω relate by the equation

$$\varepsilon^-(x_0) = \left(I + \widetilde{K}(n_0)C^{1+}(x_0)\right)\varepsilon^+(x_0)\,, \quad x_0 \in \Omega\,. \tag{4.53}$$

Here, $I = E^1$ is unit fourth rank tensor.

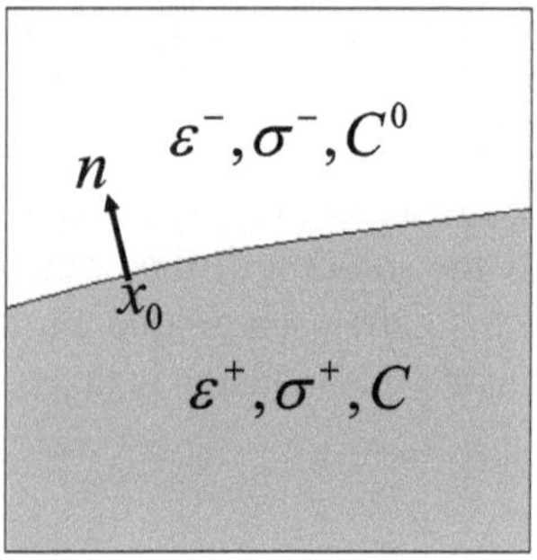

Fig. 4.1 The interface between two media

The stress tensor $\sigma(x)$ in Eq. (4.33) for the homogeneous medium with an inclusion is presented in the form

$$\sigma(x) = \sigma^0(x) + \sigma^1(x), \quad \sigma^1(x) = \int_V S(x - x')B^1(x')\sigma(x')dx'. \tag{4.54}$$

If the external field $\sigma^0(x)$ is a continuous function, the jumps of the function $\sigma(x)$ on Ω coincide with the jumps of the potential $\sigma^1(x)$. It follows from Eq. (3.52) that the potential $\sigma^1(x)$ can be interpreted as the internal stresses in a homogeneous medium with the stiffness tensor C^0 subjected to dislocation moments of the density $m(x) = B^1(x)\sigma(x)$ in the region V. The jump of $\sigma^1(x)$ on Ω is determined in Eq. (3.130) and it takes the form

$$[\sigma(x_0)] = \sigma^-(x_0) - \sigma^+(x_0) = -\widetilde{S}(n_0)B^{1+}(x_0)\sigma^+(x_0), \quad x_0 \in \Omega. \tag{4.55}$$

Here, $\sigma^-(x_0)(\sigma^+(x_0))$ are the limiting values of the stress tensor in the medium (in the inclusion) when $x \to x_0 \in \Omega$, the tensor $\widetilde{S}(n)$ is defined in Eq. (3.61). Hence the limiting values of the stress tensor outside and inside the region V at the point x_0 of the interface Ω relate by the equation

$$\sigma^-(x_0) = \left(I - \widetilde{S}(n_0)B^{1+}(x_0)\right)\sigma^+(x_0), \quad x_0 \in \Omega. \tag{4.56}$$

Consider some consequences of these equations.

1. We introduce two projectors: Π is the projector on the normal $n = n(x)$ to Ω, and Θ is the projector on the tangent plane to Ω at the point $x \in \Omega$ (don't mix with the projectors on div-free and Rot-free functional subspaces!)

$$\Theta_{\alpha\beta\lambda\mu} = \theta_{\alpha\lambda}\theta_{\mu\beta}|_{(\alpha\beta)}, \quad \theta_{\alpha\beta} = \delta_{\alpha\beta} - n_\alpha n_\beta, \tag{4.57}$$

$$\Pi = I - \Theta, \quad I_{\alpha\beta\lambda\mu} = \delta_{\alpha\lambda}\delta_{\mu\beta}|_{(\alpha\beta)}, \tag{4.58}$$

It follows from these definitions that tensors Π and Θ are orthogonal projectors:

$$\Pi_{\alpha\beta\lambda\mu}\Pi_{\lambda\mu\nu\rho} = \Pi_{\alpha\beta\nu\rho}, \quad \Theta_{\alpha\beta\lambda\mu}\Theta_{\lambda\mu\nu\rho} = \Theta_{\alpha\beta\nu\rho}, \tag{4.59}$$

$$\Pi_{\alpha\beta\lambda\mu}\Theta_{\lambda\mu\nu\rho} = \Theta_{\alpha\beta\lambda\mu}\Pi_{\lambda\mu\nu\rho} = 0. \tag{4.60}$$

These projectors split any symmetric second rank tensor function $A(x)$ on Ω into the "tangential" $\Theta_{\alpha\beta\lambda\mu}A_{\lambda\mu}$ and "normal" $\Pi_{\alpha\beta\lambda\mu}A_{\lambda\mu}$ components.

For the tensor $\widetilde{K}(k)$ in Eq. (3.16) and $k = n$, we have the identity

$$\Theta_{\alpha\beta\lambda\mu}(n)\widetilde{K}_{\lambda\mu\nu\rho}(n) \equiv 0. \tag{4.61}$$

Fig. 4.2 The ellipsoidal inhomogeneity

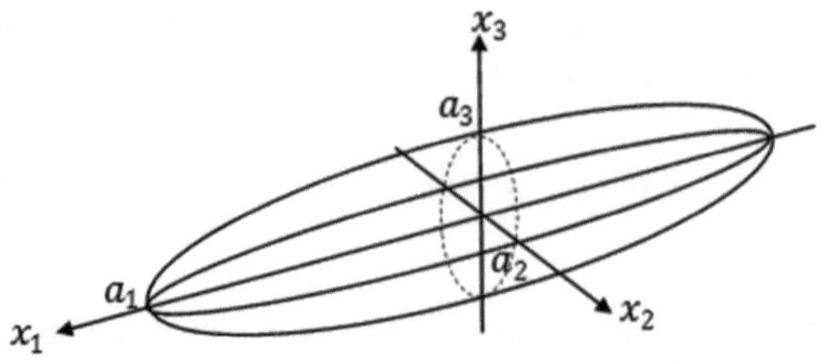

This equation and Eq. (4.53) show that the tangential component $\Theta(n)\varepsilon$ of the strain tensor is continuous on the inclusion interface Ω

$$[\Theta(n_0)\varepsilon(x_0)] = \Theta(n_0)\varepsilon^-(x_0) - \Theta(n_0)\varepsilon^+(x_0) = 0\,, \quad x_0 \in \Omega. \tag{4.62}$$

On the other hand, it follows from Eq. (3.61) that the tensor $\widetilde{S}(k)$ for $k = n$ satisfies the equation

$$n_\alpha \widetilde{S}_{\alpha\beta\lambda\mu}(n) = 0\,. \tag{4.63}$$

It follows from this equation and Eq. (4.55) that the normal component of the stress tensor is continuity on the inclusion surface

$$[n_0\sigma(x_0)] = n_0\sigma^-(x_0) - n_0\sigma^+(x_0) = 0\,, \quad x_0 \in \Omega. \tag{4.64}$$

2. Equations (4.53) and (4.55) contain the limiting values of the elastic fields and the elastic stiffness tensors of the host medium and at point $x_0 \in \Omega$ of the inclusion interface as well as the normal vector n_0 to Ω at this point. Thus, these conditions are local: they do not depend on the fields or elastic stiffness tensor at other points of the medium.

3. Equations (4.53) and (4.55) allow calculating stress and strain tensors in the medium at the surfaces of inclusions if the strain and stress fields inside inclusion are known. The limiting values of the tensors $\varepsilon^-(x)$ and $\sigma^-(x)$ at the inclusion interface $x \in \Omega$ can be presented as the following products

$$\varepsilon^-(x) = F^\varepsilon(n)\varepsilon^+(x), \quad \sigma^-(x) = F^\sigma(n)\sigma^+(x), \tag{4.65}$$

where the tensors $F^\varepsilon(n)$ and $F^\sigma(n)$ can be interpreted as the strain and stress concentration factors. The expressions for these tensors follow from Eqs. (4.53) and (4.55)

$$F^\varepsilon(n) = I + \widetilde{K}(n)C^{1+}(x), \quad F^\sigma(n) = I - \widetilde{S}(n)B^{1+}(x). \tag{4.66}$$

4.4 Ellipsoidal Inhomogeneity

Let an inclusion with the constant elastic stiffness tensor C occupy an ellipsoidal region V. In the cartesian coordinate system x_1, x_2, x_3 with the origin at the inclusion center and the axes directed along the ellipsoid semiaxes, the inclusion surface is defined by the equation

$$\frac{x_1^2}{a_1^2} + \frac{x_2^2}{a_2^2} + \frac{x_3^2}{a_3^2} = 1. \tag{4.67}$$

Here, a_1, a_2, a_3 are of the ellipsoid semiaxes (Fig. 4.2).

In this case, the solution of Eq. (4.12) has a remarkable property (the polynomial conservation theorem).

If the external field $\varepsilon^0(x)$ is a polynomial of the degree m with respect to coordinates x_1, x_2, x_3 inside the ellipsoidal region V, the strain field $\varepsilon^+(x)$ inside V is also a polynomial of the same degree m.

The proof of this theorem given in [1] is presented below.

Let the right-hand side $\varepsilon^0(x)$ of Eq. (4.12) for the strain tensor $\varepsilon(x)$ in the medium with a homogeneous ellipsoidal inclusion is a polynomial of degree m

$$\varepsilon_{\alpha\beta}(x) + \int_V K_{\alpha\beta\lambda\mu}(x - x')C^1_{\lambda\mu\nu\rho}\varepsilon_{\nu\rho}(x')dx' = \varepsilon^0_{\alpha\beta}(x). \tag{4.68}$$

Linear transformation $y = a^{-1}x$ ($a_{\alpha\beta} = a_\alpha\delta_{\alpha\beta}$ (no sum with α !)) of the x-space converts the region V into the ball of the unit radius, but the right-hand side $\varepsilon^0(x)$ remains polynomial of the same degree. Therefore, it is sufficient to give the proof for the case when V is the ball $|x| \leq 1$.

Polynomial conservation holds if the operator K with the kernel $K(x)$ transforms a polynomial of degree m into a polynomial of the same degree. Moreover, in order to prove this property, it is sufficient to consider the action of the operator K on a homogeneous polynomial of degree m or on the tensor product $x^{(m)}_{\lambda_1\lambda_2\ldots\lambda_m} = x_{\lambda_1}x_{\lambda_2}\ldots x_{\lambda_m}$.

Using regularization formula (3.22) for the operator K with $q_{\alpha\beta}(x) = \left(q^0x^{(m)}\right)_{\alpha\beta} = q^0_{\alpha\beta\lambda_1\lambda_2\ldots\lambda_m}x_{\lambda_1}x_{\lambda_2}\ldots x_{\lambda_m}$, where q^0 is a constant tensor, we yield

$$\int_V K(x - x')q(x')dx' = Aq(x) + pv\int_V K(x - x')q^0x'^{(m)}dx'. \tag{4.69}$$

Here, A is the constant tensor, and pv is Cauchy principal value of the integral. Because $K(x)$ is a homogeneous function of the order of -3, it is presented in the form

$$K(x - x') = \frac{1}{r^3}K(n), \quad n = \frac{(x - x')}{r}, \quad r = \left|x - x'\right|, \tag{4.70}$$

where $K(n)$ is an even function of unit vector n: $K(-n) = K(n)$. A consequence of existence of Cauchy principal value of the integral in Eq. (4.69) is the equation

$$\int_{\Omega_1} K(n)d\Omega_n = 0, \tag{4.71}$$

where Ω_1 is the surface of a unit sphere.

Substituting $x' = x + r\,n$ in the integral in the right-hand side of Eq. (4.69), we obtain

$$pv\int_V K(x-x')x'^{(m)}dx' = \sum_{k=0}^{m} x^{(m-k)}J^{(k)}(x)\,, \tag{4.72}$$

$$J^{(k)}(x) = pv\int_V K(n)n^{(k)}r^{k-3}dx'\,,\quad x \in V\,. \tag{4.73}$$

For the property (4.71) $J^{(0)} = 0$, and other integrals in Eq. (4.73) exist in the ordinary sense.

Then, we put $dx' = r^2 dr d\Omega_1$ in Eq. (4.73) and integrate over the elemental cone with the origin at the point $x \in V$ (see Fig. 4.3). As a result, this integral takes the form

$$J^{(k)}(x) = \frac{1}{k}\int_{\Omega_1} K(n)n^{(k)}\rho^k(x,n)d\Omega_1\,,\quad x \in V,\quad k = 1, 2, ..., m\,, \tag{4.74}$$

where $\rho(x,n)$ is the distance from the point x in the direction n to the surface of the unit sphere (Fig. 4.3)

$$\rho(x,n) = -(x\cdot n) + \sqrt{1-(x\cdot x)+(x\cdot n)^2}\,. \tag{4.75}$$

For ρ^k we have

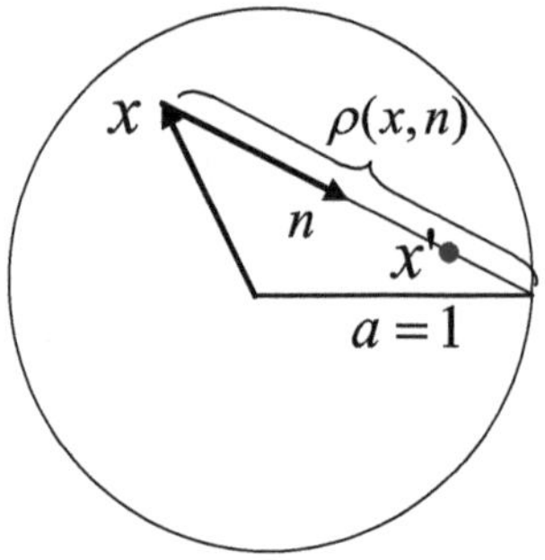

Fig. 4.3 To the proof of the polynomial conservation theorem

$$\rho^k = \sum_{l=0}^{k} (-1)^l C_k^l (x \cdot n)^l \left(1 - (x \cdot x) + (x \cdot n)^2\right)^{(k-l)/2}, \tag{4.76}$$

where C_k^l are the binomial coefficients. Because $K(n)$ is an even function of n, the terms in this series satisfying the condition $k + l = 2m$ can only contribute to the integral $J^{(k)}(x)$. These terms contain the root $\sqrt{1 - (x \cdot x) + (x \cdot n)^2}$ with an even degree, and the kth term has the form

$$\sum_{l=0}^{\left[\frac{k}{2}\right]} A_{(k-2l)}(n) x^{(k-2l)},$$

where $\left[\frac{k}{2}\right]$ is the integer part of $\frac{k}{2}$, $A_{(k-2l)}(n)$ are tensor coefficients. Therefore, $J^{(k)}(x)$ is a polynomial with respect to x of degree k with the tensor values

$$J^{(k)}(x) = \sum_{l=0}^{\left[\frac{k}{2}\right]} B_{(k-2l)}^{(k)} x^{(k-2l)}, \tag{4.77}$$

$$B_{(k-2l)}^{(k)} = \frac{1}{k} \int_{\Omega_1} K(n) n^{(k)} A_{(k-2l)}(n) d\Omega_1 . \tag{4.78}$$

Finally, substituting Eq. (4.77) into Eq. (4.72) we obtain that the second term in Eq. (4.69) is a polynomial of degree m which coefficients are tensors of the rank $(m + 4)$. This completes the proof of the polynomial conservation theorem for the ellipsoidal inclusion.

The polynomial conservation theorem can be also proved on the basis of the integral equation for stresses

$$\sigma(x) - \int_V S(x - x') B^1 \sigma(x') dx' = \sigma^0(x) \tag{4.79}$$

because the regular part of the generalized function $S(x)$ in Eq. (4.21) is proportional to $K(x)$.

4.5 An Ellipsoidal Inhomogeneity Subjected to a Constant External Field

Let x_1, x_2, x_3 be the Cartesian coordinate system with the origin at the center of an ellipsoidal inclusion. In this system, the inclusion surface is described by the equation

$$x_\alpha \left(a^{-2}\right)_{\alpha\beta} x_\beta = 1. \tag{4.80}$$

Here, a is a symmetric second rank tensor which principal values are the ellipsoid semiaxes a_1, a_2, a_3. It follows from the polynomial conservation theorem that for a constant external strain field ε^0, the strain field ε^+ inside the inclusion is constant. For calculation of the tensor ε^+, we can take $x = 0$ in Eq. (4.68)

$$\varepsilon^+ + \int K(x)V(x)dxC^1\varepsilon^+ = \varepsilon^0 \tag{4.81}$$

and obtain the solution in the form

$$\varepsilon^+ = (I + A(a)C^1)^{-1}\varepsilon^0, \;\; A(a) = \int K(x)V(x)dx. \tag{4.82}$$

Here, $V(x)$ is the characteristic function of the region occupied by the ellipsoid, $I = E^1$ is unit forth rank tensor. In the variables $y_\alpha = \left(a^{-1}\right)_{\alpha\beta} x_\beta$, the ellipsoid is transformed into unit ball with the characteristic function

$$V'(|y|) = H(1 - |y|), \tag{4.83}$$

where $H(z)$ is Heaviside's function. Transferring to the y-variables in the integral (4.82) yields

$$A(a) = \frac{1}{\det a}\int K\left(ay\right) V'(|y|)dy. \tag{4.84}$$

Using Parseval's formula and changing the integrand functions with their Fourier transforms, we obtain

$$A(a) = \frac{1}{(2\pi)^3}\int \widetilde{K}\left(a^{-1}k\right)\widetilde{V}'(|k|)dk. \tag{4.85}$$

Here, the equation

$$\int K\left(ay\right) \exp(ik \cdot y)dy =$$
$$= \det a \int K\left(x\right) \exp(\mathrm{i}\left(ka^{-1}x\right))dx = (\det a)\, \widetilde{K}(a^{-1}k) \tag{4.86}$$

is taken into account. Then, vector k can be presented in the form $k = |k|m$, where m is vector on unit sphere Ω_1 in the k-space. Integrating first over Ω_1 and then, over $|k|$ from zero to infinity, we yield

$$A(a) = \frac{1}{(2\pi)^3}\int_{\Omega_1} \widetilde{K}\left(a^{-1}m\right) d\Omega_m \int_0^\infty \widetilde{V}'(|k|)\, |k|^2 d|k|. \tag{4.87}$$

For $\widetilde{K}(k)$ is a homogeneous function of the order of 0, $\widetilde{K}\left(a^{-1}k\right) = \widetilde{K}\left(a^{-1}m\right)$. Because $V'(y)$ is the characteristic function of a unit ball centered at $y = 0$, the following equation holds

$$\frac{1}{(2\pi)^3}\int \widetilde{V}'(|k|)\exp(-ik\cdot y)dk|_{y=0} =$$

$$= \frac{1}{2\pi^2}\int_0^\infty \widetilde{V}'(|k|)\,|k|^2 d|k| = V'(|y|)\,|_{y=0} = 1\,. \tag{4.88}$$

From this equation and Eq. (4.87) we have finally

$$A(a) = \frac{1}{4\pi}\int_{\Omega_1} \widetilde{K}\left(a^{-1}k\right)d\Omega\,. \tag{4.89}$$

For $\widetilde{K}(k)$ is a homogeneous function of zero degree, tensor $A(a)$ does not depend on absolute sizes of the ellipsoid V.

Integral (4.89) can be interpreted as the average of the function $\widetilde{K}(a^{-1}k)$ over the unit sphere Ω_1 in the k-space. We denote such an average as $\prec \widetilde{K}(k)|a \succ$ and therefore, for arbitrary tensor function $T(k)$, we have

$$\prec T(k)|a \succ = \frac{1}{4\pi}\int_{\Omega_1} T\left(a^{-1}k\right)d\Omega. \tag{4.90}$$

It follows from Eq. (4.79) that for a constant external stress field σ^0, the stress tensor σ^+ inside the inclusion is also constant, and

$$\sigma^+ = (I - D(a)B^1)^{-1}\sigma^0, \tag{4.91}$$

$$D(a) = \frac{1}{4\pi}\int_{\Omega_1} \widetilde{S}\left(a^{-1}k\right)d\Omega = C^0 A(a) C^0 - C^0. \tag{4.92}$$

Note that the so-called Eshelby's tensor [2] is $\mathbb{S}(a) = A(a)C^0$.

For an ellipsoid with the semiaxes a_1, a_2, a_3, we introduce the dimensionless aspect ratios

$$\gamma_1 = \frac{a_3}{a_1},\ \ \gamma_2 = \frac{a_3}{a_2}. \tag{4.93}$$

Let x_1, x_2, x_3 be the Cartesian axes directed along the principal axes of the ellipsoid. In the conjugate basis in the k-space,

$$\widetilde{K}(a^{-1}k) = \widetilde{K}\left(\frac{k_1}{a_1}, \frac{k_2}{a_2}, \frac{k_3}{a_3}\right), \tag{4.94}$$

and the tensor $A(a)$ in Eq. (4.89) is presented as the following double integral

$$A(a) = \frac{1}{4\pi}\int_0^{2\pi} d\varphi \int_0^{\pi} \widetilde{K}\left(\frac{\cos\varphi\sin\theta}{a_1}, \frac{\sin\varphi\sin\theta}{a_2}, \frac{\cos\theta}{a_3}\right)\sin\theta d\theta. \tag{4.95}$$

Here, φ, θ are the polar and azimuthal angles of the spherical coordinate system with the polar axis k_3. Because $\widetilde{K}(k)$ is a homogeneous function of the order of zero, the equation $\widetilde{K}(\lambda k) = \widetilde{K}(k)$ holds for any scalar parameter λ. As a result, we have for the integrand function in Eq. (4.95)

$$\widetilde{K}\left(\frac{\cos\varphi\sin\theta}{a_1}, \frac{\sin\varphi\sin\theta}{a_2}, \frac{\cos\theta}{a_3}\right) =$$
$$= \widetilde{K}\left(\gamma_1\cos\varphi\sin\theta, \gamma_2\sin\varphi\sin\theta, \cos\theta\right). \tag{4.96}$$

We consider the limit expressions for the tensor $A(a)$ when the ellipsoid semiaxis a_3 tends to zero, and therefore, $\gamma_1, \gamma_2 \to 0$. From Eq. (4.96) we have

$$\widetilde{K}(a^{-1}k) \to \widetilde{K}(0, 0, \cos\theta) = \widetilde{K}(0, 0, 1) = \widetilde{K}(n). \tag{4.97}$$

Here, n is a unit vector directed along the semiaxis a_3 of the ellipsoid.

The same result can be obtain if a_3 is finite but $a_1, a_2 \to \infty$. This limit corresponds to the inclusion in the form of a plane layer (Fig. 4.4).

The strain ε^+ and stress σ^+ tensors inside the layer are defined by Eqs. (4.82) and (4.91), with the limiting tensors A and D

$$A = \widetilde{K}(n), \quad D = \widetilde{S}(n) = C^0\widetilde{K}(n)C^0 - C^0. \tag{4.98}$$

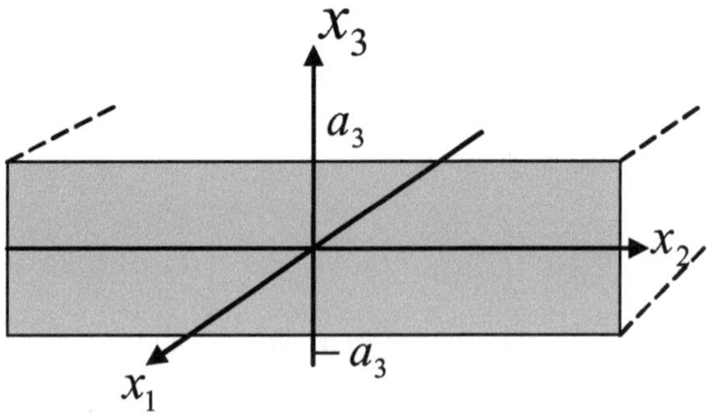

Fig. 4.4 The inclusion in the form of a plane layer

Hence for a constant external field $\varepsilon^0(\sigma^0 = C^0\varepsilon^0)$, the strain ε^+ and stress σ^+ fields inside the layer are defined by the equations

$$\varepsilon^+ = (I + \widetilde{K}(n)C^1)^{-1}\varepsilon^0, \quad \sigma^+ = (I - \widetilde{S}(n)B^1)^{-1}\sigma^0. \tag{4.99}$$

The fields σ^- and ε^- in the host medium at the layer surfaces are calculated from Eqs. (4.53), (4.56)

$$\varepsilon^- = \left(I + \widetilde{K}(n)C^1\right)\varepsilon^+ = \varepsilon^0, \tag{4.100}$$

$$\sigma^- = \left(I - \widetilde{S}(n)B^1\right)\sigma^+ = \sigma^0. \tag{4.101}$$

It follows from these equations that the strain and stress fields outside the layer V coincide with the external fields applied to the medium

$$\sigma(x) = \sigma^0, \quad \varepsilon(x) = \varepsilon^0, \quad x \notin V. \tag{4.102}$$

Let the ellipsoid semiaxis a_3 tends to infinity. In this case, we have

$$\lim_{a_3 \to \infty} \widetilde{K}\left(\frac{\cos\varphi \sin\theta}{a_1}, \frac{\sin\varphi \sin\theta}{a_2}, \frac{\cos\theta}{a_3}\right) =$$

$$= \widetilde{K}\left(\frac{\cos\varphi \sin\theta}{a_1}, \frac{\sin\varphi \sin\theta}{a_2}, 0\right) = \widetilde{K}\left(\frac{a_2}{a_1}\cos\varphi, \sin\varphi, 0\right), \tag{4.103}$$

and the limiting function $\widetilde{K}(a^{-1}k)$ does not depend on the azimuthal angle θ. As a result Eq. (4.95) takes the form of the 1D-integral

$$A(a) = \frac{1}{2\pi}\int_0^{2\pi} \widetilde{K}\,(\gamma\cos\varphi, \sin\varphi, 0)\,d\varphi, \quad \gamma = \frac{a_2}{a_1}. \tag{4.104}$$

4.6 An Ellipsoidal Inclusion in an Isotropic Medium

For an isotropic medium, the tensor $\widetilde{K}(k)$ has the following explicit form (Appendix B.1)

$$\widetilde{K}\,(k) = \frac{1}{\mu_0}\left(E^5\,(m) - E^6\,(m)\right), \quad \kappa_0 = \frac{\lambda_0 + \mu_0}{\lambda_0 + 2\mu_0}, \quad m_\alpha = \frac{k_\alpha}{|k|}, \tag{4.105}$$

where tensors E^5, E^6 are the elements of the E-basis (Appendix A.1)

$$E^5_{\alpha\beta\lambda\mu}(m) = m_\alpha \delta_{\beta\mu} m_\lambda|_{(\alpha\beta)(\lambda\mu)}, \quad E^6_{\alpha\beta\lambda\mu}(m) = m_\alpha m_\beta m_\lambda m_\mu. \tag{4.106}$$

Here, μ_0, λ_0 are Lame's parameters of the medium. As a result, the equation for $A(a)$ follows from Eqs. (4.95), (4.96) in the form

$$A(a) = \frac{1}{\mu_0}\left(\prec E^5(m)\,|a\succ - \kappa_0 \prec E^6(m)\,|a\succ\right) =$$
$$= \frac{1}{4\pi\mu_0}\int_{\Omega_1}\left(E^5\left(\frac{a^{-1}k}{|a^{-1}k|}\right) - \kappa_0 E^6\left(\frac{a^{-1}k}{|a^{-1}k|}\right)\right) d\Omega_k, \tag{4.107}$$

For calculation of the integrals over unit sphere Ω_1 in this equation, we take unit vectors n^1, n^2, n^3 of the principal axes of the ellipsoid as the basis of the Cartesian coordinate system and introduce a spherical coordinate system with the polar axis n^3 and polar φ and azimuthal θ coordinate angles. In this basis,

$$k_1 = \cos\varphi \sin\theta, \; k_2 = \sin\varphi \sin\theta, \; k_3 = \cos\theta, \tag{4.108}$$

$$\frac{(a^{-1}k)_1}{|a^{-1}k|} = \frac{\gamma_1}{\sqrt{\Delta(\varphi,\theta)}}\cos\varphi\sin\theta, \quad \frac{(a^{-1}k)_2}{|a^{-1}k|} = \frac{\gamma_2}{\sqrt{\Delta(\varphi,\theta)}}\sin\varphi\sin\theta, \tag{4.109}$$

$$\frac{(a^{-1}k)_3}{|a^{-1}k|} = \frac{\cos\theta}{\sqrt{\Delta(\varphi,\theta)}}, \; \gamma_1 = \frac{a_3}{a_1}, \; \gamma_2 = \frac{a_3}{a_2}, \tag{4.110}$$

$$\Delta(\varphi,\theta) = (\gamma_1\cos\varphi\sin\theta)^2 + (\gamma_2\sin\varphi\sin\theta)^2 + (\cos\theta)^2. \tag{4.111}$$

For presentation of the averages $\prec E^5(m)|a\succ$ and $\prec E^6(m)|a\succ$, the basis of twelve linearly independent tensors T^i (the T-basis, Appendix A.3) can be used

$$T^1_{\alpha\beta\gamma\delta} = n^1_\alpha n^1_\beta n^1_\gamma n^1_\delta,\; T^2_{\alpha\beta\gamma\delta} = n^2_\alpha n^2_\beta n^2_\gamma n^2_\delta,\; T^3_{\alpha\beta\gamma\delta} = n^3_\alpha n^3_\beta n^3_\gamma n^3_\delta,$$
$$T^4_{\alpha\beta\gamma\delta} = n^1_\alpha n^1_\beta n^2_\gamma n^2_\delta,\; T^5_{\alpha\beta\gamma\delta} = n^2_\alpha n^2_\beta n^1_\gamma n^1_\delta,\; T^6_{\alpha\beta\gamma\delta} = n^1_\alpha n^1_\beta n^3_\gamma n^3_\delta,$$
$$T^7_{\alpha\beta\gamma\delta} = n^3_\alpha n^3_\beta n^1_\gamma n^1_\delta,\; T^8_{\alpha\beta\gamma\delta} = n^2_\alpha n^2_\beta n^3_\gamma n^3_\delta,\; T^9_{\alpha\beta\gamma\delta} = n^3_\alpha n^3_\beta n^2_\gamma n^2_\delta, \tag{4.112}$$
$$T^{10}_{\alpha\beta\gamma\delta} = 4n^1_\alpha n^2_\beta n^1_\gamma n^2_\delta|_{(\alpha\beta)(\gamma\delta)},\; T^{11}_{\alpha\beta\gamma\delta} = 4n^1_\alpha n^3_\beta n^1_\gamma n^3_\delta|_{(\alpha\beta)(\gamma\delta)},$$
$$T^{12}_{\alpha\beta\gamma\delta} = 4n^2_\alpha n^3_\beta n^2_\gamma n^3_\delta|_{(\alpha\beta)(\gamma\delta)}.$$

Taking into account symmetry of the tensors $E^5(m)$ and $E^6(m)$ about index permutations, the averages $\prec E^5(m)|a \succ$ and $\prec E^6(m)|a \succ$ have the following tensor structures

$$\prec E^5(m)|a \succ = \alpha_1 \left(T^1 + \frac{1}{4}T^{10} + \frac{1}{4}T^{11}\right) + \alpha_2 \left(T^2 + \frac{1}{4}T^{10} + \frac{1}{4}T^{12}\right) +$$
$$+ \alpha_3 \left(T^3 + \frac{1}{4}T^{11} + \frac{1}{4}T^{12}\right), \tag{4.113}$$

$$\prec E^6(m)|a \succ = \alpha_4 T^1 + \alpha_5 T^2 + \alpha_6 T^3 + \alpha_7(T^4 + T^5 + T^{10}) +$$
$$+ \alpha_8(T^6 + T^7 + T^{11}) + a_9(T^8 + T^9 + T^{12}). \tag{4.114}$$

Nine scalar coefficients α_i in these equations can be obtained by convolutions of the both parts of these equations with the appropriate basic vectors n^i

$$\alpha_1 = \prec \left(m \cdot n^1\right)^2 |a \succ = \frac{(\gamma_1)^2}{4\pi} \int_0^{2\pi} d\varphi \int_0^{\pi} \frac{(\cos\varphi \sin\theta)^2}{\Delta(\varphi,\theta)} \sin\theta d\theta, \tag{4.115}$$

$$\alpha_2 = \prec \left(m \cdot n^2\right)^2 |a \succ = \frac{(\gamma_2)^2}{4\pi} \int_0^{2\pi} d\varphi \int_0^{\pi} \frac{(\sin\varphi \sin\theta)^2}{\Delta(\varphi,\theta)} \sin\theta d\theta, \tag{4.116}$$

$$\alpha_3 = \prec \left(m \cdot n^3\right)^2 |a \succ = \frac{1}{4\pi} \int_0^{2\pi} d\varphi \int_0^{\pi} \frac{(\cos\theta)^2}{\Delta(\varphi,\theta)} \sin\theta d\theta. \tag{4.117}$$

For the coefficients $\alpha_4, ..., \alpha_9$ in Eq. (4.114), we have

$$\alpha_4 = \prec \left(m \cdot n^1\right)^4 |a \succ = \frac{(\gamma_1)^4}{4\pi} \int_0^{2\pi} d\varphi \int_0^{\pi} \frac{(\cos\varphi \sin\theta)^4}{\Delta(\varphi,\theta)^2} \sin\theta d\theta, \tag{4.118}$$

$$\alpha_5 = \prec \left(m \cdot n^2\right)^4 \succ = \frac{(\gamma_2)^4}{4\pi} \int_0^{2\pi} d\varphi \int_0^{\pi} \frac{(\sin\varphi \sin\theta)^4}{\Delta(\varphi,\theta)^2} \sin\theta d\theta, \tag{4.119}$$

$$\alpha_6 = \prec \left(m \cdot n^3\right)^4 |a \succ = \frac{1}{4\pi} \int_0^{2\pi} d\varphi \int_0^{\pi} \frac{(\cos\theta)^4}{\Delta(\varphi,\theta)^2} \sin\theta d\theta, \tag{4.120}$$

$$\alpha_7 = \prec \left(m \cdot n^1\right)^2 \left(m \cdot n^2\right)^2 |a \succ =$$
$$= \frac{(\gamma_1\gamma_2)^2}{4\pi} \int_0^{2\pi} d\varphi \int_0^{\pi} \frac{(\cos\varphi \sin\varphi)^2 (\sin\theta)^4}{\Delta(\varphi,\theta)^2} \sin\theta d\theta, \tag{4.121}$$

$$\alpha_8 = \prec \left(m \cdot n^1\right)^2 \left(m \cdot n^3\right)^2 |a \succ =$$

$$= \frac{(\gamma_1)^2}{4\pi} \int_0^{2\pi} d\varphi \int_0^{\pi} \frac{(\cos\varphi \sin\theta \cos\theta)^2}{\Delta(\varphi, \theta)^2} \sin\theta d\theta, \tag{4.122}$$

$$\alpha_9 = \prec \left(m \cdot n^2\right)^2 \left(m \cdot n^3\right)^2 |a \succ =$$

$$= \frac{(\gamma_2)^2}{4\pi} \int_0^{2\pi} d\varphi \int_0^{\pi} \frac{(\sin\varphi \sin\theta \cos\theta)^2}{(\Delta(\varphi, \theta))^2} \sin\theta d\theta. \tag{4.123}$$

Note that the coefficients α_i relate by the equations

$$\alpha_1 + \alpha_2 + \alpha_3 = 1, \tag{4.124}$$

$$\alpha_4 + \alpha_5 + \alpha_6 + 2\alpha_7 + 2\alpha_8 + 2\alpha_9 = 1. \tag{4.125}$$

The double integrals (4.115)–(4.123) can be reduced to 1D-integrals because the integrals with respect to θ are expressed in terms of elementary functions. Thus, for isotropic media, the tensor $A(a)$ in Eqs. (4.107) is presented in the T-basis as follows

$$A(a) = \frac{1}{\mu_0} \sum_{l=1}^{12} A_l T^l, \tag{4.126}$$

$$A_1 = \alpha_1 - \kappa_0 \alpha_4,\ A_2 = \alpha_2 - \kappa_0 \alpha_5,\ A_3 = \alpha_3 - \kappa_0 \alpha_6, \tag{4.127}$$

$$A_4 = A_5 = -\kappa_0 \alpha_7,\ A_6 = A_7 = -\kappa_0 \alpha_8,\ A_8 = A_9 = -\kappa_0 \alpha_9, \tag{4.128}$$

$$A_{10} = \frac{1}{4} \left(\alpha_1 + \alpha_2 - 4\kappa_0 \alpha_7\right),\ A_{11} = \frac{1}{4} \left(\alpha_1 + \alpha_3 - 4\kappa_0 \alpha_8\right) \tag{4.129}$$

$$A_{12} = \frac{1}{4} \left(\alpha_2 + \alpha_3 - 4\kappa_0 \alpha_9\right). \tag{4.130}$$

The tensor $\left(I + A(a)C^1\right)^{-1}$ in Eq. (4.82) for isotropic inclusions can be calculated using the multiplication table of the tensors T^i (Appendix A.3). In this case, the tensor C^1 has the form

$$C^1 = (\lambda - \lambda_0)E^2 + 2(\mu - \mu_0)E^1, \tag{4.131}$$

where λ, μ are Lame's coefficients of the inclusion material, and the tensors E^1 and E^2 in the T-basis are

$$E^1 = T^1 + T^2 + T^3 + \frac{1}{2}\left(T^{10} + T^{11} + T^{12}\right), \tag{4.132}$$

$$E^2 = T^1 + T^2 + T^3 + T^4 + T^5 + T^6 + T^7 + T^8 + T^9. \tag{4.133}$$

For an ellipsoid of revolution around vector $n = n^3$ (spheroid), two ellipsoid semiaxes coincide $a_1 = a_2 = a$. In this case, the averages $\prec E^5(m)|a \succ$ and $\prec E^6(m)|a \succ$ in Eqs. (4.115)–(4.123) are simplified after integrating over the polar angle φ. These averages have transverse isotropic symmetry, and in the E-basis (Appendix A.1), they are presented as follows

$$\prec E^5(m)\,|a \succ = \beta_1 E^1 + (1 - 3\beta_1)E^5(n), \tag{4.134}$$

$$\prec E^6(m)\,|a \succ = \beta_2\left(2E^1 + E^2\right) +$$

$$+\,\beta_3\left(E^3(n) + E^4(n) + 4E^5(n)\right) + \beta_4 E^6(n). \tag{4.135}$$

Here, the coefficients $\beta_1, \beta_2, \beta_3, \beta_4$ are

$$\beta_1 = \frac{\gamma^2}{4}\int_0^\pi \frac{(\sin\theta)^3\, d\theta}{\Delta(\theta)},\quad \Delta(\theta) = (\gamma\sin\theta)^2 + (\cos\theta)^2,\ \gamma = \frac{a_3}{a}; \tag{4.136}$$

$$\beta_2 = \frac{\gamma^4}{16}\int_0^\pi \frac{(\sin\theta)^5}{\Delta(\theta)^2}d\theta,\ \beta_3 = \frac{\gamma^2}{16}\int_0^\pi \frac{4(\cos\theta)^2 - \gamma^2(\sin\theta)^2}{\Delta(\theta)^2}(\sin\theta)^3 d\theta, \tag{4.137}$$

$$\beta_4 = \frac{1}{16}\int_0^\pi \frac{8(\cos\theta)^4 - 6\gamma^2(\sin 2\theta)^2 + 3\gamma^4(\sin\theta)^4}{\Delta(\theta)^2}\sin\theta d\theta. \tag{4.138}$$

These integrals can be expressed in terms of elementary functions. Note that four coefficients β_i relate by the equations

$$\beta_1 - 5\beta_2 - \beta_3 = 0,\ 10\beta_1 - 35\beta_2 + \beta_4 = 1. \tag{4.139}$$

Therefore, two of four coefficients β_i are independent only. Finally, for spheroidal inclusions in an isotropic medium, the tensor $A = A(\gamma)$ is presented in the form

$$A(\gamma) = \frac{1}{\mu_0}\sum_{l=1}^{6} \mathcal{A}_l E^l(n), \tag{4.140}$$

$$\mathcal{A}_1 = \beta_1 - 2\kappa_0\beta_2,\ \mathcal{A}_2 = -\kappa_0\beta_2,\ \mathcal{A}_3 = \mathcal{A}_4 = -\kappa_0(\beta_1 - 5\beta_2), \tag{4.141}$$

$$\mathcal{A}_5 = 1 - (3 + 4\kappa_0)\beta_1 - 20\kappa_0\beta_2,\ \mathcal{A}_6 = -\kappa_0(1 - 10\beta_1 + 35\beta_2). \tag{4.142}$$

In the limit $\gamma_1, \gamma_2 \to 0$ (thin ellipsoidal disk), the expressions for the tensors A and D are simplified

$$A = \frac{1}{\mu_0}\left(E^5(n) - \kappa_0 E^6(n)\right), \tag{4.143}$$

$$D = -2\mu_0\left(P^1(n) + (2\kappa_0 - 1)\, P^2(n)\right). \tag{4.144}$$

Here, n is the normal to the disk plane, $P^1(n)$, $P^2(n)$ are the elements of P- basis (Appendix A.2).

When the semiaxis a_3 tends to infinity, $\gamma_1, \gamma_2 \to \infty$, the inclusion is transformed into an infinite elliptical cylinder with the aspect ratio $\gamma = a_2/a_1$. In this case, the tensor $A = A(\gamma)$ is defined in Eq. (4.104) and in the T-basis, it takes the form

$$A(\gamma) = A_1 T^1 + A_2 T^2 + A_4(T^4 + T^5) + A_{10} T^{10} + A_{11}(T^{11} + T^{12}), \tag{4.145}$$

$$A_1 = \frac{\gamma}{2\mu_0(1+\gamma)^2}\left(2(1+\gamma) - \kappa_0(1+2\gamma)\right), \tag{4.146}$$

$$A_2 = \frac{\gamma}{2\mu_0(1+\gamma)^2}\left(2(1+\gamma) - \kappa_0(2+\gamma)\right), \quad A_4 = -\frac{\kappa_0\gamma}{2\mu_0(1+\gamma)^2}, \tag{4.147}$$

$$A_{10} = \frac{1}{4\mu_0}\left(1 + \frac{\kappa_0\gamma}{2(1+\gamma)^2}\right), \quad A_{11} = \frac{1}{4\mu_0(1+\gamma)}. \tag{4.148}$$

For $\gamma = 1$, the inclusion is a circular cylinder, and the tensors A and D have transverse isotropic symmetry. In the P-basis with vector n directed along the cylinder generatrix, these tensors are

$$A = \frac{1}{8\mu_0}\left(2(2-\kappa_0)P^1(n) - \kappa_0 P^2(n) + 4P^5(n)\right), \tag{4.149}$$

$$D = -\frac{\mu_0}{4}\left(\kappa_0\left(2P^1(n) + P^2(n) + 8P^6(n)\right) + \right.$$
$$\left. + 2(2\kappa_0 - 1)\left(P^3(n) + P^4(n)\right) + 2P^5(n)\right). \tag{4.150}$$

For a spherical inclusion, $\gamma_1 = \gamma_2 = 1$, the tensors A and D are isotropic

$$A = \frac{1-\kappa_0}{9\mu_0}E^2 + \frac{5 - 2\kappa_0}{15\mu_0}\left(E^1 - \frac{1}{3}E^2\right), \tag{4.151}$$

$$D = \frac{4\mu_0(1-4\kappa_0)}{9}E^2 - \frac{2\mu_0(5+4\kappa_0)}{15}\left(E^1 - \frac{1}{3}E^2\right). \tag{4.152}$$

4.7 An Ellipsoidal Inclusion in the Elastic Medium with Ellipsoidal Anisotropy

For the medium with ellipsoidal anisotropy, the tensor C^0 has the following structure [3]

$$C^0_{\alpha\beta\gamma\delta} = \lambda_0 Q_{\alpha\beta} Q_{\gamma\delta} + \mu_0 (Q_{\alpha\gamma} Q_{\beta\delta} + Q_{\alpha\delta} Q_{\beta\gamma}). \tag{4.153}$$

Here, λ_0, μ_0 are elastic constants, $Q_{\alpha\beta}$ is a symmetric second rank tensor with positive eigenvalues d_1, d_2, d_3 such as

$$\det Q = d_1 d_2 d_3 = 1. \tag{4.154}$$

Thus, the medium with ellipsoidal anisotropy has four independent elastic constants.

We consider an ellipsoidal inclusion in the medium with ellipsoidal anisotropy subjected to a constant external field ε^0 at infinity. The field ε^+ inside the inclusion is also constant and

$$\varepsilon^+ = \Lambda^\varepsilon \varepsilon^0, \ \Lambda^\varepsilon = \left(E^1 + A(a)C^1\right)^{-1},$$

where tensor $A(a)$ is defined in Eq. (4.89)

$$A(a) = \frac{1}{4\pi} \int\limits_{\Omega_1} \widetilde{K}(a^{-1}k) d\Omega_k. \tag{4.155}$$

Here,

$$\widetilde{K}_{\alpha\beta\gamma\delta}(k) = k_\alpha \widetilde{G}_{\beta\gamma}(k) k_\delta |_{(\alpha\beta)(\gamma\delta)}, \tag{4.156}$$

$$\widetilde{G}_{\alpha\beta}(k) = L^{-1}(k)_{\alpha\beta}, \ L_{\alpha\beta}(k) = k_\gamma C^0_{\gamma\alpha\beta\delta} k_\delta. \tag{4.157}$$

For the medium with ellipsoidal anisotropy, the tensor $L(k)$ has the form

$$L_{\alpha\beta}(k) = (\lambda + \mu) q_\alpha q_\beta + \mu \left(q Q^{-1} q\right) Q_{\alpha\beta}, \tag{4.158}$$

$$q_\alpha = Q_{\alpha\beta} k_\beta, \ \left(q Q^{-1} q\right) = q_\alpha (Q^{-1})_{\alpha\beta} q_\beta = k_\alpha Q_{\alpha\beta} k_\beta. \tag{4.159}$$

Then, we introduce two second rank tensors

$$M^1_{\alpha\beta} = \delta_{\alpha\beta} - \frac{1}{(q Q^{-1} q)} \left(Q^{-1}\right)_{\alpha\gamma} q_\gamma q_\beta, \tag{4.160}$$

$$M^2_{\alpha\beta} = \frac{1}{(q Q^{-1} q)} \left(Q^{-1}\right)_{\alpha\gamma} q_\gamma q_\beta \tag{4.161}$$

that satisfy the equations

$$M^1_{\alpha\gamma}M^2_{\gamma\beta}=0,\ M^1_{\alpha\gamma}M^1_{\gamma\beta}=M^1_{\alpha\beta},\ M^2_{\alpha\gamma}M^2_{\gamma\beta}=M^2_{\alpha\beta}. \tag{4.162}$$

Using these tensors, the tensor $L(k)$ is presented in the form

$$L(k)=\mu_0(qQ^{-1}q)Q\left(M^1+\frac{\lambda_0+2\mu_0}{\mu_0}M^2\right), \tag{4.163}$$

and the inverse tensor is

$$L(k)^{-1}=\frac{1}{\mu_0(qQ^{-1}q)}\left(M^1+\frac{\mu_0}{\lambda_0+2\mu_0}M^2\right)Q^{-1}. \tag{4.164}$$

The explicit expression for the tensor $\widetilde{G}(k)$ follows from this equation in the form

$$\widetilde{G}(k)=\frac{1}{\mu_0(kQk)}\left(Q^{-1}-\frac{\kappa_0}{(kQk)}k\otimes k\right),\quad \kappa_0=\frac{\lambda_0+\mu_0}{\lambda_0+2\mu_0}. \tag{4.165}$$

As the result, for the tensor $\widetilde{K}(k)$ in Eq. (4.156), we have the expression

$$\widetilde{K}_{\alpha\beta\gamma\delta}(k)=$$

$$=\frac{1}{\mu_0(kQk)}\left(k_\alpha\left(Q^{-1}\right)_{\beta\gamma}k_\delta|_{(\alpha\beta)(\gamma\delta)}-\frac{\kappa_0}{(kQk)}k_\alpha k_\beta k_\gamma k_\delta\right). \tag{4.166}$$

The explicit expression for the integrand function in Eq. (4.155) allows calculating the tensor $A(a)$ numerically.

4.8 An Ellipsoidal Inclusion in an Orthorhombic Elastic Medium

The elastic stiffness tensor C^0 of orthorhombic medium has three mutually orthogonal 3-fold symmetry axes. If the T-basis in Eqs. (4.111) is formed from the unit vectors n^1, n^2, n^3 of these axes, tensor C^0 is presented the form

$$C^0= c_1T^1+c_2T^2+c_3T^3+c_4(T^4+T^5)+$$

$$+c_5(T^6+T^7)+c_6(T^8+T^9)+c_7T^{10}+c_8T^{11}+c_9T^{12}. \tag{4.167}$$

The scalar coefficients c_i in this equation relate the elastic constants c_{ij} in the matrix representation of the elastic stiffness tensors of an orthorhombic material [4] by the equations

$$c_1 = c_{11}, \;\; c_2 = c_{22}, \;\; c_3 = c_{33}, \;\; c_4 = c_{12}, \;\; c_5 = c_{13}, \tag{4.168}$$

$$c_6 = c_{23}, \;\; c_7 = c_{55}, \;\; c_8 = c_{66}, \;\; c_9 = c_{44}. \tag{4.169}$$

For an orthorhombic medium, Fourier transform $\widetilde{K}(k)$ of the function $K(x)$ in Eq. (4.13) has the following form

$$\widetilde{K}_{\gamma\alpha\beta\delta}(m) = m_\gamma \left(\mathsf{k}_1 n^1_\alpha n^1_\beta + \mathsf{k}_2 n^2_\alpha n^2_\beta + \mathsf{k}_3 n^3_\alpha n^3_\beta + 2\mathsf{k}_4 n^1_\alpha n^2_\beta |_{(\alpha\beta)} \right.$$

$$\left. + 2\mathsf{k}_5 n^1_\alpha n^3_\beta |_{(\alpha\beta)} + 2\mathsf{k}_6 n^2_\alpha n^3_\beta |_{(\alpha\beta)} \right) m_\delta |_{(\gamma\alpha)(\beta\delta)}, \quad m_\alpha = \frac{k_\alpha}{|k|}. \tag{4.170}$$

In this equation,

$$\mathsf{k}_1 = \frac{l_6^2 - l_2 l_3}{\Delta}, \;\; \mathsf{k}_2 = \frac{l_1 l_3 - l_5^2}{\Delta}, \;\; \mathsf{k}_3 = \frac{l_1 l_2 - l_4^2}{\Delta}, \tag{4.171}$$

$$\mathsf{k}_4 = \frac{l_5 l_6 - l_3 l_4}{\Delta}, \;\; \mathsf{k}_5 = \frac{l_4 l_6 - l_2 l_5}{\Delta}, \;\; \mathsf{k}_6 = \frac{l_1 l_6 - l_4 l^5}{\Delta}, \tag{4.172}$$

$$\Delta = -l_3 l_4^2 - l_5 (l_2 l_5 - 2 l_4 l_6) - l_1 (l_6^2 - l_2 l_3), \tag{4.173}$$

where the coefficients l_i are

$$l_1 = c_1 (m \cdot n^1)^2 + c_7 (m \cdot n^2)^2 + c_8 (m \cdot n^3)^2, \tag{4.174}$$

$$l_2 = c_7 (m \cdot n^1)^2 + c_2 (m \cdot n^2)^2 + c_9 (m \cdot n^3)^2, \tag{4.175}$$

$$l_3 = c_8 (m \cdot n^1)^2 + c_2 (m \cdot n^3)^2 + c_9 (m \cdot n^2)^2, \tag{4.176}$$

$$l_4 = (c_4 + c_7)(m \cdot n^1)(m \cdot n^2), \;\; l_5 = (c_5 + c_6)(m \cdot n^1)(m \cdot n^3), \tag{4.177}$$

$$l_6 = (c_6 + c_9)(m \cdot n^2)(m \cdot n^3), \;\; (m \cdot n^i) = m_\alpha n^i_\alpha, \;\; i = 1, 2, 3. \tag{4.178}$$

Let unit vectors n^1, n^2, n^3 be taken as the basis of the global Cartesian system X_1, X_2, X_3 (see Fig. 4.5), and vectors e^1, e^2, e^3 be directed along the principal axes of the ellipsoid. The vectors e^i compose the basis of the local coordinate system X'_1, X'_2, X'_3. The orientation of the local basis with respect to the global basis is defined by Euler angles (φ, ψ, θ) indicated in this figure. Relation of the e^i and n^i bases is defined by the equation $n^j = \sum_{i=1}^{3} Q^j_i e^i$, where the matrix Q^j_i has the form

$$\left[Q^j_i \right] = \begin{bmatrix} \cos\varphi\cos\psi - \sin\varphi\sin\psi\cos\theta & \sin\varphi\cos\psi + \cos\varphi\sin\psi\cos\theta & \sin\varphi\sin\theta \\ -\cos\varphi\cos\psi - \sin\varphi\cos\psi\cos\theta & -\sin\varphi\sin\psi + \cos\varphi\cos\psi\cos\theta & \cos\varphi\sin\theta \\ \sin\varphi\sin\theta & -\cos\varphi\sin\theta & \cos\theta \end{bmatrix}. \tag{4.179}$$

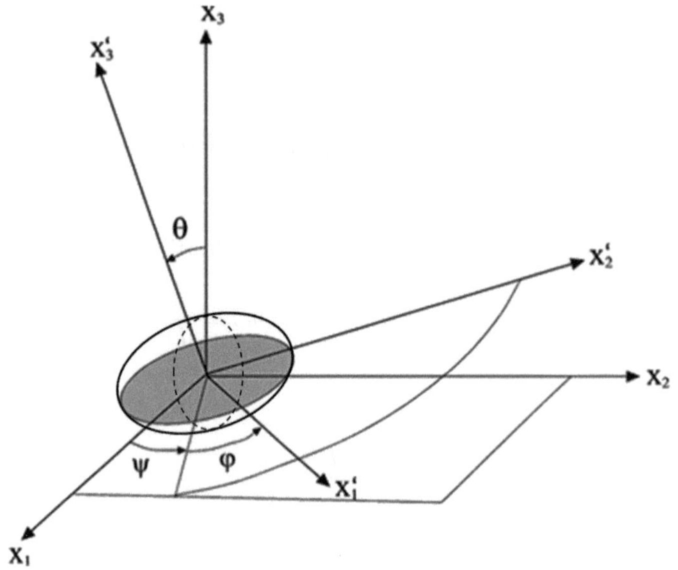

Fig. 4.5 The global and local coordinate systems for a medium with an ellipsoidal inclusion

Substituting $n^i_{\underset{\sim}{\alpha}} = \alpha^{i\alpha}$, $i = 1, 2, 3$ into Eqs. (4.170)–(4.178), we obtain presentation of the tensors $\underset{\sim}{K}(k)$ in the e-basis of the principal axes of the ellipsoid. In this basis, the components of the vector m are defined in Eqs. (4.109), (4.110). Then, the integral (4.95) can be calculated numerically, and it determines the components of the tensor $A(a)$ in the local e-basis.

Then, the tensor $\Lambda^{\varepsilon} = \left(I + A(a)C^1\right)^{-1}$ in Eq. (4.82) for the strain tensor inside the inclusion can be calculated. Note that the forth rank tensor $I + A(a)C^1$ is symmetric about permutations of the index of the first and second pairs but it is not symmetric about permutation of the index pares. Thus, this tensor has 36 independent components, and its inversion consists in solution of 36 linear algebraic equations (Appendix A.6).

4.9 Temperature Stresses in a Homogeneous Medium with an Ellipsoidal Inclusion

Let a homogeneous medium with the elastic stiffness tensor C^0 and the tensor of thermo-expansion coefficients α^0 contains an ellipsoidal inclusion with the stiffness tensor C and the tensor of thermo-expansion coefficients α. If the medium is subjected to a constant temperature field T, the integral equation for stress tensor $\sigma(x)$ in the medium follows from Eq. (4.32) in the form

$$\sigma(x) - \int S(x-x')B^1\sigma(x')V(x')dx' = \int S(x-x')\alpha(x)dx'T. \tag{4.180}$$

Here, $V(x)$ is the characteristic function of the region V occupied by the inclusion,

$$B^1 = B - B^0,\ \alpha(x) = \alpha^0 + \alpha^1 V(x),\ \alpha^1 = \alpha - \alpha^0. \tag{4.181}$$

The integral on the right-hand side of Eq. (4.180) is presented in the form

$$\int S(x-x')\alpha(x)dx' = \int S(x-x')dx'\alpha^0 + \int S(x-x')V(x')dx'\alpha^1. \tag{4.182}$$

The value of the first integral on the right-hand side depends on conditions at infinity (see Sect. 3.3) and

$$\int S(x-x')dx' = \begin{cases} -C^0, & \text{for a constrained medium} \\ 0, & \text{for a non-constrained medium} \end{cases}. \tag{4.183}$$

For an ellipsoidal region V, the second integral on the right-hand side of Eq. (4.182) is constant inside V

$$\int S(x-x')\alpha^1 V(x')dx' = D(a)\alpha^1,\ x \in V, \tag{4.184}$$

where tensor $D(a)$ is defined in Eq. (4.92). Thus, for a non-constrained medium, the stress field in the ellipsoidal region V satisfy the equation

$$\sigma(x) - \int_V S(x-x')B^1\sigma(x')dx' = D(a)\alpha^1 T,\ x \in V. \tag{4.185}$$

For a constant right-hand side, the stress tensor σ^+ inside the ellipsoidal inclusion is constant and has the form

$$\sigma^+ = \left(I - D(a)B^1\right)^{-1} D(a)\alpha^1 T. \tag{4.186}$$

For the constrained medium, an additional term appears in the right-hand side of Eq. (4.184)

$$\sigma(x) - \int_V S(x-x')B^1\sigma(x')dx' = D(a)\alpha^1 T - C^0\alpha^0 T, \tag{4.187}$$

and the field inside the inclusion changes

$$\sigma^+ = \left(I - D(a)B^1\right)^{-1}\left(D(a)\alpha^1 - C^0\alpha^0\right)T. \tag{4.188}$$

The stresses concentration in the host medium at the inclusion interface is defined in Eq. (4.65) for $\sigma^{-}(x_0)$ with the tensor σ^{+} in Eqs. (4.186) or (4.188).

4.10 An Isolated Inclusion in a Homogeneous Conductive Media

We consider a steady electric current in a homogeneous medium with the conductivity tensor C^0 containing a region V (inclusion) with the conductivity tensor $C(x)$. The vectors of the electric field $e(x)$ and the current $j(x)$ in the medium satisfy the system of differential equations

$$\nabla_\alpha j_\alpha(x) = 0\,, \quad j_\alpha(x) = \left(C^0_{\alpha\beta} + C^1_{\alpha\beta}(x)V(x)\right) e_\beta(x),$$

$$\mathrm{rot}_{\alpha\beta} e_\beta(x) = 0. \tag{4.189}$$

Here, $V(x)$ is the characteristic function of the region V, $C^1(x) = C(x) - C^0$. Because the electric field $e_\alpha(x)$ is rot-free, it is presented as gradient of a scalar potential $\varphi(x)$

$$e_\alpha(x) = -\nabla_\alpha \varphi(x), \tag{4.190}$$

and the third equation of the system (4.189) is satisfied automatically. The equation for the potential $\varphi(x)$ follows from Eq. (4.189) and takes the form

$$\nabla_\alpha \left(C^0_{\alpha\beta} + C^1_{\alpha\beta}(x)V(x)\right) \nabla_\beta \varphi(x) = 0\,. \tag{4.191}$$

The system (4.189)–(4.190) coincides with the governing system of the heat transfer problem. In this case, $\varphi(x) = -T(x)$, $T(x)$ is the temperature field, $j_\alpha(x)$ is the heat flux, and $C_{\alpha\beta}(x)$ is the tensor of thermo-conductivity.

The same procedure as in Sect. 4.1 allows deriving the integral equation for the vector fields $e(x)$ and $j(x)$ in the homogeneous host medium with an isolated heterogeneous inclusion

$$e_\alpha(x) + \int K_{\alpha\beta}(x - x')C^1_{\beta\mu}(x')e_\mu(x')V(x')dx' = e^0_\alpha(x)\,, \tag{4.192}$$

$$j_\alpha(x) - \int S_{\alpha\beta}(x - x')B^1_{\beta\mu}(x')j_\mu(x')V(x')dx' = j^0_\alpha(x)\,, \tag{4.193}$$

$$B^1(x) = B(x) - B^0, \quad B(x) = (C(x))^{-1}, \quad B^0 = (C^0)^{-1}. \tag{4.194}$$

Here, e^0 and j^0 are the fields in the host medium without the inclusion ($C^1(x) = 0$) by the same field sources. The kernels $K(x)$ and $S(x)$ are determined by the equations similar to Eqs. (4.13) and (4.31):

$$K_{\alpha\beta}(x) = -\nabla_\alpha \nabla_\beta g(x)\,, \quad S_{\alpha\beta}(x) = C^0_{\alpha\lambda} K_{\lambda\mu}(x) C^0_{\mu\beta} - C^0_{\alpha\beta}\delta(x)\,, \tag{4.195}$$

where $g(x)$ is Green's function for the infinite homogeneous medium with the property tensor C^0, and the explicit form of $g(x)$ is presented in Eq. (2.20)

The solutions of integral equations (4.192) and (4.193) have the same properties as the solutions of Eqs. (4.12) and (4.33) for the elastic fields in the homogeneous medium with a heterogeneous inclusion. For instance, the conditions for the jumps of the fields at the inclusion interface Ω have the same form as for elastic media in Eqs. (4.53), (4.56)

$$e^-(x_0) = (I + \widetilde{K}(n_0) C^{1+}(x_0)) e^+(x_0)\,, \quad n_0 = n(x_0), \tag{4.196}$$

$$j^-(x_0) = \left(I - \widetilde{S}(n_0) B^{1+}(x_0)\right) j^+(x_0)\,, \quad x_0 \in \Omega\,. \tag{4.197}$$

Here, $e^-(x_0)$ and $j^-(x_0)$ are the limiting values of the electric field and current at the point x_0 from the side of the host medium, and $e^+(x_0)$ and $j^+(x_0)$ are such limits from the side of the inclusion, $n(x_0)$ is external normal to the surface Ω at point x_0. The tensors $\widetilde{K}(k)$ and $\widetilde{S}(k)$ in Eqs. (4.196), (4.197) are Fourier transforms of the tensors $K(x)$ and $S(x)$

$$\widetilde{K}_{\alpha\beta}(k) = \frac{k_\alpha k_\beta}{\left(k_\gamma C^0_{\gamma\delta} k_\delta\right)}, \quad \widetilde{S}_{\alpha\beta}(k) = C^0_{\alpha\delta} \widetilde{K}_{\delta\gamma}(k) C^0_{\gamma\beta} - C^0_{\alpha\beta}. \tag{4.198}$$

4.10.1 An Ellipsoidal Inclusion in a Conductive Medium

The polynomial conservation theorem for ellipsoidal inclusions holds for Eqs. (4.192) and (4.193) of the conductivity problem. Therefore, if the external field e^0 (j^0) is constant, the constant field inside an ellipsoidal inclusion is determined by the expressions

$$e^+_\alpha = \Lambda^e_{\alpha\beta} e^0_\beta\,, \quad \left(\Lambda^e\right)^{-1}_{\alpha\beta} = \delta_{\alpha\beta} + A^0_{\alpha\lambda}(a) C^1_{\lambda\beta}, \tag{4.199}$$

$$A_{\alpha\beta}(a) = \frac{1}{4\pi}\int\limits_{\Omega_1} \widetilde{K}_{\alpha\beta}(a^{-1}k) d\Omega\,, \quad \widetilde{K}_{\alpha\beta}(k) = \frac{k_\alpha k_\beta}{\left(k_\lambda C^0_{\lambda\mu} k_\mu\right)}\,. \tag{4.200}$$

If the medium is isotropic ($C^0_{\alpha\beta} = c_0 \delta_{\alpha\beta}$), and the inclusion is an ellipsoid with the semiaxes a_1, a_2, a_3, the tensor $A(a)$ takes the form

$$A_{\alpha\beta} = \frac{1}{c_0}\left(\alpha_1 n^1_\alpha n^1_\beta + \alpha_2 n^2_\alpha n^2_\beta + \alpha_3 n^3_\alpha n^3_\beta\right), \tag{4.201}$$

where n^1, n^2, n^3 are unit vectors of the ellipsoid principal axes, and the coefficients $\alpha_1, \alpha_2, \alpha_3$ are the integrals

$$\alpha_1 = \frac{(\gamma_1)^2}{4\pi}\int_0^{2\pi} d\varphi \int_0^{\pi} \frac{(\cos\varphi\sin\theta)^2}{\Delta(\varphi,\theta)}\sin\theta d\theta, \tag{4.202}$$

$$\alpha_2 = \frac{(\gamma_2)^2}{4\pi}\int_0^{2\pi} d\varphi \int_0^{\pi} \frac{(\sin\varphi\sin\theta)^2}{\Delta(\varphi,\theta)}\sin\theta d\theta, \tag{4.203}$$

$$\alpha_3 = \frac{1}{4\pi}\int_0^{2\pi} d\varphi \int_0^{\pi} \frac{(\cos\theta)^2}{\Delta(\varphi,\theta)}\sin\theta d\theta, \tag{4.204}$$

$$\alpha_1 + \alpha_2 + \alpha_3 = 1. \tag{4.205}$$

Here,

$$\Delta(\varphi,\theta) = (\gamma_1\cos\varphi\sin\theta)^2 + (\gamma_2\sin\varphi\sin\theta)^2 + (\cos\theta)^2, \tag{4.206}$$

and $\gamma_1 = a_3/a_1,\ \gamma_2 = a_3/a_2$ are the ellipsoid aspect ratios. The double integrals in these equations are reduced to 1D integrals after integrating over the θ-angle.

For a spheroid with semiaxes $a_1 = a_2 = a,\ a_3,\ \gamma = a_3/a$, the tensor A in (4.201) takes the form

$$A_{\alpha\beta} = \frac{1}{c_0}\left(\beta_1\delta_{\alpha\beta} + (1-2\beta_1)n_\alpha n_\beta\right),\quad n = n^3, \tag{4.207}$$

$$\beta_1 = \frac{\gamma^2}{4}\int_0^{\pi} \frac{(\sin\theta)^3\, d\theta}{\Delta(\theta)},\quad \Delta(\theta) = (\gamma\sin\theta)^2 + (\cos\theta)^2. \tag{4.208}$$

For a spherical inclusion, $A_{\alpha\beta} = (3c_0)^{-1}\,\delta_{\alpha\beta}$, and the equations for the electric field and current inside an isotropic inclusion with conductivity coefficient c take the forms

$$e_\alpha^+ = \frac{3c_0}{2c_0 + c}e_\alpha^0,\quad j_\alpha^+ = \frac{3c}{2c_0 + c}j_\alpha^0,\quad j_\alpha^0 = c_0 e_\alpha^0. \tag{4.209}$$

Let the host conductive medium be transverse isotropic with the conductivity tensor

$$C^0_{\alpha\beta} = c_1^0\theta_{\alpha\beta} + c_3^0 n_\alpha n_\beta,\quad \theta_{\alpha\beta} = \delta_{\alpha\beta} - n_\alpha n_\beta, \tag{4.210}$$

and the inclusion be a spheroid with semiaxes $a_1 = a_2 = a$, a_3 and the rotation axis n. In this case, the tensor A in Eq. (4.200) takes the form

$$A_{\alpha\beta}(\gamma) = A_1\theta_{\alpha\beta} + A_3 n_\alpha n_\beta, \tag{4.211}$$

where the coefficients A_1, A_3 are the following integrals

$$A_1 = \frac{\gamma^2}{2}\int_0^\pi \frac{(\sin\theta)^3\, d\theta}{c_3^0 + \gamma^2 c_1^0 + (c_3^0 - \gamma^2 c_1^0)\cos 2\theta}, \tag{4.212}$$

$$A_3 = \int_0^\pi \frac{(\cos\theta)^2 \sin\theta d\theta}{c_3^0 + \gamma^2 c_1^0 + (c_3^0 - \gamma^2 c_1^0)\cos 2\theta}, \quad \gamma = \frac{a_3}{a}. \tag{4.213}$$

4.11 Numerical Solution of the One-Particle Problems for Inclusions of Arbitrary Shapes

The one-particle problems for inclusions of arbitrary shapes can be solved only numerically. The finite element method (FEM) or the boundary element method (BEM) can be used for this purpose. A specific feature of the one-particle problems is that the host medium is infinite. Therefore by using the FEM, the domain of calculation should be sufficiently large in order to avoid the influence of the boundaries. It reduces efficiency of the EFM in application to numerical solution of the one-particle problems.

In the BEM, the one-particle problem is reduced to 2D-integral equations for the unknown vector functions on the inclusion interface. Reduction of the problem dimensions is an advantage of the BEM over the FEM. But for application of the BEM, the inclusion should be homogeneous, and explicit equations for Green's functions of the host medium and inclusion materials should be known.

The volume integral equations of the one-particle problems considered in this chapter reduce the solution domain to the region occupied by an inclusion. For construction of numerical solutions of these equations, Green's function of the host medium should be known but the property of inclusion material can be arbitrary. In the conventional method of solution of the volume integral equations, the integration domain (inclusion volume) is divided into finite elements. Inside each element, the solution is approximated by standard functions, e.g., polynomial splines or non-uniform rational B-splines with unknown coefficients. Then, using the collocation method or the method of moments, the problem is reduced to a linear algebraic system for coefficients of the approximation. The components of the matrix of this system are integrals over the volume elements. For conventional approximating functions, these integrals can be calculated only numerically that requires substantial computational resources.

A numerical method proposed in [5] has several advantages over the conventional method of solution of volume integral equations. First, Gaussian quasi-interpolants (Gaussian functions concentrated at a system of nodes) are used for approximation of the fields inside the inclusion. The theory of approximation by Gaussian functions was developed in [6]. The numerical method based on such approximating functions is mesh free: the coordinates of the nodes and material properties at the nodes is the only necessary information for performing the method. Second, for isotropic host media, the elements of the matrix of the discretized problems can be calculated in explicit analytical forms, and since numerical integration is not involved, these matrices are calculated fast. Third, for periodic node grids, the matrices have Toeplitz's structure, and therefore only one matrix column should be computed. In addition, for such matrices, the Fast Fourier Transform (FFT) algorithm can be used for calculation of the matrix-vector products in the process of iterative solution of the discretized problem. The numerical algorithm is simpler and more efficient than the ones based on the conventional numerical methods of solution of volume integral equations. The details of the method are presented in [5].

4.12 Notes

In this chapter, the elasticity and conductivity problems for an isolated inclusion embedded in an infinite homogeneous host medium are reduced to solution of the volume integral equations for the fields inside the inclusion. For ellipsoidal homogeneous inclusions, the one-particle problem consists in calculation of regular 2D-integrals, multiplication and inversion of fourth rank tensors of specific symmetry. All these operations can be successfully fulfilled by the commercial computational packages like Wolfram "Mathematica" or MATLAB.

Volume integral equations for strain and stress tensors in heterogeneous elastic media are similar to Lippmann-Shwinger's equation of quantum mechanics [7]. For elastic heterogeneous media, these equations were used first in [8]. General theory of this type of equations was developed at the end of 1960s in a number of mathematical publications. Historical surveys and references to the original works can be found in [9, 10].

The conditions for the jumps of strain and stress fields on the boundary of two elastic media were obtained in [11, 12].

A homogeneous isotropic elastic medium with an ellipsoidal isotropic inclusion was considered in [2, 13–15]. For polynomial external fields and arbitrary anisotropy of the media and the inclusion material, the polynomial conservation theorem was proved in [11].

References

1. Kunin I, Sosnina E (1971) An ellipsoidal inhomogeneity in an elastic medium. Proc Acad Sci USSR 199:571–575

2. Eshelby J (1957) The determination of the elastic field of an ellipsoidal inclusion, and related problems. Proc Royal Soc A241:376–396
3. Pouya A (2011) Ellipsoidal anisotropy in linear elasticity: approximation models and analytical solutions. Int J Solids Struct 48:2245–2254
4. Nye J (1985) Physical Properties of Crystals, their Representation by Tensors and Matrices. Clarendon Press, Oxford, p 329
5. Kanaun S (2021) Heterogeneous media: local fields. Effective properties and wave propagation. Elsevier, Amsterdam
6. Maz'ya V, Schmidt G (2007) Approximate approximation, mathematical surveys and monographs, vol 141. American Mathematical Society, Providence
7. Lippmann B, Schwinger J (1950) Variational principles for scattering processes. I Phys Rev Lett 79(3):469–480
8. Lifshits I, Rosentsveig L (1946) Zur Theorie der elastischen Eigenschaften von Polykristallen. Zh Eksp Teor Fiz 16:967–975
9. Mikhlin S (1962) Multi-dimensional Singular Integrals and Integral Equations. Phys Math Lit 254 (Moscow)
10. Eskin G (1981) Boundary-value problems for elliptic pseudo-differential equations. Math Soc 32. (New York, USA)
11. Kunin I, Sosnina E (1973) Stress concentration on an ellipsoidal inhomogeneity in an anisotropic elastic medium. Appl Math Mech (PMM) 37:306–315
12. Kunin I (1983) The theory of elastic media with microstructure II. Springer, Berlin Heilderberg New York, p 272
13. Sadowsky M, Sternberg E (1949) Stress concentration around a three-axial ellipsoidal cavity. J Appl Mech 16:149–157
14. Robinson K (1951) Elastic energy of an ellipsoidal inclusion in an infinite solid. J Appl Mech 22:1045–1054
15. Eshelby (1961) Elastic inclusions and inhomogeneities. In: Sneddon I, Hill R (eds) Progress in solid mechanics, vol 2, pp 88–140

Chapter 5
Crack in a Homogeneous Elastic Medium

Abstract The elasticity problem for an anisotropic homogeneous medium containing an isolated crack is reduced to 2D-integral equation for the crack opening vector. For an elliptical crack subjected to a polynomial external stress field, the polynomial conservation theorem is proved. For an elliptical crack subjected to a constant external field, determination of the crack opening vector and the stress intensity factors at the crack edge are reduced to calculation of 2D improper integrals. For isotropic host media, these integrals are expressed in terms of complete elliptic integrals. The conductivity problem for a crack is considered.

Keywords Crack in anisotropic elastic medium · 2D-integral equation of the crack problem · Elliptical crack · Polynomial conservation for an elliptical crack · Elliptical crack in a constant external field · Stress intensity factor

5.1 Integral Equation of the Crack Problem

Let a homogeneous elastic medium with the stiffness tensor C^0 contain an inclusion in the flattened region V with the stiffness tensor C. The strain and stress tensor in the medium with the inclusion satisfy the integral equations (4.12), (4.20). Further, we consider the limiting form of these equations when tensor C tens to zero together with transversal size of the region V.

The integral equations (4.12), (4.20) for strain $\varepsilon(x)$ and stress $\sigma(x)$ tensors can be presented in the forms

$$\varepsilon(x) + \int K(x - x')(C - C^0)\varepsilon^+(x')dx' = \varepsilon^0(x)\,, \tag{5.1}$$

$$\sigma(x) - \int S(x - x')\left(I - \left(C^0\right)^{-1} C\right)\varepsilon^+(x')dx' = \sigma^0(x)\,. \tag{5.2}$$

Here, $\varepsilon^+(x) = \varepsilon(x)V(x)$ and $V(x)$ is the characteristic function of the region V. The limiting equation for the strain tensor when the tensor C tends to zero is

S. Kanaun, *Self-Consistent Methods for Composites*, Solid Mechanics and Its Applications 282, https://doi.org/10.1007/978-3-031-96739-9_5

$$\varepsilon(x) - \int K(x - x')C^0\varepsilon^+(x')dx' = \varepsilon^0(x). \tag{5.3}$$

The stress field outside the cavity is expressed in terms of the tensor $\varepsilon^+(x)$ by the equation

$$\sigma(x) - \int S(x - x')\varepsilon^+(x')dx' = \sigma^0(x). \tag{5.4}$$

For transition from a thin cavity to a crack, i.e., a cut along a smooth surface Ω inside V, we introduce the local Cartesian coordinate systems at each point $x \in \Omega$ with the z-axis directed along the normal $n(x)$ to Ω. Let $h(x)$ is the transverse size of the cavity, and $z_1(x, h)$, $z_2(x, h)$ are the coordinates of the points of intersection of the z-axis with the boundary of the cavity V, $z_1, z_2 \to 0$ as $h \to 0$. For a fixed point $x \notin V$, the kernel $S(x - x')$ in Eq. (5.4) is a smooth bounded function. Hence, the principal term of the stress field outside the cavity has the form:

$$\sigma(x) = \sigma^0(x) + \int_\Omega S(x - x')\bar{\varepsilon}(x', h)d\Omega', \tag{5.5}$$

$$\bar{\varepsilon}(x, h) = \int_{z_1(x,h)}^{z_2(x,h)} \varepsilon^+(x + zh(x))dz, \quad x \in \Omega. \tag{5.6}$$

The function $\bar{\varepsilon}(x, h)$ can be interpreted as the coefficient of the principal term of the expansion of the function $\varepsilon^+(x)$ in a series of multipoles concentrated on Ω (Sect. 3.6):

$$\varepsilon^+(x)V(x) = \bar{\varepsilon}(x, h)\Omega(x) + ... \tag{5.7}$$

Here, $\Omega(x)$ is the delta-function concentrated on Ω that is defined in Eq. (3.38). It is important that all other terms of the multipole expansion of $\varepsilon^+(x)V(x)$ vanish as $h \to 0$, i.e., the total strain field $\varepsilon(x)$ contains a singular component proportional to $\Omega(x)$. If the transverse size h of the cavity tend to zero, the cavity is transformed into an infinitely thin cut (crack). In this limit, the stress vector $t_\alpha(x) = n_\beta(x)\sigma_{\beta\alpha}(x)$ is equal to zero on Ω, and the displacement field $u(x)$ has a jump $b(x)$ on Ω

$$n_\alpha(x)\sigma_{\alpha\beta}(x) = 0, \quad [u_\alpha(x)] = b_\alpha(x), \quad x \in \Omega. \tag{5.8}$$

The vector $b(x)$ can be interpreted as the crack opening vector that should be found from the solution of the crack problem. In what follows we suppose that the crack faces do not contact. This condition imposes some restrictions on the external field $\sigma^0(x)$.

Since the displacement field is discontinuous on Ω, the strain tensor ε is the sum of the regular ε^r and singular ε^s components

$$\varepsilon(x) = \varepsilon^r(x) + \varepsilon^s(x)\,, \quad \varepsilon^s_{\alpha\beta}(x) = n_{(\alpha}(x)b_{\beta)}(x)\Omega(x)\,. \tag{5.9}$$

It follows from comparison of this equation with Eq. (5.7) that in the limit $h \to 0$ the unwritten terms in (5.7) vanish, and the limiting value $\bar{\varepsilon}(x)$ of the function $\varepsilon^+(x, h)$ takes the form

$$\bar{\varepsilon}_{\alpha\beta}(x) = \lim_{h\to 0}\bar{\varepsilon}^+_{\alpha\beta}(x, h) = n_{(\alpha}(x)b_{\beta)}(x), \;\; x \in \Omega\,. \tag{5.10}$$

This equation and Eq. (5.6) show that in the limit $h \to 0$, the normal component of the tensor $\varepsilon^+(x)$ tends to infinity, whereas the integral $\bar{\varepsilon}(x)$ is finite. Thus taking into account Eqs. (5.5), (5.10) and symmetry of the tensor $S(x)$ with respect to index permutations, the stress field outside the crack is presented in the form

$$\sigma_{\alpha\beta}(x) = \sigma^0_{\alpha\beta}(x) + \int_\Omega S_{\alpha\beta\lambda\mu}(x - x')n_\lambda(x')b_\mu(x')d\Omega'. \tag{5.11}$$

Comparison of Eq. (5.11) with Eq. (3.110) for the stress field in an elastic medium containing Somigliana dislocation show that Eq. (5.11) corresponds to the dislocation model of a crack.

From the first condition (5.8) we obtain the equation for the crack opening vector $b(x)$

$$\int_\Omega T_{\alpha\beta}(x, x')b_\beta(x')d\Omega' = n_\beta(x)\sigma^0_{\alpha\beta}(x)\,, x \in \Omega. \tag{5.12}$$

$$T_{\alpha\beta}(x, x') = -n_\lambda(x)S_{\lambda\alpha\beta\mu}(x - x')n_\mu(x')\,. \tag{5.13}$$

Note that the operator T in Eq. (5.12) can be written as the integral operator with the kernel $T(x, x')$ only symbolically, because the corresponding integral formally diverges as $x \in \Omega$ for any smooth function $b(x)$ ($T(x, x')\sim|x-x'|^{-3}$ as $x \to x'$). For revealing the regularization formula of this integral, we consider the generalized function $T(x, x')$ in Eq. (5.13) in detail.

Let Ω be a finite region in the plane $x_3 = 0$, and function $b(x_1, x_2)$ be continued by zero outside Ω. In this case, T can be presented as the convolution operator with the generalized function $T(x_1, x_2)$ that is generated by the function $S(x_1, x_2, x_3)$ in Eq. (5.13)

$$T_{\alpha\beta}(x_1, x_2) = -S_{3\alpha\beta 3}(x_1, x_2, x_3)|_{x_3=0}\,. \tag{5.14}$$

Fourier transform $\widetilde{T}_{\alpha\beta}(k_1, k_2)$ of the function $T_{\alpha\beta}(x_1, x_2)$ follows from this equation in the form

$$\widetilde{T}_{\alpha\beta}(k_1, k_2) = -\frac{1}{2\pi}\int\limits_{-\infty}^{\infty} \widetilde{S}_{3\alpha\beta 3}(k_1, k_2, k_3)dk_3\,, \tag{5.15}$$

where $\widetilde{S}(k_1, k_2, k_3)$ is Fourier transform of the function $S(x_1, x_2, x_3)$ in Eq. (5.2). This function is defined by the equations

$$\widetilde{S}_{\alpha\beta\gamma\delta}(k_1, k_2, k_3) = C^0_{\alpha\beta\lambda\mu}\widetilde{K}_{\lambda\mu\nu\rho}(k_1, k_2, k_3)C^0_{\nu\rho\gamma\delta} - C^0_{\alpha\beta\gamma\delta}\,, \tag{5.16}$$

$$\widetilde{K}_{\alpha\beta\gamma\delta}(k_1, k_2, k_3) = k_\alpha \widetilde{G}_{\beta\gamma}(k_1, k_2, k_3)k_\delta\big|_{(\alpha\beta)(\gamma\delta)} \tag{5.17}$$

$$\widetilde{G}(k_1, k_2, k_3) = L^{-1}(k_1, k_2, k_3),\, L_{\alpha\beta}(k_1, k_2, k_3) = k_\gamma C^0_{\gamma\alpha\beta\delta}k_\delta. \tag{5.18}$$

The integral (5.15) converges absolutely and defines an even homogeneous function of the argument $\bar{k} = \bar{k}(k_1, k_2)$ of the order of 1, e.g., $\widetilde{T}(\lambda\bar{k}) = \lambda\widetilde{T}(\bar{k})$, where λ is an arbitrary scalar parameter. Therefore, if the function $b(\bar{x})$ ($\bar{x} = \bar{x}(x_1, x_2)$) has Fourier transform $\widetilde{b}(\bar{k})$ that is a bounded function vanishing at infinity faster than $|\bar{k}|^{-3}$, action of the operator T on $b(\bar{x})$ is defined by the equation

$$(Tb)(\bar{x}) = \frac{1}{(2\pi)^2}\int \widetilde{T}(\bar{k})\widetilde{b}(\bar{k})e^{-\mathrm{i}(\bar{k}\cdot\bar{x})}d\bar{k}\,. \tag{5.19}$$

Here, the integral converges absolutely. This equation defines regularization of the operator T on the mentioned class of functions in the plane $x_3 = 0$.

If Ω is an arbitrary bounded Lyapunov surface with a smooth contour Γ, the regularization of the operator T in Eq. (5.12) is obtained in Appendix C.2 in the form

$$(T_{\alpha\beta}b_\beta)(x) = pv\int_\Omega T_{\alpha\beta}(x, x')\left(b_\beta(x') - b_\beta(x)\right)d\Omega' -$$
$$- n_\nu(x)\oint_\Gamma \mathrm{rot}_{\lambda\beta}Z_{\nu\alpha\lambda\mu}(x - x')d\Gamma'_\beta b_\mu(x)\,,\ x \in \Omega. \tag{5.20}$$

The function $b(x)$ is assumed to be continuous together with the first derivative on Ω, and $b(x) = 0$ on the crack contour Γ. If Ω is a domain in the plane $x_3 = 0$, the regularization formula for the operator T can be presented as follows

$$(T_{\alpha\beta}b_\beta)(\overline{x}) = pv\int T_{\alpha\beta}(\overline{x} - \overline{x}')\left(b_\beta(\overline{x}') - b_\beta(\overline{x})\right)d\overline{x}',\, \overline{x} = (x_1, x_2), \tag{5.21}$$

where integration is spread to the entire plane $x_3 = 0$, and the function $b(\overline{x})$ is continued by zero outside Ω.

5.2 Elliptical Crack

For a planar elliptical crack in a homogeneous elastic medium, there exists an analogy of the polynomial conservation theorem for ellipsoidal inclusions (Sect. 4.3).

If the external field $\sigma^0(x)$ is a polynomial of degree m in the crack surface, the crack opening vector $b(x) = [u(x)]_\Omega$ has the form

$$b(x_1, x_2) = B(x_1, x_2)\sqrt{1 - \left(\frac{x_1}{a_1}\right)^2 - \left(\frac{x_2}{a_2}\right)^2}. \tag{5.22}$$

Here, x_1, x_2 are Cartesian coordinates directed along the principal axes of the ellipse, $B(x_1, x_2)$ is a polynomial of the degree not more than m, $a_1, a_2$are the ellipse semiaxes.

For the proof, we consider an elliptical crack as a limit of an ellipsoidal inclusion with the semiaxes a_1, a_2, h when the semiaxis h and the stiffness tensor C of the inclusion tend to zero. According to Eqs. (5.7), (5.10), the vector $b(x)$ is determined by the limiting value of the integral

$$\bar{\varepsilon}^+(\bar{x}, h) = \int_{-hz(\bar{x})}^{hz(\bar{x})} \varepsilon^+(\bar{x} + \xi n)d\xi\ , \quad \bar{x} \in \Omega\ , \tag{5.23}$$

where ξ is the coordinate along the normal n to Ω,

$$z(\bar{x}) = \sqrt{1 - \left(\frac{x_1}{a_1}\right)^2 - \left(\frac{x_2}{a_2}\right)^2}. \tag{5.24}$$

It follows from the polynomial conservation theorem (Sect. 4.3) that for the polynomial external field of degree m, the field $\varepsilon^+(\bar{x} + \xi n)$ inside the inclusion is also a polynomial of degree m in the coordinates x_1, x_2, ξ . Thus this field is presented the form

$$\varepsilon^+(\bar{x} + \xi n) = \varepsilon^{(0)}(\bar{x}) + \varepsilon^{(1)}(\bar{x})\xi + ... + \varepsilon^{(m)}(\bar{x})\xi^m, \tag{5.25}$$

where $\varepsilon^{(k)}(\bar{x})$ is polynomial of a degree no higher than $m - k$ in the coordinates x_1, x_2. After substituting this equation in Eq. (5.23) and integrating over ξ, we find that the terms with odd degrees of ξ vanish, and the integrals of the terms with ξ^{2k} give terms that have the form $Q^{2k}(\bar{x})z(\bar{x})$, where $Q^{2k}(\overline{x})$ is a polynomial of degree not higher than $2k$. Hence, the tensor $\bar{\varepsilon}^+(\overline{x}, h)$ is represented in the form

$$\bar{\varepsilon}^+(\bar{x}, h) = Q_m(\bar{x}, h)z(\bar{x})\ , \tag{5.26}$$

where Q_m is a polynomial of degree not more than m in the coordinates x_1, x_2. Finally, setting $h = 0$, we obtain from Eq. (5.10) the required Eq. (5.22).

It follows from Eq. (5.22) that for an elliptical crack, the operator T in Eq. (5.12) transforms a polynomial $B(\bar{x})$ of a degree m multiplied by the function $z(\bar{x})$ (5.24) into a polynomial of the same degree m on Ω.

5.3 An Elliptical Crack Subjected to a Constant External Stress Field

If σ^0 is a constant tensor, the vector $B = B^0$ in Eq. (5.22) is constant, and the crack opening vector is

$$b(\overline{x}) = B^0 z(\overline{x}). \tag{5.27}$$

For calculation of B^0, we can take $x = 0$ in the integral equation (5.12) for a planar crack. As a result, vector B^0 is the solution of the linear algebraic equation

$$T^0_{\alpha\beta} B^0_\beta = n_\beta \sigma^0_{\beta\alpha}, \tag{5.28}$$

$$T^0 = \int_\Omega T(\bar{x}) z(\bar{x}) d\bar{x} = \int T(\bar{x}) \left(z(\bar{x}) - z(0)\right) d\bar{x}\,. \tag{5.29}$$

Here, the regularization formula (5.21) is used. The last integral in this equation is calculated over the entire (x_1, x_2)-plane, and the function $z(\bar{x})$ is continued by zero outside of the crack surface. Note that the improper integral T^0 in Eq. (5.29) exists in the ordinary sense.

For calculation of the integral T^0, we introduce dimensionless variables $y_1 = \frac{x_1}{a_1}$, $y_2 = \frac{x_2}{a_1}$ and present T^0 in the form

$$T^0 = a_1 a_2 \iint T(a_1 y_1, a_2 y_2) Z(|y|) dy_1 dy_2,$$

$$Z(|y|) = \sqrt{1 - |y|^2} - 1,\ |y| < 1;\ Z(|y|) = -1, \quad |y| > 1. \tag{5.30}$$

Using Parceval's formula and transferring to Fourier transforms of the integrand functions, we yield

$$T^0 = \frac{1}{(2\pi)^2} \int\int \widetilde{T}\left(\frac{k_1}{a_1}, \frac{k_2}{a_2}\right) \widetilde{Z}(|\bar{k}|) dk_1 dk_2 =$$

$$= \frac{1}{(2\pi)^2} \int_0^\infty \widetilde{Z}(|\bar{k}|) |\bar{k}|^2 d|\bar{k}| \int_0^{2\pi} \widetilde{T}\left(\frac{\cos\varphi}{a_1}, \frac{\sin\varphi}{a_2}\right) d\varphi. \tag{5.31}$$

Here, $\widetilde{Z}(|\bar{k}|)$ and $\widetilde{T}(k_1, k_2)$ are Fourier transforms of the functions $Z(|y|)$, $T(y_1, y_2)$, φ is polar angle in the plane (k_1, k_2). It is taken into account that $\widetilde{T}(k_1, k_2)$ is a homogeneous function of the order of 1, and $\widetilde{T}(\lambda k_1, \lambda k_2) = \lambda \widetilde{T}(k_1, k_2)$ for any scalar parameter λ. The integral over $|\bar{k}|$ in Eq. (5.31) is calculated as follows

$$\frac{1}{(2\pi)^2}\int_0^{\infty} \widetilde{Z}(|\bar{k}|)|\bar{k}|^2 d|k| = \frac{1}{(2\pi)^3}\iint \widetilde{Z}(|\bar{k}|)|\bar{k}|dk_1dk_2 =$$

$$= -\frac{1}{(2\pi)^2}\iint \frac{Z(|y|)}{|y|^3}dy_1dy_2 = -\frac{1}{2\pi}\int_0^{\infty}\frac{Z(|y|)}{|y|^2}d|y| = \frac{1}{4}. \tag{5.32}$$

Here, it is taken into account that $|\bar{k}|$ is Fourier transform of the function $-1/\left(2\pi|y|^3\right)$ [1]. As a result for the tensor T^0 in Eq. (5.31), we yield the equation

$$T^0 = \frac{1}{4}\int_0^{2\pi} \widetilde{T}\left(\frac{\cos\varphi}{a_1}, \frac{\sin\varphi}{a_2}\right)d\varphi. \tag{5.33}$$

Taking into account Eq. (5.15) for $\widetilde{T}(k_1, k_2)$, we obtain that tensor T^0 is the following double integral

$$T^0_{\alpha\beta} = -\frac{1}{8\pi}\int_0^{2\pi} d\varphi \int_{-\infty}^{\infty} \widetilde{S}_{3\alpha\beta 3}\left(\frac{\cos\varphi}{a_1}, \frac{\sin\varphi}{a_2}, k_3\right)dk_3. \tag{5.34}$$

The explicit expression for the function $\widetilde{S}(k_1, k_2, k_3)$ is given in Eqs. (5.16)–(5.18), and therefore, this absolutely converging integral can be calculated numerically for any arbitrary anisotropic medium.

For an isotropic medium, the tensor $\widetilde{S}(k_1, k_2.k_3)$ is defined in Eq. (3.96).

$$\widetilde{S}(k) = -2\mu_0\left(P^1(m) + (2\kappa_0 - 1)P^2(m)\right), \tag{5.35}$$

$$m_\alpha = \frac{k_\alpha}{|k|}, \kappa_0 = \frac{\lambda_0 + \mu_0}{\lambda_0 + 2\mu_0}. \tag{5.36}$$

Here, λ_0, μ_0 are Lame's parameters of the medium, $P^1(m)$, $P^2(m)$ are the elements of the tensor basis of forth rank tensors (P -basis, Appendix A.2)

$$P^1_{\alpha\beta\gamma\delta} = \theta_{\alpha\gamma}\theta_{\beta\delta}|_{(\alpha\beta)(\gamma\delta)}, P^2_{\alpha\beta\gamma\delta} = \theta_{\alpha\beta}\theta_{\gamma\delta}, \quad \theta_{\alpha\beta} = \delta_{\alpha\beta} - m_\alpha m_\beta. \tag{5.37}$$

In this case, the only components of the tensor $T^0_{\alpha\beta}$ that do not vanish by integration in Eq. (5.34) are T^0_{11}, T^0_{22}, T^0_{33}. The explicit forms for the components $\widetilde{S}_{3113}$, $\widetilde{S}_{3223}$, $\widetilde{S}_{3333}$ are

$$\widetilde{S}_{3113} = -\frac{\mu_0}{\left(k_1^2+k_2^2+k_3^2\right)^2}\left(k_2^2(k_2^2+k_3^2)+k_1^2(k_2^2+4\kappa_0 k_3^2\right), \tag{5.38}$$

$$\widetilde{S}_{3223} = -\frac{\mu_0}{\left(k_1^2+k_2^2+k_3^2\right)^2}\left(k_1^4+k_1^2(k_2^2+k_3^2)+4\kappa_0 k_2^2 k_3^2\right), \tag{5.39}$$

$$\widetilde{S}_{3333} = -\frac{4\kappa_0\mu_0(k_1^2+k_2^2)^2}{\left(k_1^2+k_2^2+k_3^2\right)^2}. \tag{5.40}$$

Integration of the function $\widetilde{S}_{3113}$, $\widetilde{S}_{3223}$ and $\widetilde{S}_{3333}$ over k_3 from $-\infty$ to ∞ results in the components of the tensor $\widetilde{T}_{\alpha\beta}(k_1, k_2)$ in Eq. (5.15)

$$\widetilde{T}_{11} = \frac{\pi\mu_0}{|\bar{k}|}\left(2\kappa_0 k_1^2+k_2^2\right), \widetilde{T}_{22} = \frac{\pi\mu_0}{|\bar{k}|}\left(k_1^2+2\kappa_0 k_2^2\right), \tag{5.41}$$

$$\widetilde{T}_{33} = 2\pi\kappa_0\mu_0|\bar{k}|, \quad |\bar{k}| = \sqrt{k_1^2+k_2^2}. \tag{5.42}$$

Finally, the diagonal components of the tensor T^0 in Eq. (5.34) are the following regular integrals

$$T_{11}^0 = \frac{\mu_0}{8a_2}\int_0^{2\pi}\frac{2\kappa_0(\gamma\cos\varphi)^2+(\sin\varphi)^2}{\sqrt{(\gamma\cos\varphi)^2+(\sin\varphi)^2}}d\varphi, \tag{5.43}$$

$$T_{22}^0 = \frac{\mu_0}{8a_2}\int_0^{2\pi}\frac{\left(\gamma^2\cos\varphi\right)^2+2\kappa_0(\sin\varphi)^2}{\sqrt{(\gamma\cos\varphi)^2+(\sin\varphi)^2}}d\varphi, \tag{5.44}$$

$$T_{33}^0 = \frac{\kappa_0\mu_0}{4a_2}\int_0^{2\pi}\sqrt{(\gamma\cos\varphi)^2+(\sin\varphi)^2}d\varphi, \gamma = \frac{a_2}{a_1}. \tag{5.45}$$

These integrals are expressed in terms of the complete elliptic integrals as follows

$$T_{11}^0 = \frac{\mu_0}{2(1-\nu_0)\sqrt{a_1 a_2}}\left(c_1+\nu_0 c_2\right), \tag{5.46}$$

$$T_{22}^0 = \frac{\mu_0}{2(1-\nu_0)\sqrt{a_1 a_2}}\left(c_1+\nu_0 c_3\right), \tag{5.47}$$

$$T_{33}^0 = \frac{\mu_0}{2(1-\nu_0)\sqrt{a_1 a_2}}c_1, \tag{5.48}$$

$$c_1 = \frac{E(\eta)}{\sqrt{\gamma}}, \quad c_2 = \frac{\gamma^2 K(\eta)-E(\eta)}{\sqrt{\gamma}\eta}, \tag{5.49}$$

$$c_3 = \gamma^{3/2}\frac{E(\eta)-K(\eta)}{\eta}, \quad \eta = 1-\gamma^2. \tag{5.50}$$

Here, ν_0 is Poisson's ratio of the medium, $K(\eta)$ and $E(\eta)$ are the complete elliptic integrals of the first and second kinds

$$K(\eta) = \int_0^{\pi/2} \frac{d\theta}{\sqrt{1-\eta\sin^2\theta}}, \ E(\eta) = \int_0^{\pi/2} \sqrt{1-\eta\sin^2\theta} d\theta. \tag{5.51}$$

For a penny-shape crack of the radius a,

$$T_{11}^0 = T_{22}^0 = \frac{\pi\mu_0}{8a}\frac{(2-\nu_0)}{(1-\nu_0)}, \ T_{33}^0 = \frac{\pi\mu_0}{4a\,(1-\nu_0)}. \tag{5.52}$$

Finally, the crack opening vector $b(\overline{x})$ of an elliptical crack subjected to a constant external stress field σ^0 is defined by the equation

$$b_\alpha(x_1, x_2) = B_\alpha^0 \sqrt{1 - \left(\frac{x_1}{a_1}\right)^2 - \left(\frac{x_2}{a_2}\right)^2}, \tag{5.53}$$

$$B_\alpha^0 = (T^0)^{-1}_{\alpha\beta} t_\beta^0, \ \ t_\alpha^0 = \sigma_{\alpha\beta}^0 n_\beta, \tag{5.54}$$

where the tensor T^0 is in Eq. (5.34). For an isotropic medium, the matrix of the tensor T^0 in the basis of the principal axes of the ellipse is diagonal with the component $T_{11}^0, T_{22}^0, T_{33}^0$ in Eqs. (5.46)–(5.48).

5.4 Stress Intensity Factor at the Crack Edge

For an arbitrary smooth crack surface Ω, the operator T in the integral equation (5.12) is a pseudo-differential operator which symbol $\widetilde{T}(k)$ is a homogeneous function of the order of 1. The existence and uniqueness of the solution of the equation $Tb = f$ are proved in [2]. It is shown in [2] that for an analytical right-hand side $f(x)$, the asymptotic form of the vector $b(x)$ at the crack contour Γ is

$$b(x) = \beta(x_0)\sqrt{r} + 0(r^{3/2}), \tag{5.55}$$

where r is the distance from the point $x \in \Omega$ to $x_0 \in \Gamma$ along the normal to Γ, $\beta(x_0)$ is a smooth function on Γ.

For analysis of the asymptotic form of the stress field $\sigma(x)$ in the medium near the crack edge Γ, we consider an arbitrary point $x_0 \in \Gamma$ and introduce the local Cartesian coordinates y_1, y_2, y_3 with the origin at this point. The y_3-axis is directed along the limit normal to Ω at the point x_0, and the y_2-axis is tangent to Γ. Then, the y_1-axis is in the tangent plane to Ω at the point x_0. Let the basis vectors of this system be g^1, g^2, g^3 (Fig. 5.1).

It follows from Eq. (5.55) that in the y-coordinates, the asymptotic form of the crack opening vector near the point $x_0 \in \Gamma$ is

$$b(y) = \beta(x_0)\sqrt{y_1} + 0(y_1^{3/2}). \tag{5.56}$$

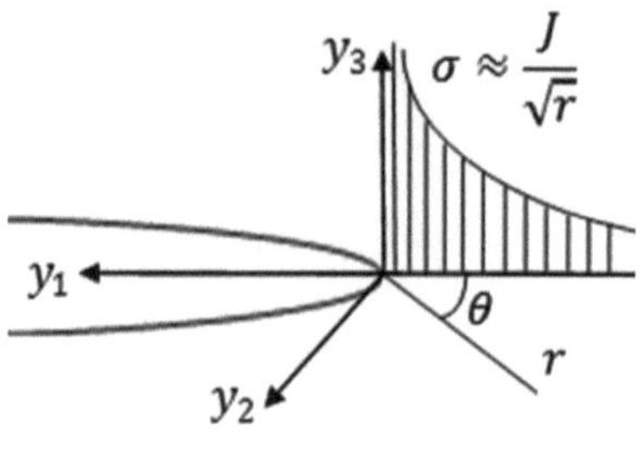

Fig. 5.1 Local coordinate system in the vicinity of the crack edge

Then, it follows from Eq. (5.11) that the stress tensor $\sigma(y_1, y_2, y_3)$ at the point with the coordinates $y_1 = -r\cos\theta$, $y_2 = 0$, $y_3 = -r\sin\theta$ is presented in the form

$$\sigma(-r\cos\theta, 0, -r\sin\theta) = \tag{5.57}$$

$$= \int_{\Omega} S\left(-r\cos\theta - y_1', -y_2', -r\sin\theta - y'_3\right) n\left(y'\right)\sqrt{y_1'}d\Omega_y'\beta\left(x_0\right) + O\left(\sigma^0\right).$$

Here, r is the distance from point y up to the origin, θ is the polar angle in the plane (y_1, y_3) (Fig. 5.1). Transferring to dimensionless variables $\xi_i = y_i/r$, $i = 1, 2, 3$ in the integral in this equation and taking into account that $d\Omega'_y = r^2 d\Omega'_\xi$ and $S(x)$ is an even homogeneous function of the order of -3, we obtain

$$\sigma(-r\cos\theta, 0, -r\sin\theta) = \tag{5.58}$$

$$= \frac{1}{\sqrt{r}}\int_{\Omega(r)} S\left(\cos\theta + \xi'_1, \xi'_2, \sin\theta + \xi'_3\right) n\left(r\xi'\right)\sqrt{\xi_1'}d\Omega_\xi'\beta\left(x_0\right) + O\left(\sigma^0\right).$$

In the limit $r \to 0$, the multiplier at $r^{-1/2}$ in this equation has a finite value

$$J(\theta, x_0) = \lim_{r\to 0}\sqrt{r}\sigma(y). \tag{5.59}$$

Because in the limit $r \to 0$ the surface $\Omega(r)$ in coordinates ξ_i is transformed into the semi-plane ($\xi_3 = 0$, $\xi_1 > 0$), we yield the equation for the tensor $J(\theta, x_0)$

$$J_{\alpha\beta}(\theta, x^0) = s_{\alpha\beta\lambda}(\theta, x^0)\beta_\lambda(x^0), \tag{5.60}$$

$$s_{\alpha\beta\mu}(\theta, x^0) = \int_0^\infty \sqrt{\xi_1}d\xi_1 \int_{-\infty}^{\infty} S_{\alpha\beta\mu\lambda}(\cos\theta + \xi_1, \xi_2, \sin\theta)n_\lambda(x^0)d\xi_2. \tag{5.61}$$

Here, $n(x^0)$ is the limiting normal to the crack surface Ω at point x^0. Therefore, the function $J(\theta, x_0)$ is the product of two terms. The first term $s(\theta, x^0)$ is independent of the shape of the surface Ω and the external field σ^0. Tensor $s(\theta, x^0)$ is determined by the local orientation of the crack edge Γ at the point $x_0 \in \Gamma$ only. The second term (vector $\beta(x^0)$) is a functional of the entire surface Ω and the external stress field $\sigma^0(x)$.

It follows from Eq. (5.59) that near the crack edge the stress $\sigma(y)$ is presented as

$$\sigma(y) = \frac{1}{\sqrt{r}} J(\theta, x_0) + O(\sigma^0), \tag{5.62}$$

and therefore, $J(\theta, x_0)$ can be interpreted as the stress intensity factor. The tensor $J(\theta, x_0)$ can be presented as a sum of three tensors that correspond to the three components of the vector $\beta(x^0)$ in the basis g^1, g^2, g^3 of the local coordinate system y_1, y_2, y_3 (not sum over i!)

$$J = J^1 + J^2 + J^3, \quad J^i_{\alpha\beta}(\theta, x^0) = s_{\alpha\beta i}(\theta, x^0)\beta_i(x^0). \tag{5.63}$$

In linear fracture mechanics, the asymptotic form of the stress field at the crack edge is characterized by the stress intensity factors K_I, K_{II}, K_{III} [3]. Relations of these factors with the components of the tensor $J^i(\theta, x_0)$ are given by the equations

$$K_I(x_0) = \sqrt{2\pi} J^3_{33}(0, x_0), \quad K_{II}(x_0) = \sqrt{2\pi} J^1_{13}(0, x_0), \tag{5.64}$$

$$K_{III}(x_0) = \sqrt{2\pi} J^2_{23}(0, x_0). \tag{5.65}$$

It follows from these equations and Eq. (5.63) that the stress intensity factors are proportional to the corresponding components of the vector $\beta(x^0)$.

The tensor $s_{\alpha\beta\gamma}(0, x^0)$ in Eq. (5.61) is the double integral

$$s_{\alpha\beta\gamma} = \int_0^\infty \sqrt{\xi_1} d\xi_1 \int_{-\infty}^\infty S_{\alpha\beta\gamma3}(1 + \xi_1, \xi_2, 0) d\xi_2. \tag{5.66}$$

Here, the components of the tensor $S_{\alpha\beta\gamma3}(\xi_1, \xi_2, \xi_3)$ are to be taken in the local basis g^i at the point x^0. The function $S_{\alpha\beta\gamma3}(1 + \xi_1, \xi_2, 0)$ in this equation can be expressed in term of Fourier transform $\widetilde{S}_{\alpha\beta\gamma3}(k)$ of the function $S(x)$

$$S_{\alpha\beta\gamma3}(1 + \xi_1, \xi_2, 0) =$$
$$= \frac{1}{(2\pi)^3} \int\int\int \widetilde{S}_{\alpha\beta\gamma3}(k_1, k_2, k_3) e^{-i(1+\xi_1)k_1 - ik_2\xi_2} dk_1 dk_2 dk_3. \tag{5.67}$$

Here integration is spread to the entire 3D-space. Using the identities [1]

$$\frac{1}{2\pi} \int_{-\infty}^\infty e^{-ik_2\xi_2} = \delta(k_2), \tag{5.68}$$

$$\int_{-\infty}^{\infty} \widetilde{S}_{ijk3}(k_1, 0, k_3)dk_3 = |k_1| \int_{-\infty}^{\infty} \widetilde{S}_{ijk3}(1, 0, k_3)dk_3, \tag{5.69}$$

$$\int_{-\infty}^{\infty} |k_1| e^{-\mathrm{i}(1+\xi_1)k_1} dk_1 = -\frac{2}{(1+\xi_1)^2}, \tag{5.70}$$

$$\int_{0}^{\infty} \frac{\sqrt{\xi_1}}{(1+\xi_1)^2} d\xi_1 = \frac{\pi}{2}, \tag{5.71}$$

the expression of the tensor $s_{\alpha\beta\gamma}(0, x^0)$ takes the form of the 1D-integral

$$s_{\alpha\beta\gamma}(0, x^0) = -\frac{1}{4\pi} \int_{-\infty}^{\infty} \widetilde{S}_{\alpha\beta\gamma 3}(1, 0, k_3)dk_3. \tag{5.72}$$

Here, the components of the tensor $\widetilde{S}_{\alpha\beta\gamma 3}(1, 0, k_3)$ are taken in the local basis g^i at the edge point x_0 of the crack.

For an isotropic medium, the tensor s in Eq. (5.72) does not depend on the point x^0 at the crack contour Γ, and the non-zero components of this tensor are

$$s_{131} = s_{311} = \frac{1}{2}\kappa_0\mu_0, \; s_{232} = s_{322} = \frac{1}{4}\mu_0, \tag{5.73}$$

$$s_{113} = s_{333} = \frac{1}{2}\kappa_0\mu_0, \; s_{223} = \frac{1}{2}(2\kappa_0 - 1)\mu_0. \tag{5.74}$$

For an elliptical crack and a polynomial external field, the asymptotic form of the vector $b(x)$ on the crack contour follows from Eq. (5.22) in the form

$$b(x) = \beta(x_0)\sqrt{r} + O(r^{3/2}), \tag{5.75}$$

$$\beta(x_0) = \sqrt{2}B(x_0)\left(\left(\frac{x_1^0}{a_1^2}\right)^2 + \left(\frac{x_2^0}{a_2^2}\right)^2\right)^{1/4}. \tag{5.76}$$

Here, r is the same as in Eq. (5.56), $B(x_0)$ is the value of the polynomial $B(x)$ at the point $x_0 = (x_1^0, x_1^0)$ of the crack contour Γ. The stress intensity tensor $J(0, x_0)$ at the elliptical crack contour follows from Eq. (5.62) in the form

$$J_{\alpha\beta}(0, x^0) = \sqrt{2}s_{\alpha\beta\lambda}(0, x^0)B_\lambda(x_0)\left(\left(\frac{x_1^0}{a_1^2}\right)^2 + \left(\frac{x_2^0}{a_2^2}\right)^2\right)^{1/4}, \tag{5.77}$$

Here, x_1^0, x_2^0 are the Cartesian coordinates of the point $x^0 \in \Gamma$ in the basis of the principal exes of the ellipse Ω.

5.5 An Elliptical Crack in a Transverse Isotropic Medium

Transverse isotropic media have a rotation symmetry axis and a mirror symmetry plane orthogonal to this axis. The elastic stiffness tensor C^0 of such media has 5 independent constants. If n is unit vector of the rotation symmetry axis, tensor C^0 can be presented in the P-basis as follows

$$C^0 = k_0 P^2(\mathsf{n}) + 2m_0 \left(P^1(\mathsf{n}) - \frac{1}{2} P^2(\mathsf{n}) \right) + l_0 \left(P^3(\mathsf{n}) + P^4(\mathsf{n}) \right) +$$
$$+ 4\mu_0 P^5(\mathsf{n}) + n_0 P^6(\mathsf{n}). \tag{5.78}$$

The constants in this equation relate to the constant c_{ij} in matrix presentation of the elastic stiffness tensor C^0 by the equation

$$k_0 = \frac{1}{2}(c_{11} + c_{12}), m_0 = \frac{1}{2}(c_{11} - c_{12}), \tag{5.79}$$

$$l_0 = c_{13}, \ \mu_0 = c_{44}, n_0 = c_{33}. \tag{5.80}$$

For a transverse isotropic medium, Fourier transform $\widetilde{K}(k)$ of the tensor $K(x)$ in Eq. (4.13) is presented in the form

$$\widetilde{K}_{\alpha\beta\gamma\delta}(k) = k_\alpha \widetilde{G}_{\beta\gamma}(k) k_\delta|_{(\alpha\beta)(\gamma\delta)}, \tag{5.81}$$

$$\widetilde{G}_{\alpha\beta}(k) = g_0(k)\theta_{\alpha\beta} + g_1(k)\overline{k}_\alpha \overline{k}_\beta + g_2(k) n_\alpha n_\beta +$$
$$+ h_3(k)(n_\alpha \overline{k}_\beta + n_\alpha \overline{k}_\beta). \tag{5.82}$$

Here, $\overline{k} = k_1 n^1 + k_2 n^2$ is the projection of the vector k on the isotropy plane, $\theta_{\alpha\beta} = \delta_{\alpha\beta} - n^3_\alpha n^3_\beta$,

$$g_0 = \left(m_0 \overline{k}^2 + \mu_0 k_3^2 \right)^{-1}, \tag{5.83}$$

$$g_1 = \frac{g_0}{\Delta} \left[k_0(\mu_0 \overline{k}^2 + n_0 k_3^2) - (l_0 + \mu_0)^2 k_3^2 \right], \tag{5.84}$$

$$g_2 = -\frac{1}{\Delta} \left[(k_0 + m_0)\overline{k}^2 + \mu_0 k_3^2 \right], g_3 = \frac{1}{\Delta}(l_0 + \mu_0) k_3, \tag{5.85}$$

where

$$\Delta = -\mu_0 n_0 \left(k_3^4 + a_0 \overline{k}^2 k_3^2 + b_0 \overline{k}^4 \right), \tag{5.86}$$

$$a_0 = \frac{1}{\mu_0 n_0} \left(n_0(k_0 + m_0) - 2l_0\mu_0 - l_0^2 \right), \quad b_0 = \frac{k_0 + m_0}{n_0}, \tag{5.87}$$

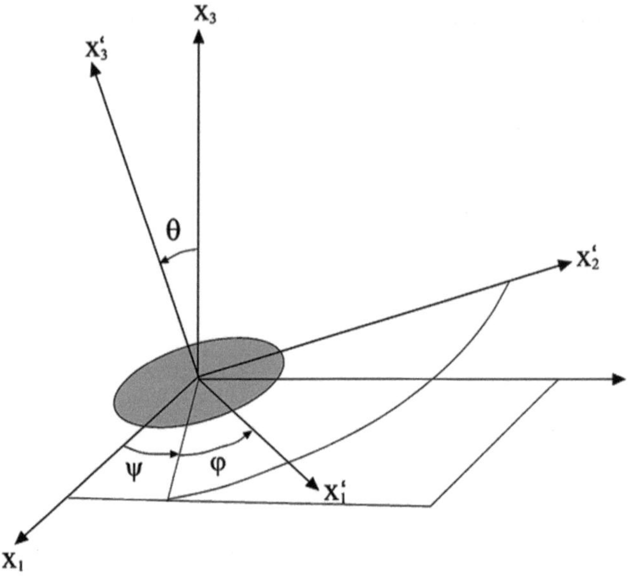

Fig. 5.2 The global and local coordinate systems for an elliptical crack

We consider an elliptical crack in a transverse isotropic medium subjected to a constant external stress field σ^0. Let the basis of a global coordinate system X_1, X_2, X_3 be composed from three orthogonal unit vectors n^i $(i = 1, 2, 3)$, and $n^3 = \mathsf{n}$ is the vector of the rotation symmetry of the medium. We introduce also the Cartesian system X'_1, X'_2, X'_3 with the axes X'_1, X'_2 directed along the principal axes of the ellipse Ω, the basis of this system is e^1, e^2, e^3 and $e^3 = n$ is the vector normal to the crack surface (Fig. 5.2). Orientation of the e-basis with respect to the n-basis is defined by the three Euler angles (φ, ψ, θ), and relation of the elements of the n and e-bases has the form $n^\alpha = Q^\alpha_\beta e^\beta$, where the matrix Q^α_β is

$$\left[Q^\alpha_\beta\right] = \tag{5.88}$$

$$\begin{bmatrix} \cos\varphi\cos\psi - \sin\varphi\sin\psi\cos\theta & \sin\varphi\cos\psi + \cos\varphi\sin\psi\cos\theta & \sin\varphi\sin\theta \\ \cos\varphi\cos\psi - \sin\varphi\cos\psi\cos\theta & -\sin\varphi\sin\psi + \cos\varphi\cos\psi\cos\theta & \cos\varphi\sin\theta \\ \sin\varphi\sin\theta & -\cos\varphi\sin\theta & \cos\theta \end{bmatrix}$$

The components $C^{0\prime}_{\alpha\beta\gamma\delta}$ of the tensor C^0 in the basis of the principal axes of the ellipse Ω is

$$C^{0\prime}_{\alpha\beta\gamma\delta} = Q^\lambda_\alpha Q^\mu_\beta Q^\nu_\gamma Q^\tau_\delta C^0_{\lambda\mu\nu\tau}. \tag{5.89}$$

The components of the crack opening vector $b(X')$ of an elliptical crack in the e-basis are (Fig. 5.2)

$$b_\alpha(X') = M_{\alpha\beta} t^0_\beta \sqrt{1 - \left(\frac{X'_1}{a_1}\right)^2 - \left(\frac{X'_2}{a_2}\right)^2}, \tag{5.90}$$

$$B^0_\alpha = M_{\alpha\beta} t^0_\beta, \quad t^0_\alpha = \sigma^0_{\alpha\beta} n_\beta. \tag{5.91}$$

Here, $M = (T^0)^{-1}$, and the components of the tensor T^0 are defined by the integral (5.34)

$$T^0_{\alpha\beta} = -\frac{1}{8\pi} \int_0^{2\pi} d\varphi \int_{-\infty}^{\infty} \widetilde{S}'_{3\alpha\beta 3}\left(\frac{\cos\varphi}{a_1}, \frac{\sin\varphi}{a_2}, k_3\right) dk_3. \tag{5.92}$$

Here, the components of the tensor $\widetilde{S}(k_1, k_2, k_3)$ in the e-basis are

$$\widetilde{S}'_{3\alpha\beta 3}(k_1, k_2, k_3) = C^{0\prime}_{3\alpha\gamma\delta} \widetilde{K}'_{\gamma\delta\lambda\mu}(k_1, k_2, k_3) C^{0\prime}_{\lambda\mu\beta 3} - C^{0\prime}_{3\alpha\beta 3}, \tag{5.93}$$

where

$$\widetilde{K}'_{\alpha\beta\gamma\delta}(k) = Q^\lambda_\alpha Q^\mu_\beta Q^\nu_\gamma Q^\tau_\delta \widetilde{K}_{\lambda\mu\nu\tau}(k). \tag{5.94}$$

For the function $\widetilde{K}_{\lambda\mu\nu\tau}(k)$ in Eqs. (5.81)–(5.87), integral (5.92) can be calculated numerically.

An elliptical crack in the isotropy plane of a transverse isotropic medium. If an elliptical crack is in the isotropy plane of a transverse isotropic medium, the components of the tensor $\widetilde{S}_{3ij3}(k)$ in the basis of the principal axes of the crack are

$$\widetilde{S}_{3113} = -\frac{\mu_0 m_0 k_2^2}{m\overline{k}^2 + \mu k_3^2} + \frac{\mu_0}{\Delta}\left(n_0(k_0 + m_0) - l_0^2\right) k_1^2 k_3^2, \tag{5.95}$$

$$\widetilde{S}_{3223} = -\frac{\mu_0 m_0 k_1^2}{m\overline{k}^2 + \mu k_3^2} + \frac{\mu_0}{\Delta}\left(n_0(k_0 + m_0) - l^2\right) k_2^2 k_3^2, \tag{5.96}$$

$$\widetilde{S}_{3333} = \frac{\mu_0}{\Delta}\left(n_0(k_0 + m_0) - l_0^2\right) \overline{k}^4, \quad \overline{k}^2 = k_1^2 + k_2^2, \tag{5.97}$$

$$\widetilde{S}_{3123} = \frac{\mu_0^2 k_1 k_2}{(m\overline{q}^2 + \mu q_3^2)\Delta}\left((m_0\overline{k}^2 - l_0 k_3^2) + k_0(m_0\overline{k}^2 - n_0 k_3^2)\right), \tag{5.98}$$

$$\widetilde{S}_{3133} = \frac{\mu_0 \overline{k}^2 k_1 k_3}{\Delta}\left(n_0(k_0 + m_0) - l_0^2\right), \quad \widetilde{S}_{3233} = \frac{k_2}{k_1}\widetilde{S}_{3133}, \tag{5.99}$$

where

$$\Delta = -\mu n \left(k_3^4 + a_0 \overline{q}^2 k_3^2 + b_0 \overline{k}^4\right), \tag{5.100}$$

$$a_0 = \frac{1}{\mu_0 n_0}\left[n_0(k_0 + m_0) - 2 l_0 \mu_0 - l_0^2\right], \quad b_0 = \frac{k_0 + m_0}{n_0}. \tag{5.101}$$

The component $\widetilde{T}_{11}(k_1, k_2)$ of the tensor $\widetilde{T}(k_1, k_2)$ in Eq. (5.15) takes in the form

$$\widetilde{T}_{11}(k_1, k_2) = -\frac{1}{2\pi}\int_{-\infty}^{\infty} \widetilde{S}_{3113}(k_1, k_2, k_3)dk_3 =$$
$$= \frac{1}{2}\left(\frac{m_0}{\sqrt{\xi_1}}\frac{k_2^2}{|\bar{k}|} + \frac{n_0(k_0+m_0)-l_0^2}{n_0(\sqrt{\xi_2}+\sqrt{\xi_3})}\frac{k_1^2}{|\bar{k}|}\right). \tag{5.102}$$

Here, $\xi_1 = m_0/\mu_0$, and ξ_2, ξ_3 are the roots of the quadratic equation

$$\xi^2 - a_0\xi + b_0 = 0, \tag{5.103}$$

where the coefficients a_0 and b_0 are determined in Eq. (5.101). Integrating over the polar angle φ in Eq. (5.34) results in the equation for the component T_{11}^0 of the tensor T^0

$$T_{11}^0 = \frac{1}{2a_2}\left(\frac{m_0}{\sqrt{\xi_1}}J_{11}^{(1)} + \frac{n_0(k_0+m_0)-l_0^2}{n_0(\sqrt{\xi_2}+\sqrt{\xi_3})}J_{11}^{(2)}\right), \tag{5.104}$$

$$J_{11}^{(1)} = \frac{E(\eta)-(1-\eta)K(\eta)}{\eta}, \eta = 1-\left(\frac{a_2}{a_1}\right)^2, \tag{5.105}$$

$$J_{11}^{(2)} = (1-\eta)\frac{K(\eta)-E(\eta)}{\eta}, \ a_2 \le a_1. \tag{5.106}$$

In these equations, $K(\eta)$ and $E(\eta)$ are the complete elliptic integrals of the first and the second kind. A similar integration results in the following equations for T_{22}^0 and T_{33}^0

$$T_{22}^0 = \frac{1}{2a_2}\left(\sqrt{\mu_0 m_0}J_{11}^{(2)} + \frac{n_0(k_0+m_0)-l_0^2}{n_0(\sqrt{\xi_2}+\sqrt{\xi_3})}J_{11}^{(1)}\right), \tag{5.107}$$

$$T_{33}^0 = \frac{1}{2a_2}\frac{\left(n_0(k_0+m_0)-l_0^2\right)}{n_0\sqrt{\xi_2\xi_3}\left(\sqrt{\xi_2}+\sqrt{\xi_3}\right)}E(\eta). \tag{5.108}$$

Other components of the tensor T^0 are proved to be zero

$$T_{12}^0 = T_{13}^0 = T_{23} = 0. \tag{5.109}$$

For a penny-shape crack of the radius a, we have

$$T_{11}^0 = T_{22}^0 = \frac{\pi}{8a}\left(\sqrt{m_0\mu_0} + \frac{(k_0+m_0)n_0 - l_0^2}{n_0\left(\sqrt{\xi_2}+\sqrt{\xi_3}\right)}\right), \tag{5.110}$$

$$T_{33}^0 = \frac{\pi}{4a}\frac{(k_0+m_0)n_0 - l_0^2}{n_0\left(\sqrt{\xi_2}+\sqrt{\xi_3}\right)\sqrt{\xi_2\xi_3}}. \tag{5.111}$$

For an elliptical crack of arbitrary orientation, tensor T^0 can be calculated only numerically. We consider an example of a transverse isotropic medium with the elastic constants: $c_{11} = 50GPa$, $c_{12} = 10GPa$, $c_{13} = 60GPa$, $c_{33} = 100GPa$, $c_{44} = 20GPa$. For a circular crack ($a_1 = a_2 = 1m$), the dependencies of the components of the tensor $M = (T^0)^{-1}$ on the Euler angle θ ($\varphi = \psi = 0$) are presented in Fig.5.3 ($M_{\alpha\beta}$ are the components of the tensor M in the e-basis, and $M_{12} = M_{13} = 0$) (Fig. 5.3).

The stress intensity factor at the crack edge is proportional to the integral (5.72)

$$s_{\alpha\beta\gamma} = -\frac{1}{4\pi}\int_{-\infty}^{\infty} \widetilde{S}_{\alpha\beta\gamma 3}(1, 0, k_3)dk_3. \tag{5.112}$$

In this equation, the components of the tensor $\widetilde{S}_{\alpha\beta\gamma 3}(1, 0, k_3)$ should be taken in the local basis g^i at the edge point of the crack. Relation of the g-basis and the e-basis of the principal axes of the elliptical crack is defined by the equations

$$e^i = \gamma^{ij}g^j,\ [\gamma^{ij}] = \begin{bmatrix} \gamma^{11} & \gamma^{12} & 0 \\ -\gamma^{12} & \gamma^{11} & 0 \\ 0 & 0 & 1 \end{bmatrix}, \tag{5.113}$$

$$\gamma^{11} = \frac{\cos t}{\sqrt{(\cos t)^2 + (a_1 \sin t/a_2)^2}}, \tag{5.114}$$

$$\gamma^{12} = \frac{\sin t}{\sqrt{(\sin t)^2 + (a_1 \cot s/a_2)^2}}. \tag{5.115}$$

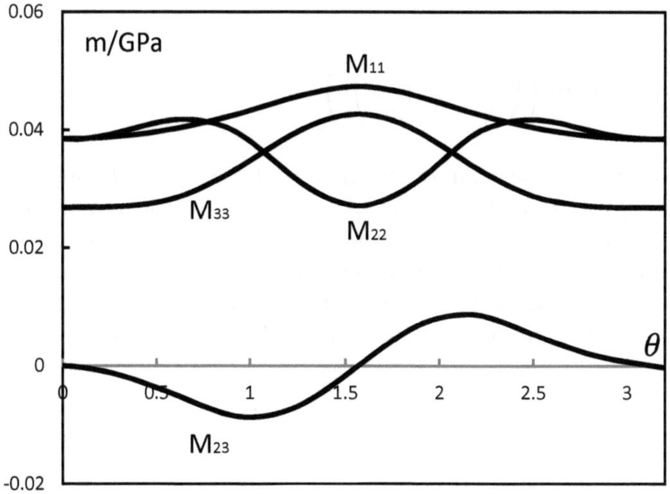

Fig. 5.3 Dependencies of the components of the tensor $M = (T^0)^{-1}$ on the Euler angle θ ($\varphi = \psi = 0$) for a transversally isotropic medium

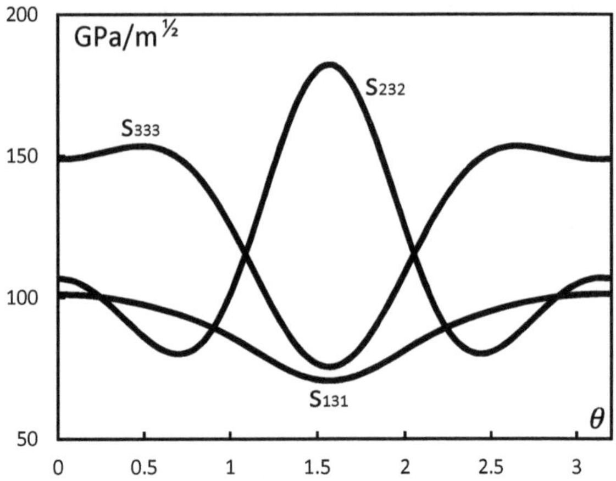

Fig. 5.4 Dependencies of the components of the tensor $s_{\alpha\beta\gamma}$ on the Euler angle θ ($\varphi = \psi = 0$) for a transversally isotropic medium

Here, parameter t relates to the coordinates of the point x^0 at the crack edge in the e-basis by the equations: $x_1^0 = a_1 \cos t,\ \ x_2^0 = a_2 \sin t$.

The stress intensity factors K_I, K_{II}, K_{II} in Eqs. (5.64), (5.65) are expressed in terms the tensor $s_{\alpha\beta\gamma}(x^0)$ and the vector $\beta_\alpha(x^0)$

$$K_I(x^0) = \sqrt{2\pi} s_{333}(x^0)\beta_3(x^0), \quad K_{II}(x^0) = \sqrt{2\pi} s_{131}(x^0)\beta_1(x^0), \tag{5.116}$$

$$K_{III}(x^0) = \sqrt{2\pi} s_{232}(x^0)\beta_2(x^0). \tag{5.117}$$

Here, vector $\beta(x^0)$ is

$$\beta(x_0) = \sqrt{2} B(x_0) \left(\left(\frac{x_1^0}{a_1^2} \right)^2 + \left(\frac{x_2^0}{a_2^2} \right)^2 \right)^{1/4}. \tag{5.118}$$

The dependencies of the components s_{333}, s_{131} and s_{232} of the tensor s on the Euler angle θ ($\varphi = \psi = 0$) for a circular crack ($a_1 = a_2 = 1m$) are presented in Fig. 5.4.

The algorithm of solution the crack problem for elliptical cracks can be extended to the anisotropic media when the explicit equation for Fourier transform of Green's function of the host medium is available. In particular, this algorithm can be applied for the media with ellipsoidal symmetry and orthorhombic media.

5.6 Notes

The surface integral equation of the crack problem of elasticity was obtained as a limit of the volume integral equation of elasticity for a cavity in [4]. The regularization formula for the integral operator in the integral equation of a crack problem was proposed in [5, 6]. In [7], the solution of the elasticity problem for an elliptical crack in an anisotropic medium subjected to a constant external field was obtained as a limit of an ellipsoidal pore by tending to zero one of its semiaxes. The same method was used in [8] for a circular crack in a transverse isotropic medium.

The analogy of the polynomial conservation theorem for an elliptical crack was proved in [9]. It is shown in [4] that this theorem is a consequence of the polynomial conservation theorem for an ellipsoidal inclusion. An elliptical crack in orthorhombic and transverse isotropic media are considered in [10, 11], and in the medium with ellipsoidal symmetry in [12].

For cracks of non-elliptical shapes, the elasticity problem can be solved only numerically. An efficient numerical method of solution of the crack problems is proposed in [12, 13]. In this method, Gaussian approximating functions are used for discretization of the integral equation of the problem. For such functions, the matrix of the discretized problem can be obtained in an explicit analytical form. Application of Fast Fourier Transform technique for calculation of matrix-vector products accelerates substantially construction of the numerical solution of the crack problem in comparison with conventional numerical methods.

References

1. Brychkov Y, Prudnikov A (1989) Integral transforms of generalized functions. CRC Press, p 344
2. Eskin G (1981) Boundary-value problems for elliptic pseudo-differential equations. Math Soc 232. (New York, USA)
3. Cherepanov G (1983) Fracture mechanics of composite materials. Nauka, Moscow, p 296
4. Kunin I (1983) The theory of elastic media with microstructure II. Springer, Berlin Heilderberg New York, p 272
5. Kanaun S (1981) To the crack problem in 3D anisotropic elastic medium. Appl Math Mech (PMM) 45:361–370
6. Kanaun S (1984) On singular models of thin inclusions in an homogeneous elastic medium. Appl Math Mech (PMM) 48:81–91
7. Huang J, Liu H (1998) On a flat ellipsoidal inclusion or crack in three-dimensional anisotropic media. Int Jour Engin Sci 36(2):143–155
8. Levin V, Michelitsch T, Sevostianov I (2000) Spheroidal inhomogeneity in a transversely isotropic piezoelectric medium. Arch Appl Mech 70:673–693
9. Willis J (1968) The stress field around an elliptical crack in an anisotropic elastic medium. Int J Eng Sci 6:253–263
10. Kanaun S (2007) An elliptical crack in an anisotropic medium subjected to a constant external field. Int J Fracture 148:95–102
11. Kanaun S, Levin V (2009) Elliptical cracks arbitrarily oriented in 3D-anisotropic elastic media. Int J Eng Sci 47:777–792
12. Kanaun S (2021) Heterogeneous media: local fields. Effective properties and wave propagation. Elsevier, Amsterdam
13. Kanaun S (2007) Fast solution of 3D-elasticity problem of a planar crack of arbitrary shape. Int J Fract 148:435–442

Chapter 6
Thin Inclusions in a Homogeneous Elastic Medium

Abstract Thin inclusions which elastic moduli differ substantially from the moduli of the host medium are considered. Such inclusions are characterized by two dimensionless small parameters: the ratio of the characteristic linear sizes of the inclusion and the ratio of the characteristic elastic moduli of the inclusion and the host medium. The principal terms of the asymptotic expansions of the elastic fields in the medium with thin inclusions with respect to these parameters are derived. The elasticity problems for thin soft and thin stiff inclusions are reduced to 2D integral equations on the inclusion middle surfaces. For thin ellipsoidal inclusions, explicit analytical solutions of these equations are presented.

Keywords Conductive and elastic media with thin inclusions · External and internal expansion of fields · Matching of asymptotic expansions · 2D integral equations for thin inclusions · Singular models of thin inclusions · Thin ellipsoidal inclusions

6.1 External Expansions of Elastic Fields in the Medium with a Thin Inclusion

Let an infinite homogeneous elastic medium contain an inclusion in a finite region V, and one characteristic size of this region be much smaller that two other sizes (a thin inclusion). We suppose that the middle surface Ω of V is smooth and Γ is the boundary contour of Ω. At each point x of Ω we introduce a local coordinate system y_1, y_2, z with the z-axis directed along the normal $n(x)$ to Ω, and $h(x)$ is the transverse size of V along the z-axis (Fig. 6.1).

For a thin inclusion, the function $h(x)$ is presented in the form

$$h(x) = \delta_1 l(x), \quad \delta_1 << 1, \tag{6.1}$$

S. Kanaun, *Self-Consistent Methods for Composites*, Solid Mechanics and Its Applications 282, https://doi.org/10.1007/978-3-031-96739-9_6

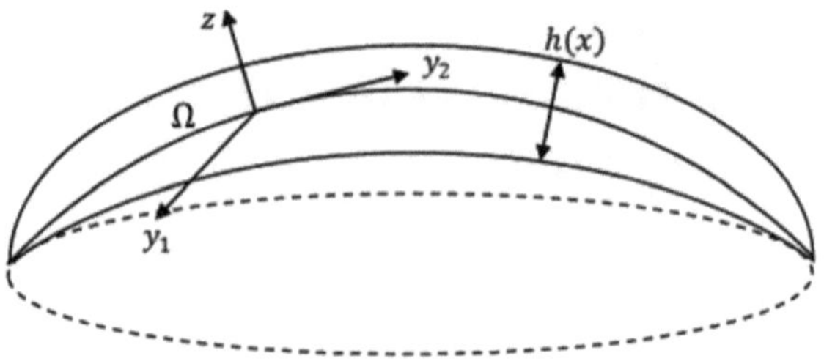

Fig. 6.1 Thin inclusion in a homogeneous medium

where $l(x)$ has the order of the maximal size of V. We assume that $h(x)$ satisfies the condition

$$|\partial_\alpha h(x)| << 1 \tag{6.2}$$

on Ω except a small vicinity of the boundary contour Γ. Here, ∂ is the gradient along the surface Ω

$$\partial_\alpha = \nabla_\alpha - n_\alpha(x)n_\beta(x)\nabla_\beta\,. \tag{6.3}$$

The integral presentations of the stress and strain fields in the medium with an inclusion has forms in Eqs. (4.36), (4.37).

$$\varepsilon(x) = \varepsilon^0(x) - \int K(x-x')C^1(x')\varepsilon^+(x')dx', \tag{6.4}$$

$$\sigma(x) = \sigma^0(x) + \int S(x-x')B^1(x')\sigma^+(x')dx'. \tag{6.5}$$

Here, $\varepsilon^+(x) = \varepsilon(x)V(x)$, $\sigma^+(x) = \sigma(x)V(x)$, and $V(x)$ is the characteristic function of the region V. For a fixed point $x \notin V$, the functions $K(x-x')$ and $S(x-x')$ in these equations are smooth and bounded functions of x as $x' \in V$. The principal terms of the asymptotic strain and stress fields far from the inclusion can be expressed in terms of the integrals over inclusion middle surface Ω

$$\varepsilon(x) = \varepsilon^0(x) - \int_\Omega K(x-x')C^1\bar{\varepsilon}^+(x',h)d\Omega', \tag{6.6}$$

$$\sigma(x) = \sigma^0(x) + \int_\Omega S(x-x')B^1\bar{\sigma}^+(x',h)d\Omega'. \tag{6.7}$$

Here, $\bar{\varepsilon}^+(x,h)$, $\bar{\sigma}^+(x,h)$ are the integral strain and stress tensors over the transverse coordinate z of the inclusion

$$\bar{\sigma}^+(x,h) = \int_{-h(x)/2}^{h(x)/2} \sigma^+(x+n(x)z)dz\,, \tag{6.8}$$

$$\bar{\varepsilon}^+(x,h) = \int_{-h(x)/2}^{h(x)/2} \varepsilon^+(x+n(x)z)dz\,. \tag{6.9}$$

If we introduce the function

$$m(x) = B^1\bar{\sigma}^+(x,h) = -B^0C^1\bar{\varepsilon}^+(x,h),\ \ x \in \Omega, \tag{6.10}$$

where $\bar{\sigma}^+(x,h)$ and $\bar{\varepsilon}^+(x,h)$ are in Eqs. (6.8), (6.9), (6.6), (6.7) are presented in the forms

$$\varepsilon(x) = \varepsilon^0(x) + \varepsilon^1(x)\,,\ \ \sigma(x) = \sigma^0(x) + \sigma^1(x)\,, \tag{6.11}$$

$$\varepsilon^1(x) = \int_{\Omega} K(x-x')C^0m(x')d\Omega', \tag{6.12}$$

$$\sigma^1(x) = \int_{\Omega} S(x-x')m(x')d\Omega'. \tag{6.13}$$

In order to find the principal terms of asymptotic expansions of the functions $\bar{\sigma}^+(x,h)$ and $\bar{\varepsilon}^+(x,h)$ for small parameters of the problem, we consider the limiting values of the potentials in Eqs. (6.12), (6.13) on the surface Ω (the internal limits of the external solutions [1]).

6.2 Properties of the Potentials (6.12) and (6.13)

Let the density $m(x)$ of the potentials $\varepsilon^1(x)$ and $\sigma^1(x)$ be an arbitrary smooth tensor field on Ω. We consider limiting values of the potentials $\sigma^1(x)$ and $\varepsilon^1(x)$ when $x \to \Omega$. To that end, at each point x of Ω, we introduce the projection operators $\Pi(n)$ and $\Theta(n)$ on the normal and on the tangent plane to Ω. In the E-basis, these operators have the forms

$$\Pi(n) = 2E^5(n) - E^6(n)\,,\ \ \Theta(n) = E^1 - 2E^5(n) + E^6(n)\,, \tag{6.14}$$

$$\Pi(n)\Pi(n) = \Pi(n), \Theta(n)\Theta(n) = \Theta(n), \Theta(n)\Pi(n) = 0. \tag{6.15}$$

Here, $n = n(x)$ is the normal to Ω at point x. The operators Π and Θ split any symmetric second rank tensor field $m(x)$ on Ω into the "normal" $m^{\pi}(x)$ and the "tangential" $m^{\theta}(x)$ components

$$m(x) = m^{\pi}(x) + m^{\theta}(x)\,, \tag{6.16}$$

$$m^{\pi}_{\alpha\beta} = \Pi_{\alpha\beta\lambda\mu}(n)m_{\lambda\mu}(x)\,,\ \ m^{\theta}_{\alpha\beta}(x) = \Theta_{\alpha\beta\lambda\mu}(n)m_{\lambda\mu}(x)\,. \tag{6.17}$$

The tensor $m^{\pi}(x)$ can be presented in the form

$$m^{\pi}_{\alpha\beta}(x) = n_{\alpha}(x)b_{\beta}(x)|_{(\alpha\beta)}, \tag{6.18}$$

where $b(x)$ is a vector field on Ω, and

$$b_{\alpha}(x) = 2m_{\alpha\beta}(x)n_{\beta}(x) - (n_{\beta}(x)m_{\beta\gamma}(x)n_{\gamma}(x))n_{\alpha}(x). \tag{6.19}$$

The tensor $m^{\theta}(x)$ satisfies the condition

$$n_{\alpha}(x)m^{\theta}_{\alpha\beta}(x) = 0, \tag{6.20}$$

and hence $m^{\theta}(x)$ is the tensor of the surface Ω.

The tensor density $C^0m(x)$ of the potential $\varepsilon^1(x)$ in Eq. (6.12) can be presented in the form of the following sum

$$C^0_{\alpha\beta\lambda\mu}m_{\lambda\mu}(x) = q_{\alpha\beta}(x) + C^0_{\alpha\beta\lambda\mu}n_{\lambda}(x)b_{\mu}(x), \tag{6.21}$$

where the vector $b(x)$ satisfies the equation

$$n_{\alpha}(x)C^0_{\alpha\beta\lambda\mu}n_{\lambda}(x)b_{\mu}(x) = n_{\alpha}(x)C^0_{\alpha\beta\lambda\mu}m_{\lambda\mu}(x). \tag{6.22}$$

For such a choice of vector $b(x)$, $q(x)$ in Eq. (6.21) is the tensor of the surface Ω

$$n_{\alpha}(x)q_{\alpha\beta}(x) = 0\,,\ \ \Theta_{\alpha\beta\lambda\mu}(n)q_{\lambda\mu}(x) = q_{\alpha\beta}(x), \tag{6.23}$$

Further, we consider the potential $\varepsilon_1(x)$ in Eq. (6.12) with the density $q_{\alpha\beta}(x)$ or $C^0_{\alpha\beta\lambda\mu}n_{\lambda}(x)b_{\mu}(x)$.

1. Let $C^0m = q$, and q be a tensor of the surface Ω. Using Gauss' theorem for surfaces (Appendix C.1), we present the potential $\varepsilon^1(x)$ in the form

$$\begin{aligned}\varepsilon^1_{\alpha\beta}(x) &= \int_{\Omega} K_{\alpha\beta\lambda\mu}(x-x')\Theta_{\lambda\mu\rho\nu}(x')q_{\rho\nu}(x')d\Omega' = \\ &= -\nabla_{(\alpha}\int_{\Omega} G_{\beta)\mu}(x-x')\partial'_{\lambda}q_{\lambda\mu}(x')d\Omega' + \\ &+ pv\int_{\Gamma} \nabla_{(\alpha}G_{\beta)\mu}(x-x')q_{\lambda\mu}(x')e_{\lambda}(x')d\Gamma',\end{aligned} \tag{6.24}$$

where $e(x)$ is the external normal to the contour Γ in the tangent plane to Ω at the point $x \in \Gamma$. Here, the equation

$$\nabla_\lambda G_{\beta\mu}(x-x')\Theta_{\lambda\mu\nu\rho}(x')q_{\nu\rho}(x') =$$
$$= -\partial'_\lambda \left(G_{\beta\mu}(x-x')q_{\lambda\mu}(x')\right) + G_{\beta\mu}(x-x')\partial'_\lambda q_{\lambda\mu}(x') \tag{6.25}$$

is taken into account. This equation follows from Eq. (6.3) and the definition (6.14) of the projector Θ, ∂'_λ is the derivative with respect to x'. The integral over the boundary contour Γ in Eq. (6.24) is a continuous function for $x \in \Omega$ ($x \notin \Gamma$), and the integral over Ω is the gradient of the potential of a simple layer of static elasticity with the density $\partial_\lambda q_{\lambda\mu}(x)$. Thus the integral over Ω in Eq. (6.24) is a continuous function in the entire space except the surface Ω. Let ε^{1+} and ε^{1-} be the limiting values of the potential ε^1 when the point x tends to Ω from the side of the normal $n(x)$ and from the opposed side, consequently. The jump $[\varepsilon^1]$ of the potential ε^1 on Ω is determined by the equation that follows from Eqs. (C.22), (C.23) of Appendix C

$$\left[\varepsilon^1_{\alpha\beta}(x)\right]_\Omega = \varepsilon^{1+}_{\alpha\beta}(x) - \varepsilon^{1-}_{\alpha\beta}(x) = n_\alpha(x)\widetilde{G}_{\beta\mu}(n)\partial_\lambda q_{\lambda\mu}(x)|_{(\alpha\beta)}, \tag{6.26}$$

where $\widetilde{G}(k)$ is Fourier transform of Green's function of the host medium.

Because $\Theta(n)n(x)\widetilde{G}(n) = 0$, the tangential component $\Theta(n)\varepsilon^1(x)$ of the potential $\varepsilon^1(x)$ is continuous on the surface Ω.

$$\left[\Theta_{\alpha\beta\lambda\mu}(n)\varepsilon^1_{\lambda\mu}(x)\right]_\Omega = 0. \tag{6.27}$$

Note that the integral that presents the value of the function $\Theta(n)\varepsilon^1(x)$ on Ω

$$\Theta_{\alpha\beta\lambda\mu}(n)\varepsilon^1_{\lambda\mu}(x) = \int_\Omega U_{\alpha\beta\lambda\mu}(x,x')q_{\lambda\mu}(x')d\Omega', \tag{6.28}$$

$$U_{\alpha\beta\lambda\mu}(x,x') = \Theta_{\alpha\beta\nu\rho}(x)K_{\nu\rho\tau\delta}(x-x')\Theta_{\tau\delta\lambda\mu}(x') \tag{6.29}$$

diverges formally since $\mathrm{K}(x-x') \sim |x-x'|^{-3}$ when $x \to x'$. The regularization formula for this integral is derived in Appendix C.3

$$\Theta_{\alpha\beta\lambda\mu}(x)\varepsilon^1_{\lambda\mu}(x) = pv\int_\Omega U_{\alpha\beta\lambda\mu}\left(x,x'\right)\left(q_{\lambda\mu}\left(x'\right) - q^0_{\lambda\mu}\left(x,x'\right)\right)d\Omega' -$$
$$- \Theta_{\alpha\beta\lambda\mu}(x)pv\oint_\Gamma \nabla_\lambda G_{\mu\nu}(x-x')q^0_{\nu\rho}(x,x')e_\rho(x')d\Gamma'. \tag{6.30}$$

Here, pv is Cauchy principal value of the integral, $q^0(x,x')$ is a constant tensor field on Ω. The function $q^0(x,x')$ satisfies the equation $\partial' q^0(x,x') = 0$ and the condition $q^0(x,x')|_{x=x'} = q(x')$. For a planar surface Ω, the regularization (6.30) is equivalent to the following one

$$\Theta_{\alpha\beta\lambda\mu}(x)\varepsilon^{1}_{\lambda\mu}(x) = pv\int U_{\alpha\beta\lambda\mu}\left(x, x'\right)\left(q_{\lambda\mu}\left(x'\right) - q_{\lambda\mu}\left(x\right)\right)d\Omega', \tag{6.31}$$

where integration is spread to the entire plane which surface Ω belongs to, the function $q(x)$ is continued by zero outside Ω.

Let the density of the potential $\sigma^1(x)$ in Eq. (6.13) have the form $m(x) = B^0 q(x)$, where $q(x)$ is a tensor of the surface Ω. It follows from definition (4.31) of the function $S(x)$ that this potential relates to the potential $\varepsilon^1(x)$ in Eq. (6.12) by the equation

$$\sigma^1(x) = \int_{\Omega} S(x - x')B^0 q(x')d\Omega' = C^0\varepsilon^1(x) - q(x)\Omega(x), \tag{6.32}$$

where $\Omega(x)$ is the delta-function concentrated on the surface Ω. Thus the limiting values of the potential $\sigma^1(x)$ are determined by the limiting values of the potential $\varepsilon^1(x)$ in Eq. (6.26) multiplied with the tensor C^0. Because $\widetilde{G}(n) = (nC^0 n)^{-1}$, the jump of the normal component $n(x)\sigma^1(x)$ of the potential $\sigma^1(x)$ on Ω is determined by the equation

$$\left[n_\beta(x)\sigma^{1}_{\alpha\beta}(x)\right]_{\Omega} = n_\beta(x)\sigma^{1+}_{\alpha\beta}(x) - n_\beta(x)\sigma^{1-}_{\alpha\beta}(x) = \partial_\beta q_{\beta\alpha}(x)\,. \tag{6.33}$$

2. Let the density of the potential $\varepsilon^1(x)$ (6.12) be the second term in the right-hand side of Eq. (6.21):

$$\varepsilon^{1}_{\alpha\beta}(x) = \int_{\Omega} K_{\alpha\beta\lambda\mu}(x - x')C^{0}_{\lambda\mu\rho\tau}n_\rho(x')b_\tau(x')d\Omega'. \tag{6.34}$$

It follows from definition (4.13) of the kernel $K(x)$ that $\varepsilon^1(x)$ is the symmetrized gradient of the double layer potential of static elasticity. If $b(x)$ in Eq. (6.34) is a smooth vector field on Ω, the potential $\varepsilon^1(x)$ can be presented in the form

$$\varepsilon^{1}_{\alpha\beta}(x) = \varepsilon^{1r}_{\alpha\beta}(x) + n_{(\alpha}(x)b_{\beta)}(x)\Omega(x), \tag{6.35}$$

where the function $\varepsilon^{1r}(x)$ is discontinuous on Ω. The limiting values of the potential $\varepsilon^1(x)$ on Ω are determined by the equation (Appendix C.2)

$$\varepsilon^{1\pm}_{\alpha\beta}(x) = pv\int_{\Omega} K_{\alpha\beta\lambda\mu}\left(x - x'\right)C^{0}_{\lambda\mu\tau\rho}n_\tau(x')\left(b_\rho(x') - b_\rho(x)\right)d\Omega' +$$

$$+ R_{\alpha\beta\lambda}(x)b_\lambda(x) \mp \frac{1}{2}\Lambda_{\alpha\beta\lambda\mu}(n)\partial_\lambda b_\mu(x)\,, \tag{6.36}$$

$$R_{\alpha\beta\lambda}(x) = B^{0}_{\alpha\beta\tau\rho}\oint_{\Gamma} \mathrm{rot}_{\nu\gamma} Z_{\tau\rho\nu\lambda}(x - x')d\Gamma'_{\gamma}, \tag{6.37}$$

$$\Lambda_{\alpha\beta\lambda\mu}(n) = B^0_{\alpha\beta\nu\rho}\widetilde{S}_{\nu\rho\lambda\mu}(n) \ , \tag{6.38}$$

where $Z(x)$ is the Green tensor for internal stresses that relates to the tensor $S(x)$ by the equation $S(x) = \mathrm{Rot}Z(x)$ (Section 3.3), $d\Gamma_\gamma$ is the infinitesimal vector element on the contour Γ, the function $\widetilde{S}(k)$ is Fourier transform of the function $S(x)$.

The potential $\sigma^1(x)$ in Eq. (6.13) with the density $m(x) = n(x)b(x)$ is presented in the form similar to (6.32)

$$\sigma^1_{\alpha\beta}(x) = C^0_{\alpha\beta\lambda\mu}\varepsilon^1_{\lambda\mu}(x) - C^0_{\alpha\beta\lambda\mu}n_\lambda(x)b_\mu(x)\Omega(x) \, . \tag{6.39}$$

It follows from this equation and Eq. (6.35) that the function $\sigma^1(x)$ has no singular components, and the jump of $\sigma^1(x)$ on Ω is determined by the equation

$$\left[\sigma^1_{\alpha\beta}(x)\right]_\Omega = -\widetilde{S}_{\alpha\beta\lambda\mu}(n)\partial_\lambda b_\mu(x). \tag{6.40}$$

Because $n_\alpha\widetilde{S}_{\alpha\beta\lambda\mu}(n) = 0$, the vector $n_\beta(x)\sigma^1_{\alpha\beta}(x)$ is continuous on Ω and

$$n_\beta(x)\sigma^1_{\beta\alpha}(x) = -\int_\Omega T_{\alpha\beta}(x, x')b_\beta(x')d\Omega', \, x \in \Omega, \tag{6.41}$$

$$T_{\alpha\beta}(x, x') = -n_\lambda(x)S_{\lambda\alpha\beta\mu}(x - x')n_\mu(x'). \tag{6.42}$$

Regularization of the formally divergent integral in Eq. (6.41) is defined in Eqs. (5.20), (5.21).

6.3 External Limiting Problems for Thin Inclusions

In this section, we consider the limiting elastic fields outside of a thin inclusion when its transverse size h (parameter δ_1 in Eq. (6.1)) tends to zero. If the elastic moduli of the inclusion are finite, in the limit $h \to 0$ ($C \neq 0, \infty$), the elastic fields inside the inclusion is finite also. The integrals $\bar{\varepsilon}^+(x, h)$ and $\bar{\sigma}^+(x, h)$ in Eqs. (6.8), (6.9) have the order of h and they vanish in the limit $h \to 0$ as well as the integrals in Eqs. (6.6), (6.7). But these integrals do not vanish if the elastic moduli of the inclusion tend to zero or to infinity when $\delta_1 \to 0$.

Let the elastic stiffness tensor C of the inclusion be presented in the form

$$C = \delta_2\overline{C}, \ \ \delta_2 << 1, \ \ \overline{C} = O(C^0), \tag{6.43}$$

where C^0 is the elastic stiffness tensor of the host medium. Thus the inclusion material is much softer than the material of the host medium. If the inclusion is much stiffer than the host medium, the elastic compliance tensor $B = C^{-1}$ of the inclusion can be presented as follows

$$B = \delta_2 \overline{B}, \ \delta_2 << 1, \quad \overline{B} = O(B^0). \tag{6.44}$$

We consider the limiting problems for thin inclusions when the "geometrical" δ_1 and "physical" δ_2 parameters simultaneously tend to zero, while the ratio δ_1/δ_2 has the order of 1.

6.3.1 *Thin Soft Inclusion*

Let the parameter δ_1 and the elastic stiffness tensor C of a thin inclusion together tend to zero. It follows from the equation

$$B^1\bar{\sigma}^+ = (B - B^0)C\bar{\varepsilon}^+ = -B^0C^1\bar{\varepsilon}^+ \tag{6.45}$$

that $B^1\bar{\sigma}^+ \to \bar{\varepsilon}^+$ when $C \to 0 \ (C^1 \to -C^0)$. Therefore, if $h, C \to 0 \ (\delta_1, \delta_2 \to 0)$, Eqs. (6.6), (6.7) take the forms

$$\varepsilon(x) = \varepsilon^0(x) + \int_{\Omega} K(x - x')C^0\bar{\varepsilon}^+(x')d\Omega', \tag{6.46}$$

$$\sigma(x) = \sigma^0(x) + \int_{\Omega} S(x - x')\bar{\varepsilon}^+(x')d\Omega', \tag{6.47}$$

$$\bar{\varepsilon}^+(x) = \lim \bar{\varepsilon}^+(x, h), \ h, \ C \to 0. \tag{6.48}$$

On the other hand, if h and C tend to zero, the stress vector $n(x)\sigma(x)$ should be continuous on the limiting surface Ω because it is continuous for any finite h and C. At the same time, continuity of the displacement vector $u(x)$ can be violated due to zero limiting value of the inclusion stiffness tensor C. Hence, the limiting values of the displacement and stress vectors on the surface Ω satisfy the conditions

$$[u(x)]_{\Omega} = b(x), \quad [n(x)\sigma(x)]_{\Omega} = 0. \tag{6.49}$$

Here, $b(x)$ is an unknown jump of the displacement vector on the surface Ω. It follows from the first Eq. (6.49) that the limiting strain field in the medium contains a singular component (delta-function) $\varepsilon^{1s}(x)$ concentrated on Ω

$$\varepsilon^{1s}_{\alpha\beta}(x) = n_\alpha(x)b_\beta(x)|_{(\alpha\beta)}\Omega(x). \tag{6.50}$$

In this case, the function $\bar{\varepsilon}^+(x)$ in Eqs. (6.9) and (6.47) is determined by the equation

$$\bar{\varepsilon}^+_{\alpha\beta}(x) = n_\alpha(x)b_\beta(x)|_{(\alpha\beta)}, \tag{6.51}$$

and in the limits $h, C \to 0$, Eqs. (6.46) and (6.47) take the forms

$$\varepsilon_{\alpha\beta}(x) = \varepsilon^0_{\alpha\beta}(x) + \int_\Omega K_{\alpha\beta\lambda\mu}(x - x')C^0_{\lambda\mu\nu\rho}n_\nu(x')b_\rho(x')d\Omega', \tag{6.52}$$

$$\sigma_{\alpha\beta}(x) = \sigma^0_{\alpha\beta}(x) + \int_\Omega S_{\alpha\beta\lambda\mu}(x - x')n_\lambda(x')b_\mu(x')d\Omega'. \tag{6.53}$$

The potentials on the right-hand sides of these equations coincide with the potentials (6.12), (6.13) with $m(x) = n(x)b(x)$. It follows from the properties of these potentials indicated in the previous Section that for the field in Eq. (6.53), the second condition (6.49) of continuity of the vector $n(x)\sigma(x)$ on Ω is fulfilled automatically. Therefore, with the accuracy to the terms of the order of δ_1, the following equation holds

$$n(x)\sigma(x) = h^{-1}(x)n(x)\overline{\sigma}^+(x, h), \;\; x \in \Omega, \tag{6.54}$$

where the function $\overline{\sigma}^+(x, h)$ is defined in Eq. (6.8). From Hooke's law for the inclusion material we have

$$\overline{\sigma}^+(x, h) = C\overline{\varepsilon}^+(x, h). \tag{6.55}$$

Then, the relation between the vector $b(x)$ and the stress vector $n(x)\sigma(x)$ on Ω follows from this equation and Eqs. (6.53), (6.54) in the form

$$n_\beta(x)\sigma_{\beta\alpha}(x) = \lambda_{\alpha\beta}(x)b_\beta(x), \;\; \lambda_{\alpha\beta}(x) = \lim_{h,C\to 0} \frac{n_\lambda(x)C_{\lambda\alpha\beta\mu}n_\mu(x)}{h(x)}. \tag{6.56}$$

Taking into account Eq. (6.53), we yield the equation for the vector $b(x)$ on Ω

$$\lambda_{\alpha\beta}(x)b_\beta(x) + \int_\Omega T_{\alpha\beta}(x, x')b_\beta(x')d\Omega' = n_\beta(x)\,\sigma^0_{\beta\alpha}(x)\,, \tag{6.57}$$

$$T_{\alpha\beta}(x, x') = -n_\lambda(x)S_{\lambda\alpha\beta\mu}(x - x')n_\mu(x). \tag{6.58}$$

The regularization of the integral operator T with the kernel $T(x, x')$ on smooth functions $b(x)$ is defined in Eqs. (5.20), (5.21). Hence, Eqs. (6.52), (6.53) present the solution of the limiting problem for a thin soft inclusion when $\delta_1, \delta_2 \to 0$. The vector $b(x)$ is to be found from Eq. (6.57). If $\lambda(x) = 0$, Eq. (6.57) coincides with the integral equation (5.12) of the crack problem.

Remark. Equations (6.54) and (6.55) can be interpreted as the asymptotic solution of the internal limiting problem for a thin inclusion. This is the problem for a planar layer of thickness $h = 1$ and the elastic stiffness tensor C subjected to a homogeneous external stress field (Sect. 4.4). The tensor $\sigma(x)$ in Eq. (6.56) is the limiting value of the stress tensor in Eq. (6.53) on the surface Ω. Therefore, conditions (6.56) perform matching the external and internal limiting solutions for a thin soft inclusion in a homogeneous elastic medium [1].

For actual thin soft inclusions, the parameters $\lambda(x)$ in Eq. (6.57) should be calculated for the actual values of h and C.The fields in Eqs. (6.52), (6.53) describe the strain and stress fields in the medium with a thin soft inclusion as better as smaller are the ratios h/l and $C(C^0)^{-1}$.

6.3.2 Thin Stiff Inclusions

We consider a thin stiff inclusion when its transverse size h and elastic compliance tensor B tend to zero. In this case,

$$C^1\bar{\varepsilon}^+ = (C - C^0)B\bar{\sigma}^+ = -C^0B^1\bar{\sigma}^+, \tag{6.59}$$

and $C^1\bar{\varepsilon}^+ \to \bar{\sigma}^+$ when $B \to 0$ $(B^1 \to -B^0)$. The limiting expressions for the strain and stress tensor in the medium follow from Eqs. (6.6), (6.7) in the forms

$$\varepsilon(x) = \varepsilon^0(x) - \int_{\Omega} K(x - x')\bar{\sigma}(x')d\Omega', \tag{6.60}$$

$$\sigma(x) = \sigma^0(x) - \int_{\Omega} S(x - x')B^0\bar{\sigma}(x')d\Omega', \tag{6.61}$$

$$\bar{\sigma}(x) = \lim \bar{\sigma}^+(x, h), \quad h, B \to 0. \tag{6.62}$$

Because the inclusion stiffness tensor $C = B^{-1}$ tends to infinity, the displacement vector and the tangential component of the strain tensor are to be continuous on the limiting surface Ω, but the stress vector can have a jump on this surface. Thus, for the limiting displacement vector and strain tensor, we have the conditions

$$[u(x)]_{\Omega} = 0, \quad [\Theta(x)\varepsilon(x)]_{\Omega} = 0. \tag{6.63}$$

The first of these conditions shows that the strain field does not contain a singular component concentrated on Ω. The potential in Eq. (6.60) coincides with the potential $\varepsilon^1(x)$ in Eq. (6.12) with $C^0m = \bar{\sigma}$. It follows from the properties (6.27) of the potential $\varepsilon^1(x)$ that the second condition (6.63) is satisfied automatically if the density $\bar{\sigma}(x)$ of the potentials (6.60), (6.61) is a tensor of the surface Ω, i.e.,

$$n_{\beta}(x)\bar{\sigma}_{\alpha\beta}(x) = 0, \quad \Theta_{\alpha\beta\lambda\mu}(x)\bar{\sigma}_{\lambda\mu}(x) = \bar{\sigma}_{\alpha\beta}(x). \tag{6.64}$$

For continuity of the tangential components of the limiting strain tensor on Ω, up to terms of the order δ_1, the strain tensor $\varepsilon(x)$ and the integral strain $\bar{\varepsilon}^+(x, h)$ in Eq. (6.9) relate by the equation

$$\Theta(x)\varepsilon(x) = h^{-1}(x)\Theta(x)\bar{\varepsilon}^{+}(x,h), \quad x \in \Omega, \tag{6.65}$$

Taking into account that the integral strain $\bar{\varepsilon}^{+}(x,h)$ and stress $\bar{\sigma}^{+}(x,h)$ relate by Hooke's law $\bar{\varepsilon}^{+}(x,h) = B\bar{\sigma}^{+}(x,h)$, in the limit $h, B \to 0$, we obtain

$$\Theta(x)\varepsilon(x) = \mu(x)\bar{\sigma}(x)\,, \quad \mu_{\alpha\beta\lambda\mu}(x) = \lim_{h,B\to 0}\frac{\Theta_{\alpha\beta\nu\rho}(x)B_{\nu\rho\tau\delta}\Theta_{\tau\delta\lambda\mu}(x)}{h(x)}\,. \tag{6.66}$$

Substituting in this equation $\varepsilon(x)$ from from Eq. (6.60), we yield the equation for the density $\bar{\sigma}(x)$ of the potentials in Eqs. (6.60), (6.61)

$$\mu_{\alpha\beta\lambda\mu}(x)\bar{\sigma}_{\lambda\mu}(x) + \int_{\Omega} U_{\alpha\beta\lambda\mu}(x,x')\bar{\sigma}_{\lambda\mu}(x')d\Omega' = \Theta_{\alpha\beta\lambda\mu}(x)\varepsilon^{0}_{\lambda\mu}(x)\,. \tag{6.67}$$

$$U_{\alpha\beta\lambda\mu}(x,x') = \Theta_{\alpha\beta\nu\rho}(x)K_{\nu\rho\tau\delta}(x-x')\Theta_{\tau\delta\lambda\mu}(x'). \tag{6.68}$$

Regularization of the operator U with the kernel $U(x,x')$ on smooth functions $\bar{\sigma}(x)$ is determined in Eqs. (6.30) or (6.31). If $\mu = 0$, Eq. (6.67) is the integral equation of the elasticity problem for a non-stretched membrane embedded into a homogeneous elastic medium and subjected to external strain field $\varepsilon^{0}(x)$.

The operators T and U in Eqs. (6.57), (6.67) belong to the class of elliptic pseudo-differential operators. It follows from the results of [2] that a unique solution of Eq. (6.57) exists if the function $b(x)$ vanishes at the contour Γ

$$b(x) = 0 \quad \text{as} \quad x \in \Gamma\,. \tag{6.69}$$

A unique solution of Eq. (6.67) exists if the function $\bar{\sigma}(x)$ satisfy the condition

$$\bar{\sigma}_{\alpha\beta}(x)e_{\beta}(x) = 0 \quad \text{as} \quad x \in \Gamma\,, \tag{6.70}$$

where $e(x)$ is the normal to Γ in the plane tangent to Ω at point x.

6.4 Singular Models of Thin Inclusions

For calculation of the elastic fields in homogeneous media with thin inclusions, it is natural to replace the 3D-problem of elasticity by an equivalent 2D-problem. In [3], an heuristic model of a thin inclusion was proposed. In this model, the inclusion is simulated by its middle surface Ω with the following boundary conditions

$$[u]_{\Omega} = b\,, \; [n\sigma]_{\Omega} = 0\,, \; n\sigma|_{\Omega} = \lambda b\,, \tag{6.71}$$

where tensor $\lambda(x)$ has the form

$$\lambda_{\alpha\beta}(x) = h^{-1}(x) n_{\lambda}(x) C_{\lambda\alpha\beta\mu} n_{\mu}(x) \ . \tag{6.72}$$

In [3], the second conditions (6.71) is called the equilibrium equation, and the third one is Hooke's law for the inclusion. The boundary conditions (6.71) coincide with (6.49), (6.56). Therefore, the solution of this problem is presented in the integral form (6.52), (6.53), where the vector $b(x)$ is to be found from Eq. (6.57). Strictly speaking the condition (6.69) should be added Eqs. (6.71). The smaller parameters δ_1, δ_2 (the thinner the inclusion and the smaller the ratio of the elastic moduli of the inclusion and the medium) the better the model with the boundary conditions (6.71) approximates the exact solution. Such inclusions can be called crack-like, and the conditions (6.71) serve for these inclusions only.

A thin stiff inclusion can be replaced by its middle-surface with the boundary conditions that follow from Eqs. (6.63)–(6.66)

$$[u(x)]_{\Omega} = 0 \, , \quad [\Theta(x)\varepsilon(x)]_{\Omega} = 0 \, , \tag{6.73}$$

$$\Theta(x)\varepsilon(x) = \mu(x)\bar{\sigma}(x) \, , \quad \Theta(x)\overline{\sigma}(x) = \bar{\sigma}(x) \, , \quad x \in \Omega \, , \tag{6.74}$$

The tensor $\bar{\sigma}(x)$ in these equations satisfies the integral equation (6.67), where the function $\mu(x)$ has the form

$$\mu(x) = h^{-1}(x)\Theta(x) B \Theta(x) \, . \tag{6.75}$$

Equations (6.73) can be interpreted as the compatibility conditions for the inclusion and medium deformations (the second condition is a consequence of the first one). The first condition in Eq. (6.74) is Hooke's law for the inclusion. Note that according to Eq. (6.8), the tensor $\bar{\sigma}(x)$ is the integral of the stress tensor over the cross-section of the inclusion. Equation of equilibrium for this tensor is a consequence of Eq. (6.33)

$$\partial_{\alpha}\bar{\sigma}_{\alpha\beta}(x) = -[n_{\alpha}(x)\sigma_{\alpha\beta}(x)] \, , \quad x \in \Omega \, , \tag{6.76}$$

where $[n_{\alpha}\sigma_{\alpha\beta}]$ is the jump of the stress vector on Ω. The condition (6.70) on the boundary contour Γ of Ω should be added to Eqs. (6.73)–(6.76). Note that the integral stresses in a thin momentless shell satisfy Eq. (6.76) [4]. Therefore, the boundary conditions (6.75)–(6.76) correspond to the problem of a momentless shell embedded into a homogeneous elastic medium which elastic stiffness is smaller than the stiffness of the shell material. The model with the conditions (6.73)–(6.76) on the inclusion middle surface Ω simulates the fields in the medium with a thin stiff inclusion as better as smaller are the parameters h/l and $C^0 C^{-1}$.

6.5 Thin Ellipsoidal Inclusions

Let an inclusion with the elastic stiffness tensor C occupy a thin ellipsoidal region V_h in a homogeneous host medium with elastic stiffness tensor C^0. The semi-axes of the ellipsoid V_h are a_1, a_2, h, and $h \ll a_1, a_2$. In this case, the middle surface Ω of the inclusion is an ellipse with semi-axes a_1, a_2, and the inclusion transverse size $h(x)$ is

$$h(x_1, x_2) = 2hz(x_1, x_2), \quad z(x_1, x_2) = \sqrt{1 - \left(\frac{x_1}{a_1}\right)^2 - \left(\frac{x_2}{a_2}\right)^2}. \tag{6.77}$$

Here, x_1, x_2 are the cartesian coordinates directed along the principal axes of the ellipse Ω.

For a soft (crack-like) inclusion $\delta_2 = |C\left(C^0\right)^{-1}| \ll 1$, and the principal terms of the elastic fields in the medium with the inclusion are defined in Eqs. (6.52), (6.53), where vector $b(x)$ is the solution of Eq. (6.57)

$$\lambda(x)b(x) + \int_{\Omega} T(x - x')b(x')d\Omega' = t^0(x), \tag{6.78}$$

$$\lambda(x) = \frac{1}{h(x)}nCn, \; T(x) = -nS(x)n, \; t^0(x) = n\sigma^0(x). \tag{6.79}$$

Here, n is the normal to the middle surface Ω. It is shown in Sect. 5.2 that for an elliptical region Ω, the operator T in this equation transfers a function $b(x) = P(x)z(x)$, where $P(x)$ is a polynomial of the power m, into a polynomial of the same power on Ω. It means that the solution of Eq. (6.78) with a polynomial right hand side can be found in the form $b(x) = P(x)z(x)$, where $P(x)$ is the polynomial of the same power as $t^0(x)$. In particular, if t^0 is a constant vector, $b(x) = b^0 z(x)$, and b^0 is a constant vector. Substituting this function in Eq. (6.78) and taking $x = 0$, we obtain the equation for b^0

$$\left(\lambda^0 + T^0\right) b^0 = t^0, \; b^0 = \left(\lambda^0 + T^0\right)^{-1} t^0, \tag{6.80}$$

$$\lambda^0 = \frac{1}{2h}nCn, \; T^0 = \int_{\Omega} T(x)z(x)d\Omega. \tag{6.81}$$

The integral T^0 for a 2D-elliptical region Ω is considered in Sect. 5.3, and it is presented in the form of the double integral in Eq. (5.34)

$$T^0_{\alpha\beta} = -\frac{1}{8\pi}\int_0^{2\pi} d\varphi \int_{-\infty}^{\infty} \widetilde{S}_{3\alpha\beta 3}\left(\frac{\cos\varphi}{a_1}, \frac{\sin\varphi}{a_2}, k_3\right) dk_3, \tag{6.82}$$

where the function $\widetilde{S}(k_1, k_2, k_3)$ is defined in Eqs. (5.16)–(5.18). For an isotropic host media, T^0 is expressed in terms of elliptic integrals in Eqs. (5.46)–(5.48). Note that the elastic stiffness tensor C of the inclusion in Eq. (6.81) for λ^0 can be anisotropic.

For an elastic medium with a thin stiff ellipsoidal inclusion, the principal terms of the elastic fields are in Eqs. (6.60), (6.61), where tensor $\overline{\sigma}(x)$ is the solution of the integral equation

$$\mu(x)\bar{\sigma}(x) + \int_{\Omega} U(x - x')\bar{\sigma}(x')d\Omega' = \zeta^0(x), x \in \Omega, \tag{6.83}$$

$$\mu(x) = \frac{1}{h(x)}\Theta C^{-1}\Theta, U(x) = \Theta K(x)\Theta, \ \zeta^0(x) = \Theta\varepsilon^0(x). \tag{6.84}$$

Here, $\Theta = \Theta(n)$ is the projector on the inclusion middle-surface Ω, n is the normal to Ω. In this equation, $U(x_1, x_2)$ is a homogeneous function of the power -3, and its Fourier transform $\widetilde{U}(k_1, k_2)$ is a homogeneous function of the power 1. For an elliptical region, the integral operator U in Eq. (6.83) have the same properties as the operator T in Eqs. (6.78). In particular, U transforms a polynomial multiplied with the function $z(x)$ in Eq. (6.77) into a polynomial of the same power on Ω. Therefore, for a constant right hand side ζ^0, the solution of Eq. (6.83) can be found in the form

$$\bar{\sigma}(x) = M^0 z(x), \tag{6.85}$$

where M^0 is a constant tensor. This tensor is the solution of the equation

$$\left(\mu^0 + U^0\right) M^0 = \zeta^0, \ M^0 = \left(\mu^0 + U^0\right)^{-1} \zeta^0, \tag{6.86}$$

$$\mu^0 = \frac{1}{2h}\Theta(n)C^{-1}\Theta(n), U^0 = \int_{\Omega} U(x)z(x)d\Omega. \tag{6.87}$$

Similar to tensor T^0 in Eq. (6.82), the tensor U^0 can be presented in the form of the double integral

$$U^0_{\alpha\beta\gamma\delta} = \frac{1}{4}\int_0^{2\pi} \widetilde{U}_{\alpha\beta\gamma\delta}\left(\frac{\cos\varphi}{a_1}, \frac{\sin\varphi}{a_2}\right) d\varphi, \tag{6.88}$$

where

$$\widetilde{U}_{\alpha\beta\gamma\delta}(k_1, k_2) = \frac{1}{2\pi}\int_{-\infty}^{\infty} \Theta(n)_{\alpha\beta\mu\nu}\widetilde{K}_{\mu\nu\lambda\rho}(k_1, k_2, k_3)\Theta_{\lambda\rho\gamma\delta}(n)dk_3. \tag{6.89}$$

Here the function $\widetilde{K}(k_1, k_2, k_3)$ is defined in Eqs. (5.17), (5.18), and the normal n to Ω is the unit vector of the k_3-axis.

For an isotropic host medium, the integral (6.89) is calculated explicitly, and non-zero components of the tensor $\widetilde{U}\,(k_1, k_2)$ are

$$\widetilde{U}_{1111} = \frac{k_1^2(2k_2^2 + (2-\kappa_0)k_1^2)}{4\mu_0(k_1^2+k_2^2)^{3/2}}, \quad \widetilde{U}_{1122} = -\frac{\kappa_0 k_1^2 k_2^2}{4\mu_0(k_1^2+k_2^2)^{3/2}}, \tag{6.90}$$

$$\widetilde{U}_{2222} = \frac{k_2^2(2k_1^2 + (2-\kappa_0)k_2^2)}{4\mu_0(k_1^2+k_2^2)^{3/2}}, \quad \widetilde{U}_{1212} = \frac{\left(k_1^2+k_2^2\right)^2 - 2\kappa_0 k_1^2 k_2^2}{8\mu_0(k_1^2+k_2^2)^{3/2}}, \tag{6.91}$$

where λ_0, μ_0 are Lame's parameters of the medium. The integral (6.88) can be expressed in terms of elliptic integrals. For a spheroidal inclusion ($a_1 = a_2 = a$), non-zero components of the tensor U^0 are

$$U^0_{1111} = U^0_{2222} = \frac{\pi(8-3\kappa_0)}{64a\mu_0}, \quad \kappa_0 = \frac{\lambda_0+\mu_0}{\lambda_0+2\mu_0}, \tag{6.92}$$

$$U^0_{1122} = -\frac{\pi\kappa_0}{64a\mu_0}, \quad U^0_{1212} = \frac{\pi(4-\kappa_0)}{64a\mu_0}. \tag{6.93}$$

6.6 A Thin Inclusion in Homogeneous Conductive Media

We consider an anisotropic homogeneous conductive medium with the conductivity tensor C^0 containing an isolated inclusion with the conductivity tensor C. The inclusion occupies a finite region V, and one of its characteristic sizes h is much smaller than other two sizes (of the order of l). Therefore, $\delta_1 = h/l \ll 1$ is a small parameter. The second dimensionless parameter δ_2 is the ratio of the characteristic conductivities of the medium and the inclusion. Further it is assumed that $\delta_2 \ll 1$ or $\delta_2 \gg 1$ but the ratio δ_1/δ_2 or the product $\delta_1\delta_2$ have the order of 1. The objectives of this Section are the principal terms of the asymptotic expansions of the electric field $e(x)$ and current $j(x)$ in the medium with a thin inclusion with respect to the parameters δ_1, δ_2. These terms have the prime interest for applications. As in the case of elastic media with thin inclusions, the method of matching of asymptotic expansions of the external and internal problems for a thin inclusion can be used for construction of these terms. Here, we present the final results omitting details of the derivations that are similar to those in the elastic case.

First we consider a thin inclusion which conductivity is much smaller than the conductivity of the host medium, e.g., $\delta_2 \ll 1$. In this case, the principal terms of the fields $e_\alpha(x)$ and $j_\alpha(x)$ have the forms

$$e_\alpha(x) = e^0_\alpha(x) + \int_\Omega K_{\alpha\beta}(x-x')C^0_{\beta\lambda}n_\lambda(x')b(x')d\Omega', \tag{6.94}$$

$$j_\alpha(x) = j^0_\alpha(x) + \int_\Omega S_{\alpha\beta}(x-x')n_\beta(x')b(x')d\Omega', \tag{6.95}$$

where the functions $K_{\alpha\beta}(x)$ and $S_{\alpha\beta}(x)$ are defined in in Eqs. (4.195), $n(x)$ is the normal to the inclusion middle surface Ω. $e^0(x)$ and $j^0(x)$ are the external electric field and the current applied to the medium. The scalar function $b(x)$ in Eqs. (6.94), (6.95) is the solution of the integral equation

$$\lambda(x)b(x) + \int_{\Omega} T(x, x')b(x')d\Omega' = n_\alpha(x) j^0_\alpha(x), \tag{6.96}$$

$$T(x, x') = -n_\alpha(x) S_{\alpha\beta}(x - x')n_\beta(x'). \tag{6.97}$$

The coefficient $\lambda(x)$ in Eq. (6.96) is

$$\lambda(x) = \frac{1}{h(x)} n_\alpha(x) C_{\alpha\beta} n_\beta(x), \tag{6.98}$$

where C is the conductivity tensor of the inclusion, and $h(x)$ is its transverse size. The regularization of the integral operator with the kernel $T(x, x')$ is defined in Sect. 5.1, Eq. (5.20).

For a thin inclusion which conductivity is much larger than the conductivity of the host medium $\delta_2 \gg 1$, the principal terms of the electric field and the current in the medium have forms

$$e_\alpha(x) = e^0_\alpha(x) + \int_{\Omega} K_{\alpha\beta}(x - x')\zeta_\beta(x')d\Omega', \tag{6.99}$$

$$j_\alpha(x) = j^0_\alpha(x) + \int_{\Omega} S_{\alpha\beta}(x - x')B^0_{\beta\lambda}\zeta_\lambda(x')d\Omega', \tag{6.100}$$

where $B^0 = \left(C^0\right)^{-1}$, $\zeta_\alpha(x)$ is the vector of the middle inclusion surface Ω

$$\theta_{\alpha\beta}(x)\zeta_\beta(x) = \zeta_\alpha(x), \quad \theta_{\alpha\beta}(x) = \delta_{\alpha\beta} - n_\alpha(x)n_\beta(x). \tag{6.101}$$

This vector is the solution of the integral equation

$$\mu_{\alpha\beta}(x)\zeta_\beta(x) + \int_{\Omega} U_{\alpha\beta}(x, x')\zeta_\beta(x')dx' = \theta_{\alpha\beta}(x)e^0_\beta(x), \tag{6.102}$$

where the functions $U_{\alpha\beta}(x, x')$ and $\mu_{\alpha\beta}(x)$ are

$$U_{\alpha\beta}(x, x') = \theta_{\alpha\lambda}(x)K_{\lambda\mu}(x - x')\theta_{\mu\beta}(x'), \tag{6.103}$$

$$\mu_{\alpha\beta}(x) = \frac{1}{h(x)}\theta_{\alpha\lambda}(x)\left(C^{-1}\right)_{\lambda\mu}\theta_{\mu\beta}(x). \tag{6.104}$$

For a thin ellipsoidal inclusion with the semi-axes be a_1, a_2, h $(h/a_1, h/a_2 \ll 1)$, Ω is a planar elliptical surface with the normal n. In the Cartesian coordinates x_1, x_2 directed along the ellipsoid axes, the function $h(x)$ has the form

$$h(x) = 2hz(x), \quad z(x) = \sqrt{1 - \left(\frac{x_1}{a_1}\right)^2 - \left(\frac{x_2}{a_2}\right)^2}. \tag{6.105}$$

For a thin inclusion of small conductivity ($\delta_2 \ll 1$) and a constant external current j^0, the function $b(x)$ in Eq. (6.96) has the form

$$b(x) = b_0 z(x). \tag{6.106}$$

where a scalar b_0 is constant

$$b_0 = \frac{\left(n \cdot j^0\right)}{\lambda_0 + T_0}, \lambda_0 = \frac{1}{2h} n \cdot C \cdot n, \tag{6.107}$$

$$T_0 = -\frac{1}{4} \int_0^{2\pi} \widetilde{T}\left(\frac{\cos\varphi}{a_1}, \frac{\sin\varphi}{a_2}\right) d\varphi. \tag{6.108}$$

Here, the function $\widetilde{T}(k_1, k_2)$ is the absolutely converging improper integral

$$\widetilde{T}(k_1, k_2) = \frac{1}{2\pi} \int_{-\infty}^{\infty} \widetilde{S}_{33}(k_1, k_2, k_3)\, dk_3, \tag{6.109}$$

where

$$\widetilde{S}_{\alpha\beta}(k_1, k_2, k_3) = \frac{C^0_{\alpha\gamma} k_\gamma k_\delta C^0_{\delta\beta}}{k_\lambda C^0_{\lambda\mu} k_\mu} - C^0_{\alpha\beta}. \tag{6.110}$$

For an isotropic host medium $C^0_{\alpha\beta} = c_0 \delta_{\alpha\beta}$, and we have

$$\widetilde{S}_{33}(k_1, k_2, k_3) = c_0 \left(\frac{k_3^2}{k_1^2 + k_2^2 + k_3^2} - 1\right), \tag{6.111}$$

$$\widetilde{T}(k_1, k_2) = -c_0 \sqrt{k_1^2 + k_2^2}. \tag{6.112}$$

As the result, the integral T_0 takes the form

$$T^0 = \frac{c_0}{4a_2} \int_0^{2\pi} \sqrt{(\gamma\cos\varphi)^2 + (\sin\varphi)^2} d\varphi = \frac{c_0}{\sqrt{a_1 a_2}} \frac{E\left(1 - \gamma^2\right)}{\sqrt{\gamma}}, \tag{6.113}$$

$$\gamma = \frac{a_2}{a_1},\ a_1 \geq a_2. \tag{6.114}$$

Here, $E(\varsigma)$ is the elliptic integral of the second kind.

In the case of a thin ellipsoidal inclusion of large conductivity, the solution of Eq. (6.102) has the form

$$\zeta(x) = \zeta^0 z(x). \tag{6.115}$$

where the constant vector ζ^0 is the constant vector and

$$\zeta^0_\beta = \left(\mu^0_{\alpha\beta} + U^0_{\alpha\beta}\right)^{-1} e^0_\beta, \tag{6.116}$$

$$\mu^0_{\alpha\beta} = \frac{1}{2h}\theta_{\alpha\lambda}\left(C^{-1}\right)_{\lambda\mu}\theta_{\mu\beta},\ U^0_{\alpha\beta}(x) = \frac{1}{4}\int_0^{2\pi}\widetilde{U}_{\alpha\beta}\left(\frac{\cos\varphi}{a_1}, \frac{\sin\varphi}{a_2}\right)d\varphi. \tag{6.117}$$

where the tensor $\widetilde{U}(k_1, k_2)$ is

$$\widetilde{U}_{\alpha\beta}(k_1, k_2) = \frac{1}{2\pi}\int_{-\infty}^{\infty}\theta_{\alpha\gamma}\widetilde{K}_{\gamma\delta}(k_1, k_2, k_3)\,\theta_{\delta\beta}dk_3, \tag{6.118}$$

$$\widetilde{K}_{\alpha\beta}(k_1, k_2, k_3) = \frac{k_\alpha k_\beta}{k_\lambda C^0_{\lambda\mu}k_\mu}. \tag{6.119}$$

For an isotropic host medium,

$$\widetilde{U}_{\alpha\beta}(k_1, k_2) = \frac{k_\alpha k_\beta}{2c_0\sqrt{k_1^2+k_2^2}},\ \alpha, \beta = 1, 2, \tag{6.120}$$

$$\widetilde{U}_{\alpha 3}(k_1, k_2) = \widetilde{U}_{3\alpha}(k_1, k_2) = 0. \tag{6.121}$$

Then, from Eq. (6.118) we obtain non-zero components of the tensor U^0 in the form of 1D-integrals

$$U^0_{11} = \frac{\gamma^2}{8c_0a_2}\int_0^{2\pi}\frac{(\cos\varphi)^2}{\sqrt{(\gamma\cos\varphi)^2 + (\sin\varphi)^2}}d\varphi,\ \gamma = \frac{a_2}{a_1}, \tag{6.122}$$

$$U^0_{22} = \frac{1}{8c_0a_2}\int_0^{2\pi}\frac{(\sin\varphi)^2}{\sqrt{(\gamma\cos\varphi)^2 + (\sin\varphi)^2}}d\varphi,\ U^0_{12} = 0. \tag{6.123}$$

These integrals are expressed in terms of the elliptic integrals of the first $K(\eta)$ and second $E(\eta)$ kinds

$$U_{11}^{0} = \frac{\gamma^2}{2c_0a_2}\frac{K(\eta) - E(\eta)}{\tau}, \quad \eta = 1 - \gamma^2. \tag{6.124}$$

$$U_{22}^{0} = \frac{1}{2c_0a_2}\frac{E(\eta) - \gamma^2 K(\eta)}{\eta}. \tag{6.125}$$

6.7 Notes

Heuristic models of thin inclusions were proposed in [5–7].

Asymptotic methods for solutions of partial differential equations in thin domains were developed in [1, 8, 9]. Thin inclusions which properties differ substantially from the properties of the host medium are considered in [10–14].

The contents of this Chapter is based on the works [15, 16].

In the case of a non-ellipsoidal thin inclusions, solution of the integral equations (6.78), (6.83) can be obtained only numerically. For the numerical solutions, the same methods as for the solution of the integral equation of the crack problem can be used. In particular, application of Gaussian approximation functions result in explicit analytical expression of the matrix of the discretized problem. For thin inclusions with planar middle surfaces, Fast Fourier Transform technique can be used in the process of iterative solution of the discretized problem. This method allows us to obtain the solution with accuracy less than 1% by personal computers in reasonable time. The details of the method are described in [17].

References

1. Dyke V (1964) Perturbation methods in fluid mechanics. Academic Press, New York, London, p 310
2. Eskin G (1981) Boundary-value problems for elliptic pseudo-differential equations. Math Soc 232. (New York, USA)
3. Cherepanov G (1983) Fracture mechanics of composite materials. Nauka, Moscow, p 296
4. Novozgilov V (1959) Thin shell theory. Noordhoff, Amsterdam, p 376
5. Aleksandrov V, Mhitaryan S (1983) Contact problems for bodies with thin layers and inclusions. Nauka, Moscow, p 487
6. Sotkilava O, Cherepanov G (1974) Some problems of the heterogeneous elasticity theory. J Appl Math Mech (PMM) 38:537–550
7. Kit G, Hay M (1989) The method of potentials in 3d thermo-elastic problems for bodies with cracks. Naukova Dumka, Kiev, p 283
8. Ilyin A (1989) Matching of asymptotic expansions of the solutions of boundary-value problems. Nauka, Moscow, p 335
9. Maz'ya V, Nazarov S, Plamenevskii B (2000) Asymptotic theory of elliptic boundary value problems in singularly perturbed domains. Birkhauser, Berlin, p 460
10. Sanches-Palencia E (1984) Heterogeneous media and oscillation theory. Mir, Moscow, p 472
11. Simonenko I (1974) Electrostatic problem for an inhomogeneous medium. The case of thin dielectric inclusion with high dielectric constants.-I. Differ Equ 10:301–310

12. Simonenko I (1975) Electrostatic problem for an inhomogeneous medium. The case of thin dielectric inclusion with high dielectric constants.-II. Differ Equ 11:1870–1878
13. Poddubnyak A (1986) Integral equations of torsion of an elastic body with a thin disk-like inclusion. App Math Mech (PMM) 50:644–650
14. Ya K (1987) Axisymmetric torsion of the elastic space with a thin elastic inclusion. Appl Math Mech (PMM) 51:638–645
15. Kanaun S (1984) On singular models of thin inclusions in an homogeneous elastic medium. Appl Math Mech (PMM) 48:81–91
16. Kanaun S (1984) Thin defect in an homogeneous elastic medium. Proc Acad Sci USSR Mech Solids 3:74–83
17. Kanaun S (2021) Heterogeneous media: local fields. Effective properties and wave propagation. Elsevier, Amsterdam

Chapter 7
A Stiff Axisymmetric Fiber in a Homogeneous Elastic Medium

Abstract The elasticity problem for a homogeneous elastic medium containing a stiff axisymmetric fiber is considered. The solution depends on two small parameters: the ratio of the transverse size of the fiber and its length, and the ratio of the elastic moduli of the host medium and fiber materials. Construction of the principal terms of the elastic field in the medium with a fiber is reduced to a differential equation for the axial component of the stress field inside the fiber. The asymptotic solution for cylindrical, ellipsoidal, and double cone files is presented.

Keywords Elasticity problem for a stiff fiber · 1D-integral equation of the fiber problem · Differential equation of the fiber problem · Cylindrical fiber · Ellipsoidal fiber · Double cone fiber

In this chapter, the equilibrium of a homogeneous elastic medium containing a stiff axisymmetric fiber of a finite length is considered. The solution of this problem depends on two small parameters: the ratio of the transverse size of the fiber and its length, and the ratio of the elastic moduli of the host medium and fiber materials. The construction of the principal terms of the elastic field in the medium with a fiber with respect to the small parameters is the objective of this chapter.

7.1 An Ellipsoidal Stiff Fiber

We consider a homogeneous isotropic elastic medium with the stiffness tensor C^0 containing a fiber with the stiffness tensor C. The fiber is a prolate axisymmetric ellipsoid with the semiaxes $a_1 = a_2 = a$ and $a_3 = l \gg a$. The aspect ratio $\delta_1 = a/l$ is a small parameter of the problem. The second small parameter is the ratio of characteristic elastic moduli of the host medium and fiber materials. Thus the fiber elastic stiffness tensor C can be presented in the form

$$C = \frac{1}{\delta_2}\overline{C},\ \delta_2 \ll 1,\ \overline{C} = O(C^0). \tag{7.1}$$

S. Kanaun, *Self-Consistent Methods for Composites*, Solid Mechanics and Its Applications 282, https://doi.org/10.1007/978-3-031-96739-9_7

For a constant external stress field σ^0 applied to the medium, the stress field σ inside an ellipsoidal fiber is constant, and it is defined in Eqs. (4.81), (4.82) that can be presented in the form

$$C^{-1}\sigma + A(\delta_1)(C - C^0)C^{-1}\sigma = \left(C^0\right)^{-1}\sigma^0. \tag{7.2}$$

For an isotropic host medium, the tensor $A(\delta_1)$ is defined in Eq. (4.141), and its presentation in the P-basis is

$$A(\delta_1) = A_1 P^2(n) + A_2\left(P^1(n) - \frac{1}{2}P^2(n)\right) + \\ + A_3\left(P^3(n) + P^4(n)\right) + A_5 P_5(n) + A_6 P^6(n). \tag{7.3}$$

Here, n is the unit vector of the symmetry axis of the ellipsoid, the scalar coefficients $A_i = A_i(\delta_1)$ are

$$A_1 = \frac{1}{4\mu_0}\left(I_1 + (1-\kappa_0)I_2\right),\ A_2 = \frac{1}{4\mu_0}\left(2I_1 + (2-\kappa_0)I_2\right), \tag{7.4}$$

$$A_3 = -\frac{\kappa_0}{2\mu_0}I_1,\ A_5 = \frac{1}{2\mu_0}\left(2 - (1+4\kappa_0)I_1 - I_2\right), \tag{7.5}$$

$$A_6 = \frac{1}{\mu_0}\left((2\kappa_0 - 1)I_1 - (1-\kappa_0)(I_2 - 1)\right),\ \kappa_0 = \frac{\lambda_0 + \mu_0}{\lambda_0 + 2\mu_0}. \tag{7.6}$$

In these equations, λ_0, μ_0 are Lame's parameters of the host medium,, I_1, I_2 are the 1D-integrals

$$I_1 = \frac{\delta_1^2}{2}\int_0^{\pi}\frac{(\cos\theta)^2(\sin\theta)^2}{\Delta(\theta)}\sin\theta d\theta,\ I_2 = \frac{1}{2}\int_0^{\pi}\frac{(\sin\theta)^4}{\Delta(\theta)}\sin\theta d\theta, \tag{7.7}$$

$$\Delta(\theta) = \left(\delta_1^2\cos\theta^2 + (\sin\theta)^2\right)^2. \tag{7.8}$$

These integrals are expressed in terms of elementary functions

$$I_1 = \frac{\delta_1^2}{2(1-\delta_1^2)^2}\left(-3 + \frac{2+\delta_1^2}{\sqrt{1-\delta_1^2}}\operatorname{arctan h}\sqrt{1-\delta_1^2}\right), \tag{7.9}$$

$$I_2 = \frac{1}{2(1-\delta_1^2)^2}\left(2 + \delta_1^2 - \frac{\delta_1^2(4-\delta_1^2)}{\sqrt{1-\delta_1^2}}\operatorname{arctan h}\sqrt{1-\delta_1^2}\right). \tag{7.10}$$

If $\delta_1 \ll 1$, expansions of these integral in terms of δ_1 with accuracy to $\delta_1^2\ln\delta_1$ are

$$I_1 = -\delta_1^2\ln\delta_1 + O(\delta_1^2),\ I_2 = 1 + 2\delta_1^2\ln\delta_1 + O(\delta_1^2). \tag{7.11}$$

As a result, the principal terms of the tensor $A(\delta_1)$ in Eq. (7.3) takes the form

$$A(\delta_1) = A^0 + \delta_1^2 \ln \delta_1 A^1 + O(\delta_1^2). \tag{7.12}$$

Here, tensor A^0 is defined by the equation

$$A^0 = \frac{1-\kappa_0}{4\mu_0} P^2(n) + \frac{2-\kappa_0}{4\mu_0}\left(P^1(n) - \frac{1}{2}P^2(n)\right) + \frac{1}{2\mu_0} P_5(n), \tag{7.13}$$

and tensor A_1 is

$$A^1 = \frac{1-2\kappa_0}{2\mu_0} P^2(n) + \frac{3-4\kappa_0}{8\mu_0}\left(P^1(n) - \frac{1}{2}P^2(n)\right) +$$
$$+ \frac{\kappa_0}{2\mu_0}\left(P^3(n) + P^4(n)\right) + \frac{4\kappa_0 - 1}{4\mu_0} P_5(n) - \frac{1}{\mu_0} P^6(n). \tag{7.14}$$

As a result, Eq. (7.2) for the stress tensor inside the fiber with accuracy to the terms of the order of $\delta_1^2 \ln \delta_1$ is presented in the form

$$C^{-1}\sigma + A^0\sigma - A^0 C^0 C^{-1}\sigma + \delta_1^2 \ln \delta_1 A^1 (I - C^0 C^{-1})\sigma = \left(C^0\right)^{-1}\sigma^0. \tag{7.15}$$

In this equation, $C^0 C^{-1} = O(\delta_2)$ and $\delta_2 \ll 1$. To specify the structure of the stress tensor σ inside the fiber, we introduce two orthogonal projectors $\widehat{\Theta}(n)$ and $\widehat{\Pi}(n)$

$$\widehat{\Theta}(n) = P^1(n) + 2P^5(n), \quad \widehat{\Pi}(n) = P^6(n), \quad \widehat{\Theta} + \widehat{\Pi} = I, \tag{7.16}$$

$$\widehat{\Theta}\widehat{\Theta} = \widehat{\Theta}, \quad \widehat{\Pi}\widehat{\Pi} = \widehat{\Pi}, \quad \widehat{\Theta}\widehat{\Pi} = \widehat{\Pi}\widehat{\Theta} = 0. \tag{7.17}$$

These projectors split the linear shell of the P-basis into two orthogonal subspaces ($\widehat{\Theta}$ and $\widehat{\Pi}$). The tensor A^0 in Eq. (7.13) belongs to the $\widehat{\Theta}$ - subspace, and in this subspace, it has the inverse tensor $(A^0)^{-1}$

$$(A^0)^{-1} = \frac{\mu_0}{1-\kappa_0} P^2(n) + \frac{4\mu_0}{2-\kappa_0}\left(P^1(n) - \frac{1}{2}P^2(n)\right) + 8\mu_0 P_5(n). \tag{7.18}$$

The tensors A^0, $(A^0)^{-1}$ satisfy the equations

$$\widehat{\Theta} A^0 = A^0, \ \widehat{\Theta}\left(A^0\right)^{-1} = \left(A^0\right)^{-1}, \ A^0\left(A^0\right)^{-1} = \widehat{\Theta}, \tag{7.19}$$

$$\widehat{\Pi} A^0 = \widehat{\Pi}\left(A^0\right)^{-1} = 0. \tag{7.20}$$

The projectors $\widehat{\Theta}$ and $\widehat{\Pi}$ split the stress tensor σ inside the fiber in two parts

$$\sigma = \sigma^\theta + \sigma^\pi, \ \sigma^\theta = \widehat{\Theta}\sigma, \tag{7.21}$$

$$\sigma^{\pi} = \widehat{\Pi}\sigma = \sigma_n n \otimes n,\ \sigma_n = n_{\alpha}\sigma_{\alpha\beta}n_{\beta}. \tag{7.22}$$

After multiplication of Eq. (7.15) with the projector $\widehat{\Theta}(n)$, we yield

$$\widehat{\Theta}C^{-1}\sigma + A^0\sigma^{\theta} - A^0C^0C^{-1}\sigma + \delta_1^2 \ln\delta_1 \widehat{\Theta}A^1\sigma = \widehat{\Theta}\left(C^0\right)^{-1}\sigma^0. \tag{7.23}$$

Here, it is taken into account that $A^0\sigma = A^0\sigma^{\theta}$. Because $C^0C^{-1} = O(\delta_2)$, the principal term of the right-hand side of this equation with respect to δ_1, δ_2 is the tensor $A^0\sigma^{\theta}$ only. Keeping this term, we obtain the principal term of the component σ^{θ} of the tensor σ:

$$\sigma^{\theta} = \left(A^0\right)^{-1}\left(C^0\right)^{-1}\sigma^0 + O(\delta_1^2 \ln\delta_1, \delta_2). \tag{7.24}$$

After multiplication of Eq. (7.15) with the projector $\widehat{\Pi}$, and taking into account Eq. (7.20), we obtain

$$\widehat{\Pi}C^{-1}\sigma + \delta_1^2 \ln\delta_1 \widehat{\Pi}A^1\sigma = \widehat{\Pi}\left(C^0\right)^{-1}\sigma^0. \tag{7.25}$$

Because both terms in the left-hand side of this equation are of the order of small parameters δ_2 and $\delta_1^2 \ln\delta_1$, tensor σ has the order of $\delta_2^{-1}\sigma^0$ or $\left(\delta_1^2 \ln\delta_1\right)^{-1}\sigma^0$. For $\sigma^{\theta} = O(\sigma^0)$, the principal term of σ is the tensor σ^{π}

$$\sigma = \sigma^{\pi} + O(\sigma^0),\ \sigma^{\pi} = O(\delta_2^{-1}, \left(\delta_1^2 \ln\delta_1\right)^{-1}). \tag{7.26}$$

Taking into account these equations, we obtain

$$\widehat{\Pi}C^{-1}\sigma \approx \widehat{\Pi}C^{-1}\sigma^{\pi},\ \widehat{\Pi}A^1\sigma \approx \widehat{\Pi}A^1\sigma^{\pi}. \tag{7.27}$$

Because $\widehat{\Pi}\sigma^{\pi} = \sigma^{\pi}$, for other terms in Eq. (7.25) we have

$$\widehat{\Pi}C^{-1}\widehat{\Pi} = \frac{1}{E}\widehat{\Pi},\quad \widehat{\Pi}A^1\widehat{\Pi} = -\frac{1}{\mu_0}\widehat{\Pi}, \tag{7.28}$$

$$\widehat{\Pi}\left(C^0\right)^{-1} = -\frac{\lambda_0}{2\mu_0(3\lambda_0 + 2\mu_0)}P^4(n) + \frac{1}{E_0}P^6(n). \tag{7.29}$$

Here, E_0, E are Young's moduli of the host medium and the fiber. It follows from these equations and Eq. (7.25) that the axial component σ_n of the stress tensor in Eq. (7.22) satisfies the equation

$$\frac{1}{E}\sigma_n - \frac{\delta_1^2 \ln\delta_1}{\mu_0}\sigma_n = \frac{1}{E_0}\widehat{\sigma}_n^0, \tag{7.30}$$

$$\widehat{\sigma}_n^0 = \sigma_n^0 - \frac{\lambda_0}{2(\lambda_0 + \mu_0)}(tr(\sigma^0) - \sigma_n^0),\ tr(\sigma^0) = \sigma_{\alpha\alpha}^0. \tag{7.31}$$

Finally, the principal term σ_n of the axial component of the stress tensor σ inside the prolate ellipsoidal fiber takes the form

$$\sigma_n = \left(1 - \frac{E}{\mu_0}\delta_1^2 \ln \delta_1\right)^{-1} \frac{E}{E_0}\widehat{\sigma}_n^0. \tag{7.32}$$

The parameter $\left(\delta_1^2 \ln \delta_1\right) E/\mu_0$ in this equation has the order of the ratio of the small parameters of the problem $\left(\delta_1^2 \ln \delta_1\right)/\delta_2$, and this ratio can be not small.

7.2 An Axisymmetric Stiff Fiber of Arbitrary Shape

In this Section, we consider an axisymmetric fiber of the length $2l$ which radius a is a function of the axial coordinate z (Fig. 7.1). We assume that $a(z)$ is a smooth function that satisfies the condition

$$\left|\frac{da}{dz}\right| \ll 1, \; -l < z < l \tag{7.33}$$

except the regions of the fiber ends. For $a(z)$ much smaller than the fiber length l, the function $a(z)$ can be presented in the form

$$a(z) = \delta_1 l \, \alpha(z), \quad \delta_1 << 1, \; \alpha(z) = O(1). \tag{7.34}$$

The integral equation for the stress tensor $\sigma(x)$ inside the fiber follows from Eq. (4.12) in the form

$$C^{-1}\sigma(x) + \int_V K(x - x')C^1C^{-1}\sigma(x')dx' = (C^0)^{-1}\sigma^0, \; x \in V, \tag{7.35}$$

where V is the region occupied by the fiber, $C^1 = C - C^0$, and the function $K(x)$ is defined in Eq. (4.13).

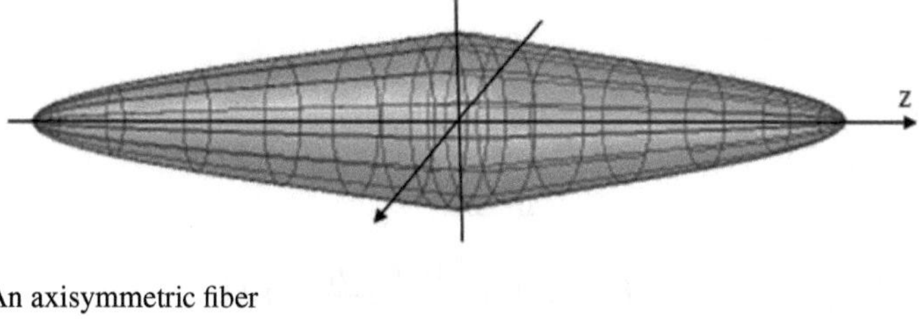

Fig. 7.1 An axisymmetric fiber

In the dimensionless local coordinates $\xi_i = a(z_0)^{-1}x_i, i = 1, 2$ and $\xi_3 = a(z_0)^{-1}(z - z_0)$ with the origin at an arbitrary point z_0 of the fiber axis, Eq. (7.35) takes the form

$$C^{-1}\sigma(\xi, z_0) + \int_{V'} K(\xi - \xi')C^1C^{-1}\sigma(\xi', z_0)d\xi' = \left(C^0\right)^{-1}\sigma^0. \tag{7.36}$$

Here, V' is the fiber region in the coordinates ξ_i. If $a(z_0) \to 0$, V' is transformed into a circular cylinder of unit radius with the generatrix ξ_3. Since σ^0 is constant, the solution of Eq. (7.36) inside the cylinder is also constant. Therefore, the principal term of the stress field inside the fiber is constant in each fiber cross section

$$\sigma(\overline{\xi}, z) \approx \sigma(z), \ \left(\overline{\xi}, z\right) \in V, \ \overline{\xi} = (\xi_1, \xi_2). \tag{7.37}$$

Then, we introduce the dimensionless coordinates $\zeta_i = x_i/l$, $i = 1, 2$, $\overline{\zeta} = (\zeta_1, \zeta_2)$, $\varsigma = z/l$ and present Eq. (7.35) in the form

$$C^{-1}\sigma(\varsigma) + \int_{V'} K(\overline{\zeta} - \overline{\zeta}', \varsigma - \varsigma')C^1C^{-1}\sigma(\varsigma')d\overline{\zeta}'d\varsigma = (C^0)^{-1}\sigma^0, \ \overline{\zeta}, \varsigma \in V', \tag{7.38}$$

where V' is the fiber region in the $\overline{\zeta}$, ς-coordinates. Because σ is constant in each fiber cross section $\Omega(\varsigma)$, this equation can be considered on the fiber axis $\overline{\zeta} = 0$ only

$$C^{-1}\sigma(\varsigma) + \int_{-1}^{1} A(\varsigma - \varsigma', \varsigma')C^1C^{-1}\sigma(\varsigma')d\Omega' = (C^0)^{-1}\sigma^0, \ -1 < \varsigma < 1. \tag{7.39}$$

$$A(\varsigma - \varsigma', \varsigma') = \int_{\Omega(\varsigma')} K(-\overline{\zeta}', \varsigma - \varsigma')d\Omega'. \tag{7.40}$$

The integral over the fiber cross section $\Omega(\varsigma')$ in this equation can be calculated explicitly using the Fourier transform $\widetilde{K}(\overline{k}, k_3)$ of the function $K(\overline{\zeta}, \varsigma)$

$$\int_{\Omega(\varsigma')} K(-\overline{\zeta}', \varsigma)d\Omega' =$$

$$= \frac{1}{(2\pi)^3}\int \widetilde{K}(\overline{k}, k_3)e^{-i\varsigma k_3}d\overline{k}dk_3 \int_{\Omega(\varsigma')} \exp\left(-i\left(\overline{\zeta}' \cdot \overline{k}\right)\right)d\Omega' =$$

$$= \frac{\delta_1\alpha(\varsigma)}{(2\pi)^2}\int_{-\infty}^{\infty} e^{(-i\varsigma k_3)}dk_3 \int_0^{\infty} \widetilde{K}(\overline{k}, k_3)\frac{J_1(\delta_1\alpha(\varsigma)|\overline{k}|)}{|\overline{k}|}d|\overline{k}|. \tag{7.41}$$

Here, $\overline{k} = (k_1, k_2)$, $J_1(y)$ is Bessel's function. For an isotropic host medium, the function $\widetilde{K}(\overline{k}, k_3)$ has the form

$$\widetilde{K}(\overline{k}, k_3) = \frac{1}{\mu_0}\left(E^5(m) - \kappa_0 E^6(m)\right), \; m_\alpha = \frac{k_\alpha}{\sqrt{|\overline{k}|^2 + k_3^2}}, \tag{7.42}$$

and the explicit form of the integral over $|\overline{k}|$ in Eq. (7.41) is

$$\widetilde{A}(k_3, \varsigma') = \frac{\delta_1 \alpha(\varsigma')}{2\pi} \int_0^\infty \widetilde{K}(\overline{k}, k_3) \frac{J_1(\delta_1 \alpha(\varsigma')|\overline{k}|)}{|\overline{k}|} d|\overline{k}| =$$
$$= \widetilde{A}_1 P^2(n) + \widetilde{A}_2 \left(P^1(n) - \frac{1}{2} P^2(n)\right) + \widetilde{A}_3 \left(P^3(n) + P^4(n)\right) +$$
$$+ \widetilde{A}_5 P^5(n) + \widetilde{A}_6 P^6(n). \tag{7.43}$$

Here, n is unit vector of the z-axis, the scalar coefficients $\widetilde{A}_i(k_3, \varsigma')$ are

$$\widetilde{A}_1 = \frac{1}{8\mu_0}\left(\widetilde{T}_2 - 2(1-\kappa_0)\widetilde{T}_1\right), \; \widetilde{A}_2 = \frac{1}{8\mu_0}\left(\widetilde{T}_2 + 2(2-\kappa_0)\widetilde{T}_1\right), \tag{7.44}$$

$$\widetilde{A}_3 = -\frac{\kappa_0}{4\mu_0}\widetilde{T}_2, \; \widetilde{A}_5 = \frac{1}{2\mu_0}\left(2 - \widetilde{T}_1 - 2\widetilde{T}_2\right), \tag{7.45}$$

$$\widetilde{A}_6 = \frac{1}{2\mu_0}\left(2(1-\kappa_0)(1-\widetilde{T}_1) + \widetilde{T}_2\right). \tag{7.46}$$

The functions $\widetilde{T}_1 = \widetilde{T}_1(k_3, \varsigma')$ and $\widetilde{T}_2 = \widetilde{T}_2(k_3, \varsigma')$ in these equations are

$$\widetilde{T}_1(k_3, \varsigma') = \delta_1 \alpha(\varsigma') \mathrm{K}_1\left(\delta_1 \alpha(\varsigma')|k_3|\right), \tag{7.47}$$

$$\widetilde{T}_2(k_3, \varsigma) = \delta_1^2 \alpha^2(\varsigma') \mathrm{K}_0\left(\delta_1 \alpha(\varsigma')|k_3|\right), \tag{7.48}$$

where $\mathrm{K}_0(z)$, $\mathrm{K}_1(z)$ are the modified Bessel functions of zero and first kinds.

In order to obtain the principal terms of the left-hand side of Eq. (7.39), we expand the modified Bessel functions into series with respect to the parameter δ_1 and keep the principal terms of these expansions only [1]

$$\mathrm{K}_0(z) = -\ln z + O(z), \; \mathrm{K}_1(z) = \frac{1}{z} + \frac{z}{2}\ln z + O(z). \tag{7.49}$$

As a result, the asymptotic expression for the tensor $\widetilde{A}(k_3, \varsigma')$ in Eq. (7.43) takes the form

$$\widetilde{A}(k_3, \varsigma') = A^0 + \delta_1^2 \ln \delta_1 k_3^2 \alpha(\varsigma')^2 A^1 + O(\delta_1^2). \tag{7.50}$$

Here, the tensor A^0 is the same as in Eqs. (7.13) and the tensor A^1 is

$$A^1 = \frac{(1-2\kappa_0)}{8\mu_0} P^2(n) + \frac{(1-\kappa_0)}{4\mu_0}\left(P^1(n) - \frac{1}{2}P^2(n)\right) + \\ + \frac{\kappa_0}{4\mu_0}\left(P^3(n) + P^4(n)\right) + \frac{(4\kappa_0 - 1)}{4\mu_0} P^5(n) - \frac{1}{2\mu_0} P^6(n). \tag{7.51}$$

Because product with k_3^2 in the Fourier transform space corresponds to the differential operator $-d^2/d\varsigma^2$ in the original ς-space, Eq. (7.39) for the stress tensor $\sigma(\varsigma)$ inside the stiff fiber takes the form

$$C^{-1}\sigma(\varsigma) + A^0\sigma^\theta(\varsigma) - (\delta_1^2 \ln \delta_1) A^1 \frac{\mathrm{d}^2}{\mathrm{d}\varsigma^2}\left(\alpha^2(\varsigma)\sigma(\varsigma)\right) = \left(C^0\right)^{-1}\sigma^0 . \tag{7.52}$$

Here, it is taken into account that $C^1C^{-1} = I + O(\delta_2)$. Because tensor A^0 belongs to $\widehat{\Theta}$-subspace, similar to the case of ellipsoidal inclusion, the $\widehat{\Theta}$-component $\sigma^\theta(\varsigma) = \widehat{\Theta}\sigma(\varsigma)$ of the tensor $\sigma(\varsigma)$ has the order of the external stress tensor σ^0 and

$$\sigma_\theta(\varsigma) = \left(A^0\right)^{-1}\left(C^0\right)^{-1}\sigma^0. \tag{7.53}$$

Meanwhile the $\widehat{\Pi}$-component $\sigma^\pi(\varsigma) = \widehat{\Pi}\sigma(\varsigma) = \sigma_n(\varsigma) n \otimes n$ has the order of δ_2^{-1} or $(\delta_1^2 \ln \delta_1)^{-1}$. Multiplying Eq. (7.52) with the projector $\widehat{\Pi}$ and taking into account that $\sigma = \sigma^\pi + O(\sigma^0)$, we yield the equation for the axial component of the stress tensor in the fiber

$$\sigma_n(\varsigma) + \frac{E}{2\mu_0}(\delta_1^2 \ln \delta_1)\frac{d^2}{d\xi^2}\left(\alpha^2(\xi)\sigma_n(\xi)\right) = \frac{E}{E_0}\widehat{\sigma}_n^0. \tag{7.54}$$

Here, E and E_0 are Young's moduli of the inclusion and the host medium, $E/E_0 = O(\delta_2^{-1})$, $\widehat{\sigma}_n^0(\varsigma)$ is defined in Eq. (7.27), and the equation

$$\widehat{\Pi} A^1 \widehat{\Pi} = -\frac{1}{2\mu_0}\widehat{\Pi} \tag{7.55}$$

is taken into account.

Let us return to Eqs. (7.43)–(7.48) for the tensor $\widetilde{A}(k_3, \varsigma')$. The inverse Fourier transforms the function $\widetilde{T}_1(k_3, \varsigma')$ and $\widetilde{T}_2(k_3, \varsigma')$ in Eqs. (7.48) with respect to k_3 have the following explicit expressions

$$T_1(\varsigma, \varsigma') = \frac{\delta_1^2 \alpha(\varsigma')^2}{2\left(\varsigma^2 + \delta_1^2\alpha(\varsigma')^2\right)^{3/2}}, \tag{7.56}$$

$$T_2(\varsigma, \varsigma') = -\frac{1}{2}\frac{d^2}{d\varsigma^2}\left(\frac{\delta_1^2\alpha(\varsigma')^2}{\sqrt{\varsigma^2 + \delta_1^2\alpha(\varsigma')^2}}\right). \tag{7.57}$$

As a result, Eq. (7.39) for the stress tensor inside the fiber takes the form

$$C^{-1}\sigma(\varsigma) + \int_{-1}^{1} A(\varsigma - \varsigma', \varsigma')\sigma(\varsigma')d\Omega' = (C^0)^{-1}\sigma^0, \ -1 < \varsigma < 1. \tag{7.58}$$

Multiplying Eq. (7.58) with the projector $\widehat{\Pi}$ and taking into account that $\sigma_\theta(\varsigma) \ll \sigma_\pi(\varsigma)$, we obtain the integral equation for the axial component $\sigma_n(\varsigma)$ of the stress tensor inside the fiber

$$\rho\sigma_n(\varsigma) - \int_{-1}^{1} T(\varsigma - \varsigma', \varsigma')\sigma_n(\varsigma')d\varsigma' = \frac{\mu_0}{E_0(1-\kappa_0)}\widehat{\sigma}_n^0, \ |\varsigma| < 1, \tag{7.59}$$

$$\rho = 1 + \frac{\mu_0}{E(1-\kappa_0)}, \ T(\varsigma, \varsigma') = T_1(\varsigma, \varsigma') - \frac{\kappa_0}{2(1-\kappa_0)}T_2(\varsigma, \varsigma'). \tag{7.60}$$

Here, the functions $T_1(\varsigma, \varsigma')$ and $T_2(\varsigma, \varsigma')$ are defined in Eqs. (7.56), (7.57).

Differential equation (7.54) follows from the integral equation (7.59) after presenting $T(\varsigma, \varsigma')$ in terms of the Fourier transform with respect to variable ς and expanding the result in the series over the parameter δ_1. Keeping in this series the terms of the order of $\delta_1^2 \ln \delta_1$, we yield Eq. (7.54) after returning to ς-space.

7.3 Differential Equation for the Axial Component of the Stress Tensor Inside a Stiff Fiber

The differential equation (7.54) for the principal component of the stress tensor inside a stiff fiber can be present in the form

$$\frac{d^2}{d\varsigma^2}\left(\alpha^2(\varsigma)\sigma_n(\varsigma)\right) - q^2\sigma_n(\varsigma) = -q^2\frac{E}{E_0}\widehat{\sigma}_n^0(\varsigma). \tag{7.61}$$

Here, q^2 is a positive dimensionless parameter

$$q^2 = -\frac{2\mu_0}{E(\delta_1^2 \ln \delta_1)} = O\left(\frac{\delta_2}{\delta_1^2 \ln \delta_1}\right). \tag{7.62}$$

We consider Eq. (7.61) for different relations between the parameters δ_1 and δ_2.

1. $\delta_2 \gg \delta_1^2 ln\delta_1, \ q^2 \gg 1$.

In this case, the term proportional to $\delta_1^2 ln\delta_1$ in Eq. (7.52) can be neglected, and application of the projector $\widehat{\Pi}$ to (7.52) results in the equation

$$\widehat{\Pi} \cdot C^{-1}\sigma(\varsigma) = \widehat{\Pi} \cdot \left(C^0\right)^{-1}\sigma^0(\varsigma). \tag{7.63}$$

Thus the axial component $\sigma_n(\varsigma)$ of the stress field inside the fiber has the order of δ_2^{-1}, and

$$\sigma_n(\varsigma) = \frac{E}{E_0}\widehat{\sigma}_n^0(\xi). \tag{7.64}$$

2. $\delta_2 = O(\delta_1^2 \ln \delta_1),\ q^2 = O(1)$.

In this case, the principal term of stress field $\sigma_n(\varsigma)$ inside the fiber is the solution of the differential equation of the second order (7.61). The general solution of this equation contains two arbitrary constants, and the conditions for the determination of these constants should be formulated. To that end we take into account that the function $\sigma_n(\varsigma)$ satisfies the integral equation (7.59). If we substitute the general solution of Eq. (7.61) into the left-hand side of Eq. (7.59), it will be satisfied with precision to discrepancy $R(\sigma_n, \varsigma)$

$$\rho\sigma_n(\varsigma) - (T\sigma_n)(\varsigma) = \frac{\mu_0}{E_0(1-\kappa_0)}\widehat{\sigma}_n^0(\varsigma) + R(\sigma_n, \varsigma). \tag{7.65}$$

The constants in the general solution of Eq. (7.61) should be taken to minimize the discrepancy $R(\sigma_n, \varsigma)$ in this equation.

3. $\delta_2 \ll \delta_1^2 ln\delta_1,\ q^2 \ll 1$.

In this case, the differential equation (7.61) takes the form

$$\frac{d^2}{d\varsigma^2}\left(\alpha^2(\varsigma)\sigma_n(\varsigma)\right) = \frac{2\mu_0}{(\delta_1^2 \ln \delta_1)E_0}\widehat{\sigma}_n^0(\varsigma), \tag{7.66}$$

and as in case 2, two constants in the general solution of this equation should be taken to minimize the discrepancy $R(\sigma_n, \varsigma)$ in the integral equation (7.65) after substitution in its left-hand side the general solution of Eq. (7.66).

7.4 Stress Tensor Inside the Fibers of Particular Forms

1. *Cylindrical fiber*

For a cylindrical fiber of radius a, we have

$$\delta_1 = a/l,\quad \alpha(\varsigma) = 1. \tag{7.67}$$

The differential equation (7.61) for the principal term of the axial component of the stress tensor inside the fiber takes the form

$$\frac{d^2}{d\varsigma^2}\sigma_n(\varsigma) - q^2\sigma_n(\varsigma) = -q^2\frac{E}{E_0}\widehat{\sigma}_n^0,\ q^2 = O(1). \tag{7.68}$$

The general solution of this equation is

$$\sigma_n(\varsigma) = \frac{E}{E_0}\widehat{\sigma}_n^0 + c_1 e^{q\varsigma} + c_2 e^{-q\varsigma}, \tag{7.69}$$

where c_1, c_2 are arbitrary constant.

In order to obtain the discrepancy R in Eq. (7.65) for this case, we have to evaluate the result of the action of the integral operators with the kernels $T_1(\varsigma)$ and $T_2(\varsigma)$ in Eqs. (7.59)–(7.57) on a smooth bounded functions $\sigma(\varsigma)$. To that end we present the function $(T_1\sigma)(\varsigma)$ in the form

$$(T_1\sigma)(\varsigma) = T_1^0(\varsigma)\sigma(\varsigma) + T_1^1(\varsigma)D\sigma(\varsigma) + \frac{1}{2}T_1^2 D^2\sigma(\varsigma) + \\ + \int_{-1}^{1} T_1\left(\varsigma - \varsigma'\right)Q\left(\varsigma, \varsigma'\right)d\varsigma',\ D = \frac{d}{d\varsigma}, \tag{7.70}$$

$$Q\left(\varsigma, \varsigma'\right) = \sigma\left(\varsigma'\right) - \sigma\left(\varsigma\right) - D\sigma\left(\varsigma\right)\left(\varsigma - \varsigma'\right) - \frac{1}{2}D^2\sigma\left(\varsigma\right)\left(\varsigma' - \varsigma\right)^2, \tag{7.71}$$

$$T_1^k(\varsigma) = \int_{-1}^{1} T_1(\varsigma - \varsigma')(\varsigma - \varsigma')^k d\varsigma',\quad k = 0, 1, 2, \tag{7.72}$$

where the kernel $T_1(\varsigma)$ is defined in (7.56) with $\alpha(\varsigma') = 1$. Calculating integrals $T_1^k(\varsigma)$ and substituting the results in (7.70), we obtain

$$(T_1\sigma)(\varsigma) = \sigma(\varsigma) - \left(\Phi_0\left(\frac{1-\varsigma}{\delta_1}\right) + \Phi_0\left(\frac{1+\varsigma}{\delta_1}\right)\right)\sigma(\varsigma) - \\ - \delta_1\left(\Phi_1\left(\frac{1-\varsigma}{\delta_1}\right) - \Phi_1\left(\frac{1+\varsigma}{\delta_1}\right)\right)D\sigma(\varsigma) + O(\delta_1^2 \ln \delta_1). \tag{7.73}$$

Here, the functions Φ_0, Φ_1 are of the boundary layer type localized at the vicinities of the fiber ends $\varsigma = \pm 1$:

$$\Phi_0(t) = \frac{1}{2}\operatorname{sign} t\left(1 - \frac{|t|}{\sqrt{1+t^2}}\right),\quad \Phi_1(t) = \frac{1}{2\sqrt{1+t^2}}. \tag{7.74}$$

In the same way, we can show that the integral operator with the kernel $T_2(\varsigma)$ in Eq. (7.57) with $\alpha(\varsigma) = 1$ is evaluated as follows

$$(T_2\sigma)(\varsigma) = -\delta_1\left(\Phi_1\left(\frac{1-\varsigma}{\delta_1}\right) - \Phi_1\left(\frac{1+\varsigma}{\delta_1}\right)\right)D\sigma(\varsigma) + O(\delta_1^2 \ln \delta_1). \tag{7.75}$$

Thus for the function $\sigma_n(\varsigma)$ in Eq. (7.69), the principal term of the discrepancy R in Eq. (7.65) takes the form

$$R = \left(\Phi_0 \left(\frac{1-\varsigma}{\delta_1} \right) + \Phi_0 \left(\frac{1+\varsigma}{\delta_1} \right) \right) \sigma_n(\varsigma) + O(\delta_1). \tag{7.76}$$

Because $\Phi_0 \left(\frac{1-\varsigma}{\delta_1} \right)$ and $\Phi_0 \left(\frac{1+\varsigma}{\delta_1} \right)$ are the functions of boundary layer types localized at the ends of the fiber $\varsigma = \pm 1$, the principal term of R is minimal if the function $\sigma_n(\varsigma)$ satisfies the conditions

$$\sigma_n(-1) = \sigma_n(1) = 0. \tag{7.77}$$

For these conditions, the function $\sigma_n(\varsigma)$ in Eq. (7.69) takes the form

$$\sigma_n(\varsigma) = \left(1 - \frac{\cosh(q\varsigma)}{\cosh(q)} \right) \frac{E}{E_0} \widehat{\sigma}_n^0. \tag{7.78}$$

To visualize the character of the approximation of the solution of Eq. (7.59) by the function $\sigma_n(\varsigma)$ in (7.78), we consider a model equation similar to Eq. (7.59):

$$\beta\sigma(\varsigma) - (T_1\sigma)(\varsigma) = -\frac{1}{2}\delta_1^2 \ln \delta_1 , \quad \beta = 1 - \frac{1}{2} q^2 \delta_1^2 \ln \delta_1 , \tag{7.79}$$

$$(T_1\sigma)(\varsigma) = \int_{-1}^{1} T_1(\varsigma - \varsigma')\sigma(\varsigma')d\varsigma', \ T_1(\varsigma) = \frac{\delta_1^2}{2\left(\varsigma^2 + \delta_1^2\right)^{3/2}}. \tag{7.80}$$

Here, $\delta_1 \ll 1$, $q = O(1)$. Because expansion of the Fourier transform $\widetilde{T}_1(k)$ of the function $T_1(\varsigma)$ in the series with respect to δ_1 is

$$\widetilde{T}_1(k) = \delta_1 |k| \mathrm{K}_1(\delta_1 |k|) = 1 + \frac{1}{2}\delta_1^2 (\ln \delta_1) k^2 + O(\delta_1^2), \tag{7.81}$$

the operator T_1 can be approximated as follows

$$(T_1\sigma)(\varsigma) = \sigma(\varsigma) - \frac{1}{2}\delta_1^2 (\ln \delta_1) \frac{\mathrm{d}^2}{\mathrm{d}\varsigma^2}\sigma(\varsigma) + O(\delta_1^2), \tag{7.82}$$

and the equation for $\sigma(\varsigma)$ takes the form

$$\frac{\mathrm{d}^2}{\mathrm{d}\varsigma^2}\sigma(\varsigma) - q^2\sigma(\varsigma) = -1 . \tag{7.83}$$

For the boundary conditions $\sigma(\pm 1) = 0$, the solution of this equation is

$$\sigma(\varsigma) = \frac{1}{q^2}\left(1 - \frac{\cosh (q\varsigma)}{\cosh (q)} \right). \tag{7.84}$$

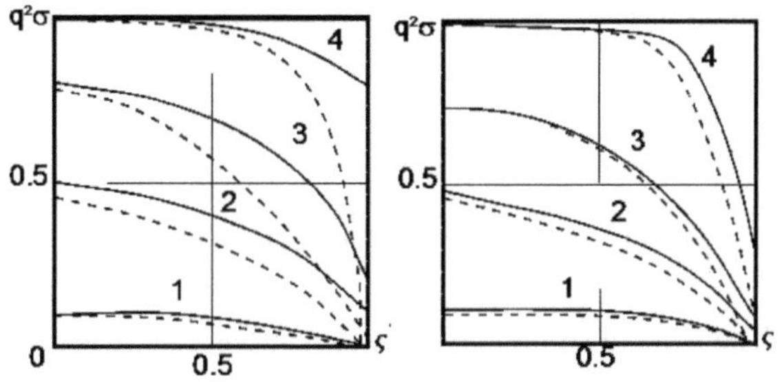

Fig. 7.2 Comparison of the results of the numerical solution of Eq. (7.81) (sold lines) with the approximation (7.84) (dashed lines), left figure for $\delta_1 = 0.1$, right figure for $\delta_1 = 0.01$, lines 1, 2, 3, 4 correspond to the values of the parameter $q = 0.4;\ 1.2;\ 2;\ 10$

Comparison of $\sigma(\varsigma)$ in Eq. (7.84) with the results of the numerical solution of Eq. (7.79) is presented on Fig. 7.2 (left figure for $\delta_1 = 0.1$, right figure for $\delta_1 = 0.01$) by the solid lines; the dashed lines correspond to the function $\sigma(\varsigma)$ in (7.84). Curves 1–4 are for the values of the parameter $q = 0.4;\ 1.2;\ 2;\ 10$, correspondingly. When δ_1 decreases, the difference between the numerical solution and the approximate solution is substantial only in the δ_1-vicinity of the fiber ends.

2. *Prolate ellipsoidal fiber*
For the ellipsoidal fiber $\alpha(\varsigma) = \sqrt{1-\varsigma^2}$, and differential equation (7.61) for the axial stress $\sigma_n(\varsigma)$ inside the fiber takes the form

$$\frac{d^2}{d\varsigma^2}\left(\left(1-\varsigma^2\right)\sigma_n\left(\varsigma\right)\right) - q^2\sigma_n\left(\varsigma\right) = -q^2\frac{E}{E_0}\widehat{\sigma}_n^0. \tag{7.85}$$

The general solution of this equation is

$$\sigma_n\left(\varsigma\right) = \left(\frac{q^2}{q^2+2}\right)\frac{E}{E_0}\widehat{\sigma}_n^0 + c_1\varphi_1(\varsigma) + c_2\varphi_2(\varsigma). \tag{7.86}$$

Here, c_1, c_2 are arbitrary constants, $\varphi_1(\varsigma)$, $\varphi_2(\varsigma)$ are hypergeometric functions that have non-integrable singularities $(1 \mp \varsigma)^{-1}$ at the fiber ends $\varsigma = \pm 1$. Since the elastic field inside a fiber with a smooth boundary is finite, the constants c_1, c_2 in Eq. (7.86) should be equal to zero. Thus, for a constant external stress field, the principal term of the stress field inside the fiber is

$$\sigma_n = \frac{q^2}{2+q^2}\frac{E}{E_0}\widehat{\sigma}_n^0 = \left(1 - \frac{E}{\mu_0}\delta_1^2 \ln \delta_1\right)^{-1}\frac{E}{E_0}\widehat{\sigma}_n^0. \tag{7.87}$$

This equation coincides with the principal term of the stress field in a stiff prolate ellipsoidal fiber indicated in Eq. (7.32).

3. *Fiber in the form of a double cone*
For the fiber in the form of a double cone, the function $\alpha(\varsigma)$ has the form

$$\alpha(\varsigma) = 1 - |\varsigma|. \tag{7.88}$$

Equation (7.61) for the principal term of the stress field inside the fiber takes the form

$$\frac{d^2}{d\varsigma^2}\left((1+\varsigma)^2\sigma_n(\varsigma)\right)-q^2\sigma_n(\varsigma)=-q^2\frac{E}{E_0}\widehat{\sigma}_n^0,\ -1<\varsigma<0, \tag{7.89}$$

$$\frac{d^2}{d\varsigma^2}\left((1-\varsigma)^2\sigma_n(\varsigma)\right)-q^2\sigma_n(\varsigma)=-q^2\frac{E}{E_0}\widehat{\sigma}_n^0,\ \ 0<\varsigma<1. \tag{7.90}$$

The general solution of these equations can be presented in the form

$$\sigma_n(\varsigma)=\overline{\sigma}_n+c_1(1-|\varsigma|)^{\beta_-}+c_2(1-|\varsigma|)^{\beta_+},\ q^2\neq 2. \tag{7.91}$$

$$\overline{\sigma}_n=\left(\frac{q^2}{q^2-2}\right)\frac{E}{E_0}\widehat{\sigma}_n^0,\ \beta_\pm=-\frac{1}{2}\left(3\pm\sqrt{1+4q^2}\right), \tag{7.92}$$

where c_1, c_2 are arbitrary constants. Since the elastic field near a conical point on the inclusion interface is square-integrable [2], the constant c_2 in front of the non-integrable function $(1-|\varsigma|)^{\beta_+}$ should be equal to zero, and therefore,

$$\sigma_n(\varsigma)=\overline{\sigma}_n+c(1-|\varsigma|)^{\beta},\ \beta=\beta_-,\ c=c_1. \tag{7.93}$$

For $q^2=2$, the general solution of Eqs. (7.89), (7.90) changes and takes the form

$$\sigma_n(\varsigma)=\overline{\sigma}_n(\varsigma)+c_1+\frac{c_2}{(1-|\varsigma|)^3},\ q^2=2, \tag{7.94}$$

$$\overline{\sigma}_n(\varsigma)=\frac{2E\widehat{\sigma}_n^0}{9E_0(1-|\varsigma|)^3}-\frac{2E\widehat{\sigma}_n^0}{3E_0}\ln(1-|\varsigma|). \tag{7.95}$$

Because $(1-|\varsigma|)^{-3}$ is a non-integrable function, the constant c_2 should be $c_2=-2E\sigma_n^0/(9E_0)$, and for $c_1=c$, we have

$$\sigma_n(\varsigma)=c-\frac{2E}{3E_0}\sigma_n^0\ln(1-|\varsigma|),\ \ q^2=2. \tag{7.96}$$

For the fiber in the form of a double cone, the integral operators T_1 and T_2 in the integral equation Eqs. (7.56), (7.57) are

$$(T_1\sigma)(\varsigma)=\frac{1}{2}\int_{-1}^{1}\frac{\delta_1^2(1-|\varsigma'|)^2}{\left((\varsigma-\varsigma')^2+\delta_1^2(1-|\varsigma'|)^2\right)^{3/2}}\sigma(\varsigma')\,d\varsigma', \tag{7.97}$$

$$(T_2\sigma)(\varsigma)=-\frac{1}{2}\frac{d^2}{d\varsigma^2}\int_{-1}^{1}\frac{\delta_1^2(1-|\varsigma'|)^2}{\left((\varsigma-\varsigma')^2+\delta_1^2(1-|\varsigma'|)^2\right)^{1/2}}\sigma(\varsigma')d\varsigma_1'. \tag{7.98}$$

Using Eqs. (7.70)–(7.72), the function $(T_1\sigma)(\varsigma)$ near the fiber middle point $\varsigma = 0$ are evaluated as follows

$$(T_1\sigma)(\varsigma) = \sigma(\varsigma) + \Phi_0\left(\frac{\varsigma}{\delta_1}\right)(\sigma_-(\varsigma) - \sigma_+(\varsigma)) -$$

$$- \delta_1\Phi_1\left(\frac{\varsigma}{\delta_1}\right)(\sigma(\varsigma) + D\sigma(\varsigma)) + O(\delta_1^2 \ln \delta_1), \tag{7.99}$$

$$(T_2\sigma)(\varsigma) = O(\delta_1^2 \ln \delta_1). \tag{7.100}$$

Here the functions $\Phi_0(t)$, $\Phi_1(t)$ are defined in Eq. (7.74), the singular terms arising by the differentiation of $|\varsigma|$ should be omitted. As a result, near the middle point, the discrepancy R in Eq. (7.65) is evaluated as follows

$$R = -\Phi_0\left(\frac{\varsigma}{\delta_1}\right)\left(\sigma^-(\varsigma) - \sigma^+(\varsigma)\right) +$$

$$+ \delta_1\Phi_1\left(\frac{\varsigma}{\delta_1}\right)\left(\sigma^-(\varsigma) + \sigma^+(\varsigma) + D\sigma^-(\varsigma) - D\sigma^+(\varsigma)\right) + O(\delta_1^2 \ln \delta_1). \tag{7.101}$$

For $q^2 \neq 2$, the functions $\sigma^-(\varsigma)$, $\sigma^+(\varsigma)$ and $D\sigma^-(\varsigma)$, $D\sigma^+(\varsigma)$ in this equation are

$$\sigma^-(\varsigma) = \overline{\sigma}_n + c(1+\varsigma)^\beta, \sigma^+(\varsigma) = \overline{\sigma}_n + c(1-\varsigma)^\beta, \tag{7.102}$$

$$D\sigma^-(\varsigma) = c\beta(1+\varsigma)^{\beta-1}, \; D\sigma^+(\varsigma) = c\beta(1-\varsigma)^{\beta-1}. \tag{7.103}$$

Therefore, in the vicinity of the point $\varsigma = 0$, we have

$$\sigma^-(\varsigma) - \sigma^+(\varsigma) = 2c\beta + O(\varsigma^2), \tag{7.104}$$

$$\sigma^-(\varsigma) + \sigma^+(\varsigma) + D\sigma^-(\varsigma) - D\sigma^+(\varsigma) =$$

$$= 2\overline{\sigma}_n + 2c(1+\beta) + O(\varsigma^2), \tag{7.105}$$

and the discrepancy R takes the form

$$R(\delta_1, \varsigma) = 2\delta_1\left(\sqrt{\left(\frac{\varsigma}{\delta_1}\right) + 1} - \frac{|\varsigma|}{\delta_1}\right)c\beta +$$

$$+ \frac{\delta_1}{\sqrt{\left(\frac{\varsigma}{\delta_1}\right) + 1}}(\overline{\sigma}_n + c) + O(\varsigma^2) =$$

$$= \delta_1\left(\frac{1}{2}c\beta + \overline{\sigma}_n + c\right)\left(\frac{\delta_1}{|\varsigma|}\right) + o\left(\frac{\delta_1^3}{|\varsigma|^3}\right). \tag{7.106}$$

If the constant c is taken from the condition

$$\frac{1}{2}c\beta + \overline{\sigma}_n + c = 0, \tag{7.107}$$

the discrepancy decays as $(|\varsigma|/\delta_1)^{-3}$ by moving away from the fiber middle point. For $c = -2\overline{\sigma}_n/(2+\beta)$, the expression for $\sigma_n(\varsigma)$ takes the form

$$\sigma_n(\varsigma) = \left(1 - \frac{2}{2+\beta}\left(1 - |\varsigma|\right)^{\beta}\right)\bar{\sigma}_n. \tag{7.108}$$

For $q^2 = 2$,

$$\sigma^-(\varsigma) = c - \frac{2E\sigma_n^0}{3E_0}\ln(1+\varsigma),\ \sigma^+(\varsigma) = c - \frac{2E\sigma_n^0}{3E_0}\ln(1-\varsigma), \tag{7.109}$$

and in the vicinity of the fiber middle point $\varsigma = 0$, we have

$$\sigma^-(\varsigma) - \sigma^+(\varsigma) = -\frac{4E}{3E_0}|\varsigma| + O(\varsigma^3), \tag{7.110}$$

$$\sigma^-(\varsigma) + \sigma^+(\varsigma) + D\sigma^-(\varsigma) - D\sigma^+(\varsigma) = 2c - \frac{4E\sigma_n^0}{3E_0} + O(\varsigma^2). \tag{7.111}$$

As a result, the discrepancy R is

$$\begin{aligned} R(\delta_1, \varsigma) &= \frac{\delta_1}{2}\left(\sqrt{\left(\frac{\varsigma}{\delta_1}\right) + 1} - \frac{|\varsigma|}{\delta_1}\right)\frac{4E\sigma_n^0}{3E_0} + \\ &+ \frac{\delta_1}{2\sqrt{\left(\frac{\varsigma}{\delta_1}\right) + 1}}\left(2c - \frac{4E\sigma_n^0}{3E_0}\right) = \\ &= \delta_1\left(c - \frac{E\sigma_n^0}{3E_0}\right)\left(\frac{\delta_1}{|\varsigma|}\right) + O\left(\frac{\delta_1^3}{|\varsigma|^3}\right). \end{aligned} \tag{7.112}$$

For the constant c in the form

$$c = \frac{E\sigma_n^0}{3E_0}, \tag{7.113}$$

the discrepancy decays as $(\delta_1/|\varsigma|)^3$ by moving away from the fiber middle point. Finally, the principal term of the stress field inside a stiff double cone fiber takes the form

$$\sigma_n(\varsigma) = \frac{q^2}{q^2-2}\left(1 - \frac{2}{2+\beta}\left(1 - |\varsigma|\right)^{\beta}\right)\frac{E}{E_0}\widehat{\sigma}_n^0,\quad q^2 \neq 2, \tag{7.114}$$

$$\sigma_n(\varsigma) = \frac{1}{3}\left(1 - 2\ln\left(1 - |\varsigma|\right)\right)\frac{E}{E_0}\widehat{\sigma}_n^0,\quad q^2 = 2. \tag{7.115}$$

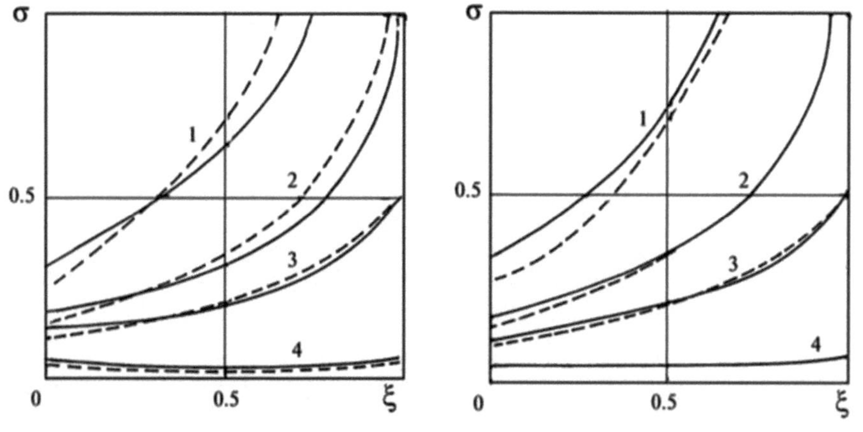

Fig. 7.3 Comparison of the results of the numerical solution of Eq. (7.116) and approximation in Eq. (7.114) (left part: $\delta_1 = 0.1$, right part: $\delta_1 = 0.01$), lines 1–4 corresponds to the parameters $q^2 = 0.8, 1.4, 4, 20$

In order to visualize the difference between the asymptotic solution (7.114), (7.115) and the solution of integral equation (7.59) we consider a model integral equation similar to (7.59) for a fiber in the form of a double cone

$$\alpha_m \sigma(\varsigma) - (T_1\sigma)(\varsigma) = -\frac{1}{2}\delta_1^2 \ln \delta_1 \,, \quad \alpha_m = 1 - \frac{q^2}{2}\delta_1^2 \ln \delta_1 \,, \tag{7.116}$$

where the operator T_1 is defined in (7.97). The principal term $\sigma(\varsigma)$ of the asymptotic solution of this equation has the form (7.115) or (7.115) with $E\widehat{\sigma}_n^0/E_0 = 1/2$.

The results of the numerical solution of Eq. (7.59) are presented in Fig. 7.3 (left figure: $\delta_1 = 0.1$, right figure: $\delta_1 = 0.01$) by solid lines, the dashed line are the functions $\sigma_n(\varsigma)$ in Eq. (7.115). Lines 1–4 correspond to the values of the parameter $q^2 = 0.8, 1.4, 4, 20$. The discrepancies between the solid and dashed lines can be reduced by adding to $\sigma_n(\varsigma)$ functions of boundary layer types localized at the middle $\varsigma = 0$ and at the ends $\varsigma = \pm 1$ of the fiber.

Thus for a constant external field σ^0, the principal part of the stress tensor inside a stiff axisymmetric fiber is its axial component $\sigma_n(\varsigma)$ that is presented the form

$$\sigma_n(\varsigma) = f(\varsigma)\frac{E}{E_0}\widehat{\sigma}_n^0. \tag{7.117}$$

Here, the function $f(\varsigma)$ depends on the shape of the fiber. For the considered fibers (Fig. 7.4), these functions are:

Cylinder:

$$f(\varsigma) = 1 - \frac{\cosh q\varsigma}{\cosh q}\,, \tag{7.118}$$

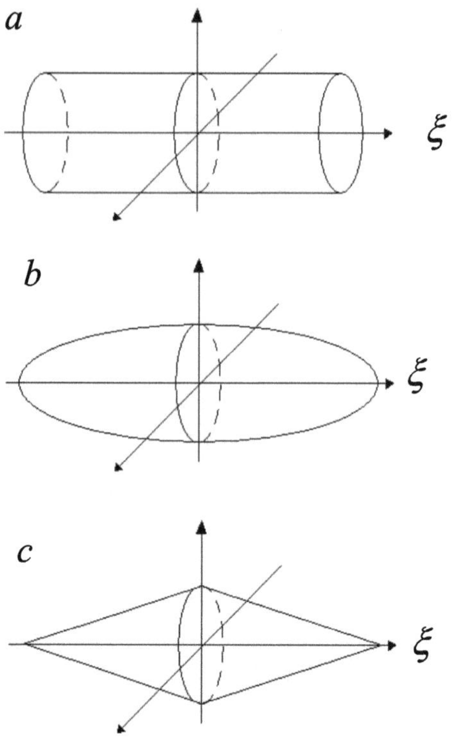

Fig. 7.4 Fibers of the three different forms

Ellipsoid:

$$f(\varsigma) = \frac{q^2}{2+q^2}, \tag{7.119}$$

Double cone:

$$f(\varsigma) = \frac{q^2}{2-q^2}\left[1 - \frac{2}{2+\beta}(1-|\varsigma|)^{\beta}\right], \tag{7.120}$$

$$\beta = \frac{1}{2}\left(\sqrt{1+4q^2} - 3\right), \; q^2 \neq 2; \tag{7.121}$$

$$f(\varsigma) = \frac{1}{3}(1 - 2\ln(1-|\varsigma|)), \; q^2 = 2. \tag{7.122}$$

Note that from Eqs. (7.54) and (7.117) follows the equation

$$\frac{d^2}{d\varsigma^2}\left(\alpha^2(\varsigma)\sigma_n(\varsigma)\right) = -\frac{2\mu_0}{E_0(\delta_1^2 \ln \delta_1)}(f(\varsigma)-1)\,\widehat{\sigma}_n^0. \tag{7.123}$$

This equation allows refining the equation for the transversal component σ^θ for the stress field inside the fiber.

Finally, the principal term of the stress tensor inside the fiber is presented in the form

$$\sigma(\varsigma) = \Lambda(\varsigma, n)\varepsilon^0, \tag{7.124}$$

$$\Lambda(\varsigma, n) = \left(A^0\right)^{-1} + \gamma(\varsigma)\left(P^3(n) + P^4(n)\right) + E_n f(\varsigma) P^6(n). \tag{7.125}$$

Here, the function $f(\varsigma)$ in Eqs. (7.118)–(7.121) depends on the fiber shape,

$$\gamma(\varsigma) = \lambda_0 + \mu_0\left(1 - \frac{\mu}{\lambda + \mu} f(\varsigma)\right). \tag{7.126}$$

7.5 An Axisymmetric Fiber of Large Conductivity in a Homogeneous Conductive Medium

In the case of the conductivity problem, the analogy of a stiff elastic fiber is a fiber which conductivity is larger than the conductivity of the host medium. The integral equation (4.192) for the current vector $j(x)$ inside the fiber can be presented in the form

$$C^{-1} j(x) + \int_V K(x - x') C^1 C^{-1} j(x') \mathrm{d}x' = \left(C^0\right)^{-1} j^0(x). \tag{7.127}$$

Here, V is the region occupied by the fiber, C is the conductivity tensor of the fiber material, C^0 is the same for the host medium, $C^1 = C - C^0$. As in the elastic case, the principal term of the current is constant in each fiber cross section. Therefore, in the Cartesian coordinate system (x_1, x_2, z) with the origin at the fiber center and the z-axis directed along its symmetry axis, we have

$$j(x_1, x_2, z) \approx j(z), \ -l < z < l, \tag{7.128}$$

where $2l$ is the fiber length.

We introduce dimensionless coordinates $\zeta_i = x_i/l, \ i = 1, 2; \ \varsigma = z/l$ and the dimensionless fiber radius $\alpha(\varsigma)$ by the equation

$$a(\varsigma) = \delta_1 l \alpha(\varsigma), \ \delta_1 \ll 1, \ \alpha(\varsigma) = O(1). \tag{7.129}$$

In the coordinates $\overline{\zeta}(\zeta_1, \zeta_2)$ and ς, the integral equation (7.127) on the fiber axis takes the form

$$C^{-1}j(\varsigma)+\int_{-1}^{1}d\varsigma'\int_{\Omega(\varsigma')}K(-\overline{\zeta}',\varsigma-\varsigma')j(\varsigma')d\Omega'd\varsigma'=\left(C^0\right)^{-1}j^0(\varsigma). \tag{7.130}$$

Here, we assume that $C^0C^{-1}=O(\delta_2)$, where δ_2 is a small parameter, and therefore, $\left(C-C^0\right)C^{-1}=I+O(\delta_2)$, $\Omega(\varsigma)$ is the area of the fiber cross section with coordinate ς. The integral over fiber cross section $\Omega(\varsigma')$ in Eq. (7.130) is calculated as follows

$$\begin{aligned}\widehat{K}(\overline{\zeta},\varsigma')&=\int_{\Omega(\varsigma')}K(-\overline{\zeta}',\varsigma)d\overline{\zeta}'=\\&=\frac{1}{(2\pi)^3}\int\widetilde{K}(\overline{k},k_3)\exp\left(-ik_3\varsigma\right)dk_3d\overline{k}\int_{\Omega(\varsigma')}\exp(i\overline{k}\cdot\overline{\zeta}')d\overline{\zeta}'.\end{aligned} \tag{7.131}$$

Here, integration over the Fourier parameters $\overline{k}, k_3$ is performed from $-\infty$ to ∞. The integral over the fiber cross section $\Omega(\varsigma')$ is

$$\frac{1}{2\pi}\int_{\Omega(\varsigma')}\exp(i\overline{k}\cdot\overline{\zeta}')d\overline{\zeta}'=\delta_1\alpha(\varsigma')\frac{J_1(\delta_1\alpha(\varsigma')|\overline{k}|)}{|\overline{k}|}, \tag{7.132}$$

where $J_1(z)$ is the Bessel function of the first kind. For an isotropic host medium $C^0_{\alpha\beta}=c_0\delta_{\alpha\beta}$, and $\widetilde{K}(\overline{k},k_3)$ has the form

$$\widetilde{K}_{\alpha\beta}(\overline{k},k_3)=\frac{1}{c_0}m_\alpha m_\beta,\ m_\alpha=\frac{k_\alpha}{\sqrt{|\overline{k}|^2+k_3^2}},\ |\overline{k}|^2=k_1^2+k_2^2. \tag{7.133}$$

After integrating over $\overline{k}$ in Eq. (7.131), we obtain the non-zero components of the tensor $\widehat{K}_{\alpha\beta}(\varsigma,\varsigma')$ in the forms

$$\widehat{K}_{11}(\varsigma,\varsigma')=\widehat{K}_{22}(\varsigma,\varsigma')=\frac{1}{2\pi c_0}\int_{-\infty}^{\infty}\widetilde{T}(k_3,\varsigma')e^{-ik_3\varsigma}dk_3, \tag{7.134}$$

$$\widehat{K}_{33}(\varsigma,\varsigma')=\frac{1}{2\pi c_0}\int_{-\infty}^{\infty}\left(\frac{1}{2}-\widetilde{T}(k_3,\varsigma')\right)e^{-ik_3\varsigma}dk_3. \tag{7.135}$$

Here, the explicit equation for $\widetilde{T}(k_3,\varsigma')$ is

$$\widetilde{T}(k_3,\varsigma')=\delta_1\alpha(\varsigma')|k_3|\mathrm{K}_1\left(\delta_1\alpha(\varsigma')|k_3|\right), \tag{7.136}$$

where $K_1(z)$ is the modified Bessel function of the first kind. After calculation of the integral in Eq. (7.134), we obtain finally

$$\widehat{K}_{11}(\varsigma,\varsigma') = \frac{1}{c_0} T(\varsigma,\varsigma'),\ \widehat{K}_{33}(\varsigma,\varsigma') = \frac{1}{c_0}\left(\delta(\varsigma) - T(\varsigma,\varsigma')\right), \tag{7.137}$$

$$T(\varsigma,\varsigma') = \frac{\delta_1^2\alpha(\varsigma')^2}{2\left(\delta_1^2\alpha(\varsigma')^2 + \varsigma^2\right)^{3/2}}. \tag{7.138}$$

Note that the integral of the function $T(\varsigma,\varsigma')$ with respect to ς over the interval $(-\tau,\tau)$, $\tau > 0$ is evaluated as follows

$$\int_{-\tau}^{\tau} \frac{\delta_1^2\alpha(\varsigma')^2}{2\left(\delta_1^2\alpha(\varsigma')^2 + \varsigma^2\right)^{3/2}} d\varsigma = 1 + o(\delta_1). \tag{7.139}$$

It means that the kernel $T(\varsigma,\varsigma')$ contains Dirac's delta-function and other terms vanish when $\delta_1 \to 0$ or

$$(Tj)(\varsigma) = \int_{-1}^{1} T(\varsigma - \varsigma', \varsigma') j(\varsigma') d\varsigma' = j(\varsigma) + o(\delta_1). \tag{7.140}$$

For isotropic fiber material $C_{\alpha\beta} = c\delta_{\alpha\beta}$, Eq. (7.130) takes the form

$$\frac{1}{c} j(\varsigma) + \int_{-1}^{1} \widehat{K}(\varsigma - \varsigma', \varsigma') j(\varsigma') d\varsigma' = \frac{1}{c_0} j^0(\varsigma), \quad -1 < \varsigma < 1. \tag{7.141}$$

The equations for the components of the vector $j_\alpha(\varsigma)$ follow from this equation

$$\frac{1}{c} j_\alpha(\varsigma) + \frac{1}{c_0}\int_{-1}^{1} T(\varsigma - \varsigma', \varsigma') j_\alpha(\varsigma') d\varsigma' = \frac{1}{c_0} j_\alpha^0(\varsigma),\ \alpha = 1, 2; \tag{7.142}$$

$$\left(\frac{1}{c} + \frac{1}{c_0}\right) j_n(\varsigma) - \int_{-1}^{1} T(\varsigma - \varsigma', \varsigma') j_n(\varsigma') d\varsigma' = \frac{1}{c_0} j_n^0(\varsigma). \tag{7.143}$$

Here, $j_n(\varsigma) = j_3(\varsigma)$ is the current along the fiber symmetry axis, $j_n^0(\varsigma) = j_3^0(\varsigma)$ is the component of the external current in the medium. Taking into account the property (7.140) of the operator T, we obtain

$$\frac{1}{c} + \frac{1}{c_0} j_\alpha(\varsigma) + o(\delta_1) = \frac{1}{c_0} j_\alpha^0(\varsigma),\ \alpha = 1, 2. \tag{7.144}$$

Because $c^{-1}c_0 = O(\delta_2)$ is small, the principal terms of $j_\alpha(\varsigma)$ with respect to the parameters δ_1, δ_2 are

$$j_\alpha(\varsigma) \approx j_\alpha^0(\varsigma), \ \alpha = 1, 2. \tag{7.145}$$

For Eq. (7.143), both terms in the left-hand side are small, and therefore,

$$j_n(\varsigma) = O(\delta_2^{-1} j_n^0) \gg j_n^0 \tag{7.146}$$

Thus $j_n(\varsigma)$ is the principal component of the current vector inside the fiber of large conductivity.

In order to render the principal terms of the asymptotic solution for the current $j_n(\varsigma)$ inside the fiber, we expand the Fourier transform $\widetilde{T}(k_3, \varsigma')$ of the kernel $T(\varsigma, \varsigma')$ in series with respect to the small parameter δ_1

$$\widetilde{T}(k_3, \varsigma') = 1 + \frac{1}{2}\delta_1^2 \ln(\delta_1)\alpha^2(\varsigma')k_3^2 + O(\delta_1^2). \tag{7.147}$$

For multiplication with k_3^2 in the Fourier transform space corresponds to the operator $-d^2/d\varsigma^2$ in the original space, the principal terms of $j_n(\varsigma)$ satisfy the differential equation

$$\frac{1}{c} j_n(\varsigma) + \frac{1}{2c_0}\delta_1^2 \ln(\delta_1)\frac{d^2}{d\varsigma^2}\left(\alpha^2(\varsigma) j_n(\varsigma)\right) = \frac{1}{c_0} j_n^0(\varsigma). \tag{7.148}$$

This equation can be written in the form

$$\frac{d^2}{d\varsigma^2}\left(\alpha^2(\varsigma) j_n(\varsigma)\right) d\varsigma' - q^2 j_n(\varsigma) = -q^2 \frac{c}{c_0} j_n^0(\varsigma), \tag{7.149}$$

$$q^2 = -\frac{2c_0}{c\delta_1^2 \ln(\delta_1)} = O\left(\frac{\delta_2}{\delta_1^2 \ln(\delta_1)}\right). \tag{7.150}$$

The general solution of Eq. (7.149) depends on 2 arbitrary constants. For determination of these constants, we substitute this solution in the left-hand side of Eq. (7.143)

$$\left(\frac{c_0}{c} + 1\right) j_n(\varsigma) - (T j_n)(\varsigma) = j_n^0(\varsigma) + R(j_n, \varsigma). \tag{7.151}$$

Here, $R(j_n, \varsigma)$ is the discrepancy of the left and right sides of this equation. Then, the values of the constants have to be taken to minimize $R(j_n, \varsigma)$. Similar to the case of elastic fibers, the principal component of the current inside the fiber can be presented in the form

$$j_n(\varsigma) = \frac{c}{c_0} f(\varsigma) j_n^0(\varsigma) = c f(\varsigma) e_n^0(\varsigma), \tag{7.152}$$

where the function $f(\varsigma)$ depends on the shape of the fiber, and for the fibers in the form of a cylinder, ellipsoid, and double cone, this functions are defined in Eqs. (7.118)–(7.122).

7.6 Notes

The first approximate solution for the elastic medium reinforced with a stiff cylindrical fiber was proposed in [3], where simplifications typical for the technical theory of elastic systems were accepted. A more accurate approach was presented in [4, 5]. Some mistakes in these works were corrected in [6].

The method described in this chapter differs from the mentioned approaches, and it is based on the works [7, 8].

References

1. Abramowitz M, Stigan I (1964) Handbook of mathematical functions. Nat Buro Stand, 1046
2. Kondratiev V (1967) Boundary value problems for regions with conical and angle points. Proc Moscow Math Soc 16:209–292
3. Cox H (1952) The elasticity and strength of paper and other fibrous materials. Brit J Appl Phys 3:72–79
4. Cherepanov G (1983) Fracture mechanics of composite materials. Nauka, Moscow, p 296
5. Nikishkov G, Cherepanov G (1984) Tension of the elastic space with an isolated rigid fiber. Appl Math Mech (PMM) 48:460–465
6. Eshelby (1982) The stress on and in a thin inextensible fiber in a stretched elastic medium. Eng Fracture Mech 16:453–460
7. Kanaun S (1987) Static fields in an homogeneous medium disturbed by an inclusion in the form of a curvilinear fiber. Appl Math Mech (PMM) 51:293–304
8. Kanaun S (1988) Equilibrium of a homogeneous elastic medium reinforced with a rectilinear rigid fiber. Appl Math Mech (PMM) 52:789–800

Chapter 8
A Radially Transverse Isotropic Inclusion in an Isotropic Homogeneous Medium

Abstract Radially transverse isotropic inclusions in homogeneous isotropic conductive and elastic media are considered. The Mellin transform method is used for the solution of the volume integral equations of the problems. The method reveals the tensor structure of the solution with precision to three scalar functions of the radial coordinate. The system of ordinary differential equations for these functions arises in the process of realization of the method. For multilayered radially transverse isotropic inclusions with constant elastic and conductive coefficients inside the layers, explicit equations for the fields inside the layers are obtained. An efficient numerical algorithm of solution for solving inclusions with an arbitrary number of transverse isotropic layers is presented.

Keywords Radially transverse isotropic inclusion · Conductive and elastic host media · Volume integral equations · Mellin transform method · Multilayered transverse isotropic inclusions · Numerical algorithms of efficient solutions

8.1 A Radially Transverse Isotropic Inclusion in a Homogeneous Elastic Medium

Let a homogeneous isotropic elastic medium contain an isolated radially transverse isotropic spherical inclusion. For a spherical radially transverse isotropic inclusion, the elastic stiffness tensor $C(x)$ is presented the form

$$C(r,n) = c_1(r)E^1 + c_2(r)E^2 + c_3(r)\left(E^3(n) + E^4(n)\right) +$$

$$+ c_5(r)E^5(n) + c_6(r)E^6(n). \tag{8.1}$$

Here, $r = |x|$ is distance from the inclusion center to point x, $n_\alpha = x_\alpha/r$, $c_k(r)$ are scalar coefficients, and $E^k(n)$ $(k = 1, 2, ..., 6)$ are the elements of the E-basis

$$E^1_{\alpha\beta\lambda\mu} = \delta_{\alpha\lambda}\delta_{\beta\mu}|_{(\alpha\beta)}\,, \quad E^2_{\alpha\beta\lambda\mu} = \delta_{\alpha\beta}\delta_{\lambda\mu}\,, \quad E^3_{\alpha\beta\lambda\mu} = \delta_{\alpha\beta}n_\lambda n_\mu\,,$$

S. Kanaun, *Self-Consistent Methods for Composites*, Solid Mechanics and Its Applications 282, https://doi.org/10.1007/978-3-031-96739-9_8

$$E^4_{\alpha\beta\lambda\mu} = n_\alpha n_\beta \delta_{\lambda\mu}, \quad E^5_{\alpha\beta\lambda\mu} = n_\alpha \delta_{\beta\mu} n_\lambda|_{(\alpha\beta)(\lambda\mu)}, \tag{8.2}$$

$$E^6_{\alpha\beta\lambda\mu} = n_\alpha n_\beta n_\lambda n_\mu .$$

This basis is closed with respect to the product operation, and the multiplication table of the tensors E^i is presented in Appendix A.1.

In the local Cartesian coordinate system with the x_3-axis directed along the vector n, tensor C can be associated with a 6×6 matrix c_{ij} (Voigt's notations for elastic stiffness tensors, [1])

$$\left[c_{ij}\right] = \begin{bmatrix} c_{11} & c_{12} & c_{13} & 0 & 0 & 0 \\ c_{12} & c_{11} & c_{13} & 0 & 0 & 0 \\ c_{13} & c_{13} & c_{33} & 0 & 0 & 0 \\ 0 & 0 & 0 & c_{44} & 0 & 0 \\ 0 & 0 & 0 & 0 & c_{44} & 0 \\ 0 & 0 & 0 & 0 & 0 & c_{66} \end{bmatrix}. \tag{8.3}$$

The coefficients $c_{ij} = c_{ij}(r)$ of this matrix are expressed in terms of the coefficients $c_i(r)$ in Eq. (8.1) as follows

$$c_{11} = c_1 + c_2, \; c_{44} = \frac{1}{4}(2c_1 + c_5), \; c_{66} = \frac{1}{2}c_1, \tag{8.4}$$

$$c_{12} = c_2, \; c_{13} = c_2 + c_3, \; c_{33} = c_1 + c_2 + 2c_3 + c_5 + c_6. \tag{8.5}$$

Because $C(x)$ is a definitely positive tensor, the matrix c_{ij} is definitely positive also, and the coefficients c_{ij} satisfy the inequalities

$$c_{11} > 0, c_{33} > 0, c_{44} > 0, c_{66} > 0, c_{11}c_{33} - c_{13}^2 > 0,$$

$$(c_{11} + c_{12})c_{33} - 2c_{13}^2 > 0. \tag{8.6}$$

The integral equation (4.12) for the strain tensor $\varepsilon(x)$ in the medium with a radially transverse isotropic inclusion subjected to a constant external strain field ε^0 has the form

$$\varepsilon(x) + \int K(x - x')C^1(x')\varepsilon(x')dx' = \varepsilon^0 \tag{8.7}$$

$$C^1(x) = C(r, n) - C^0. \tag{8.8}$$

For linearity of the problem, the tensor $\varepsilon(x)$ can be presented as the sum

$$\varepsilon(x) = \varepsilon^0 + \varepsilon^1(x), \quad \varepsilon^1(x) = A(x)\varepsilon^0, \tag{8.9}$$

where $A(x)$ is a vanishing at infinity fourth rank tensor: $A(x) \to 0$ as $|x| \to \infty$. Substituting Eq. (8.9) into Eq. (8.7) and taking into account that ε^0 is an arbitrary tensor, we yield the equation for the tensor $A(x)$

$$A(x) + \int_V K(x - x')C^1(x')A(x')dx' + \int_V K(x - x')C^1(x')dx' = 0. \tag{8.10}$$

A unique solution of this equation exists if C^0 and $C(x)$ are not degenerated tensors.

8.2 Special Presentation of the K-Operator

Let $f(r, n)$ be a tensor function of scalar $r = |x|$ and vector $n_\alpha = x_\alpha / r$ arguments. We denote by $f^*(s, n)$ the Mellin transform of $f(r, n)$ with respect to the variable r [2]

$$f^*(s, n) = \int_0^\infty r^{s-1} f(r, n)\, dr\,, \quad f(r, n) = \frac{1}{2\pi \mathrm{i}} \int_{\tau - \mathrm{i}\infty}^{\tau + \mathrm{i}\infty} r^{-s} f(s, n)\, ds\,. \tag{8.11}$$

In terms of the Mellin transform, the Fourier transform $(Ff)(k)$ of a function $f(r, n)$ is presented in the form ($m_\alpha = k_\alpha / |k|$)

$$(Ff)(k) = \int f(r, n) \exp(\mathrm{i}k \cdot x) dx = \frac{1}{2\pi \mathrm{i}} e^{\mathrm{i}\frac{\pi}{2}(d-s)} \Gamma(d - s)\, |k|^{s-d} \times$$

$$\times \int_{\tau - \mathrm{i}\infty}^{\tau + \mathrm{i}\infty} ds \int_{\Omega_1(d)} f^*(s, n)\, ((n \cdot m) + \mathrm{i}0)^{s-d} d\Omega_n,\ m_\alpha = \frac{k_\alpha}{|k|}. \tag{8.12}$$

Here, d is the dimension of the $x(k)$-space, and in what follows $d = 2, 3$; $\Omega_1(d)$ is the surface of the unit sphere in the space of the dimension d, $\Gamma(s)$ is the Euler Gamma function. To prove Eq. (8.12) we consider the Fourier transform of the function $f(r, n)$ using the Mellin transform inversion formula (8.11)

$$(Ff)(k) = \int_0^\infty r^{d-1} dr \int_{\Omega_1(d)} e^{\mathrm{i}r|k|(m \cdot n)} d\Omega_n \frac{1}{2\pi \mathrm{i}} \int_{\tau - \mathrm{i}\infty}^{\tau + \mathrm{i}\infty} r^{-s} f^*(s, n)\, ds\,. \tag{8.13}$$

To justify the change of the integration order in this equation, we introduce the parameter $\varepsilon > 0$ and present $(Ff)(k)$ as follows

$$(Ff)(k) = \lim_{\varepsilon \to 0} \int_0^\infty r^{d-1} dr \int_{\Omega_1(d)} e^{\mathrm{i}r|k|(m \cdot n) - r\varepsilon} d\Omega \frac{1}{2\pi \mathrm{i}} \int_{\tau - \mathrm{i}\infty}^{\tau + \mathrm{i}\infty} r^{-s} f^*(s, n)\, ds =$$

$$= \lim_{\varepsilon \to 0} \frac{1}{2\pi \mathrm{i}} \int_{\tau - \mathrm{i}\infty}^{\tau + \mathrm{i}\infty} ds \int_{\Omega_1(d)} f^*(s, n)\, d\Omega_n \int_0^\infty e^{\mathrm{i}|k| r (m \cdot n) - \varepsilon r} r^{d-s-1} dr\,. \tag{8.14}$$

The last integral in this equation is calculated explicitly

$$\int_0^\infty e^{\mathrm{i}|k| r (m \cdot n) - \varepsilon r} r^{d-s-1} dr = e^{\mathrm{i}\frac{\pi}{2}(d-s)} \left(|k|\,(n \cdot m) + \mathrm{i}\varepsilon\right)^{s-d} \Gamma(d-s)\,, \tag{8.15}$$

and for the function $(Ff)(k)$ we yield the equation

$$(Ff)(k) = \lim_{\varepsilon \to 0} \frac{1}{2\pi \mathrm{i}} \int_{\tau - \mathrm{i}\infty}^{\tau + \mathrm{i}\infty} e^{\mathrm{i}\frac{\pi}{2}(p-s)} \Gamma(d-s)\, |k|^{s-d} ds \times$$

$$\times \int_{\Omega_1(p)} f(s, n) \left((n \cdot m) + \mathrm{i}\varepsilon\right)^{s-p} d\Omega_n =$$

$$= \frac{1}{2\pi \mathrm{i}} \int_{\tau - i\infty}^{\tau + i\infty} e^{\mathrm{i}\frac{\pi}{2}(d-s)} \Gamma(d-s)\, |k|^{s-d} ds \times$$

$$\times \int_{\Omega_1(d)} f^*(s, n) \left((n \cdot m) + \mathrm{i}0\right)^{s-d} d\Omega_n\,, \tag{8.16}$$

where

$$\left((n \cdot m) + \mathrm{i}0\right)^{\mu} = \lim_{\varepsilon \to 0} \left((n \cdot m) + \mathrm{i}\varepsilon\right)^{\mu}. \tag{8.17}$$

The left-hand side of Eq. (8.16) is a generalized function of vector m on the unit sphere $\Omega_1(d)$. Hence, Eq. (8.12) is proved.

Then, we consider the integral operator K in Eq. (8.7) which symbol $\widetilde{K}(k)$ (the Fourier transform of $K(x)$) is a homogeneous function of the order of 0. Using the convolution properties, we obtain

$$(Kf)(x) = \int K\left(x - x'\right) f\left(x'\right) dx' = \left(F^{-1}\left(\widetilde{K}(k)\, Ff(k)\right)\right)(x)\,. \tag{8.18}$$

Here, F and $F^{-1}$are the direct and inverse Fourier transform operators (Appendix B.2). After introducing the parameter $\varepsilon > 0$ and using Eq. (8.12), the right-hand side of Eq. (8.18) is presented as follows

$$(Kf)(x) = \lim_{\varepsilon \to +0} (2\pi)^{-d} \int_0^{\infty} d|k| \int_{\Omega_1(p)} e^{-\mathrm{i}|k|r(n \cdot m) - r\varepsilon} |k|^{d-1} \widetilde{K}(m)\, d\Omega_m \times$$

$$\times \frac{1}{2\pi\mathrm{i}} \int_{\tau - i\infty}^{\tau + i\infty} \Gamma(d-s)\, e^{\mathrm{i}\frac{\pi}{2}(d-s)} |k|^{s-d} ds \int_{\Omega_1(p)} f^*(s,l)\, ((m \cdot l) + \mathrm{i}0)^{s-d} dl\,. \quad (8.19)$$

Changing the order of integration in this equation, we yield

$$(Kf)(x) = \lim_{\varepsilon \to 0} \frac{1}{(2\pi)^d} \int_{\tau - i\infty}^{\tau + i\infty} ds \int_{\Omega_1(p)} \widetilde{K}(m)\, \Gamma(p-s)\, e^{\mathrm{i}\frac{\pi}{2}(d-s)} d\Omega_m \times$$

$$\times \int_{\Omega_1(d)} f^*(s,l)\, ((l \cdot m) + \mathrm{i}0)^{s-d} d\Omega_l \int_0^{\infty} e^{-\mathrm{i}r|k|(n \cdot m) - \varepsilon|k|} |k|^{s-1} d|k|\,. \quad (8.20)$$

Here, the last integral is similar to the integral in Eq. (8.15)

$$\int_0^{\infty} e^{\mathrm{i}r|k|(n \cdot m) - \varepsilon|k|} |k|^{s-1} d|k| = e^{\mathrm{i}\frac{\pi}{2}s}\, \Gamma(s)\, (-r\,(n \cdot m) + \mathrm{i}\varepsilon)^{-1}. \quad (8.21)$$

Transition to the limit $\varepsilon \to +0$ in Eq. (8.20) results in presentation of the integral operator K in terms of its symbol $\widetilde{K}(m)$ and the Mellin transform $f^*(s,n)$ of the integrated function

$$(Kf)(r,n) = \frac{1}{2\pi\mathrm{i}} \int_{\tau - \mathrm{i}\infty}^{\tau + \mathrm{i}\infty} r^{-s} \left(K^s f^*\right)(s,n)\, ds, \quad (8.22)$$

$$\left(K^s f^*\right)(s,n) = \frac{e^{\mathrm{i}\frac{\pi}{2}d}}{(2\pi)^d} \Gamma(d-s)\, \Gamma(s) \times$$

$$\times \int_{\Omega_1(d)} (-n \cdot m + \mathrm{i}0)^{-s} \widetilde{K}(m)\, d\Omega_m \int_{\Omega_1(d)} (m \cdot l + \mathrm{i}0)^{s-d} f^*(s,l)\, d\Omega_l\,. \quad (8.23)$$

8.3 The Mellin Transform Method

In this section, the Mellin transform method is used for the solution of the volume integral equation (8.7) for an isotropic host medium containing a spherical radially transverse isotropic inclusion. For an isotropic medium, the tensor $\widetilde{K}(m)$ in Eq. (8.23) has the form (Appendix B.1)

$$\widetilde{K}(k) = \frac{1}{\mu_0}(E^5(m) - \kappa_0 E^6(m)),\ \kappa_0 = \frac{\lambda_0 + \mu_0}{\lambda_0 + 2\mu_0},\ m_\alpha = \frac{k_\alpha}{|k|}. \tag{8.24}$$

First we consider action of the operator K^s in Eq. (8.23) on the elements of the E-basis in Eqs. (8.2). Because of the following formulas for the integrals over unit sphere Ω_1 (Re$s < 1$)

$$\int_{\Omega_1} (n \cdot m)^{-s} d\Omega_m = I(s),\ I(s) = 2\pi \frac{1 + e^{-i\pi s}}{1-s}, \tag{8.25}$$

$$\int_{\Omega_1} (n \cdot m)^{-s}\, m_\alpha m_\beta d\Omega_m = \frac{I(s)}{3-s}\left(\delta_{\alpha\beta} - s n_\alpha n_\beta\right), \tag{8.26}$$

$$\int_{\Omega_1} (n \cdot m)^{-s} E^6(m) d\Omega_m = \frac{I(s)}{(3-s)(5-s)}\left(E^2 + 2E^1 - \right.$$

$$\left. - s\left(E^3(n) + E^4(n) + 4E^5(n)\right) + s(2+s)E^6(n)\right), \tag{8.27}$$

actions of the operator K^s on the elements of the E-basis are defined by the equations

$$K^s E^1 = \frac{1}{\mu_0(3-s)(5-s)}\left(T^1 + (1-\kappa_0)T^3\right), \tag{8.28}$$

$$K^s E^2 = \frac{1-\kappa_0}{\mu_0} \cdot \frac{1}{(3-s)} T^2, \tag{8.29}$$

$$K^s E^3 = \frac{1-\kappa_0}{\mu_0 s(3-s)(5-s)}\left((5-s)T^2 - (3-s)T^3\right), \tag{8.30}$$

$$K^s E^4 = -\frac{(1-\kappa_0)(2-s)}{\mu_0 s(3-s)} T^2, \tag{8.31}$$

$$K^s E^5 = -\frac{1}{2\mu_0 s(3-s)(5-s)}\left((1-s)T^1 + 2(1-\kappa_0)(2-s)T^3\right), \tag{8.32}$$

$$K^s E^6 = \frac{(s-2)}{\mu_0 s(2+s)(3-s)(5-s)} \left(2T^1 + \right.$$

$$\left. +(1-\kappa_0) \left((5-s)T^2 - (3-1)T^3\right)\right). \tag{8.33}$$

Here, T^1, T^2, T^3 are the linearly independent fourth rank tensors

$$T^1(s,n) = (5-s)(E^1 - sE^5(n)) - T^3(s,n), \tag{8.34}$$

$$T^2(s,n) = E^2 - sE^4(n), \tag{8.35}$$

$$T^3(s,n) = E^2 + 2E^1 - s\left(E^3(n) + E^4(n) + 4E^5(n)\right) +$$

$$+ s(s+2)E^6(n). \tag{8.36}$$

Note that these tensors are the eigen-elements of the operator K^s

$$K^s T^1 = \frac{1}{2\mu_0} T^1, \quad K^s T^2 = \frac{1-\kappa_0}{\mu_0} T^2, \quad K^s T^3 = \frac{1-\kappa_0}{\mu_0} T^3. \tag{8.37}$$

Because E-basis (8.2) is closed with respect to the product operation, the solution $A(r,n)$ of Eq. (8.10) can be presented as a linear combination of the tensors of the E-basis with scalar coefficients that depend on the radial coordinate r. As a result, the product $C^1 A$ under the integral in Eq. (8.10) is presented in the similar form

$$(C^1 A)(r,n) = \sum_{i=1}^{6} S_i(r) E^i(n), \tag{8.38}$$

where $S_i(r)$ are scalar functions. After substituting this equation into Eq. (8.10) and applying the Mellin transform operator, we yield

$$A^*(s,n) + \sum_{i=1}^{6} S_i^*(s)(K^s E^i)(s,n) + (K^s C^{1*})(s,n) = 0, \tag{8.39}$$

$$C^{1*}(s,n) = (c_1 - 2\mu_0)^*(s)E^1 + (c_2 - \lambda_0)^*(s)E^2 +$$

$$+ c_3^*(s)(E^3(n) + E^4(n)) + c_5^*(s)E^5(n) + c_6^*(s)E^6(n). \tag{8.40}$$

Here, $S_i^*(s)$ ($i = 1, ..., 6$) are the Mellin transforms of the scalar coefficients $S_i(r)$ in Eq. (8.38), $c_m^*(s)$ are the Mellin transforms of the elastic coefficients in Eq. (8.1). For Eqs. (8.28)–(8.33), the tensors $K^s E^i$ and $K^s C^{1*}$ are linear combinations of three linearly independent tensors T^j in Eqs. (8.34)–(8.36). Therefore, tensor $A^*(s,n)$ can be presented as a linear combination of these three tensors

$$A^*(s,n)=\sum_{j=1}^{3}T^j(s,n)\,\alpha_j^*(s). \tag{8.41}$$

Here, $\alpha_j^*(s)$ $(j=1,2,3)$ are the Mellin transforms of scalar functions $\alpha_j(r)$. Because the multiplier $(-s)$ in the s-space of the Mellin transforms corresponds to the differential operation D in the r-space

$$D=r\frac{d}{dr}, \tag{8.42}$$

the r-presentation of the tensor $A(r,n)$ follows from Eqs. (8.28)–(8.33) and (8.41) in the form

$$\begin{aligned}A(r,n)&=\left(E^1+E^5(n)D\right)(D+5)\alpha_1(r)+\left(E^2+E^4(n)D\right)\alpha_2(r)+\\&+\left(E^2+2E^1+\left(E^3(n)+E^4(n)+4E^5(n)\right)D+\right.\\&\left.+E^6(n)\,D\,(D-2)\right)(\alpha_3(r)-\alpha_1(r)).\end{aligned} \tag{8.43}$$

After substituting Eq. (8.41) into Eq. (8.39), we yield the equation which left hand side is a linear combination of the tensors T^1, T^2, T^3. Equating the coefficients of the tensors T^1 and T^3 to zero, we obtain the system for the functions $\alpha_1^*(s)$ and $\alpha_3^*(s)$:

$$2\mu_0 s(s+2)(s-3)(s-5)\alpha_1^*(s)+\Phi_1^*(s)+F_1^*(s)=0\,, \tag{8.44}$$

$$\frac{2\mu_0}{1-\kappa_0}s(s+2)(s-3)(s-5)\alpha_3^*(s)+\Phi_3^*(s)+F_3^*(s)=0. \tag{8.45}$$

In these equations, $\Phi_1^*(s)$ and $\Phi_3^*(s)$ are functions of the Mellin transforms of $\alpha_1(r)$, $\alpha_3(r)$ only

$$\Phi_1^*(s)=2s(s+2)S_1^*(s)-(s+2)(1-s)S_5^*(s)-4(2-s)S_6^*(s)\,, \tag{8.46}$$

$$\begin{aligned}\Phi_3^*(s)&=2s(s+2)S_1^*(s)+2(s+2)(s-3)S_3^*(s)+2(s^2-4)S_5^*(s)+\\&+2(s-1)(s-2)S_6^*(s)\,.\end{aligned} \tag{8.47}$$

The functions $F_1^*(s)$ and $F_3^*(s)$ are expressed in terms of the Mellin transforms $c_m^*(s)$ of the elastic coefficients $c_m(r)$

$$\begin{aligned}F_1^*(s)&=2s(s+2)\,(c_1-2\mu_0)^*\,(s)+\\&+(s+2)(s-1)c_5^*(s)-4(s+2)c_6^*(s)\,,\end{aligned} \tag{8.48}$$

$$F_3^*(s) = 2s^2((c_1 - 2\mu_0)^*(s) + c_3^*(s) + c_5^*(s) + c_6^*(s)) +$$

$$+ 2s(2(c_1 - 2\mu_0)^*(s) - 2c_3^*(s) - 6c_6^*(s)) +$$

$$+ 12c_3^*(s) + 8c_5^*(s) - 4c_6^*(s)) \,. \tag{8.49}$$

In Eqs. (8.46), (8.47), $S_i^*(s)$ are the Mellin transforms of the coefficients $S_i(r)$ in Eq. (8.38). After substitution of Eq. (8.43) into Eq. (8.38), we obtain these coefficients in the forms

$$S_1(r) = (c_1(r) - 2\mu_0)\,((D+3)\alpha_1(r) + 2\alpha_3(r))\,, \tag{8.50}$$

$$S_3(r) = (c_1(r) - 2\mu_0)\,D(\alpha_3(r) - \alpha_1(r)) +$$

$$+ (c_2(r) - \lambda_0)D(D+5)\alpha_3(r) +$$

$$+ c_3(r)(D+1)(3\alpha_1(r) + (D+2)\alpha_3(r))\,, \tag{8.51}$$

$$S_5(r) = (c_1(r) - 2\mu_0)\,D((D+1)\alpha_1(r) + 4\alpha_3(r)) +$$

$$+ \frac{1}{2}c_5(r)((6 + D(D+2))\alpha_1(r) + 4(D+1)\alpha_3(r)), \tag{8.52}$$

$$S_6(r) = (c_1(r) - 2\mu_0)D(D-2)(\alpha_3(r) - \alpha_1(r)) +$$

$$+ c_3(r)D(5+D)\alpha_3(r) + \frac{1}{2}c_5(r)((3-D)\alpha_1(r) +$$

$$+ 2D(1+D)\alpha_3(r)) + c_6(r)(D+1)(3\alpha_1(r) + (D+2)\alpha_3(r)). \tag{8.53}$$

Equating to zero the coefficient of the tensor T^2 in Eq. (8.39) for $A^*(s,n)$ results in an equation for the function $\alpha_2(r)$. But instead of α_2, it is convenient to introduce the function

$$\beta(r) = 3\alpha_2(r) + (5+D)\alpha_3(r)\,. \tag{8.54}$$

The equation for $\beta^*(s)$ follows from Eq. (8.39) after multiplying it with the tensor E^2 and taking into account Eq. (8.54). As a result, we obtain the equation that contains the function $\beta^*(s)$ only

$$\frac{\mu_0}{1-\kappa_0}s(s-3)\beta^*(s) + \Phi_2^*(s) + F_2^*(s) = 0\,, \tag{8.55}$$

$$\Phi_2^*(s) = sS_2^*(s) + (s-2)S_4^*(s), \tag{8.56}$$

$$F_2^*(s) = s((c_1 - 2\mu_0)^*(s) + 3(c_2 - \lambda_0)^*(s)) +$$

$$+ 2(2s + 3)c_3^*(s) + (s + 2)(c_5^*(s) + c_5^*(s)). \quad (8.57)$$

Here, $S_2^*(s)$, $S_4^*(s)$ are the Mellin transforms of the functions

$$S_2(r) = (c_1(r) - 2\mu_0 + 3(c_2(r) - \lambda_0) + c_3(r) +$$

$$+ (c_2(r) - \lambda_0 + c_3(r))D)\beta(r), \quad (8.58)$$

$$S_4(r) = (3c_3(r) + c_5(r) + c_6(r) +$$

$$+ (c_1(r) - 2\mu_0 + c_3(r) + c_5(r) + c_6(r))D)\beta(r). \quad (8.59)$$

After applying the inverse Mellin transform operator to Eqs. (8.46), (8.47) and (8.55), we yield differential equations for the functions $\alpha_1, \alpha_3, \beta$ in the r-presentation. By this transition, the Mellin transforms of the functions should be changed with their originals and the parameter s with the differential operator $-D$ in Eq. (8.42).

In terms of the scalar functions $\alpha_1(r), \alpha_3(r), \beta(r)$, the strain $\varepsilon(x)$ and stress $\sigma(x)$ tensors in the isotropic medium with a radially transverse isotropic inclusion subjected to a constant external strain ε^0 (stress $\sigma^0 = C^0\varepsilon^0$) field takes the forms

$$\varepsilon(x) = \Lambda^{\varepsilon}(x)\varepsilon^0, \; \Lambda^{\varepsilon}(x) = \sum_{j=1}^{6} \epsilon_j(r)E^j(n), \quad (8.60)$$

$$\sigma(x) = \Lambda^{\sigma}(x)\sigma^0, \; \Lambda^{\sigma}(x) = C(x)\Lambda^{\varepsilon}(x)\left(C^0\right)^{-1}, \quad (8.61)$$

where the scalar coefficients $\epsilon_j(r)$ are

$$\epsilon_1 = 1 + (3 + D)\alpha_1 + 2\alpha_3, \; \epsilon_2 = \frac{1}{3}\left(\beta - 3\alpha_1 - (2 + D)\alpha_3\right), \quad (8.62)$$

$$\epsilon_3 = D(\alpha_3 - \alpha_1), \quad \epsilon_4 = \frac{1}{3}D(\beta - 3\alpha_1 - (2 + D)\alpha_3), \quad (8.63)$$

$$\epsilon_5 = D(4\alpha_3 + (1 + D)\alpha_1), \; \epsilon_6 = D(2 - D)(\alpha_1 - \alpha_3). \quad (8.64)$$

Note that for hydrostatic external strain field $\varepsilon_{ij}^0 = \varepsilon_0\delta_{ij}$ (ε_0 is a scalar), the strain tensor $\varepsilon(x)$ takes the form

$$\varepsilon_{ij}(x) = \left((1+\beta(r))\delta_{ij} + n_i n_j D\beta(r)\right)\varepsilon_0. \tag{8.65}$$

Thus, the only function $\beta(r)$ is defined for elastic fields in the medium with a radially transverse isotropic inclusion by hydrostatic loading.

Further, the elastic constants $c_k(r)$ in Eq. (8.1) are changed with the dimensionless constants $c_k(r)/c_1^0$, $c_1^0 = 2\mu_0$, where μ_0 is the shear modulus of the host medium. Thus, for the host medium we have $c_1 = 1, c_2 = \nu_0/(1-2\nu_0)$, where ν_0 is Poisson's ratio of the host medium.

8.4 A Multilayered Radially Transverse Isotropic Inclusion in Homogeneous Isotropic Elastic Medium

Let the coefficients $c_j(r)$ in Eq. (8.1) for the elastic stiffness tensor of the inclusion be piece-wise constant functions of r with jumps at points $r = a_i$, $i = 0, 1, 2..., N$, $(a_0 = 0 < a_1 < ... < a_N)$. In this case, the inclusion consists of a spherical core and $N-1$ spherical layers with constant elastic coefficients $c_j^{(i)}$, $j = 1, 2, 3, 5, 6$, and $i = 1, 2, ..., N$ (Fig. 8.1). Because in each layer the derivatives $Dc_k^{(i)}$ vanish, the functions $\alpha_1, \alpha_3, \beta$ inside layers satisfy ordinary differential equations with constant coefficients. These equations are considered in the next subsections.

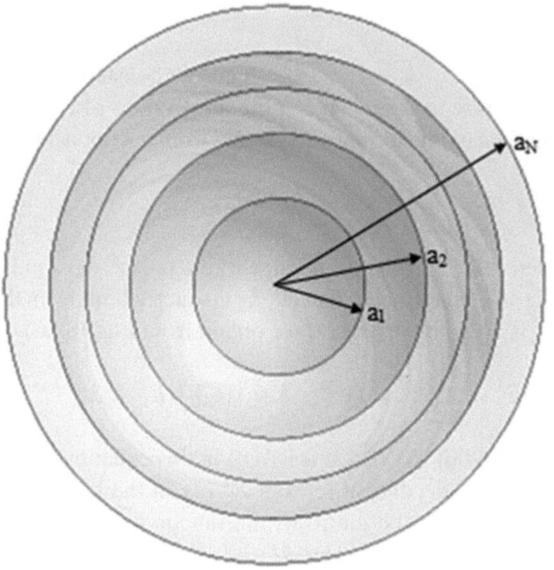

Fig. 8.1 A multilayered inclusion

8.4.1 Conditions for the Jumps of the Functions $\alpha_1(r)$, $\alpha_2(r)$ and $\beta(r)$ at the Layer Boundaries

Let the layers of a radial transverse isotropic inclusion be perfectly bonded along the boundaries Ω_i of the layers. Normal vector n to the surface Ω_i coincides with the direction of the isotropy axis of the materials of both layers. Let the elastic stiffness tensors of the ith and $(i+1)$th layers be $C^{(i)}(n)$ and $C^{(1+i)}(n)$. It follows from the conditions for strain tensor jumps at the boundary of two elastic media (Sect. 4.2, Eq. (4.53)) that the limit values of the tensor $\varepsilon(x)$ at point $x \in \Omega_i$ from the side of the normal $\varepsilon^{(1+i)}(n)$ and from the opposite side $\varepsilon^{(i)}(n)$ relate by the equation

$$\varepsilon^{(1+i)}(n) = (E^1 - \widetilde{K}^{(1+i)}(n)\,[C]_i)\varepsilon^{(i)}(n), \;\; [C]_i = C^{(1+i)}(n) - C^{(i)}(n). \tag{8.66}$$

Here, the tensor $\widetilde{K}^{(1+i)}(k)$ is the Fourier transform of the function $K(x)$ in Eqs. (5.81), (5.81) for the transverse isotropic medium with the elastic stiffness tensor $C^{(i+1)}(n)$. The explicit expression for this tensor is

$$\widetilde{K}^{(1+i)}(n) = \frac{1}{k_1}E^5(n) - \frac{k_2}{k_1(k_1+k_2)}E^6(n) \tag{8.67}$$

$$k_1 = \frac{1}{4}\left(2c_1^{(1+i)} + c_5^{(1+i)}\right), \tag{8.68}$$

$$k_2 = \frac{1}{2}c_1^{(1+i)} + c_2^{(1+i)} + 2c_3^{(1+i)} + \frac{3}{4}c_5^{(1+i)} + c_6^{(1+i)}. \tag{8.69}$$

Here, $c_k^{(i+1)}$ $(k = 1, ..., 5)$ are elastic constants of the $(i+1)$th layer. For radially transverse isotropic layers, the limit values of the strain tensor $\varepsilon(x)$ at point x from the side of the normal n to Ω_i and from the opposite side are defined by the equations

$$\varepsilon^{(j)} = (E^1 + A^{(j)})\varepsilon^0, \; j = i, 1+i. \tag{8.70}$$

Here, $A^{(1+i)}$ is the limit value of the tensor $A(x)$ in Eq. (8.43) at point $x \in \Omega_i$ from the side of normal n, and $A^{(i)}(x)$ is such a limit from the opposite side of Ω_i. Because ε^0 is an arbitrary tensor, we obtain from Eqs. (8.66)–(8.70)

$$E^1 + A^{(1+i)} = (E^1 - \widetilde{K}(n)\,[C]_i)(E^1 + A^{(i)}). \tag{8.71}$$

Substituting Eq. (8.43) for $A(x)$ in this equation and equating the coefficients of the linearly independent tensors $E^m(n)$ in the left- and right-hand sides, we yield six equations for the jumps of the functions $\alpha_1(r)$, $\alpha_2(r)$, $\beta(r)$, their first and second derivatives at the surface Ω_i

$$[\alpha_1]_i = [D\alpha_1]_i = 0, \;\; [\alpha_3]_i = [D\alpha_3]_i = 0, \tag{8.72}$$

$$\left[(2c_1 + c_5)D^2\alpha_1\right]_i =$$

$$= -[2c_1 + c_5]_i(2 + 3(2\alpha_1(a_i) + D\alpha_1(a_i)) + 4(\alpha_3(a_i) + D\alpha_3(a_i)), \tag{8.73}$$

$$\left[b_0 D^2\alpha_3\right]_i =$$

$$= -[c_1 + c_3 + c_5 + c_6]_i(1 + 3(\alpha_1(a_i) + D\alpha_1(a_i)) + 2\alpha_3(a_i)) -$$

$$- [3b_0 + 2c_2 + 2c_3]_i D\alpha_3(a_i), \tag{8.74}$$

$$[\beta]_i = 0,\ [b_0 D\beta]_i + [b_1 + b_3]_i(1 + \beta^{(i)}(a_i)) = 0. \tag{8.75}$$

Here,

$$b_0 = c_1 + c_2 + 2c_3 + c_5 + c_6,\ b_1 = 3c_3 + c_5 + c_6, \tag{8.76}$$

$$b_2 = c_2 + c_3,\ \ b_3 = c_1 + 3c_2 + c_3. \tag{8.77}$$

Symbol $[f]_i$ means the jump of $f(r)$ at the ith layer boundary

$$[f]_i = f(a_i + 0) - f(a_i - 0),\ f(a_i \pm 0) = \lim_{\delta \to 0} f(a_i \pm \delta), \tag{8.78}$$

and $\delta > 0$.

8.4.2 *Radially Transverse Isotropic Inclusion Subjected to a Hydrostatic External Field*

1. Differential equation for the function $\beta(r)$.
Let external hydrostatic pressure p_0 be applied to the medium with a radially transverse isotropic inclusion. The corresponding external strain tensor is

$$\varepsilon^0_{\alpha\beta} = \frac{p_0}{k_0}\delta_{\alpha\beta},\ k_0 = \lambda_0 + \frac{2}{3}\mu_0, \tag{8.79}$$

and the strain field $\varepsilon_{ij}(x)$ in the medium with the inclusion depends on a scalar function $\beta(r)$ only

$$\varepsilon_{\alpha\beta}(x) = \frac{p_0}{k_0}\left(\delta_{\alpha\beta} + 3(\delta_{\alpha\beta} + n_\alpha n_\beta D)\beta(r)\right). \tag{8.80}$$

The differential equation for the function $\beta(r)$ inside the ith layer follows from Eqs. (8.55)–(8.57) in the form

$$b_0^{(i)} D(3+D)\beta + 2b_1^{(i)}\beta + 2b_1^{(i)} = 0, i = 1, 2, ..., N, \tag{8.81}$$

$$b_0^{(i)} = c_1^{(i)} + c_2^{(i)} + 2c_3^{(i)} + c_5^{(i)} + c_6^{(i)}, \; b_1^{(i)} = 3c_3^{(i)} + c_5^{(i)} + c_6^{(i)}. \tag{8.82}$$

Equation (8.81) is of Euler's type, and its general solution is

$$\beta(r) = y_1^{(i)} r^{\lambda_1(i)} + y_2^{(i)} r^{\lambda_2(i)} - 1, \;\; a_{i-1} < r < a_i, i = 1, 2, ..., N+1. \tag{8.83}$$

Here, $y_1^{(i)}$, $y_2^{(i)}$ are arbitrary constants, and $\lambda_1(i)$, $\lambda_2(i)$ are the roots of the algebraic equation

$$b_0^{(i)} \lambda(3+\lambda) + 2b_1^{(i)} = 0, \tag{8.84}$$

$$\lambda_1(i) = -\frac{3}{2}\left(1 - \sqrt{1 - \frac{8b_1^{(i)}}{9b_0^{(i)}}}\right), \; \lambda_2(i) = -\frac{3}{2}\left(1 + \sqrt{1 - \frac{8b_1^{(i)}}{9b_0^{(i)}}}\right). \tag{8.85}$$

In terms of the elements c_{ij} in the matrix presentation of the tensor $C(x)$ in Eq. (8.1), the expression under the root in these equations is

$$\Delta^{(i)} = 1 - \frac{8b_1^{(i)}}{9b_0^{(i)}} = 1 - \frac{8\left(c_{33}^{(i)} - c_{11}^{(i)} + c_{13}^{(i)} - c_{12}^{(i)}\right)}{9c_{33}^{(i)}}. \tag{8.86}$$

Numerical search of minimum of $\Delta^{(i)}$ in the region of possible values of c_{ij} restricted by the inequalities (8.6) shows that $\Delta^{(i)} \geq 0$, and therefore, the roots $\lambda_1(i)$, $\lambda_2(i)$ are real numbers. The root $\lambda_2(i)$ is always negative but $\lambda_1(i)$ can be positive or negative and

$$-\frac{3}{2} < \lambda_1(i) < \infty, \;\; -\infty < \lambda_2(i) < -\frac{3}{2}. \tag{8.87}$$

Two branches of the roots of Eq. (8.85) as functions of the ratio b_1/b_0 are shown in Fig. 8.2.

In the case of an isotropic ith layer $b_1^{(i)} = 0$, and Eq. (8.84) is simplified

$$\left(c_1^{(i)} + c_2^{(i)}\right) D(3+D)\beta = 0. \tag{8.88}$$

Its general solution can be taken in the form similar to Eq. (8.83)

$$\beta(r) = y_1^{(i)} + y_2^{(i)} r^{-3} - 1, a_{i-1} < r < a_i, \tag{8.89}$$

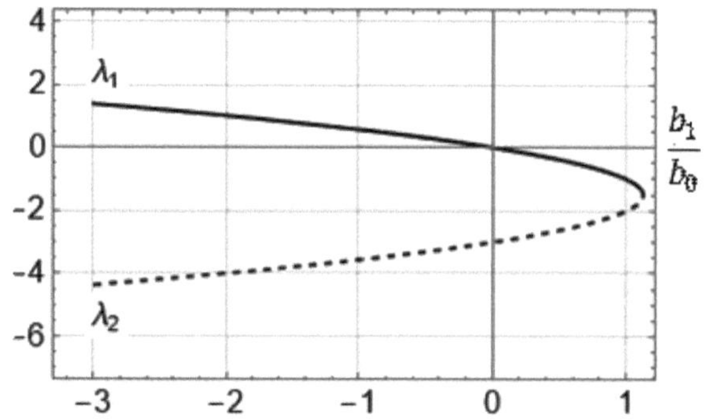

Fig. 8.2 Two branches of the solution of Eq. (8.84)

where $y_1^{(i)}$, $y_2^{(i)}$ are arbitrary constants.

2. *Jumps of the function $\beta(r)$ at the layer boundaries.*
Jumps of the function $\beta(r)$ and its first derivative at the layer boundaries are defined in Eq. (8.75) that is a consequence of the general conditions for the strain tensor discontinuity at the boundary of two elastic media. In this section, we show that these conditions follow also from Eqs. (8.55)–(8.57) for the Mellin transform of the function $\beta(r)$.

For multilayered inclusions with perfectly bonded layers, the strain tensor $\varepsilon(x)$ has no singularities of the delta-function type at the layer boundaries, and therefore, the function $\beta(r)$ in Eq. (8.65) is continuous. The derivative $D\beta(r)$ can be discontinuous at $r = a_i$, and in order to determine the jumps of $D\beta(r)$, we consider Eqs. (8.55)–(8.57) for the Mellin transform $\beta^*(s)$ of the function $\beta(r)$. This equation can be written in the form

$$s(s-2)\beta^* + (s-2)\,(b_1\beta)^* + s\,(b_3\beta)^* + s\,(b_2 D\beta)^* -$$

$$-\,(s-2)\,(b_4 D\beta)^* + s(b_2+b_3)^* - 2b_1^* = 0, \tag{8.90}$$

where

$$b_1 = 3c_3 + c_5 + c_6,\;\; b_2 = c_2 + c_3,\, b_3 = c_1 + 3c_2 + c_3, \tag{8.91}$$

$$b_4 = c_1 + c_3 + c_5 + c_6. \tag{8.92}$$

Here, $b_m = b_m(r)$ $(m = 1, 2, 3, 4)$ are piece-wise constant functions of r with jumps at the layer boundaries, $b_m^* = b_m^*(s)$ are the Mellin transforms of these functions. Using integration by parts, the Mellin transforms of a piece-wise constant functions $b_m(r)$ and $b_m(r)D\beta(r)$ in Eq. (8.90) can be presented in the forms

$$b_m^*(s) = -\frac{1}{s}\sum_{i=1}^{N}[b_m]_i\, a_i^s, \tag{8.93}$$

$$(b_m D\beta)^*(s) = -\sum_{i=1}^{N}[b_m]_i\, \beta(a_i)a_i^s - s\,(b_m\beta)^*(s). \tag{8.94}$$

Taking into account these equations, Eq. (8.60) is transformed as follows

$$((s+\lambda_1)(s+\lambda_2)b_0\beta)^*(s) +$$

$$+\sum_{i=1}^{N}\left((s\,[b_0]_i + 2\,[b_4]_i)\beta(a_i) + [b_3]_i + \left(1+\frac{2}{s}\right)[b_1]_i\right)a_i^s = 0. \tag{8.95}$$

Here, $\lambda_j = \lambda_j(r) = \lambda_j(i)$ for $a_{i-1} < r < a_i$, $i = 1, 2, ..., N+1$, and $\lambda_j(i)$ for $j = 1, 2$ are defined in Eq. (8.85). In this equation, the function $((s+\lambda_1)(s+\lambda_2)b_0\beta)^*(s)$ is the following integral

$$((s+\lambda_1)(s+\lambda_2)b_0\beta)^*(s) =$$

$$= \int_0^\infty (s+\lambda_1(r))(s+\lambda_2(r))b_0(r)\beta(r)r^{s-1}dr =$$

$$= \sum_{i=1}^{N+1} b_0^{(i)}(s+\lambda_1(i))(s+\lambda_2(i))\int_{a_{i-1}}^{a_i}\beta(r)r^{s-1}dr. \tag{8.96}$$

It follows from Eq. (8.83) that the functions $\beta(r)$ and $D\beta(r)$ in the ith and $(i+1)$th layers have the forms

$$\beta(r) = y_1^{(j)}r^{\lambda_1(i)} + y_2^{(j)}r^{\lambda_2(j)} - 1,\ \ a_{j-1} < r < a_j \tag{8.97}$$

$$D\beta(r) = \lambda_1(j)y_1^{(j)}r^{\lambda_1(j)} + \lambda_2(j)y_2^{(j)}r^{\lambda_2(j)},\ \ j = i, i+1. \tag{8.98}$$

Calculating the integrals in Eq. (8.96), we obtain

$$((s+\lambda_1)(s+\lambda_2)b_0\beta)^* = \sum_{i=1}^{N}\left(b_0^{(i)}\Psi^{(i)}(s) - b_0^{(i+1)}\Psi^{(i+1)}(s)\right)a_i^s, \tag{8.99}$$

$$\Psi^{(i)}(s) = (s+\lambda_1(i))y_1^{(i)}a_i^{\lambda_1(i)} +$$

$$+ (s+\lambda_2(i))y_2^{(i)}a_i^{\lambda_2(i)} - \frac{(s+\lambda_2(i))(s+\lambda_2(i)}{s}. \tag{8.100}$$

Then, we denote

$$\beta^{(i)}(a_i) = \beta(a_i - 0), \quad \beta^{(i+1)}(a_i) = \beta(a_i + 0), \tag{8.101}$$

$$D\beta^{(i)}(a_i) = D\beta(a_i - 0), D\beta^{(i+1)}(a_i) = D\beta(a_i + 0). \tag{8.102}$$

It follows from Eqs. (8.97), (8.98) that the constants $y_1^{(j)}$ and $y_2^{(j)}$ are expressed in terms of $\beta^{(j)}(a_i)$ and $D\beta^{(j)}(a_i)$ $(j = i, i+1)$ as follows

$$y_1^{(j)} = \frac{a_i^{-\lambda_1(j)}}{\lambda_1(j) - \lambda_2(j)} \left(D\beta^{(j)}(a_i) - \lambda_2(j)(1 + \beta^{(j)}(a_i))\right), \tag{8.103}$$

$$y_2^{(j)} = \frac{a_i^{-\lambda_2(j)}}{\lambda_1(j) - \lambda_2(j)} \left(-D\beta^{(j)}(a_i) + \lambda_1(j)(1 + \beta^{(j)}(a_i))\right). \tag{8.104}$$

For these $y_{1,2}^{(i)}$ and $y_{1,2}^{(i+1)}$, Eq. (8.99) takes the form

$$((s + \lambda_1)(s + \lambda_2)b_0\beta)^* =$$

$$= \sum_{i=1}^{N} \frac{1}{s}(s(s-3)\,[b_0\beta]_i + 2[b_1]_i + s[b_0 D\beta]_i)a_i^s, \tag{8.105}$$

$$[b_0\beta]_i = b_0^{(i+1)}\beta^{(i+1)}(a_i) - b_0^{(i)}\beta^{(i)}(a_i), \tag{8.106}$$

$$[b_0 D\beta]_i = b_0^{(i+1)} D\beta^{(i+1)}(a_i) - b_0^{(i)} D\beta^{(i)}(a_i). \tag{8.107}$$

After substituting Eq. (8.105) into Eq. (8.95) and equating to zero the coefficients of linearly independent functions a_i^s $(i = 1, 2, ..., N)$, we yield the equations

$$s^2\left([b_0\beta]_i - [b_0]_i\beta^{(i)}\right) + s([b_0 D\beta]_i + [b_1 + b_3]_i(1 + \beta^{(i)}(a_i))) = 0. \tag{8.108}$$

This equation holds for any s, and thus the coefficients of s and s^2 should be equal to zero. As a result, we obtain the conditions for the jumps of the functions β and $D\beta$ at the layer boundaries

$$[b_0\beta]_i - [b_0]_i\beta^{(i)} = 0 \text{ or } [\beta]_i = 0, \tag{8.109}$$

$$[b_0 D\beta]_i + [b_1 + b_3]_i(1 + \beta^{(i)}(a_i)) = 0, i = 1, 2, ..., N. \tag{8.110}$$

These conditions coincide with the conditions (8.75) that follow from the general equations for the discontinuity of the strain tensor at the boundary of two elastic media.

3. Numerical algorithm of construction of the function $\beta(r)$ for multilayered inclusions

We introduce the vector of solution $\mathbf{x}^{(i)}(r)$ which components $x_1^{(i)}(r), x_2^{(i)}(r)$ inside each ith layer are

$$x_1^{(i)}(r) = \beta(r) + 1 = y_1^{(i)} r^{\lambda_1(i)} + y_2^{(i)} r^{\lambda_2(i)}, \tag{8.111}$$

$$x_2^{(i)}(r) = D\beta(r) = \lambda_1(i) y_1^{(i)} r^{\lambda_1(i)} + \lambda_2(i) y_2^{(i)} r^{\lambda_2(i)}. \tag{8.112}$$

If $\mathbf{x}^{(i)}(r)$ is known at $r = a_{i-1}$ (the left end of the ith interval a_{i-1}, a_i), its value at the right end ($r = a_i$) follows from Eqs. (8.97), (8.98)

$$\mathbf{x}^{(i)}(a_i) = \mathbf{r}^{(i)} \mathbf{x}^{(i)}(a_{i-1}), \ \mathbf{x}^{(i)}(r) = \begin{bmatrix} x_1^{(i)}(r) \\ x_2^{(i)}(r) \end{bmatrix}, \quad i = 2, 3, ...N, \tag{8.113}$$

$$\mathbf{r}^{(i)} = \frac{1}{\lambda_1 - \lambda_2} \begin{bmatrix} \lambda_1 \rho^{\lambda_2} - \lambda_2 \rho^{\lambda_1}, & \rho^{\lambda_1} - \rho^{\lambda_2} \\ \lambda_1 \lambda_2 \left(\rho^{\lambda_2} - \rho^{\lambda_1}\right), & \lambda_1 \rho^{\lambda_1(i)} - \lambda_2 \rho^{\lambda_2} \end{bmatrix}, \tag{8.114}$$

$$\rho = \rho(i) = \frac{a_i}{a_{i-1}}, \ \lambda_1 = \lambda_1(i), \ \lambda_2 = \lambda_2(i). \tag{8.115}$$

The matrix $\mathbf{r}^{(i)}$ performs transition of the solution through the ith interval (layer). Relation between the vectors $\mathbf{x}^{(i)}(a_i)$ and $\mathbf{x}^{(i+1)}(a_i)$ at point a_i of the boundary of ith and $(i+1)$th layers follows from conditions (8.109), (8.110) for the jumps of the functions $\beta(r)$ and $D\beta(r)$ at $r = a_i$. This relation can be presented in the matrix form

$$\mathbf{x}^{(i+1)}(a_i) = \gamma^{(i)} \mathbf{x}^{(i)}(a_i), i = 1, 2, ..., N, \tag{8.116}$$

$$\gamma^{(i)} = \begin{bmatrix} 1, & 0 \\ -\frac{[b_1]_i + [b_3]_i}{b_0^{(i+1)}}, & \frac{b_0^{(i)}}{b_0^{(i+1)}} \end{bmatrix}. \tag{8.117}$$

The algorithm of calculation of the constants $y_1^{(i)}, y_2^{(i)}$ in Eq. (8.83) inside layers follows from Eqs. (8.113), (8.116) and the conditions for $\beta(r)$ at $r = 0$ and $r \to \infty$. Inside the core ($i = 1$), the constant $y_2^{(1)}$ is equal to zero because the solution $y_2^{(i)} r^{\lambda_2(i)}$ can be non-integrable at $r = 0$ for possible values of the elastic coefficients (Fig. 8.2). Thus, in the core, $\beta(r)$ has the form

$$\beta(r) = y_1^{(1)} r^{\lambda_1(1)} - 1, \ 0 < r < a_1. \tag{8.118}$$

Then, using transition matrices $\mathbf{r}^{(i)}$ and relations (8.116) between $\mathbf{x}^{(i)}(r)$ and $\mathbf{x}^{(i+1)}(r)$ at $r = a_i$, we present the solution in the $(N+1)$th layer (in the host medium) in the form

$$\mathbf{x}^{(N+1)}(a_N) = \mathbf{g}^N \mathbf{x}^{(1)}(a_1), \ \mathbf{x}^{(1)}(a_1) = \begin{bmatrix} y_1^{(1)} a_1^{\lambda_1(1)} \\ \lambda_1(1) y_1^{(1)} a_1^{\lambda_1(1)} \end{bmatrix}, \tag{8.119}$$

$$\mathbf{g}^N = \gamma^{(N)}\mathbf{r}^{(N)}\gamma^{(N-1)}\mathbf{r}^{(N-1)}...\gamma^{(1)}\mathbf{r}^{(1)} = \prod_{k=0}^{N-1} \gamma^{(N-k)}\mathbf{r}^{(N-k)}. \tag{8.120}$$

Here, the transition matrices $\mathbf{r}^{(i)}$ are defined in Eq. (8.114) for $i > 1$, and $\mathbf{r}^{(1)}$ is unit matrix. Because the host medium is isotropic, in the $(N+1)$th layer we have from Eq. (8.89)

$$\beta(r) = y_1^{(N+1)} + y_2^{(N+1)} r^{-3} - 1, \quad a_N < r < \infty. \tag{8.121}$$

For the function $\beta(r)$ should vanish at infinity, $y_1^{(N+1)} = 1$ and vector $\mathbf{x}^{(N+1)}(a_N)$ takes the form

$$\mathbf{x}^{(N+1)}(a_N) = \begin{bmatrix} x_1^{(N+1)}(a_N) \\ x_2^{(N+1)}(a_N) \end{bmatrix} = \begin{bmatrix} 1 + y_2^{(N+1)} a_N^{-3} \\ -3y_2^{(N+1)} a_N^{-3} \end{bmatrix}. \tag{8.122}$$

Therefore, the components of the vector $\mathbf{x}^{(N+1)}(a_N)$ relate by the equation

$$x_2^{(N+1)}(a_N) + 3\left(x_1^{(N+1)}(a_N) - 1\right) = 0. \tag{8.123}$$

Substituting in this equation $x_1^{(N+1)}(a_N)$, $x_2^{(N+1)}(a_N)$ from Eqs. (8.119), we yield the equation for the constant $y_1^{(1)}$in Eq. (8.118) for the function $\beta(r)$ in the core. After solution of this equation, the vector $\mathbf{x}^{(i+1)}(a_i)$ for the $(1+i)$th layer is calculated from the equation that is similar to Eq. (8.119)

$$\mathbf{x}^{(i+1)}(a_i) = \mathbf{g}^i \mathbf{x}^{(1)}(a_1), i = 2, 3, ..., N. \tag{8.124}$$

Then, the vectors of constants $\mathbf{y}^{(i)}$ $(i = 2, 3, ..., N+1)$ for each layer can be found from the equations

$$\mathbf{y}^{(i)} = \begin{bmatrix} y_1^{(i)} \\ y_2^{(i)} \end{bmatrix} = \mathbf{m}^{(i)}(a_i)\mathbf{x}^{(i)}(a_i), \quad \mathbf{x}^{(i)}(a_i) = \mathbf{r}^{(i)}\mathbf{x}^{(i)}(a_{i-1}), \tag{8.125}$$

$$\mathbf{m}^{(i)}(r) = \frac{1}{\lambda_1(i) - \lambda_2(i)} \begin{bmatrix} -\lambda_2(i) r^{-\lambda_1(i)}, & r^{-\lambda_1(i)} \\ \lambda_1(i) r^{-\lambda_2(i)}, & -r^{-\lambda_2(i)} \end{bmatrix}. \tag{8.126}$$

8.4.3 Functions $\alpha_1(r)$ and $\alpha_3(r)$ for Multilayered Inclusions

1. The ordinary differential equations for the functions $\alpha_1(r)$, $\alpha_3(r)$ inside layers
The system of differential equations for the functions $\alpha_1(r)$ and $\alpha_3(r)$ in the layers follows from the system (8.44), (8.45) for the Mellin transforms of these functions.

Further, index i for indication of the constants c_k in the ith layer is omitted. Changing in Eqs. (8.44), (8.45) the functions $\alpha_1^*(s)$ and $\alpha_3^*(s)$ with their originals and the parameter s with the differential operator $-D$, we yield the system of ordinary differential equations for $\alpha_1(r)$ and $\alpha_3(r)$ inside layers ($a_{i-1} < r < a_i, i = 1, 2, ..., N+1$)

$$P_{11}(D)\alpha_1 + P_{12}(D)\alpha_3 = F_1, \tag{8.127}$$

$$P_{21}(D)\alpha_1 + P_{22}(D)\alpha_3 = F_2, \tag{8.128}$$

where

$$P_{11}(D) = \frac{c_1}{2}D(D-2)(D+3)(D+5) +$$

$$+\frac{c_5}{4}(D(D+3)(D(D+3)-12)-12) - 6c_6(D+1)(D+2), \tag{8.129}$$

$$P_{12}(D) = -c_5(D(D+5)+2)(D+1) -$$

$$-2(D+2)(c_6(D+1)(D+2) + c_3D(D+5)), \tag{8.130}$$

$$P_{21}(D) = 3c_3(D-2)(D+1)(D+3) +$$

$$+\frac{3}{2}(D+2)(2c_6(D+1)^2 + c_5(D^2+D-4)), \tag{8.131}$$

$$P_{22}(D) = (c_1+c_2)(D-2)(D+3)(D+5)D +$$

$$+2c_3(D+1)(D+2)(D(D+3)-3) +$$

$$+c_5(D^2-1)(D+2)(D+4) + c_6(D^2+3D+2)^2, \tag{8.132}$$

$$F_1 = c_5 + 4c_6, \quad F_2 = 2(3c_3 + 2c_5 + c_6). \tag{8.133}$$

The system (8.127), (8.128) is of Euler's type, and the solution of the corresponding homogeneous system (for $F_1 = F_2 = 0$) can be presented in the form

$$\alpha_1(r) = Yr^\lambda, \alpha_3(r) = Zr^\lambda, \tag{8.134}$$

where Y, Z, λ are constants. After substitution, these expressions into homogeneous system of Eqs. (8.127), (8.128), we get two linear algebraic equations for Y and Z

$$P_{11}(\lambda)Y + P_{12}(\lambda)Z = 0, \tag{8.135}$$

$$P_{21}(\lambda)Y + P_{22}(\lambda)Z = 0. \tag{8.136}$$

A non-trivial solution of this system exists if its determinant $Q(\lambda)$ is equal to zero

$$Q(\lambda) = P_{11}(\lambda)P_{22}(\lambda) - P_{12}(\lambda)P_{21}(\lambda) = 0. \tag{8.137}$$

It follows from Eqs. (8.129)–(8.132) that $Q(\lambda)$ is a polynomial of the order of 8,

$$Q(\lambda) = \lambda(\lambda-2)(\lambda+3)(\lambda+5)Q^{(1)}(\lambda), \tag{8.138}$$

$$Q^{(1)}(\lambda) = B_0\lambda^4 - B_1\lambda^3 + B_2\lambda^2 - B_3\lambda + B_4. \tag{8.139}$$

Here, the coefficients $B_m = B_m^{(i)}$ $(m = 0, 1, 2, 3, 4)$ are expressed in terms of the elastic constants $c_k = c_k^{(i)}$ of the ith layer

$$B_0 = (2c_1 + c_5)(c_1 + c_2 + 2c_3 + c_5 + c_6), \quad B_1 = 6B_0, \tag{8.140}$$

$$B_2 = -2c_1^2 + 24c_3^2 + 36c_3c_5 + 11c_5(c_5 + c_6) +$$

$$+ c_1^2(-2c_1 + 32c_3 + 11c_5 + 2c_6) - 3c_2(c_5 + 8c_6), \tag{8.141}$$

$$B_3 = -6(10c_1^2 - 12c_3^2 - c_5(9c_3 + c_5 + c_6) + 6c_2(c_5 + 2c_6) +$$

$$+ 2c_1(5c_2 + c_3 + 4(c_5 + c_6))), \tag{8.142}$$

$$B_4 = 4(12c_3^2 + 9c_3c_5 - 3c_2(c_5 + 4c_6) - c_1(6c_3 + 7c_5 + 10c_6)). \tag{8.143}$$

The algebraic equation (8.137) has 8 roots, the first four roots do not depend on the elastic constants of the layers

$$\lambda_1 = 0,\ \lambda_2 = 2,\ \lambda_3 = -3,\ \lambda_4 = -5. \tag{8.144}$$

Other roots $(\lambda_5, \lambda_6, \lambda_7, \lambda_8)$ are the solutions of the algebraic equation of the fourth order

$$B_0\lambda^4 - B_1\lambda^3 + B_2\lambda^2 - B_3\lambda + B_4 = 0. \tag{8.145}$$

Here, the values of the coefficients B_m in Eqs. (8.140)–(8.143) are restricted by the inequalities (8.6). In the region of possible values of $c_k^{(i)}$, the roots of Eq. (8.145) can be real or complex numbers. Because the coefficients B_m $(m = 0, 1, 2, 3, 4)$ in Eq. (8.145) are real, two complex conjugate roots or two pairs of complex conjugate roots can arise only. In the case of two complex conjugate roots $\lambda = \gamma \pm i\delta$, the corresponding constants Y and Z in the solution (8.107) of the homogeneous system (8.135), (8.136) should be also complex conjugate numbers to guarantee real values of the functions $\alpha_1(r)$, $\alpha_3(r)$. Thus for $Y = Y_1 \pm iY_2$, $Z = Z_1 \pm iZ_2$, contributions of these terms in the function $\alpha_1(r)$, $\alpha_3(r)$ are as follows

$$\alpha_1(r) = r^\gamma \left((Y_1 + \mathrm{i}Y_2)r^{\mathrm{i}\delta} + (Y_1 - \mathrm{i}Y_2)r^{-\mathrm{i}\delta}\right) =$$

$$= 2r^\gamma \left(Y_1 \cos(\delta \ln r) - Y_2 \sin(\delta \ln r)\right), \tag{8.146}$$

$$\alpha_3(r) = r^\gamma \left((Z_1 + \mathrm{i}Z_2)r^{\mathrm{i}\delta} + (Z_1 - \mathrm{i}Z_2)r^{-\mathrm{i}\delta}\right) =$$

$$= 2r^\gamma \left(Z_1 \cos(\delta \ln r) - Z_2 \sin(\delta \ln r)\right). \tag{8.147}$$

Here, the equation

$$r^{\pm \mathrm{i}\delta} = \exp\left(\pm \mathrm{i}\delta \ln r\right) = \cos(\delta \ln r) \pm \mathrm{i}\sin(\delta \ln r)] \tag{8.148}$$

is taken into account. Further, the roots λ_5, .., λ_8 are numerated as follows

$$\mathrm{Re}\lambda_5 \geq \mathrm{Re}\lambda_6 \geq \mathrm{Re}\lambda_7 \geq \mathrm{Re}\lambda_8. \tag{8.149}$$

Let λ_j $(j = 1, 2, ..., 8)$ be one of the roots of Eq. (8.145). It follows from Eqs. (8.135), (8.136) that the constants Y_j and Z_j relate by the equation

$$Z_j = g_j Y_j,\ g_j = -\frac{P_{11}(\lambda_j)}{P_{12}(\lambda_j)} = -\frac{P_{21}(\lambda_j)}{P_{22}(\lambda_j)},\ j = 1, 2, ..., 8. \tag{8.150}$$

Because F_1, F_2 on the right-hand sides of Eqs. (8.127), (8.128) are constant inside each layer, a partial solution $(\widehat{\alpha}_1, \widehat{\alpha}_3)$ of this system is constant also. For constant $\widehat{\alpha}_1, \widehat{\alpha}_3$, the system (8.127), (8.128) is reduced to two identical equations

$$3\widehat{\alpha}_1 + 2\widehat{\alpha}_3 + 1 = 0, \tag{8.151}$$

and $\widehat{\alpha}_1 = 0$, $\widehat{\alpha}_3 = -1/2$ can be taken as a partial solution. Finally, the general solution of the system (8.127), (8.128) takes the form

$$\alpha_1(r) = \sum_{j=1}^{8} Y_j(i) r^{\lambda_j(i)},\ \alpha_3(r) = \sum_{j=1}^{8} g_j(i) Y_j(i) r^{\lambda_j(i)} - \frac{1}{2}, \tag{8.152}$$
$$a_{i-1} < r < a_i,\ i = 1, 2, ..., N.$$

Here, $Y_j(i)$ are arbitrary constants to be determined for each layer in the process of solution.

In the case of an isotropic layer $(c_3 = c_5 = c_6 = 0)$, the system (8.127), (8.128) is split into two independent equations for the functions $\alpha_1(r)$ and $\alpha_3(r)$

$$D(D-2)(D+3)(D+5)\alpha_j = 0\,,\quad j = 1, 3\,. \tag{8.153}$$

The general solutions of these equations inside the ith layer can be taken in the forms similar to Eqs. (8.152)

$$\alpha_1(r) = Y_1(i) + Y_2(i)r^2 + Y_3(i)r^{-3} + Y_4(i)r^{-5}, \tag{8.154}$$

$$\alpha_3(r) = Y_5(i) + Y_6(i)r^2 + Y_7(i)r^{-3} + Y_8(i)r^{-5} - \frac{1}{2}, \tag{8.155}$$

where $Y_j(i)$ are arbitrary constants for the ith laeyr.

2. Jumps of the functions α_1,α_3 at the layer boundaries
We consider discontinuities of the functions α_1, α_3 and their derivatives at the boundaries of the layers. For perfectly bonded boundaries, the tensor $A(r, n)$ in Eq. (8.9) does not contain singular components (delta-functions and their derivatives) concentrated at the inclusion boundaries. As a result, the functions $\alpha_1(r)$ and $\alpha_3(r)$ in Eq. (8.43) for $A(r, n)$ should be continuous together with the first derivatives. But the second and third derivatives of these functions can be discontinuous on the layer boundaries.

Let $c(r)$ be a piece-wise constant function with jumps at the points $r = a_i$, $i = 1, 2, ..., N$, and $c(r) = 0$ when $r > a_N$. The Mellin transform $c^*(s)$ of $c(r)$ has the form

$$c^*(s) = -\frac{1}{s}\sum_{i=1}^{N}[c]_i a_i^s, \tag{8.156}$$

where $[c]_i$ is the jump of the function at the point a_i.Using integration by parts, it can be proved that for a piece-wise constant function $c(r)$ and a continuous function together with the first derivative function $\alpha(r)$, the following equations hold

$$(cD\alpha)^*(s) = -\sum_{i=1}^{N}[c\alpha]_i a_i^s - s(c\alpha)^*(s), \tag{8.157}$$

$$(cD^2\alpha)^*(s) = -\sum_{i=1}^{N}([cD\alpha]_i - s[c\alpha]_i)\, a_i^s + s^2(c\alpha)^*(s). \tag{8.158}$$

Using these equations, Eqs. (8.44)–(8.45) for the Mellin transforms of the functions α_1, α_3 are presented as follows

$$P_{11}^*(\alpha_1) + P_{12}^*(\alpha_3) + \sum_{i=1}^{N} F_1^{(i)}(s) a_i^s = 0, \tag{8.159}$$

$$P_{21}^*(\alpha_1) + P_{22}^*(\alpha_3) + \sum_{i=1}^{N} F_2^{(i)}(s) a_i^s = 0, \tag{8.160}$$

where

$$P_{11}^*(\alpha_1) = \frac{1}{4}(s^4 - 6s^3)((2c_1 + c_5)\alpha_1)^* - \frac{1}{4}s^2((2c_1 + 3c_5 + 24c_6)\alpha_1)^* +$$

$$+ 3s((5c_1 + 3c_5 + 6c_6)\alpha_1)^* - 3((c_5 + 4c_6)\alpha_1)^*, \tag{8.161}$$

$$P_{12}^*(\alpha_3) = s^3((2c_3 + c_5 + c_6)\alpha_3)^* - 2s^2((7c_3 + 3c_5 + 5c_6)\alpha_3)^* +$$

$$+ s((20c_3 + 7c_5 + 16c_6)\alpha_3)^* - 2((c_5 + 4c_6)\alpha_3)^*, \tag{8.162}$$

$$F_1^{(i)}(s) = f_{10}^*(s) + f_{11}^*(s)\alpha_1(a_i) + f_{12}^*(s)D\alpha_1(a_i) +$$

$$+ f_{13}^*(s)\alpha_3(a_i) + f_{14}^*(s)D\alpha_3(a_i), \tag{8.163}$$

$$f_{10}^* = \frac{1}{2s}[2c_5 + 8c_6]_i - \frac{s}{2}[2c_1 + c_5]_i - \frac{1}{2}[4c_1 + c_5 + 4c_6]_i\,. \tag{8.164}$$

$$f_{11}^* = \frac{s^3}{4}[2c_1 + c_5]_i - \frac{3}{2}s^2[2c_1 + c_5]_i -$$

$$- \frac{s}{4}[14c_1 + 9c_5 + 24c_6]_i + \frac{3}{2}[6c_1 + 5c_5 + 8c_6]_i, \tag{8.165}$$

$$f_{12}^* = -\frac{1}{4}(6 + s(3 - s))[2c_1 + c_5]_i\,, \tag{8.166}$$

$$f_{13}^* = s^2[2c_3 + c_5 + 2c_6]_i - s[2c_1 + 14c_3 + 7c_5 + 10c_6]_i -$$

$$- 2[2c_1 - 10c_3 - 3c_5 - 6c_6]_i\,, \tag{8.167}$$

$$f_{14}^* = -2(s - 2)[c_1 + c_3 + c_5 + c_6]_i\,. \tag{8.168}$$

$$P_{21}^*(\alpha_1) = -\frac{3}{2}s^3((2c_3 + c_5 + 2c_6)\alpha_1)^* + \frac{3}{2}s^2((4c_3 + 3c_5 + 8c_6)\alpha_1)^* +$$

$$+ 3s((5c_3 + c_5 - 5c_6)\alpha_1)^* - 6((3c_3 + 2c_5 - c_6)\alpha_1)^*, \tag{8.169}$$

$$P_{22}^*(\alpha_3) = (s^4 - 6s^3)((c_1 + c_2 + 2c_3 + c_5 + c_6)\alpha_3)^* -$$

$$- s^2((c_1 + c_2 - 16c_3 - 7c_5 - 13c_6)\alpha_3)^* +$$

$$+ 6s((5c_1 + 5c_2 + c_3 + c_5 - 2c_6)\alpha_3)^* - 4((3c_3 + c_5 - c_6)\alpha_3)^*, \tag{8.170}$$

$$F_2^{(i)}(s) = f_{20}^*(s) + f_{21}^*(s)\alpha_1(a_i) + f_{22}^*(s)D\alpha_1(a_i) +$$

$$+ f_{23}^*(s)\alpha_3(a_i) + f_{24}^*(s)D\alpha_3(a_i), \tag{8.171}$$

$$f_{20}^* = \frac{2}{s}[3c_3 + c_5 - c_6]_i - s[c_1 + c_3 + c_5 + c_6]_i -$$

$$- [2c_1 - c_3 - 3c_6]_i, \tag{8.172}$$

$$f_{21}^* = -\frac{3}{2}s^2 [2c_3 + c_5 + c_6]_i - \frac{3}{2}s [2c_1 - 2c_3 - c_5 - 6c_6]_i +$$

$$+ 3 [-2c_1 + 6c_3 + c_5 - 2c_6]_i , \tag{8.173}$$

$$f_{22}^* = -\frac{3}{2}(s - 2) [2c_1 + c_5]_i , \tag{8.174}$$

$$f_{23}^* = s^2(s - 6) [c_1 + c_2 + 2c_3 + c_5 + c_6]_i -$$

$$- s [3c_1 + c_2 - 14c_3 - 5c_5 - 11c_6]_i -$$

$$- 2[c_1 - 3c_2 - 2c_3 + c_5 + c_6]_i, \tag{8.175}$$

$$f_{24}^* = -s^2 [c_1 + c_2 + 2c_3 + c_5 + c_6]_i + s [3c_1 + c_2 + 4c_3 + 3c_5 + 3c_6]_i -$$

$$- 2[c_1 - 3c_2 - 2c_3 + c_5 + c_6]_i. \tag{8.176}$$

Then, using Eq. (8.152) for $\alpha_1(r)$, the Mellin transform $(c_m\alpha_1)^*(s)$ of the product $c_m(r)\alpha_1(r)$ is presented in the form

$$(c_m\alpha_1)^* (s) = \sum_{i=1}^{N} c_m^{(i)} \int_{a_{i-1}}^{a_i} \alpha_1(r)r^{s-1}dr =$$

$$= \sum_{i=1}^{N} c_m^{(i)} \sum_{j=1}^{8} \frac{Y_j(i)}{(s + \lambda_j(i))} \left(a_i^{\lambda_j(i)+s} - a_{i-1}^{\lambda_j(i)+s}\right). \tag{8.177}$$

Inside each ith layer, we introduce the eight-dimensional vector $\mathbf{X}^{(i)}(r)$ with the components

$$X_1^{(i)}(r) = \alpha_1(r), \quad X_2^{(i)}(r) = D\alpha_1(r), \tag{8.178}$$

$$X_3^{(i)}(r) = D^2\alpha_1(r), \quad X_4^{(i)}(r) = D^3\alpha_1(r), \tag{8.179}$$

$$X_5^{(i)}(r) = \alpha_3(r) + \frac{1}{2}, \quad X_6^{(i)}(r) = D\alpha_3(r), \tag{8.180}$$

$$X_7^{(i)}(r) = D^2\alpha_3(r)\,,\quad X_8^{(i)}(r) = D^3\alpha_3(r)\,, \tag{8.181}$$
$$a_{i-1} < r < a_i,\quad i = 1, 2, ..., N+1.$$

Vector $\mathbf{Y}^{(i)}$ with the components $Y_j^{(i)}$ $(j = 1, 2, ..., 8)$ and vector $\mathbf{X}^{(i)}(r)$ relate by the equations that follow from Eqs. (8.152)

$$\mathbf{X}^i(r) = \mathbf{M}^{(i)}(r)\mathbf{Y}^{(i)},\quad \mathbf{Y}^{(i)} = \left(\mathbf{M}^{(i)}(r)\right)^{-1}\mathbf{X}^{(i)}(r), \tag{8.182}$$

where $\mathbf{M}^{(i)}(r)$ is the matrix of the dimensions 8×8 with the components

$$\mathbf{M}^{(i)}(r) = \left[M_{pq}^{(i)}(r)\right],\quad p, q = 1, 2, ..., 8; \tag{8.183}$$

$$M_{pq}^{(i)}(r) = \left(\lambda_q(i)\right)^{p-1} r^{\lambda_q(i)},\ p = 1, 2, 3, 4; \tag{8.184}$$

$$M_{pq}^{(i)}(r) = g_q(i)\left(\lambda_q(i)\right)^{p-5} r^{\lambda_q(i)},\ p = 5, 6, 7, 8. \tag{8.185}$$

Because a_i^s $(i = 1, 2, ..., N)$ are linearly independent functions of s, the coefficients of a_i^s in Eqs. (8.159), (8.160) are equal to zero, and these conditions provide the system of equations that relates vectors $\mathbf{X}^{(i+1)}(a_i)$ and $\mathbf{X}^{(i)}(a_i)$ at the layer boundaries $(r = a_i, i = 1, 2, ..., N)$

$$\mathbf{Q}^1(s, i+1)\left(\mathbf{M}^{(i+1)}(a_i)\right)^{-1}\mathbf{X}^{(i+1)}(a_i) =$$

$$= \mathbf{Q}^1(s, i)\left(\mathbf{M}^{(i)}(a_i)\right)^{-1}\mathbf{X}^{(i)}(a_i) + F_1(s, i), \tag{8.186}$$

$$\mathbf{Q}^2(s, i+1)\left(\mathbf{M}^{(i+1)}(a_i)\right)^{-1}\mathbf{X}^{(i+1)}(a_i) =$$

$$= \mathbf{Q}^2(s, i)\left(\mathbf{M}^{(i)}(a_i)\right)^{-1}\mathbf{X}^{(i)}(a_i) + F_2(s, i), \tag{8.187}$$

$$F_1(s, i) = f_{10}^*(s) + f_{11}^*(s)X_1^{(i)}(a_i) + f_{12}^*(s)X_2^{(i)}(a_i) +$$

$$+ f_{13}^*(s)X_5^{(i)}(a_i) + f_{14}^*(s)X_6^{(i)}(a_i), \tag{8.188}$$

$$F_2(s, i) = f_{20}^*(s) + f_{21}^*(s)X_1^{(i)}(a_i) + f_{22}^*(s)X_2^{(i)}(a_i) +$$

$$+ f_{23}^*(s)X_5^{(i)}(a_i) + f_{24}^*(s)X_6^{(i)}(a_i). \tag{8.189}$$

In these equations, $\mathbf{Q}^1(s, i)$, $\mathbf{Q}^2(s, i)$ are eight-dimensional vectors with the components $(j = 1, 2, ..., 8)$

$$Q_j^1(s, i) = \left.\frac{P_{11}^*(s, i)P_{22}^*(-\lambda, i) - P_{12}^*(s, i)P_{21}^*(-\lambda, i)}{P_{22}^*(-\lambda, i)(s+\lambda)}\right|_{\lambda=\lambda_j(i)}, \tag{8.190}$$

$$Q_j^2(s,i) = \left.\frac{P_{21}^*(s,i)P_{12}^*(-\lambda,i) - P_{22}^*(s,i)P_{11}^*(-\lambda,i)}{P_{12}^*(-\lambda,i)(s+\lambda)}\right|_{\lambda=\lambda_j(i)}. \tag{8.191}$$

Here, the functions $P_{mn}^*(s,i)$ are defined in terms of $P_{mn}(D)$ in Eqs. (8.129)–(8.132)

$$P_{11}^*(s,i) = P_{11}(D)|_{D=-s},\ P_{12}^*(s,i) = P_{12}(D)|_{D=-s}, \tag{8.192}$$

$$P_{21}^*(s,i) = P_{21}(D)|_{D=-s},\ P_{22}^*(s,i) = P_{22}(D)|_{D=-s}. \tag{8.193}$$

Because Eq. (8.137) holds for $\lambda = \lambda_j(i)$, the numerators of the functions $Q_j^1(s,i)$ and $Q_j^2(s,i)$ turn to zero for $s = -\lambda_j(i)$

$$P_{21}^*(-\lambda_j(i),i)P_{12}^*(-\lambda_j(i),i) - P_{22}^*(-\lambda_j(i),i)P_{11}^*(-\lambda_j(i),i) = 0, \tag{8.194}$$

$$j = 1, 2, ..., 8,\ i = 1, 2, ..., N.$$

It means that the polynomials $P_{21}^*(s,i)P_{12}^*(-\lambda_j(i),i) - P_{22}^*(s,i)P_{12}^*(-\lambda_j(i),i)$ of the order of 4 are divided without remainder by $s + \lambda_j(i)$. Therefore the components of vectors $\mathbf{Q}^1(s,i)$ and $\mathbf{Q}^2(s,i)$ are polynomials of the order of 3. Equating to zero the coefficient of s^3, s^2, s^1, s^0 in Eqs. (8.186), (8.187), we yield eight equations that relate the components of the vectors $\mathbf{X}^{(i)}(a_i)$ and $\mathbf{X}^{(i+1)}(a_i)$ at the layer boundaries. A part of these equations coincide with the equations that follow from the general conditions for the jumps of the strain tensor at the boundary of two homogeneous media (Sect. 5.3). In particular, equating to zero the coefficients of s^3 and s^2 in Eqs. (8.186), (8.187), result in the equations that relate the following components of the vectors $\mathbf{X}^{(1+i)}(r)$ and $\mathbf{X}^{(i)}(r)$ at $r = a_i$, $i = 1, 2, ..., N$

$$X_1^{(i+1)}(a_i) = X_1^{(i)}(a_i),\ X_2^{(i+1)}(a_i) = X_2^{(i)}(a_i), \tag{8.195}$$

$$X_5^{(i+1)}(a_i) = X_5^{(i)}(a_i),\ X_6^{(i+1)}(a_i) = X_6^{(i)}(a_i). \tag{8.196}$$

These conditions mean that the functions $\alpha_1(r), \alpha_3(r)$ and their derivatives $D\alpha_1(r), D\alpha_3(r)$ are continuous at the layer boundaries. Equating to zero the coefficient of s in Eqs. (8.186), (8.187), provides the conditions for the jumps of the functions $X_3^{(i)}(r)$ and $X_7^{(i)}(r)$ (the second derivatives of $\alpha_1(r)$ and $\alpha_3(r)$) at the boundaries $r = a_i$

$$X_3^{(i+1)} = \frac{1}{b_4^{(i+1)}}\left(b_4^{(i)}X_3^{(i)} - [b_4]_i(3(2X_1^{(i)} + X_2^{(i)}) + 4(X_5^{(i)} + X_6^{(i)})\right), \tag{8.197}$$

$$X_7^{(i+1)} = \frac{1}{b_0^{(i+1)}}\left(b_0^{(i)}X_7^{(i)} - [b_5]_i(3(X_1^{(i)} + X_2^{(i)}(a_i)) + \right.$$

$$\left. +2X_5^{(i)}) - [b_6]_iX_6^{(i)}\right), \tag{8.198}$$

$$b_0^{(i)} = c_1^{(i)} + c_2^{(i)} + 2c_3^{(i)} + c_5^{(i)} + c_6^{(i)}, \; b_4^{(i)} = 2c_1^{(i)} + c_5^{(i)}, \tag{8.199}$$

$$b_5^{(i)} = c_1^{(i)} + c_3^{(i)} + c_5^{(i)} + c_6^{(i)}, \; b_6^{(i)} = 3b_0^{(i)} + 2\left(c_2^{(i)} + c_3^{(i)}\right). \tag{8.200}$$

Here, $X_m^{(i)} = X_m^{(i)}(a_i)$, $X_m^{(i+1)} = X_m^{(i+1)}(a_i)$. If $r = a_i$ is the boundary of two isotropic layers, we have

$$X_3^{(i+1)} = \frac{1}{c_1(i+1)}\left(c_1^{(i)}X_3^{(i)} - [c_1]_i(3(2X_1^{(i)} + X_2^{(i)}) + \right.$$
$$\left. +4(X_5^{(i)} + X_6^{(i)})\right), \tag{8.201}$$

$$X_7^{(i+1)} = \frac{1}{c_1(1+i) + c_2(1+i)}\left((c_1^{(i)} + c_2^{(i)})X_7^{(i)} - \right.$$
$$\left. -[c_1]_i(3(X_1^{(i)} + X_2^{(i)}(a_i)) + 2X_5^{(i)}) - [5c_1 + 3c_2]_i X_6^{(i)}\right). \tag{8.202}$$

Conditions for the jumps of the third derivatives $D^3\alpha_1 = X_4$, $D^3\alpha_3 = X_8$ of the functions α_1, α_3 at the layer boundaries follow from equating to zero the coefficients of $s^0 = 1$ in Eqs. (8.186), (8.187),. These conditions are consequences of specific forms of the functions $\alpha_1(r)$ and $\alpha_3(r)$ inside each layer indicated in Eq. (8.152), and they do not follow from the general conditions for the jump of the strain tensor at the boundary of two elastic media

$$\mathbf{Q}^1(0, i+1)\left(\mathbf{M}^{(i+1)}(a_i)\right)^{-1}\mathbf{X}^{(i+1)}(a_i) =$$

$$= \left(\mathbf{Q}^1(0, i)\left(\mathbf{M}^{(i)}(a_i)\right)^{-1} + \mathbf{F}^1(0)\right)\mathbf{X}^{(i)}(a_i), \tag{8.203}$$

$$\mathbf{Q}^2(0, i+1)\left(\mathbf{M}^{(i+1)}(a_i)\right)^{-1}\mathbf{X}^{(i+1)}(a_i) =$$

$$= \left(\mathbf{Q}^2(0, i)\left(\mathbf{M}^{(i)}(a_i)\right)^{-1} + \mathbf{F}^2(0)\right)\mathbf{X}^{(i)}(a_i). \tag{8.204}$$

Here, $\mathbf{F}^1(0)$ and $\mathbf{F}^2(0)$ are the eight-dimensional vectors with the components

$$F_1^1(0) = \frac{3}{2}[6c_1 + 5c_5 + 8c_6]_i\,, \; F_2^1(0) = -\frac{3}{2}[b_4]_i\,, \tag{8.205}$$

$$F_5^1(0) = -2\,[2c_1 - 10c_3 - 3c_5 - 6c_6]_i\,, \tag{8.206}$$

$$F_6^1(0) = 4\,[b_5]_i\,, \; F_3^1(0) = F_4^1(0) = F_7^1(0) = F_8^1(0) = 0; \tag{8.207}$$

$$F_1^2(0) = -3\,[2c_1 - 6c_3 - c_5 + c_6]_i\,, \; F_2^2(0) = 3\,[b_4]_i\,, \tag{8.208}$$

$$F_5^2(0) = 2\,[13c_1 + 15c_2 + 4c_3 + 3c_5 - 3c_6]_i\,, \tag{8.209}$$

$$F_6^2(0) = -2\,[c_1 - 3c_2 - 2c_3 + c_5 + c_6]_i\,, \tag{8.210}$$

$$F_3^1(0) = F_4^1(0) = F_7^1(0) = F_8^1(0) = 0. \tag{8.211}$$

Finally, the jumps of the function $\mathbf{X}^{(i)}(r)$ at the boundary of the ith and $(i+1)$th layers are presented in the matrix form

$$\mathbf{X}^{(i+1)}(a_i) = \Gamma^{(i)}\mathbf{X}^{(i)}(a_i),\, i = 1, 2, ..., N. \tag{8.212}$$

where the matrices $\Gamma^{(i)}$ of the dimensions 8×8 can be reconstructed from Eqs. (8.195)–(8.211).

3. Jumps of the functions α_1 and α_3 at the boundary of two isotropic layers
For an isotropic ith layer, $c_3^{(i)} = c_5^{(i)} = c_6^{(i)} = 0$, and the elements $Q_j^1(s,i)$ and $Q_j^2(s,i)$ take the forms

$$Q_j^1(s,i) = \frac{P_{11}(s)}{(s+\lambda_j(i))} = \frac{c_1^{(i)}}{2(s+\lambda_j(i))}\prod_{k=1}^{4}(s+\lambda_k(i)), \tag{8.213}$$
$$j = 1, 2, 3, 4;$$

$$Q_j^2(s,i) = \frac{P_{22}(s)}{(s+\lambda_{j-4}(i))} = \frac{c_1^{(i)}+c_2^{(i)}}{(s+\lambda_{j-4}(i))}\prod_{k=1}^{4}(s+\lambda_{k-4}(i)), \tag{8.214}$$
$$j = 5, 6, 7, 8;$$

$$Q_j^1(s,i) = 0,\, j > 4; \quad Q_j^2(s,i) = 0,\ j < 5. \tag{8.215}$$

Because for isotropic layers, $\lambda_1(i) = 0$, the vectors $\mathbf{Q}_j^1(0,i)$ and $\mathbf{Q}_j^2(0,i)$ are

$$\mathbf{Q}^1(0,i) = [15c_1^{(i)}, 0, 0, 0, 0, 0, 0, 0]^T,$$

$$\mathbf{Q}^2(0,i) = [0, 0, 0, 0, 30(c_1^{(i)} + c_2^{(i)}), 0, 0, 0,]^T, \tag{8.216}$$

where $[..]^T$ is transposition operator. Then, it follows from Eq. (8.203) that the components $X_4^{(1+i)}(a_i)$ are expressed in terms of the components of the vector $\mathbf{X}^{(i)}(a_i)$ as follows

$$X_4^{(1+i)}(a_i) = \frac{1}{c_1^{(1+i)}}\Big(c_1^{(i)}X_4^{(i)}(a_i) +$$
$$+\,[c_1]_i\,(48X_1^{(i)}(a_i) + 25X_2^{(i)}(a_i) + 32X_5^{(i)}(a_i) + 16X_6^{(i)}(a_i). \tag{8.217}$$

Similarly, from Eq. (8.204) we obtain $X_8^{(1+i)}$ in terms of $X_j^{(i)}$

$$X_8^{(1+i)}(a_i) = \frac{1}{c_1^{(1+i)} + c_2^{(1+i)}} \Big((c_1^{(i)} + c_2^{(i)}) X_8^{(i)}(a_i) +$$
$$+4 \left(6[c_1]_i X_1^{(i)}(a_i) + 3[c_1]_i X_2^{(i)}(a_i) + 4[c_1]_i X_5^{(i)}(a_i) \right) +$$
$$+[21c_1 + 25c_2]_i X_6^{(i)}(a_i) \Big) . \tag{8.218}$$

The function $\mathbf{X}^{(i)}(r)$ inside isotropic layers and the constants $\mathbf{Y}^{(i)}$ in Eqs. (8.154), (8.155) relate the equations

$$\mathbf{X}^{(i)}(r) = \mathbf{M}^0(r)\mathbf{Y}^{(i)}, \; \mathbf{Y}^{(i)} = \mathbf{M}^0(r)^{-1}\mathbf{X}^{(i)}(r); \tag{8.219}$$

$$\mathbf{M}^0(r) = \begin{bmatrix} \mathbf{m}_0(r) & \mathbf{0} \\ \mathbf{0} & \mathbf{m}_0(r) \end{bmatrix}, \; \mathbf{m}_0(r) = \begin{bmatrix} 1 & r^2 & r^{-3} & r^{-5} \\ 0 & 2r^2 & -3r^{-3} & -5r^{-5} \\ 0 & 4r^2 & 9r^{-3} & 25r^{-5} \\ 0 & 8r^2 & -27r^{-3} & -125r^{-5} \end{bmatrix}, \tag{8.220}$$

where $\mathbf{0}$ is zero-matrix of the dimensions 4×4.

4. Numerical algorithm of construction of the function $\alpha_1(r)$ and $\alpha_3(r)$ for multilayered transverse isotropic inclusions
If the vector $X^i(r)$ is known at the point $r = a_{i-1} + 0$ (at the left end of the ith interval), its value at the right end $r = a_i - 0$ is defined from Eqs. (8.182) and takes the form

$$\mathbf{X}^{(i)}(a_i) = \mathbf{R}^{(i)} X^{(i)}(a_{i-1}) \, , \quad \boldsymbol{R}^{(i)} = \mathbf{M}^{(i)}(a_i) \left(\mathbf{M}^{(i)}(a_{i-1}) \right)^{-1} , \tag{8.221}$$
$$i = 1, 2, ..., N,$$

where $\boldsymbol{R}^{(i)}$ is the matrix of transition of the solution $\mathbf{X}^{(i)}(r)$ through the ith layer. The vectors $\mathbf{X}^{(i)}$ and $\mathbf{X}^{(i+1)}$ at point $r = a_i$ of the boundary between the ith and $(i+1)$th layers relate by Eq. (8.212), where the matrix $\Gamma^{(i)}$ is reconstructed from Eqs. (8.195)–(8.211). Schematically, the process of transition of the solution through the layers and the boundary between the layers is shown in Fig. 8.3.

Let the vector $\mathbf{X}^{(1)}(a_1)$ for the first layer (core) be known. Then, the vector $\mathbf{X}^{(i+1)}(a_i)$ that defines the solution inside the $(i+1)$th layer relates with the vector

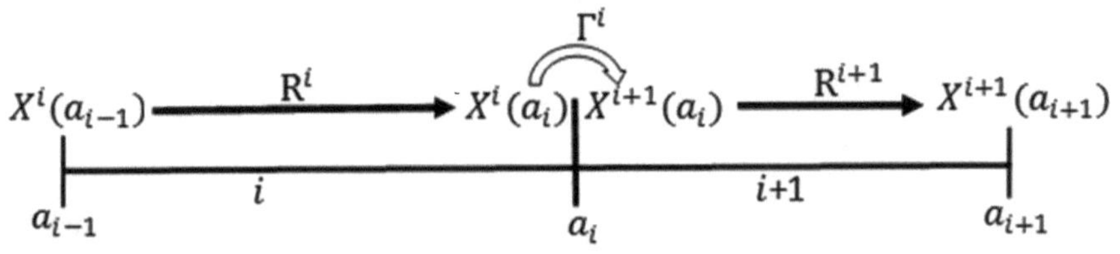

Fig. 8.3 Transition of the solution through the layers and layer boundaries for radially transverse isotropic inclusions

$\mathbf{X}^{(1)}(a_1)$ by the equation

$$\mathbf{X}^{i+1}(a_i) = \mathbf{G}^i \mathbf{X}^1(a_1), \ \mathbf{G}^i = \prod_{k=0}^{i-1} \Gamma^{(i-k)} \mathbf{R}^{(i-k)}, \quad i = 2, 3, ..., N, \tag{8.222}$$

where $\mathbf{R}^{(1)}$ is unit matrix, and the matrices $\mathbf{R}^{(i)}(i > 1)$ are defined in Eq. (8.221). For the construction of the vector $\mathbf{X}^{(1)}(a_1)$ in the core, the conditions for the functions $\alpha_1(r)$, $\alpha_3(r)$ at $r = 0$ and $r \to \infty$ should be taken into account. Unlike inclusions with isotropic layers, the strain tensor at the center of a transverse isotropic inclusion can be unbounded. To derive a unique solution, the condition of strain bounding should be changed with the condition of integrability of the squares of the strain tensor components at $r = 0$ (finite elastic energy). As a result, the roots $\lambda_j(1)$ with maximal real parts should be kept in Eqs. (8.152) for the functions $\alpha_1(r)$, $\alpha_3(r)$ in the first layer. Thus for the root numeration rule in Eq. (8.149), the functions $\alpha_1(r)$ and $\alpha_3(r)$ in the core have forms

$$\alpha_1(r) = Y_1^{(1)} + Y_2^{(1)} r^2 + Y_5^{(1)} r^{\lambda_5(1)} + Y_6^{(1)} r^{\lambda_6(1)}, \tag{8.223}$$

$$\alpha_3(r) = g_1(1) Y_1^{(1)} + g_2(1) Y_2^{(1)} r^2 + g_5(1) Y_5^{(1)} r^{\lambda_5(1)} +$$

$$+ g_6(1) Y_6^{(1)} r^{\lambda_6(1)} - \frac{1}{2}. \tag{8.224}$$

If we introduce the four-dimensional vector $\widehat{\mathbf{Y}}^{(1)} = \left[Y_1^{(1)}, Y_2^{(1)}, Y_5^{(1)}, Y_6^{(1)}\right]^T$, the vector $\mathbf{X}^{(1)}(a_1)$ is expressed in terms of $\widehat{\mathbf{Y}}^{(1)}$ as follows

$$\mathbf{X}^{(1)}(a_1) = \mathbf{M}^{(1)}(a_1) \mathbf{P} \widehat{\mathbf{Y}}^{(1)}, \ \mathbf{P} = \begin{bmatrix} \mathbf{m}^1 & \mathbf{0} \\ \mathbf{0} & \mathbf{0} \\ \mathbf{0} & \mathbf{m}^1 \\ \mathbf{0} & \mathbf{0} \end{bmatrix}, \tag{8.225}$$

$$\mathbf{m}^1 = \begin{bmatrix} 1 & 0 \\ 0 & 1 \end{bmatrix}, \mathbf{0} = \begin{bmatrix} 0 & 0 \\ 0 & 0 \end{bmatrix}. \tag{8.226}$$

For the host medium is isotropic, the functions $\alpha_1(r)$, $\alpha_3(r)$ in the $(N+1)$th layer have forms in Eqs. (8.154), (8.155), and because $\alpha_1(r), \alpha_3(r) \to 0$ when $r \to \infty$, the constants in these equations are

$$Y_1^{(N+1)} = Y_2^{(N+1)} = 0, \ Y_5^{(N+1)} = \frac{1}{2}, \ Y_6^{(N+1)} = 0. \tag{8.227}$$

Thus in the host medium $(r > a_N)$, the equations for the functions $\alpha_1(r)$, $\alpha_3(r)$ are

$$\alpha_1(r) = Y_3^{(N+1)} r^{-3} + Y_4^{(N+1)} r^{-5}, \tag{8.228}$$

$$\alpha_3(r) = Y_7^{(N+1)} r^{-3} + Y_8^{(N+1)} r^{-5}. \tag{8.229}$$

From these equation follows the equation for the vector $\mathbf{X}^{(N+1)}(a_N)$

$$\mathbf{B}\mathbf{X}^{(N+1)}(a_N) = \mathbf{f},\ \mathbf{B} = (\mathbf{P}^2 - \mathbf{L}\mathbf{P}^1), \tag{8.230}$$

where

$$\mathbf{P}^1 = \begin{bmatrix} \mathbf{m}^1 & \mathbf{0} & \mathbf{0} & \mathbf{0} \\ \mathbf{0} & \mathbf{0} & \mathbf{m}^1 & \mathbf{0} \end{bmatrix},\ \mathbf{P}^2 = \begin{bmatrix} \mathbf{0} & \mathbf{m}^1 & \mathbf{0} & \mathbf{0} \\ \mathbf{0} & \mathbf{0} & \mathbf{0} & \mathbf{m}^1 \end{bmatrix},\ \mathbf{L} = \begin{bmatrix} \mathbf{1} & \mathbf{0} \\ \mathbf{0} & \mathbf{1} \end{bmatrix}, \tag{8.231}$$

$$\mathbf{l} = \begin{bmatrix} -15 & -8 \\ 120 & 49 \end{bmatrix},\ \mathbf{B} = \begin{bmatrix} \mathbf{1} & \mathbf{m}^1 & \mathbf{0} & \mathbf{0} \\ \mathbf{0} & \mathbf{0} & \mathbf{1} & \mathbf{m}^1 \end{bmatrix},\ \mathbf{f} = \begin{bmatrix} 0 \\ 0 \\ 15/2 \\ -60 \end{bmatrix}. \tag{8.232}$$

The equation for the vector $\widehat{\mathbf{Y}}^{(1)} = \left[Y_1^{(1)}, Y_2^{(1)}, Y_5^{(1)}, Y_6^{(1)}\right]^T$ of the constant in the core is a consequence of Eqs. (8.228)–(8.232)

$$\mathbf{B}\mathbf{G}^N\mathbf{M}^{(1)}(a_1)\mathbf{P}\widehat{\mathbf{Y}}^{(1)} = \mathbf{f}. \tag{8.233}$$

After solution of this equation, the vectors $\mathbf{X}^{(i)}(r)$ and $\mathbf{Y}^{(i)}$, $i = 1, 2, ..., N+1$ are calculated from Eqs. (8.221) and (8.222).

Note that in the algorithm of construction of the functions α_1 and α_3, the transition matrices $\Gamma^{(i)}$ in Eq. (8.212) are calculated numerically except the boundaries of two isotropic layers, where the analytical expressions for the matrix $^{(i)}$ is available. The proposed numerical algorithm is robust and stable except in the cases when the roots of Eq. (8.137) have very large values ($\lambda = O(10^2)$). For such roots, inversion of the matrices $\mathbf{M}^{(i)}(a_i)$ in Eq. (8.221) for $\mathbf{R}^{(i)}$ can result in substantial loss of accuracy.

5. Multilayered spherical inclusion with isotropic components

For the inclusion with isotropic layers, the tensor $A(r, n)$ inside the $i$the layer takes the form

$$A(r, n) = \sum_{k=1}^{6} A_k(r) E^k(n), \tag{8.234}$$

For $a_{i-1} < r < a_i$, we have for the coefficients in this equation

$$A_1(r) = 3Y_1^i + 2Y_5^i + (5Y_2^i + 2Y_6^i)r^2 +$$

$$+ 2Y_7^i r^{-3} + 2(-Y_4^i + Y_8^i) r^{-5}, \tag{8.235}$$

$$A_2(r) = \frac{1}{3}((-3Y_1^i - 2Y_5^i + Y_9^i) - (3Y_2^i + 4Y_6^i) r^2 +$$

$$+ (Y_{10}^i - 3Y_3^i + Y_7^i) r^{-3} - 3(Y_4^i - Y_8^i) r^{-5}), \tag{8.236}$$

$$A^3(r) = -2(Y_2^i - Y_6^i) r^2 + 3(Y_3^i - Y_7^i) r^{-3} + 5(Y_4^i - Y_8^i) r^{-5}, \tag{8.237}$$

$$A_4(r) = -\frac{2}{3}(3Y_2^i + 4Y_6^i) r^2 - (Y_{10}^i - 3Y_3^i + Y_7^i) r^{-3} +$$

$$+ 5(Y_4^i - Y_8^i) r^{-5}, \tag{8.238}$$

$$A_5(r) = (6Y_2^i + 8Y_6^i) r^2 + 6(Y_3^i - 2Y_7^i) r^{-3} + 20(Y_4^i - Y_8^i) r^{-5}, \tag{8.239}$$

$$A_6(r) = -15(Y_3^i - Y_7^i) r^{-3} - 35(Y_4^i - Y_8^i) r^{-5}. \tag{8.240}$$

The strain tensor $\varepsilon(r, n)$ in the medium and the inclusion is calculated from Eq. (8.9) for an arbitrary external strain field ε_0.

As an example of the calculations, we consider an inclusion of unit radius, containing N spherical layers of the thickness $1/N$. Let Young modulus E_i inside the ith layer be given by the equation

$$E_i = E(a_i)\,, \quad i = 1, 2, ..., N+1\,, \tag{8.241}$$

$$E(r) = E_0\left(1 + \delta \exp\left(\frac{\lambda r^2}{r^2 - 1}\right)\right), \quad r < 1\,, \tag{8.242}$$

$$E(r) = E_0, \quad r \geq 1\,,$$

where a_i is external radius of ith layer. The Poisson ratio for the medium and all the layers was taken to be 0.4. External loading is uniaxial tension along the x_3-axis.

Figures 8.4 and 8.5 show the stress component σ_{33} along the axis orthogonal to x_3, with the origin at the center of the inclusion. The calculations were performed for $\delta = -1$, $\lambda = 0, 0.1, 1, 10$ (soft inclusion, Fig. 8.4) and $\delta = 100$ with the same value of λ (stiff inclusion, Fig. 8.5), and $N = 60$ for both cases.

When $N \to \infty$, the Young modulus inside the inclusion tends to the continuous distribution (8.242). For the construction of the plots in Figs. 8.4, 8.5, the calculations were performed for an increasing number of layers N until a stable picture of the stress distribution inside the inclusion was obtained. It turns out that for $N > 40$ the distribution of the stress component σ_{33} did not change as N grows.

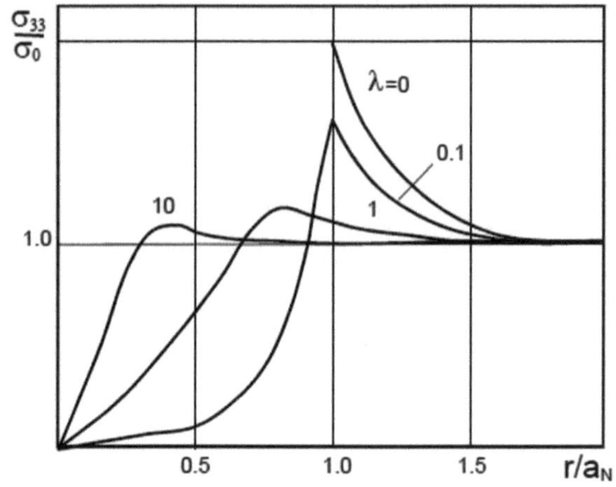

Fig. 8.4 Stress distribution in the medium with a soft spherically layered inclusion

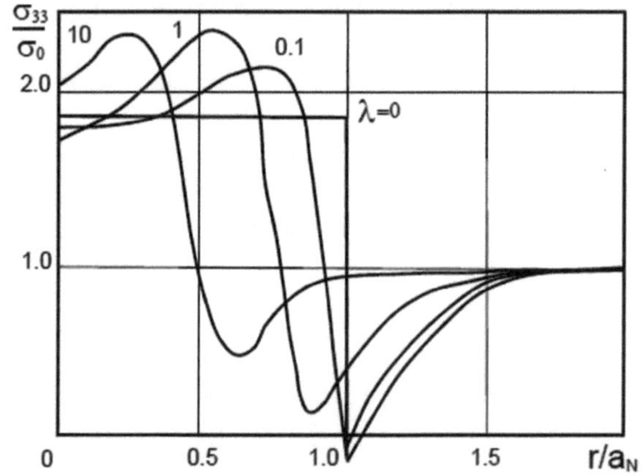

Fig. 8.5 Stress distribution in the medium with a stiff spherically layered inclusion

8.5 A Radially Transverse Isotropic Inclusion in an Isotropic Conductive Medium

We consider a spherical region V with the center at $x = 0$ in a homogeneous isotropic conductive medium with conductivity tensor $C^0_{\alpha\beta} = c_0\delta_{\alpha\beta}$, c_0 is a scalar. A radially transverse isotropic conductivity tensor $C_{\alpha\beta}(x)$ in the region V is defined by the equation

$$C_{\alpha\beta}(x) = C_{\alpha\beta}(r, n) = c(r)\delta_{\alpha\beta} + c_n(r)n_\alpha(x)n_\beta(x),$$

$$r = |x|, n_\alpha(x) = \frac{x_\alpha}{r}.$$

Here, $c(r)$ and $c_n(r)$ are scalar piecewise continuous functions. Because $C_{\alpha\beta}(x)$ is a definitely positive tensor, the coefficients $c(r)$ and $c_n(r)$ satisfy the inequalities

$$c(r) > 0, c(r) + c_n(r) > 0. \tag{8.243}$$

The electric field $e(x)$ in the medium with the inclusion is the solution of the integral equation (4.192)

$$e_\alpha(x) + \int K_{\alpha\beta}(x - x')C^1_{\beta\mu}(x')e_\mu(x')V(x')dx' = e^0_\alpha(x), \tag{8.244}$$

Here, $C^1(x) = C(x) - C^0$, $V(x)$ is the characteristic function of the region V, $e^0(x)$ is the external electric field applied to the medium, the kernels $K(x)$ is determined by Eq. (4.13)

$$K_{\alpha\beta}(x) = -\nabla_\alpha\nabla_\beta g(x)\,, \tag{8.245}$$

where $g(x)$ is the Green function of the homogeneous host medium.

Let the medium with the inclusion be subjected to a constant external electric field e^0 at infinity. Because of the linearity of the problem, electric field $e(x)$ in the entire space can be presented in the form

$$e_\alpha(x) = e^0_\alpha + e^1_\alpha(x)\,, \quad e^1_\alpha(x) = A_{\alpha\beta}(x)e^0_\beta\,, \tag{8.246}$$

where $A(x) \to 0$ when $|x| \to \infty$. Substituting Eq. (8.246) into Eq. (8.244) and taking into account that e^0 is an arbitrary vector, we yield the integral equation for the tensor $A(x)$

$$A_{\alpha\beta}(x) + \int_V K_{\alpha\gamma}(x - x')C^1_{\gamma\delta}(x')A_{\delta\beta}(x')dx' +$$

$$+ \int_V K_{\alpha\gamma}(x - x')C^1_{\gamma\beta}(x')dx' = 0. \tag{8.247}$$

Further, the Mellin transform technique is used for the solution of this equation.

Application of the Mellin transform operator to Eq. (8.247) results in the equation

$$\overline{A}^*(s,n) + K^s\overline{C^1A}^*(s,n) + K^s\overline{C^1}^*(s,n) = 0, \tag{8.248}$$

where $\overline{A}^*(s,n)$, $\overline{C^1}^*(s,n)$ are the Mellin transforms of the functions $A(r,n)$, $C^1(r,n)$ with respect to r-variable. In this equation, the operator K^s is defined in Eq. (8.23), where for an isotropic host medium, the tensor $\widetilde{K}(m)$ has the form

$$\widetilde{K}_{\alpha\beta}(m) = \frac{1}{c_0}m_\alpha m_\beta\,, \quad m_\alpha = \frac{k_\alpha}{|k|}\,. \tag{8.249}$$

It follows from Eqs. (8.24) and (8.25) that actions of operator K^s on the unit tensors $I_{\alpha\beta} = \delta_{\alpha\beta}$ and the tensor $(nn)_{\alpha\beta}(x) = n_\alpha(x)n_\beta(x)$ are defined by the equations

$$K^s I = \frac{1}{c_0(3-s)} H(s), \quad K^s(nn) = \frac{(s-2)}{c_0 s(3-s)} H(s), \tag{8.250}$$

where

$$H(s) = I - s(nn). \tag{8.251}$$

It follows from these equations that the Mellin transform of the tensor $A(r, n)$ in Eq. (8.248) can be found in the form

$$\overline{A}^*(s, n) = (I - s(nn))\, \overline{\beta}^*(s), \tag{8.252}$$

where $\overline{\beta}^*(s)$ is the Mellin transform of a scalar function $\beta(r)$. Because multiplication with the parameter s in the s-space of the Mellin transforms corresponds to the differential operator $(-D)$ in the r-space, tensor $A(r, n)$ takes the form

$$A(r, n) = (I + (nn)D)\, \beta(r), \quad D = r\frac{d}{dr}. \tag{8.253}$$

For such $A(r, n)$, the product $C^1 A$ in Eq. (8.248) is

$$(C^1 A)(r, n) = S_1(r)I + S_2(r)nn, \tag{8.254}$$

$$C^1(r) = c_1(r)I + c_n(r)nn, \; S_1(r) = c_1(r)\beta(r), \tag{8.255}$$

$$S_2(r) = c_n(r)\alpha(r) + (c_1(r) + c_n(r))D\beta(r), \; c_1(r) = c(r) - c_0. \tag{8.256}$$

After substitution of these equations into Eq. (8.247) and application of the Mellin transform operator, we yield the equation

$$\overline{A}^*(s, n) + \overline{S}_1^*(s)K^s I(s, n) + \overline{S}_2^*(s)\left(K^s nn\right)(s, n) +$$

$$+\,\overline{c}_1^*(s)\left(K^s I\right)(s, n) + \overline{c}_n^*(s)\left(K^s nn\right)(s, n) = 0. \tag{8.257}$$

Here, $\overline{S}_i^*(s)$ is the Mellin transform of scalar coefficients $S_i(r)$ in Eq. (8.256). It follows from Eq. (8.250) that tensors $K^s I$ and $K^s nn$ in Eq. (8.257) are proportional to the tensor H. After substituting Eqs. (8.250) into Eq. (8.257), and equating to zero the coefficient of the tensor H, we obtain

$$c_0 s(3-s)\overline{\beta}^*(s) + s\overline{S}_1^*(s) + (s-2)\overline{S}_2^*(s) + s\overline{c}_1^*(s) + (s-2)\overline{c}_n^*(s) = 0, \tag{8.258}$$

$$\overline{S}_1^*(s) = \overline{(c_1\beta)}^*(s), \overline{S}_2^*(s) = \overline{(c_n\beta)}^*(s) + \overline{(c_1 + c_n)D\beta}^*(s). \tag{8.259}$$

Application of the inverse Mellin transform operator to Eq. (8.258) results in the differential equation for the functions $\beta(r)$. By this transition, the Mellin transforms of functions should be changed for their r-presentations (originals) and parameter $(-s)$ for the differential operator D. As a result, we yield

$$D\left((c + c_n)(1 + D)\beta\right) + 2(c + c_n)D\beta + 2c_n(1 + \beta) + D(c + c_n) = 0. \tag{8.260}$$

Let $c(r)$, $c_n(r)$ be bounded functions with piece-wise continuous second derivatives. For the conditions at $r = 0$ and $r \to \infty$

$$r\beta(r) \to 0,\ r \to 0; \quad \beta(r) \to 0, r \to \infty, \tag{8.261}$$

there exists a unique bounded solution of the differential equation (8.260).

If $c(r)$ and $c_n(r)$ are piece-wise continuous functions, Eq. (8.260) should be understood in the generalized (weak) sense: integral of product of any smooth finite function with the left-hand side of Eq. (8.260) is equal to zero.

8.5.1 *A Multilayered Radially Transverse Isotropic Inclusion in Homogeneous Isotropic Conductive Medium*

Let the coefficients $c(r)$ and $c_n(r)$ in Eq. (8.260) for the conductivity tensor of the inclusion be piece-wise constant functions of r with jumps at the points $r = a_i,\ i = 1, 2..., N,\ (a_0 = 0,\ a_0 < a_1 < ... < a_N)$. In this case, the inclusion consists of a spherical core and $N - 1$ spherical layers with constant conductivity coefficients $c = c^{(i)}$ and $c_n = c_n^{(i)}$.

The function $\beta(r)$ in Eq. (8.260) determines the field $e(x)$ in the host medium and in the inclusion. Because in the regions $a_{i-1} < r < a_i$ (inside the layers) the derivatives $Dc^{(i)}$, $Dc_n^{(i)}$ vanish, Eq. (8.260) takes the form

$$(c^{(i)} + c_n^{(i)})D(3 + D)\beta(r) + 2c_n^{(i)}\beta(r) + 2c_n^{(i)} = 0, \tag{8.262}$$

$$a_{i-1} < r < a_i, i = 1, 2, ..., N + 1.$$

Here, $a_{N+1} = \infty$, $c^{(N+1)} = c_0$, $c_n^{(N+1)} = 0$. This equation is of Euler's type, and its general solution has the form

$$\beta(r) = y_1^{(i)}r^{\lambda_1(i)} + y_2^{(i)}r^{\lambda_2(i)} - 1, i = 1, 2, ..., N + 1. \tag{8.263}$$

Here, $y_1^{(i)}$, $y_2^{(i)}$ are arbitrary constants, $\lambda_1(i)$, $\lambda_2(i)$ are the roots of the square equation

$$(c^{(i)} + c_n^{(i)})\lambda(3+\lambda) + 2c_n^{(i)} = 0, \tag{8.264}$$

$$\lambda_1(i) = -\frac{1}{2}\left(3 - \sqrt{1 + \frac{8c^{(i)}}{c^{(i)} + c_n^{(i)}}}\right), \tag{8.265}$$

$$\lambda_2(i) = -\frac{1}{2}\left(3 + \sqrt{1 + \frac{8c^{(i)}}{c^{(i)} + c_n^{(i)}}}\right). \tag{8.266}$$

Because coefficient $c^{(i)}$, $c_n^{(i)}$ satisfy the constraints (8.243), $\lambda_1(i)$ and $\lambda_2(i)$ are real numbers, and the root $\lambda_2(i)$ is always negative, meanwhile $\lambda_1(i)$ can be positive or negative and

$$-1 < \lambda_1(i) < \infty, \quad -\infty < \lambda_1(i) < -1. \tag{8.267}$$

For isotropic layers ($c_n^{(i)} = 0$), the general solution of Eq. (8.262) can be taken the form

$$\beta(r) = y_1^{(i)} + y_2^{(i)} r^{-3} - 1. \tag{8.268}$$

Note that the solution in the host medium (in the $(N+1)$th layer) is $\beta(r) = y_2^{(N+1)} r^{-3}$ because function $\beta(r)$ should vanish at infinity, and therefore, $y_1^{(N+1)} = 1$.

8.5.2 *Conditions for Jumps of the Functions β and $D\beta$ at the Layer Boundaries*

Because the electric field $e(x)$ in a multilayered inclusion does not contain singular components (delta-functions), the function $\beta(r)$ in Eq. (8.253) is continuous, meanwhile $D\beta(r)$ can have jumps at the layer boundaries. For the determination of the values of the jumps, the general Eq. (4.196) for the field discontinuities at the boundary between two different homogeneous conductive media can be used. Let x^0 be an arbitrary point of the boundary Ω between two conductive media with conductivity tensors $C^{(i)}$ and $C^{(i+1)}$, $n(x^0)$ be the normal to Ω at point x^0 in the direction to the $(i+1)$th medium. We denote by $e^{(i+1)}$ the limit of $e(x)$ when x tends to x^0 from the side of the normal, and $e^{(i)}$ is such a limit from the opposite side. Then, the relation between these limits has the form

$$e^{(i+1)} = \left(I - \widetilde{K}^{(i+1)}(n^0)\,[C]_i\right) e^{(i)}, \quad [C]_i = C^{(i+1)} - C^{(i)}, \tag{8.269}$$

$$\widetilde{K}_{\alpha\beta}^{(i+1)}(n) = \frac{n_\alpha n_\beta}{c^{(i+1)} + c_n^{(i+1)}}, \quad n^0 = n(x^0). \tag{8.270}$$

After substituting into Eq. (8.269) expressions for $e^{(i+1)}$ and $e^{(i)}$ in terms of the tensor $A(r, n)$ in Eq. (8.253), we obtain

$$\left(I + A^{(i+1)}\right) = (I - K^{*(1+i)}(n)\,[C]_i)\left(I + A^{(i)}\right), \tag{8.271}$$

$$A^{(j)} = \beta^{(j)} I + nnD\beta^{(j)},\ j = i, 1+i, \tag{8.272}$$

$$\beta^{(i)} = \lim_{\varepsilon\to 0} \beta(a_i - \varepsilon),\quad D\beta^{(i)} = \lim_{\varepsilon\to 0} D\beta(a_i - \varepsilon), \tag{8.273}$$

$$\beta^{(i+1)} = \lim_{\varepsilon\to 0} \beta(a_i + \varepsilon),\ D\beta^{(i+1)} = \lim_{\varepsilon\to 0} D\beta(a_i + \varepsilon), \varepsilon > 0. \tag{8.274}$$

All terms in Eq. (8.271) are proportional either to the tensor I or to the tensor nn. Equating the expressions in front of these tensors in the left- and right-hand sides of Eq. (8.271), we obtain the conditions for jumps of the functions $\beta(r)$ and $D\beta(r)$ at the layer boundaries

$$[\beta]_i = 0, \qquad [(c + c_n)D\beta]_i = -\left[c + c_n\right]_i \left(\beta^{(i)} + 1\right), i = 1, 2, ..., N. \tag{8.275}$$

In these equations, symbol $[f]_i = f(a_i + 0) - f(a_i - 0)$ is the jump of a function $f(r)$ at point $r = a_i$.

8.5.3 *The Numerical Algorithm of Solution of the Conductivity Problem for a Multilayered Transverse Isotropic Inclusion*

We introduce two functions $x_1^{(i)}(r), x_2^{(i)}(r)$ that define the field $e(x)$ inside each layer of a multilayered inclusion ($i = 1, 2, ..., N+1$)

$$x_1^{(i)}(r) = \beta(r) + 1 = y_1^{(i)} r^{\lambda_1(i)} + y_2^{(i)} r^{\lambda_2(i)}, \tag{8.276}$$

$$x_2^{(i)}(r) = D\beta(r) = \lambda_1(i) y_1^{(i)} r^{\lambda_1(i)} + \lambda_2(i) y_2^{(i)} r^{\lambda_2(i)}. \tag{8.277}$$

Here, $\lambda_1(i), \lambda_2(i)$ are the roots in Eqs. (8.265), (8.266) for the ith layer, $y_1^{(i)}, y_2^{(i)}$ are constants to be determined. If functions $x_1^{(i)}(r), x_2^{(i)}(r)$ are known at point $r = a_{i-1}$ (the left end of the ith interval (a_{i-1}, a_i)), their values at the right end ($r = a_i$) are defined from the equation that follows from Eqs. (8.276)–(8.277)

$$\mathbf{x}^{(i)}(a_i) = \mathbf{r}^{(i)}\mathbf{x}^{(i)}(a_{i-1}),\ \mathbf{x}^{(i)}(r) = \begin{bmatrix} x_1^{(i)}(r) \\ x_2^{(i)}(r) \end{bmatrix},\ i = 2, 3, ...N, \tag{8.278}$$

$$\mathbf{r}^{(i)} = \frac{1}{\lambda_1 - \lambda_2}\begin{bmatrix} \lambda_1\rho^{\lambda_2} - \lambda_2\rho^{\lambda_1}, & \rho^{\lambda_1} - \rho^{\lambda_2} \\ \lambda_1\lambda_2\left(\rho^{\lambda_2} - \rho^{\lambda_1}\right), & \lambda_1\rho^{\lambda_1(i)} - \lambda_2\rho^{\lambda_2} \end{bmatrix}, \tag{8.279}$$

$$\rho = \rho(i) = \frac{a_i}{a_{i-1}}, \ \lambda_1 = \lambda_1(i), \ \lambda_2 = \lambda_2(i). \tag{8.280}$$

Here, the matrix $\mathbf{r}^{(i)}$ performs transition of the solution through the ith layer.

Relation between vectors $\mathbf{x}^{(i)}(a_i)$ and $\mathbf{x}^{(i+1)}(a_i)$ at the boundary point a_i between ith and $(i+1)$th layers follows from the conditions (8.275) for the jumps of $\beta(r)$ and $D\beta(r)$ at the layer boundaries. These relations can be presented in the form

$$\mathbf{x}^{(i+1)}(a_i) = \gamma^{(i)}\mathbf{x}^{(i)}(a_i), i = 1, 2, ..., N, \tag{8.281}$$

where the matrix $\gamma^{(i)}$ depend on the properties of the ith and $(i+1)$th layers.

$$\gamma^{(i)} = \begin{bmatrix} 1, & 0 \\ -\frac{c^{(i+1)}+c_n^{(i+1)}-c^{(i)}-c_n^{(i)}}{c^{(i+1)}+c_n^{(i+1)}}, & \frac{c^{(i)}+c_n^{(i)}}{c^{(i+1)}+c_n^{(i+1)}} \end{bmatrix}. \tag{8.282}$$

The algorithm of calculation of the constants $y_1^{(i)}$, $y_2^{(i)}$ in Eq. (8.276) inside the layers follows from Eqs. (8.278), (8.281) and the conditions for $\beta(r)$ at $r = 0$ and $r \to \infty$. Inside the core ($i = 1$), the constant $y_2^{(1)}$ is equal to zero because the solution $y_2^{(i)}r^{\lambda_2(i)}$ can be non-integrable at $r = 0$ for possible values of the coefficients c and c_n. Thus in the core, $\beta(r)$ has the form

$$\beta(r) = y_1^{(1)}r^{\lambda_1(1)} - 1, \ 0 < r < a_1. \tag{8.283}$$

Then, using transition matrices $\mathbf{r}^{(i)}$ and relations (8.281) between $\mathbf{x}^{(i)}(r)$ and $\mathbf{x}^{(i+1)}(r)$ at $r = a_i$, the solution in the $(N+1)$th layer (in the host medium) is presented in the form

$$\mathbf{x}^{(N+1)}(a_N) = \mathbf{g}^N\mathbf{x}^{(1)}(a_1), \ \mathbf{x}^{(1)}(a_1) = \begin{bmatrix} y_1^{(1)}a_1^{\lambda_1(1)} - 1 \\ \lambda_1(1)y_1^{(1)}a_1^{\lambda_1(1)} \end{bmatrix}, \tag{8.284}$$

$$\mathbf{g}^N = \gamma^{(N)}\mathbf{r}^{(N)}\gamma^{(N-1)}\mathbf{r}^{(N-1)}...\gamma^{(1)}\mathbf{r}^{(1)} = \prod_{k=0}^{N-1}\gamma^{(N-k)}\mathbf{r}^{(N-k)}. \tag{8.285}$$

Here, the transition matrices $\mathbf{r}^{(i)}$ are defined in Eq. (8.279) for $i > 1$, and $\mathbf{r}^{(1)}$ is unit matrix. Because the host medium is isotropic, in the $(N+1)$th layer we have

$$\beta(r) = y_1^{(N+1)} + y_2^{(N+1)}r^{-3} - 1, \ \ a_N < r < \infty. \tag{8.286}$$

For the function $\beta(r)$ vanishes at infinity, $y_1^{(N+1)} = 1$, and vector $\mathbf{x}^{(N+1)}(a_N)$ takes the form

$$\mathbf{x}^{(N+1)}(a_N) = \begin{bmatrix} x_1^{(N+1)}(a_N) \\ x_2^{(N+1)}(a_N) \end{bmatrix} = \begin{bmatrix} 1 + y_2^{(N+1)}a_N^{-3} \\ -3y_2^{(N+1)}a_N^{-3} \end{bmatrix}. \tag{8.287}$$

Therefore, the components of the vector $\mathbf{x}^{(N+1)}(a_N)$ relate by the equation

$$x_2^{(N+1)}(a_N) + 3\left(x_1^{(N+1)}(a_N) - 1\right) = 0. \tag{8.288}$$

Substituting in this equation $x_1^{(N+1)}(a_N)$, $x_2^{(N+1)}(a_N)$ from Eqs. (8.284), we yield the equation for the constant $y_1^{(1)}$ in Eq. (8.283) for the function $\beta(r)$ in the core. After solution of this equation, the vector $\mathbf{x}^{(i+1)}(a_i)$ for the $(1+i)$th layer is calculated from the equation that is similar to Eq. (8.284)

$$\mathbf{x}^{(i+1)}(a_i) = \mathbf{g}^i\mathbf{x}^{(1)}(a_1),\ i = 2, 3, ..., N. \tag{8.289}$$

Then, the vectors of constants $\mathbf{y}^{(i)}$ $(i = 2, 3, ..., N+1)$ for each layer can be found from the equations

$$\mathbf{y}^{(i)} = \begin{bmatrix} y_1^{(i)} \\ y_2^{(i)} \end{bmatrix} = \mathbf{m}^{(i)}(a_i)\mathbf{x}^{(i)}(a_i),\quad \mathbf{x}^{(i)}(a_i) = \mathbf{r}^{(i)}\mathbf{x}^{(i)}(a_{i-1}), \tag{8.290}$$

$$\mathbf{m}^{(i)}(r) = \frac{1}{\lambda_1(i) - \lambda_2(i)}\begin{bmatrix} -\lambda_2(i)r^{-\lambda_1(i)}, & r^{-\lambda_1(i)} \\ \lambda_1(i)r^{-\lambda_2(i)}, & -r^{-\lambda_2(i)} \end{bmatrix}. \tag{8.291}$$

8.5.4 A Radially Transverse Isotropic Inclusion with Constant Conductivity Coefficients

We consider a spherical radially transverse isotropic inclusion of radius a with constant conductivity coefficients. In this case, vector $\mathbf{x}^{(2)}(r)$ of the solution in the host medium at point $r = a$ is defined by the equation

$$\mathbf{x}^{(2)}(a) = \gamma^{(1)}\mathbf{x}^{(1)}(a), \tag{8.292}$$

where matrix $\gamma^{(1)}$ is in Eq. (8.282).Vector $\mathbf{x}^{(1)}(a)$ at point $r = a - 0$ is

$$x_1^{(1)}(r) = y_1^{(1)}r^{\lambda_1(1)},\ x_2^{(1)} = \lambda_1(1)y_1^{(1)}r^{\lambda_1(1)},\ 0 < r < a. \tag{8.293}$$

Constant $y_1^{(1)}$ is to be found from the equation

$$x_2^{(2)}(a) + 3(x_1^{(2)}(a) - 1) = 0, \tag{8.294}$$

where $x_1^{(2)}(a)$, $x_2^{(2)}(a)$ are components of vector $\mathbf{x}^{(2)}(a)$ in the host medium. If $y_1^{(1)}$ is known, constants $y_1^{(2)}$, $y_2^{(2)}$ can be found from Eqs. (8.292) and (8.290)

$$y_1^{(1)} = \frac{3c_0a^{-\lambda_1(2)}}{2c_0 + (c + c_n)(1 + \lambda_1(1))},\ y_2^{(1)} = 0, \tag{8.295}$$

$$y_1^{(2)} = 1,\ y_2^{(2)} = \left(\frac{3c_0}{2c_0 + (c + c_n)(\lambda_1(1) + 1)} - 1 \right) a^3, \tag{8.296}$$

$$\lambda_1(1) = -\frac{1}{2}\left(3 - \sqrt{1 + \frac{8c}{c + c_n}}\right). \tag{8.297}$$

We consider an example of a radially transverse isotropic inclusion with conductivity coefficients $c = 0.1c_0$, $c_n = 10c_0$, where c_0 is the conductivity of the host medium. In this case, the roots λ_1 and λ_2 in the inclusion are

$$\lambda_1 = -0.9806,\ \lambda_2 = -2.0194. \tag{8.298}$$

Distributions of the component $e_1(x_1, x_2)$ of the electric field in section $x_3 = 0$ in the inclusion vicinity are presented in Fig. 8.6, external field e^0 is the unit vector directed along the x_1-axis of the Cartesian coordinate system with the origin at the inclusion center. In Fig. 8.7, directions and sizes of the field vectors in the vicinity of the inclusion are shown.

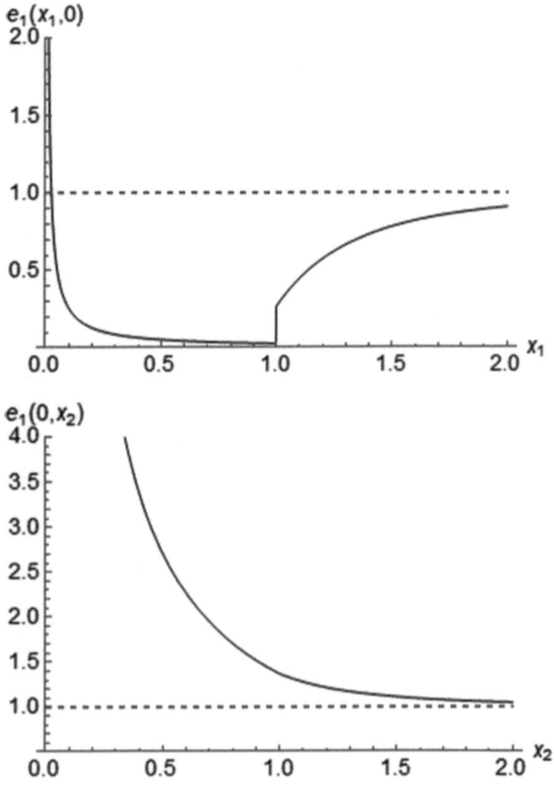

Fig. 8.6 Distributions of the electric field component $e_1(x_1, x_2)$ along the x_1 and x_2-axes ($x_3 = 0$) in the vicinity of the radially transverse isotropic inclusion of the radius $a = 1$; $c = 0.1$, $cn = 10$ are the conductivities of the inclusion, and $c_0 = 1$ is the same for the host medium; dashed lines are the component $e_1^0 = 1$ ($e_2^0 = e_3^0 = 0$) of the external filed

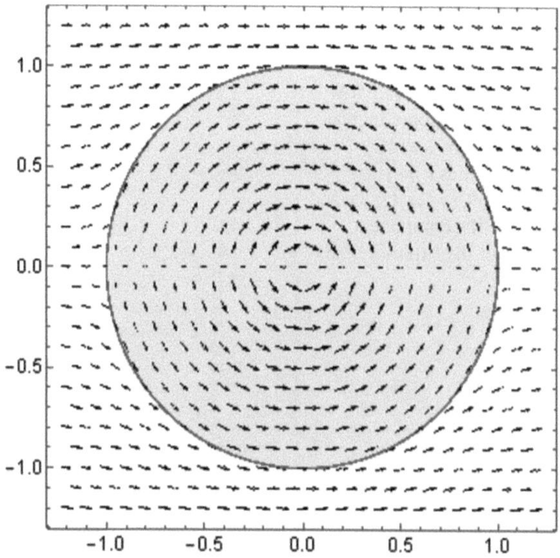

Fig. 8.7 Electric field vectors in section $x_3 = 0$ in the vicinity of the radially transverse isotropic inclusion with the parameters indicated in Fig. 8.6

In the next example, we consider the inclusion of the conductivity coefficients

$$c = 10c_0, c_n = -9.9c_0. \tag{8.299}$$

In this case, the roots λ_1 and λ_2 are

$$\lambda_1 = 12.651, \lambda_2 = -15.651. \tag{8.300}$$

Distributions of the field component $e_1(x_1, x_2)$ in the section $x_3 = 0$ are presented in Fig. 8.8, and field vectors $e(x)$ in the inclusion vicinity are shown in Fig. 8.9.

8.6 Notes

A spherical isotropic inclusion with an isotropic layer in a homogeneous isotropic elastic medium was considered in [3–5]. The solution of this problem for a cylindrical inclusion with a layer was obtained in [6]. The Mellin transform method was used for the solution of the elasticity and thermo-elasticity problem for multilayered spherical and cylindrical inclusions with isotropic layers in [7–10]. In [11], the problem for a spherical multilayered inclusion with isotropic layers in a homogeneous isotropic host medium was solved by conventional methods of elasticity.

Radially transverse isotropic inclusions with constant elastic and conductive coefficients in a homogeneous isotropic host medium were considered in [12]. The solution was found in the form of a series of spherical harmonics and radial exponential

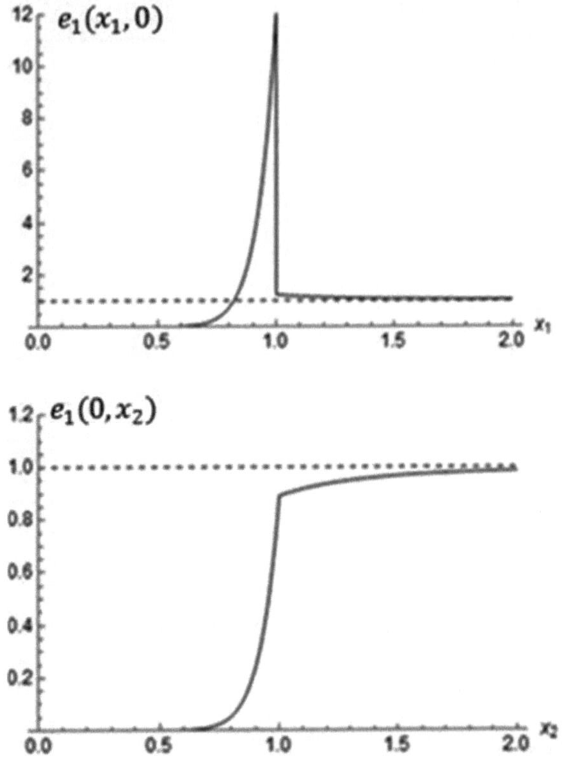

Fig. 8.8 Distributions of the field component $e_1(x_1, x_2)$ along the x_1 and x_2-axes in the vicinity of the radially transverse isotropic inclusion of radius $a = 1$ with conductivities $c = 10$, $c_n = -9.9$, ($c_0 = 1$); dashed lines are the component $e_1^0 = 1$ of the external filed

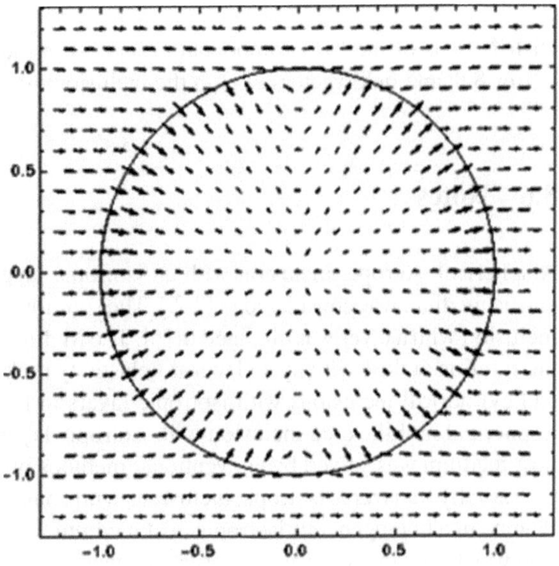

Fig. 8.9 Electric field vectors in section $x_3 = 0$ in the vicinity of the radially transverse isotropic inclusion with the parameters indicated in Fig. 8.8

functions. In [13], the solution of the conductivity problem for a periodic array of multilayered spherical inclusions was obtained in the form of a series of spherical harmonics, and the effective conductivity of the composites with a cubic lattice of the inclusions was calculated. An approximate solution of the conductivity problem for multilayered radially transverse isotropic inclusions was proposed in [14]. The elasticity and thermo-elasticity problems for cylindrical multilayered inclusion with orthorhombic layers were solved by conventional methods of elasticity in [15, 16].

The results of this chapter are based on the works [17, 18].

References

1. Nye J (1985) Physical properties of crystals, their representation by tensors and matrices. Clarendon Press, Oxford, p 329
2. Fricoris G (2022) Mellin-transform method for integral evaluation. Springer Nature Switzerland AG
3. Kerner E (1956) The elastic and thermoelastic properties of composite media. Proc Phys Soc 69(B):808–813
4. Christensen R, Loo L (1979) Solutions for effective shear properties in three phase and cylinder models. J Mech Phys Solids 27:315–330
5. Matonis V, Small N (1969) A macroscopic analysis of composites containing layered spherical inclusions. Polymer Eng Sci 9:90–99
6. Hashin Z, Rosen B (1964) The elastic moduli of fiber-reinforced materials. J Appl Mech 31:223–2
7. Kanaun S, Kudryavtzeva L (1986) Multi-layered spherical inclusions in an homogeneous elastic medium. Appl Math Mech (PMM) 50:633–643
8. Kanaun S, Kudryavtzeva L (1987) Thermal stresses in composites with multi-layered spherical inclusions. Proc Acad Sci USSR Mech Solids 4:113–121
9. Kanaun S, Kudrjavtzeva L (1989) Thermo elastic properties of composite materials reinforced with unidirectional multi-layered fibers. Appl Math Mech (PMM) 53:798–807
10. Kanaun S, Levin V (2008) Self-consistent methods for composites, v I static problems, 1st edn. Springer, Dordrecht
11. Hervé E, Zaoui A (1993) N-layered inclusion-based micromechanical modelling. Int J Eng Sci 31(1):1–10
12. Chen T (1993) Thermoelastic properties and conductivity of composites reinforced by spherically anisotropic particles. Mech Mater 14:257–268
13. Kuo H, Chen T (2006) Effective transport properties of arrays of multicoated or graded spheres with spherically transversely isotropic constituents. J. Appl. Phys. 99 (093702)
14. Pham D (2018) Solutions for the conductivity of multi-coated spheres and spherically-symmetric inclusion problems. Z Angew Math Phys 69:13
15. Tsukrov I, Drach B (2010) Elastic deformation of composite cylinders with cylindrically orthotropic layers. Int J Solids Struct 47:25–33
16. Tsukrov I, Drach B, Gross T (2012) Effective stiffness and thermal expansion coefficients of unidirectional composites with fibers surrounded by cylindrically orthotropic matrix layers. Int J Eng Sci 58:129–143
17. Kanaun S (2024) Radially transverse isotropic inclusions in isotropic conductive media: local fields, effective properties, neutral inclusions. Int J Eng Sci 194:103965
18. Kanaun S (2024) Radially transverse isotropic inclusions in isotropic elastic media: local fields, neutral inclusions, effective elastic properties. Int J Eng Sci 200:104078

Chapter 9
Neutral Inclusions

Abstract Spherical inclusions with a radially transverse isotropic layer in conductive and elastic media are considered. The objectives are revealing the properties of the interface layers that make inclusions neutral for homogeneous external fields. Neutral inclusions do not disturb external fields applied to the medium and therefore, they are "invisible" for external observers.

Keywords Strong and weak neutral inclusions · Neutral radially transverse isotropic inclusions · Neutral isotropic inclusions with transverse isotropic layers at the interfaces · Equivalent inclusions

The elasticity and conductivity problems for multilayered inclusions relate to the problem of neutral inclusions or cloaking inclusions from external fields in a homogeneous host medium. Neutral inclusions do not disturb external fields applied to the medium and therefore, they are "invisible" for external observers. A homogeneous isotropic inclusion cannot be neutral except in the case of hydrostatic pressure. But a radially transverse isotropic inclusion or isotropic inclusions with external layers can be neutral by an appropriate choice of conductive and elastic constants of the layers. In this chapter, spherical inclusions with a radially transverse isotropic layer in conductive and elastic media are considered. The objectives are revealing the properties of the interface layers that make inclusions neutral for homogeneous external fields.

9.1 Neutral Radially Transverse Isotropic Inclusions in an Isotropic Homogeneous Conductive Media

We consider the electric field in an isotropic homogeneous conductive medium containing a radially transverse isotropic inclusion. The conductivity tensor $C(x)$ of the inclusion is defined by the equation

$$C_{\alpha\beta}(x) = c(r)\delta_{\alpha\beta} + c_n(r)n_\alpha n_\beta, r = |x|, n_\alpha = \frac{x_\alpha}{r}, \tag{9.1}$$

S. Kanaun, *Self-Consistent Methods for Composites*, Solid Mechanics and Its Applications 282, https://doi.org/10.1007/978-3-031-96739-9_9

where $c(r)$, $c_n(r)$ are scalar coefficients, the conductivity tensor of the host medium is $C^0_{\alpha\beta} = c_0\delta_{\alpha\beta}$. If the medium is subjected to a constant external field e^0 at infinity, the electric field $e(x)$ in the medium is defined in Eq. (8.253)

$$e(x) = (I + A(x))\, e^0, \tag{9.2}$$

$$A_{\alpha\beta}(x) = \big(I + n_\alpha n_\beta D\big)\, \beta(r),\;\; D = r\frac{d}{dr}, \tag{9.3}$$

where the scalar function $\beta(r)$ is the solution of the differential equation (8.260). For a spherical inclusion of the radius a, the function $\beta(r)$ in the host medium is given by Eq. (8.286)

$$\beta(r) = y_2^{(2)} r^{-3},\;\; r > a, \tag{9.4}$$

where $y_2^{(2)}$ is a constant. For a neutral inclusion, $\beta(r)$ is equal to zero in the host medium, and therefore, $y_2^{(2)} = 0$.

In the case of an isotropic homogeneous inclusion, $c_n = 0$ and the expression for the constant $y_2^{(2)}$ follows from Eq. (8.296) in the form

$$y_2^{(2)} = \frac{c_0 - c}{2c_0 + c}. \tag{9.5}$$

The condition $y_2^{(2)} = 0$ leads to a trivial result $c = c_0$ (conductivities of the inclusion and the host medium coincide). But for a radially transverse isotropic inclusion, there exist non-trivial neutral inclusions with the coefficients $c \neq c_0$, $c_n \neq 0$.

For an inclusion of the radius a with the constant conductivity coefficients c and c_n, the constant $y_2^{(2)}$ in Eq. (8.296) takes the form

$$y_2^{(2)} = \left(\frac{3c_0}{2c_0 + (c + c_n)(\lambda_1 + 1)} - 1\right) a^3, \tag{9.6}$$

$$\lambda_1 = -\frac{1}{2}\left(3 - \sqrt{1 + \frac{8c}{c + c_n}}\right). \tag{9.7}$$

The condition $y_2^{(2)} = 0$ can be considered as a relation between the coefficients c and c_n that guaranties the absence of the external field disturbance in the host medium. This relation can be presented in the form

$$\frac{c + c_n}{c_0} = \left(2\frac{c}{c_0} - 1\right)^{-1}. \tag{9.8}$$

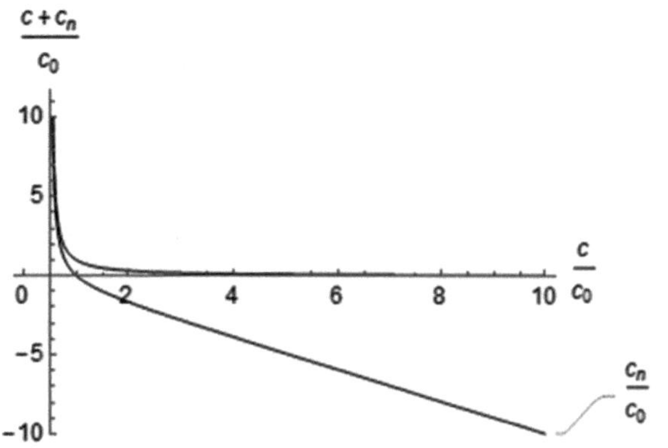

Fig. 9.1 Dependencies of the conductivities $(c + c_n)/c_0$ (upper line) and c_n/c_0 (low line) of a neutral radially transverse isotropic inclusion on the isotropic part c/c_0 of the inclusion conductivity tensor, c_0 is the conductivity of the host medium

For the conductivity coefficients c and c_n that satisfy this equation, the function $\beta(r)$ takes the form

$$\beta(r) = \left(\frac{r}{a}\right)^{\lambda_1} - 1, r < a; \beta(r) = 0, r > a. \tag{9.9}$$

Because $c + c_n > 0$, the ratio c/c_0 in Eq. (9.8) should be more than 0.5. For neutral inclusions, dependences of $(c + c_n)/c_0$ and c_n/c_0 on the isotropic part c of the inclusion conductivity tensor are presented in Fig. 9.1.

9.1.1 An Isotropic Inclusion with an Isotropic Layer at the Interface

Let a spherical inclusion of radius a consist of an isotropic core and an isotropic layer at the boundary with the host medium. The conductivity coefficient of the core is $c^{(1)} = c$, and the same for the layer is $c^{(2)}$. The core radius is $a - h$, h is the layer thickness, $c^{(3)} = c_0$ is the conductivity coefficient of the host medium. If the medium is subjected to a constant electric field at infinity, the electric field in the medium is defined in Eq. (9.3), where the function $\beta(r)$ is

$$\beta(r) = Y_1^{(1)} - 1, \ r < a - h, \tag{9.10}$$

$$\beta(r) = Y_1^{(2)} + Y_2^{(2)} r^{-3} - 1, a - h < r < a, \tag{9.11}$$

$$\beta(r) = Y_2^{(3)} r^{-3}, \ r > a. \tag{9.12}$$

In this case, the vector $\mathbf{x}^{(3)}(a) = [\beta(a+0)+1,\, D\beta(a+0)]^T$ of the solution in the host medium relates to the vector of the solution in the core $\mathbf{x}^{(1)}$ by the equation

$$\mathbf{x}^{(3)}(a) = \gamma^{(2)}\mathbf{r}^{(2)}\gamma^{(1)}\mathbf{x}^{(1)}, \tag{9.13}$$

where the matrices $\gamma^{(1)}, \gamma^{(2)}$ are defined in Eq. (8.282), the matrix $\mathbf{r}^{(2)}$ and vector $\mathbf{x}^{(1)}$ are

$$\mathbf{r}^{(2)} = \begin{bmatrix} 1, & \frac{1}{3}\left(1 - \frac{(a-h)^3}{a^3}\right) \\ 0, & \frac{(a-h)^3}{a^3} \end{bmatrix}, \mathbf{x}^{(1)} = \begin{bmatrix} Y_1^{(1)} - 1 \\ 0 \end{bmatrix}. \tag{9.14}$$

For a neutral inclusion, the components of the vector $\mathbf{x}^{(3)}(a)$ have the values

$$x_1^{(3)}(a) = 1, x_2^{(3)}(a) = 0. \tag{9.15}$$

This is the system of equations for the layer conductivity $c^{(2)}$ and the constant $Y_1^{(1)}$ in the solution inside the core. We consider two particular cases when the core conductivity c is much smaller or much larger than the conductivity c_0 of the host medium.

1. $c = 0.1c_0$, $h/a = 0.1$. For such parameters, the constant $Y_1^{(1)}$ and the layer conductivity $c^{(2)}$ of the neutral inclusion are

$$Y_1^{(1)} = 1.0969, c^{(2)} = 4.5924c_0. \tag{9.16}$$

The corresponding function $\beta(r)$ is as follows

$$\begin{aligned} &\beta(r) = 0.0969, && 0 < r < a-h, \\ &\beta(r) = -0.2608 + 0.2608\left(\tfrac{r}{a}\right)^{-3}, && a-h < r < a, \\ &\beta(r) = 0, && r > a. \end{aligned} \tag{9.17}$$

Distributions of the components e_1 and j_1 of the electric field and current vectors along x_1 and x_2-axes ($x_3 = 0$) for the neutral inclusion are presented in Figs. 9.2 and 9.3. The directions of the electric field vectors in the vicinity of the inclusion are shown in Fig. 9.4. Note that in the layer, the electric field and current vectors are almost parallel to the layer's middle surface.

We introduce the parameter η that is an integral characteristic of the layer

$$\eta = \frac{h}{a}\frac{c^{(2)}}{c_0}. \tag{9.18}$$

In Fig. 9.5, dependencies of the layer conductivity $c^{(2)}/c_0$ and the parameter η for neutral inclusions on the dimensionless core radius $(a-h)/a$ are shown. It is seen that when $h \to 0$, the layer conductivity $c^{(2)}$ tends to infinity, meanwhile the parameter η remains finite, and this parameter tends to its limit value $\eta = 0.45$.

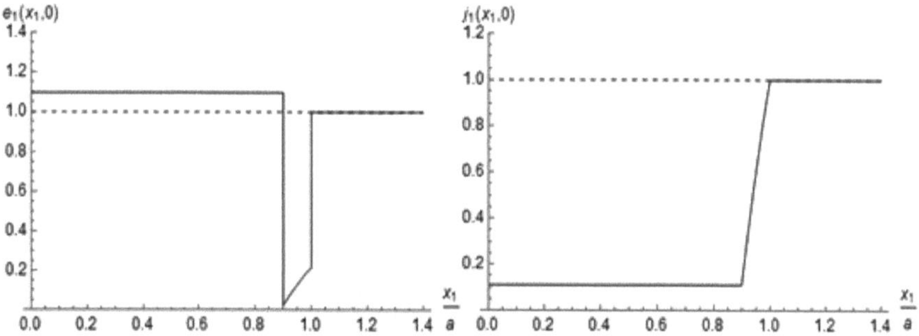

Fig. 9.2 Distributions of the components $e_1(x_1, 0)$ and $j_1(x_1, 0)$ of the field and flux vectors along the x_1-axes for the neutral inclusion with an isotropic core of the radius $a - h$ and conductivity $c = 0.1c_0$ and an isotropic layer of thickness $h = 0.1a$ and conductivity $c^{(2)} = 4.592c_0$, c_0 is conductivity of the host medium; dashed lines are the components e_1^0, j_1^0 of the external electric field and current vectors acting along the x_1-axis

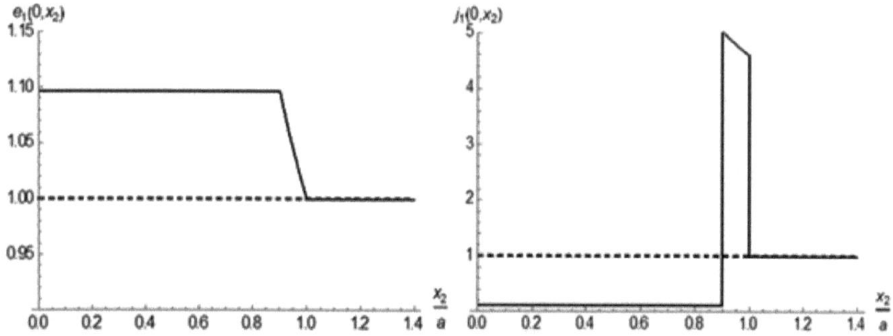

Fig. 9.3 The same as in Fig. 9.2 for the components of the electric field and current vectors along the x_2-axis

When $h \to 0$, $c^{(2)} \to \infty$, and $\eta = O(1)$, the equation for the limiting vector $\mathbf{x}^{(3)}(a)$ in Eq. (9.13) takes the form

$$\mathbf{x}^{(3)}(a) = \widehat{\gamma}^{(3)}\mathbf{x}^{(1)}(a), \widehat{\gamma}^{(3)} = \lim_{h\to 0} \gamma^{(2)}\mathbf{R}^{(2)}\gamma^{(1)}, \tag{9.19}$$

The explicit expressions for the matrix $\widehat{\gamma}^{(3)}$ is

$$\widehat{\gamma}^{(3)} = \begin{bmatrix} 1, & 0 \\ 2\eta - 1 + \frac{c}{c_0}, & \frac{c}{c_0} \end{bmatrix}, \tag{9.20}$$

and the conditions similar to (8.275) for the jumps of the functions $\beta(r)$ and $D\beta(r)$ at $r = a$ are

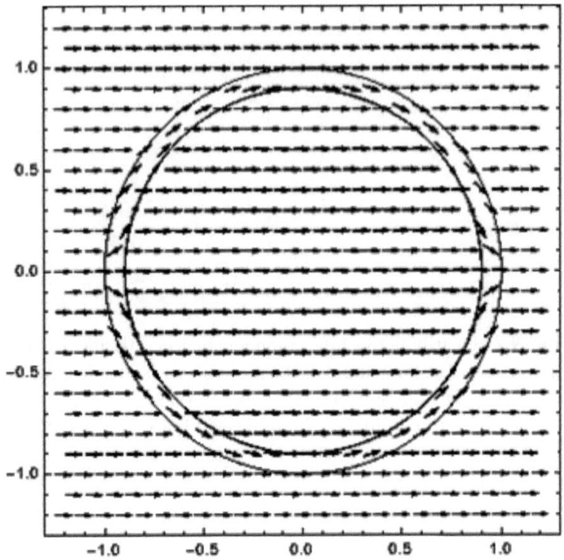

Fig. 9.4 Electric field vectors in the vicinity of the neutral inclusion with the parameters indicated in Fig. 9.2

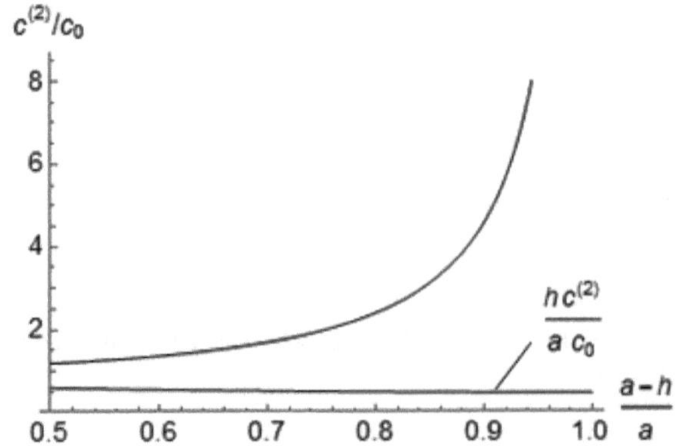

Fig. 9.5 Dependencies of the conductivity of the layer $c^{(2)}$ and the parameter η in Eq. (9.18) on the core radius $(a-h)/a$

$$[\beta]_{r=a} = \beta(a+0) - \beta(a-0) = 0, \tag{9.21}$$

$$[cD\beta]_{r=a} = c_0 D\beta(a+0) - cD\beta(a-0) =$$

$$= ((2\eta - 1)c_0 + c)(\beta(a) + 1). \tag{9.22}$$

It follows from Eq. (9.2) that projection of the electric field on the plane orthogonal to the vector n is

$$\theta_{\alpha\beta}(n)e_j(x) = (1 + \beta(r))\theta_{\alpha\beta}(n)e_j^0,\ \theta_{\alpha\beta}(n) = \delta_{\alpha\beta} - n_\alpha n_\beta. \tag{9.23}$$

For $\beta(r)$ is continuous at $r = a$, the vector $\theta(n)e(x)$ is continuous at the limiting inclusion boundary.

In the limit $h \to 0$, the vector $\mathbf{x}^{(1)}(a)$ for the neutral inclusion has the components $x_1^{(1)}(a) = y_1^{(1)} - 1$, $x_2^{(1)}(a) = 0$, and the conditions $x_1^{(3)}(a) = 1$, $x_2^{(3)}(a) = 0$ result in the following values of the constant $y_1^{(1)}$ and the parameter η

$$y_1^{(1)} = 1, \eta = \frac{c_0 - c}{2c_0}. \tag{9.24}$$

As a result, $\beta(r) = 0$ everywhere, and the external field e^0 is disturbed neither in the host medium nor in the inclusion. Meanwhile the current $j(x)$ has a jump at the inclusion surface $r = a$

$$[j]_{r=a} = j(a+0) - j(a-0) = (c_0 - c)e^0. \tag{9.25}$$

For the neutral inclusion with the limiting layer, the graphs of the components $e_1(x_1, 0)$ and $j_1(x_1, 0)$ of the electric field and current vectors in the medium are shown in Fig. 9.6.

2. $c = 10c_0, h = 0.1a$. In this case, for the neutral inclusion, the layer conductivity is $c^{(2)} = 0.1236c_0$, and the function $\beta(r)$ is defined by the equations

$$\begin{aligned} &\beta(r) = -0.87828, && 0 < r < a - h, \\ &\beta(r) = 2.36261 - 2.36261 \left(\tfrac{r}{a}\right)^{-3}, && a - h < r < a, \\ &\beta(r) = 0, && r > a. \end{aligned} \tag{9.26}$$

For the neutral inclusion, the distributions of the components e_1, j_1 of the electric field and current vectors along the x_1 and x_2-axes in section $x_3 = 0$ are presented in Figs. 9.7 and 9.8. The electric field vectors in the vicinity of the inclusion are shown in Fig. 9.9. It is seen that in the layer, the field vectors are almost orthogonal to the layer middle surface.

We introduce the integral characteristic δ of the layer

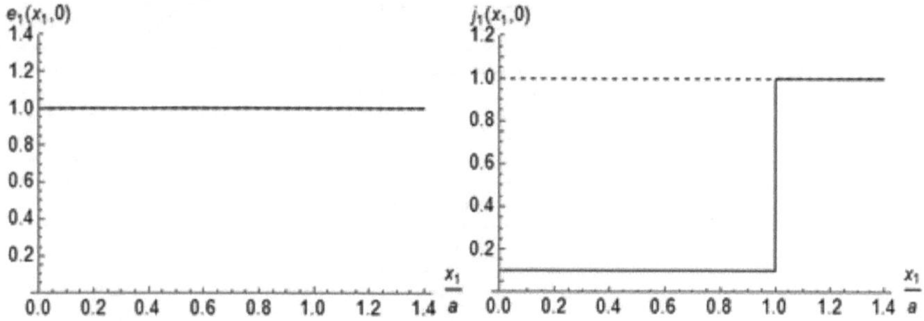

Fig. 9.6 Limit distributions of the components $e_1(x_1, 0)$ and $j_1(x_1, 0)$ of the electric field and current vectors along the x_1-axis for the neutral inclusion with a thin layer at the inclusion boundary ($x_1 = a$). The layer thickness $h \to 0$, layer conductivity $c^{(2)} \to \infty$, but parameter $\varsigma = hc^{(2)}/ac_0 \to 0.45$, $c = 0.1c_0$ is conductivity of the inclusion, c_0 is conductivity of the host medium

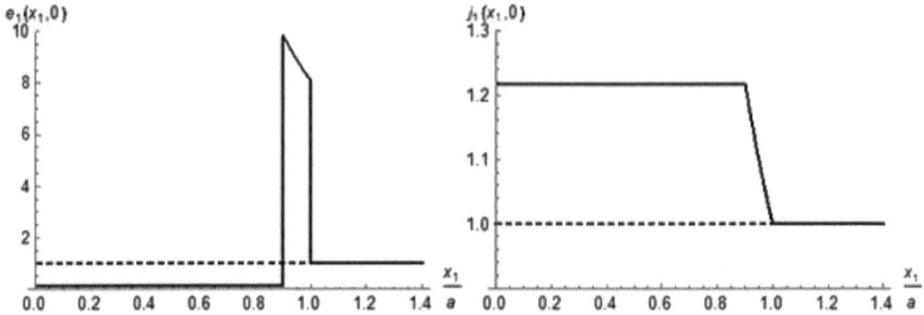

Fig. 9.7 Distributions of the components $e_1(x_1, 0)$ and $j_1(x_1, 0)$ of the field and flux vectors along the x_1-axes for the neutral inclusion with an isotropic core of the radius $a - h$ and conductivity $c = 0.1c_0$ and an isotropic layer of thickness $h = 0.1a$ and conductivity $c^{(2)} = 4.592c_0$, c_0 is conductivity of the host medium; dashed lines are the components e_1^0, j_1^0 of the external electric field and current vectors acting along the x_1-axis

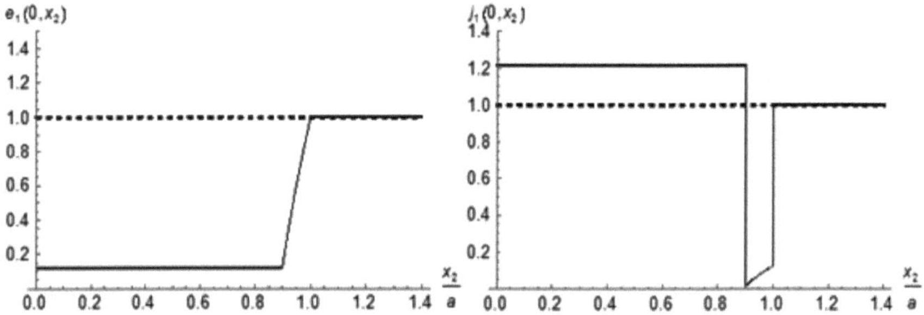

Fig. 9.8 The same as in Fig. 9.7 for the components of electric field and current vectors along the x_2-axis

Fig. 9.9 Field vectors in the vicinity of the neutral inclusion with the parameters indicated in Fig. 9.8

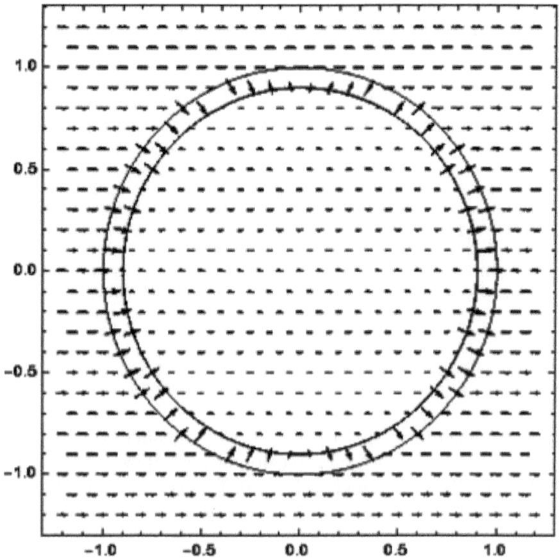

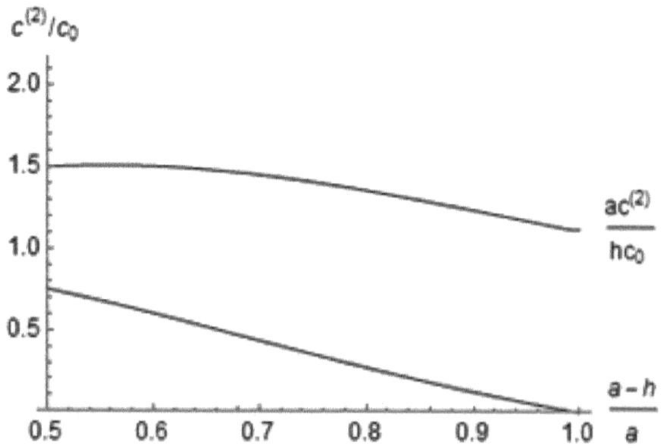

Fig. 9.10 Dependencies of the layer conductivity $c^{(2)}/c_0$ (low line) and the parameter $((ac^{(2)})/(hc_0))$ (upper line) on the kernel radius $(a-h)/a$ for isotropic neutral inclusions with an isotropic layer; the conductivity of the kernel is $c = 10c_0$, c_0 is the conductivity of the host medium

$$\delta = \frac{a}{h}\frac{c^{(2)}}{c_0} \tag{9.27}$$

In Fig. 9.10, dependences of the layer conductivity $c^{(2)}/c_0$ and the parameter δ on the dimensionless core radius $(a-h)/a$ are shown. When $h \to 0$, the conductivity $c^{(2)}$ tends to zero, meanwhile the parameter δ remains finite and tends to $\delta = 1.11$.

When $h \to 0$, $c^{(2)} \to 0$, and $\delta = O(1)$, the limiting vector $\mathbf{x}^{(3)}(a)$ in Eq. (9.73) takes the form

$$\mathbf{x}^{(3)}(a) = \widehat{\gamma}^{(3)}\mathbf{x}^{(1)}(a), \tag{9.28}$$

where the matrix $\widehat{\gamma}^{(3)}$ is

$$\widehat{\gamma}^{(3)} = \begin{bmatrix} 1 + \frac{c}{\delta c_0}, & \frac{c}{\delta c_0} \\ -1 + \frac{(\delta-1)c}{\delta c_0}, & \frac{(\delta-1)c}{\delta c_0} \end{bmatrix}. \tag{9.29}$$

In the limiting layer, the jumps of the functions $\beta(r)$ and $D\beta(r)$ at $r = a$ are defined by the equations

$$[\beta]_{r=a} = \frac{c}{\delta c_0}(1 + \beta(a-0) + D\beta(a-0)), \tag{9.30}$$

$$[D\beta]_{r=a} = \left(\frac{(\delta-1)c}{\delta c_0} - 1\right)(1 + \beta(a-0) + D\beta(a-0)). \tag{9.31}$$

In the core, the vector $\mathbf{x}^{(1)}(a)$ has components $(y_1^{(1)} - 1, 0)$. For the neutral inclusion, the constants $y_1^{(1)}$ and δ are to be found from the equations $x_1^{(3)}(a) = 1, x_2^{(3)}(a) = 0$. From this system and Eq. (9.28), and we obtain

$$y_1^{(1)} = \frac{c_0 - c}{c} - 1, \delta = \frac{c}{c - c_0}. \tag{9.32}$$

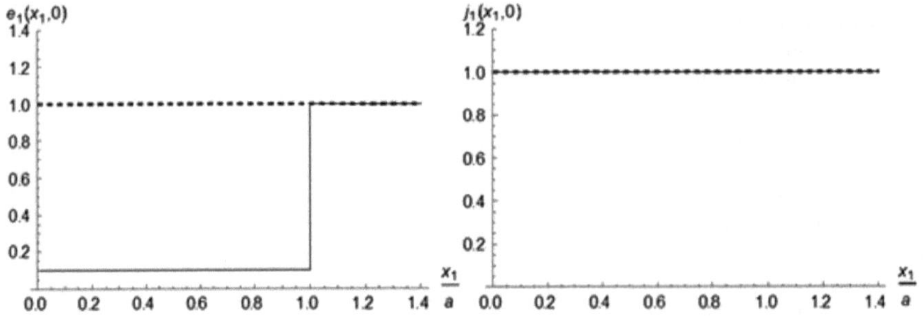

Fig. 9.11 Distributions of the components $e_1(x_1, 0)$ and $j_1(x_1, 0)$ of the electric field and current vectors along x_1-axis for the limit inclusion when $h \to 0$, $c^{(2)} \to 0$ and $\delta = (c_0 - c)/c$

As a result, function $\beta(r)$ takes the form

$$\beta(r) = \frac{c_0 - c}{c}, r < a, \beta(r) = 0, r > a. \tag{9.33}$$

For the limiting neutral inclusion, distributions of the components $e_1(x_1, 0)$ and $j_1(x_1, 0)$ of the electric field and current vectors along x_1-axis are shown in Fig. 9.11. In this case, vector $e(x)$ has a jump at the inclusion boundary

$$[e]_{r=a} = \frac{c - c_0}{c} e^0, \tag{9.34}$$

meanwhile the external current j^0 is disturbed neither in the host medium nor in the inclusion.

It follows from Eq. (9.30) that the function $\beta(r)$ has a jump at the inclusion boundary. As a result, vectors $e(r)$ and $j(r)$ contain the delta-functions concentrated at $r = a$. These delta-functions are not shown in Fig. 9.11.

Remark. The limiting layers that make the inclusions neutral are equivalent to the singular models of thin inclusions of small or large conductivity considered in Sect. 6.6.

9.2 Neutral Radially Transverse Isotropic Inclusion in a Homogeneous Elastic Medium

We consider an isotropic host medium containing a radially transverse isotropic inclusion subjected to a constant external strain field ε^0 at infinity. The strain tensor $\varepsilon(x)$ in the medium with the inclusion has form in Eq. (8.9)

$$\varepsilon(x) = (I + A(x))\, \varepsilon^0, \tag{9.35}$$

where tensor $A(x)$ is presented in the E-basis as follows

$$A(x) = \sum_{j=1}^{6} A_j(r) E^j(n), r = |x|, n_\alpha = \frac{x_\alpha}{|x|}. \tag{9.36}$$

The scalar coefficients $A_j(r)$ in this equation are expressed in terms of 3 scalar functions $\beta(r), \alpha_1(r), \alpha_3(r)$ of the radial coordinate r

$$A_1 = (3 + D)\alpha_1, A_2 = \frac{1}{3}(\beta - 3\alpha_1 + (2 + D)\alpha_3), \tag{9.37}$$

$$A_3 = D(\alpha_1 + \alpha_3), A_4 = \frac{1}{3} D(\beta - 3\alpha_1 - (2 + D)\alpha_3), \tag{9.38}$$

$$A_5 = D(4\alpha_3 + (1 + D)\alpha_1), A_6 = D(2 - D)(\alpha_1 - \alpha_3). \tag{9.39}$$

Here, $D = rd/dr$ is the differential operator. The system of differential equations for these functions is derived in Sect. 8.4, and the solution of this system depends on the coefficients $c_i(r)$ of the transverse isotropic inclusion in Eq. (8.1). In the isotropic host medium, the functions $\beta(r), \alpha_1(r), \alpha_3(r)$ have the forms

$$\beta(r) = \frac{y_2}{r^3}, \alpha_1(r) = \frac{Y_3}{r^3} + \frac{Y_4}{r^5}, \alpha_3(r) = \frac{Y_7}{r^3} + \frac{Y_8}{r^5}, \tag{9.40}$$

where y_2, Y_3, Y_4, Y_7, Y_8 are constants. A neutral inclusion does not disturb the external field applied to the medium, and therefore, all these constants should be equal to zero. According to the definition of Norris and Parnell [1], such inclusions are called strong neutral. For weak neutral inclusions, the principal terms of the far-field disturbances of the external field are equal to zero. These terms are proportional to r^{-3}, and therefore, for weak neutral inclusions, the constants y_2, Y_3, Y_7 in Eqs. (9.40) are equal to zero only. Thus, for radially transverse isotropic inclusions, we can introduce two functionals

$$F^s(c_1, c_2, c_3, c_5, c_6) = |y_2| + |Y_3| + |Y_4| + |Y_7| + |Y_8|, \tag{9.41}$$

$$F^w(c_1, c_2, c_3, c_5, c_6) = |y_2| + |Y_3| + |Y_7|, \tag{9.42}$$

where $c_m = c_m(r)$, $m = 1, 2, 3, 5, 6$ are elastic coefficients of the inclusion material. For a strong neutral inclusion, $F^s = 0$, and for a weak neutral one, $F^w = 0$.

Because the algorithm for solving the elasticity problem for the radially transverse isotropic inclusions is numerical, the problem of determination of elastic constants of strong and weak neutral inclusions is reduced to numerical search of zero-minima of the functionals F^s or F^w. In the case of multilayered inclusions, the functionals F^s, F^w become multivariable functions of the elastic constants of the layers, and the problem is reduced to numerical searching minima of such functions. The coordinate descent method can be used for this purpose [2, 3].

9.2.1 The Inclusions Neutral for Hydrostatic Pressure

We consider a transverse isotropic inclusion with constant coefficients c_m in Eq. (8.1) for the stiffness tensor $C(x)$ of the inclusion. By hydrostatic external loading, the external strain tensor is $\varepsilon^0_{\alpha\beta} = \varepsilon_0\delta_{\alpha\beta}$, and the strain tensor $\varepsilon(x)$ in the medium with a radially transverse isotropic inclusion is expressed in terms of a scalar function $\beta(r)$ only

$$\varepsilon_{\alpha\beta}(x) = \varepsilon_0\left(\delta_{\alpha\beta} + 3(\delta_{\alpha\beta} + n_\alpha n_\beta D)\beta(r)\right). \tag{9.43}$$

For a spherical inclusion with constant elastic coefficients $c^{(1)}_m$ in Eq. (8.1), the solution inside the inclusion is defined by Eq. (8.116). Vector $\mathbf{x}^{(1)}(r) = (\beta(r)+1, D\beta(r))^T$ of the solution at the inclusion boundary $r = a - 0$ has the components

$$x^{(1)}_1(a) = \beta(a-0) + 1 = y^{(1)}_1 a^{\lambda_1(1)}, \tag{9.44}$$

$$x^{(1)}_2(a) = D\beta(a-0) = \lambda_1(1) y^{(1)}_1 a^{\lambda_1(1)}, \tag{9.45}$$

$$\lambda_1(1) = -\frac{3}{2}\left(1 - \sqrt{1 - \frac{8b^{(1)}_1}{9b^{(1)}_0}}\right). \tag{9.46}$$

Here, the coefficients $b^{(i)}_0$ and $b^{(i)}_1$ are expressed in terms of the elastic constants $c^{(i)}_m$ of the inclusion ($i = 1$) and the host medium ($i = 2$)

$$b^{(i)}_0 = c^{(i)}_1 + c^{(i)}_2 + 2c^{(i)}_3 + c^{(i)}_5 + c^{(i)}_6,\, b^{(i)}_1 = 3c^{(i)}_3 + c^{(i)}_5 + c^{(i)}_6. \tag{9.47}$$

The expression for the vector $\mathbf{x}^{(2)}(a+0)$ in the host medium follows from Eq. (8.116)

$$\mathbf{x}^{(2)}(a+0) = \gamma^{(1)}\mathbf{x}^{(1)}(a-0), \tag{9.48}$$

where the matrix $\gamma^{(1)}$ is defend in Eq. (8.117). The function $\beta(r)$ in the entire space is defined by the equations

$$\beta(r) = y^{(1)}_1 r^{\lambda_1(1)} - 1, r < a;\ \beta(r) = y^{(2)}_2 r^{-3}, r > a. \tag{9.49}$$

From Eqs. (8.123)–(8.126) for $N = 1$, we obtain the constants $y^{(1)}_1$ and $y^{(2)}_2$ in Eq. (9.49)

$$y^{(1)}_1 = \frac{3b^{(2)}_0 a^{-\lambda_1(1)}}{3b^{(2)}_0 + \lambda_1(1)b^{(1)}_0 - [b_1 + b_3]_1}, \tag{9.50}$$

$$y_2^{(2)} = \left(\frac{3b_0^{(2)}}{3b_0^{(2)} + \lambda_1(1)b_0^{(1)} - [b_1 + b_3]_1} - 1 \right) a^3, \tag{9.51}$$

where $b_3^{(i)} = c_1^{(i)} + 3c_2^{(i)} + c_3^{(i)}$, $[b_m]_1 = b_m^{(2)} - b_m^{(1)}$ is the jump of the constant b_m at the inclusion boundary. For an isotropic host medium, $b_1^{(2)} = 0$.

Because for a neutral inclusion, the function $\beta(r)$ is equal to zero in the host medium ($r > a$), the constant $y_2^{(2)}$ in Eqs. (9.49) and (9.51) is equal to zero. The condition $y_2^{(2)} = 0$ results in the expression for the constant $b_0^{(1)}$ of the inclusion material in terms of the constants $b_1^{(1)}$, $b_3^{(1)}$ and $b_3^{(2)}$

$$b_0^{(1)} = \frac{\left(b_1^{(1)} + [b_3]_1\right)^2}{b_1^{(1)} + 3[b_3]_1} = c_{33}^{(1)}. \tag{9.52}$$

Here, $c_{33}^{(1)}$ is the elastic constant of the inclusion material in the matrix notations if the local x_3-axis is directed along the vector n in Eq. (8.1) for the elastic stiffness tensor $C(x)$. Because $c_{33} > 0$, the neutral inclusion exists if

$$b_1^{(1)} + 3[b_3]_1 > 0. \tag{9.53}$$

The function $\beta(r)$ for an inclusion neutral to hydrostatic loading takes the form

$$\beta(r) = \left(\frac{r}{a}\right)^{\lambda_1(1)} - 1, r < a; \beta(r) = 0, r > a, \tag{9.54}$$

$$\lambda_1(1) = -\frac{1}{2}\left(3 - \frac{|b_1^{(1)} + 3[b_3]_1|}{b_1^{(1)} + [b_3]_1}\right). \tag{9.55}$$

In the case of an isotropic inclusion, $b_1^{(1)} = 0$, $\lambda_1(1) = 0$, and the condition $y_2^{(2)} = 0$ results in the equation

$$[b_3]_1 = 0 \text{ or } 2\mu + 3\lambda = 2\mu_0 + 3\lambda_0. \tag{9.56}$$

Here, λ, μ and λ_0, μ_0 are Lame's parameters of the inclusion and the host medium. As a result, for an isotropic neutral inclusion subjected to hydrostatic loading, $\beta(r) = 0$ in the entire medium.

9.2.2 Radially Transverse Isotropic Inclusions Neutral for Arbitrary Homogeneous External Fields

Let the elastic coefficients of a radially transverse isotropic inclusion be constant. In this case, F^s and F^w in Eqs. (9.41), (9.42) are functions of 5 variables, and strong or weak neutral inclusions exist if there exist zero-minima of these functions for physically acceptable values of the constants c_m. The constant c_m that correspond to the neutral inclusions compose branches in the 5D-space of the elastic constants. It is obvious that one of these branches passes through the point

$$c_1 = c_1^0, c_2 = c_2^0, c_3 = c_5 = c_6 = 0 \tag{9.57}$$

(the elastic stiffness tensors of the inclusion and the host medium coincide). Then, let the coefficient c_1 be changed as $c_1 = c_1^0 + k\delta c_1$, where δc_1 be the step of change, $k = 1, 2, ...$ The other coefficients c_m are to be found to minimize the functions F^s (strong neural inclusion) or F^w (weak neutral inclusion). Numerical searching minima of the function F^w for weak neutral inclusions in Eq. (9.42) with respect to c_2, c_3, c_5, c_6 shows that this function has a clearly defined zero-minima for all possible values of the constant c_1. The function F^s for strong neutral inclusions in Eq. (9.41) has minima in the vicinity of the minima of F^w but these minima are nor equal to zero. Thus, the corresponding inclusion is not a strong neutral one. In Fig. 9.12, dependencies of the coefficients c_2, c_3, c_5, c_6 on the coefficient c_1 for weak neutral inclusions are shown in the region $0 < c_1 < 10$, the elastic coefficients of the host medium are $c_1^0 = 1$, $c_2^0 = 0.75$ ($\nu_0 = 0.3$).

Note that combinations of the elastic coefficients that satisfy the condition for weak neutral inclusions are not unique. In Fig. 9.13, the dashed lines indicate a branch of elastic constants of weak neutral inclusions that does not pass through the point indicated in Eq. (9.57).

The graphs of the functions $\alpha_1(r), \alpha_2(r), \beta(r)$ for weak neutral inclusions with $c_1 = 0.1$ are shown in Fig. 9.14.

The values of the other constants are indicated in the figure. The distributions of radial $\sigma_{rr}(r, \theta)$ and hoop $\sigma_{\theta\theta}(r, \theta)$ stresses along the x_3-axes ($\theta = 0$) for such inclusions subjected to hydrostatic pressure are in Fig. 9.15 (above) and such distributions for 1D-tension in the direction of the x_3-axis are shown in Fig. 9.15 (below) (x_3 is the polar axis of a spherical coordinate system with the origin at the inclusion center).

In Fig. 9.16, the functions α_1, α_3 and β for a weak neutral inclusion with the elastic coefficient $c_1 = 10$ are shown. The distributions of radial and hoop stresses in the medium with the inclusion subjected to hydrostatic pressure are shown in Fig. 9.17 (above), and the same stresses for the 1D-tension in the direction of the x_3-axis are in Fig. 9.17.

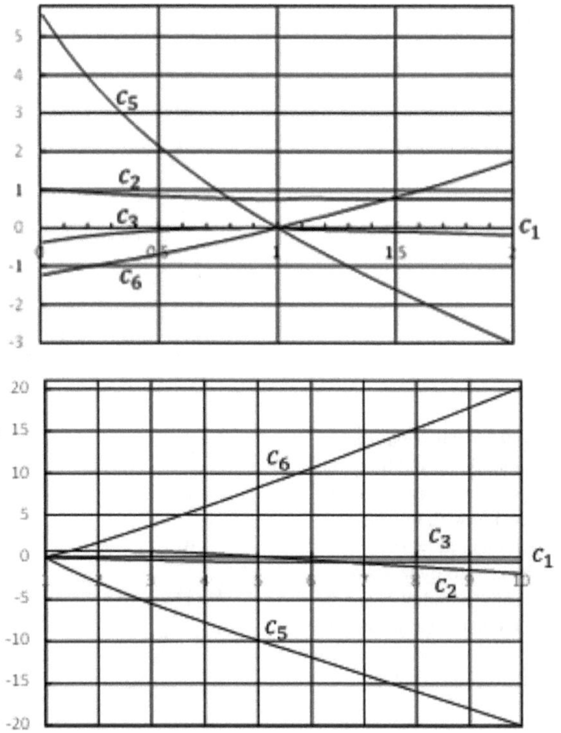

Fig. 9.12 Dependencies of the elastic coefficients of a radially transverse isotropic weak neutral inclusion on the elastic constant c_1; the elastic coefficients of the host medium are $c_1^0 = 1$, $c_2^0 = 0.75$

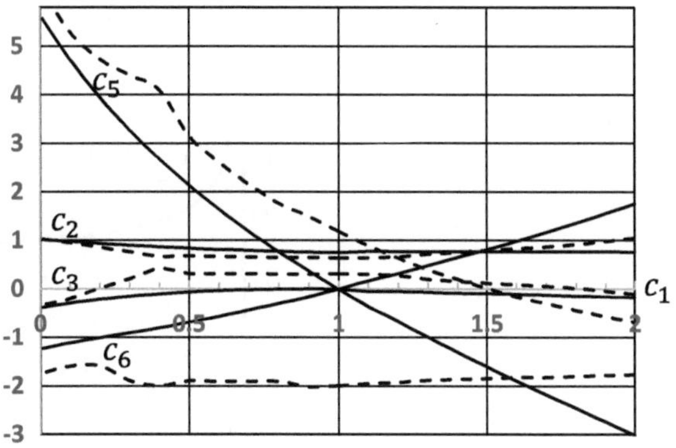

Fig. 9.13 The branches of the elastic constants for weak neutral transverse isotropic inclusions; solid lines correspond to the branch that passes through the point $C = C^0$, dashed lines correspond to another branch that does not pass through this point

9.2.3 A Neutral Isotropic Spherical Inclusion with a Transverse Isotropic Layer

In this section, we consider inclusions that consist of an isotropic core and a transverse isotropic layer of thickness h at the boundary with the host medium. The objective is to find the elastic coefficients of the layer material that make the inclusion neutral. These coefficients depend on the layer thickness h, and for $h = a$ (a is the inclusion

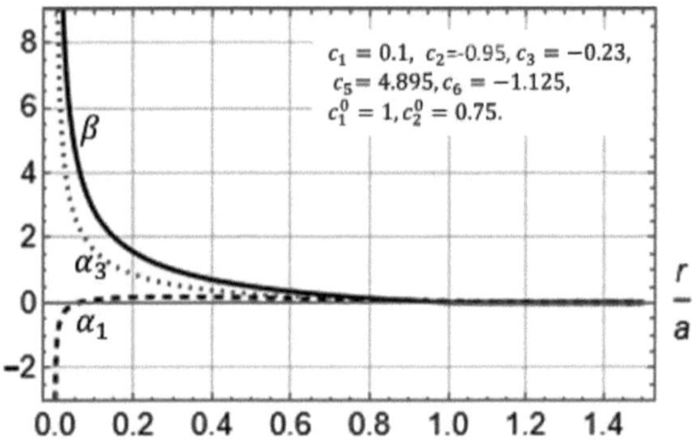

Fig. 9.14 The functions $\alpha_1(r)$, $\alpha_3(r)$ and $\beta(r)$ for a weak neutral inclusion with $c_1 = 0.1$ in the host medium with the elastic coefficients $c_1^0 = 1$, $c_2^0 = 0.75$

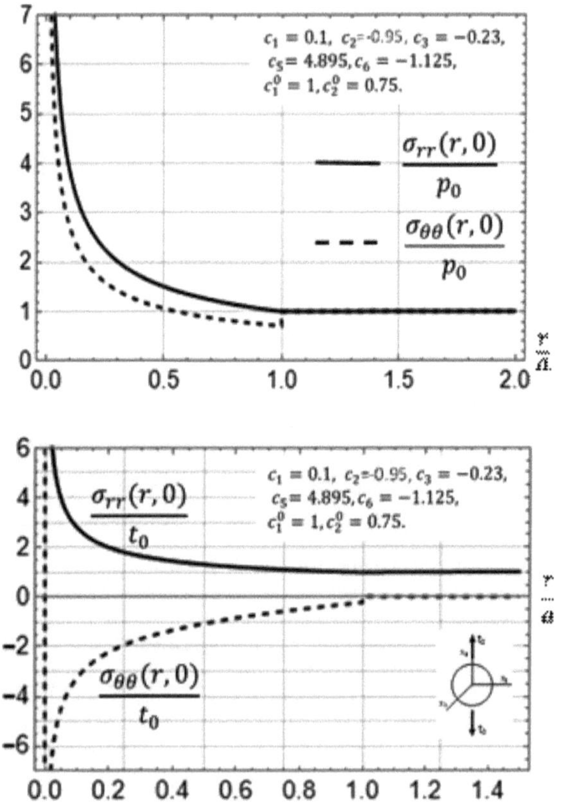

Fig. 9.15 Distributions of the radial σ_{rr} and hoop $\sigma_{\theta\theta}$ stresses in a weak neutral inclusion with the elastic coefficient $c_1 = 0.1$ subjected to hydrostatic pressure p_0 (above) and in the inclusion subjected to 1D-tension t_0 in the direction of the x_3-axis (below)

radius), the evident layer elastic coefficients of the neutral inclusion are in Eq. (9.57). Then, decreasing the layer thickness h, a branch of elastic coefficients of neutral inclusions as functions of the layer thickness can be constructed. In order to find the elastic coefficients of the layer for a fixed h, the minimum of the functions F^s

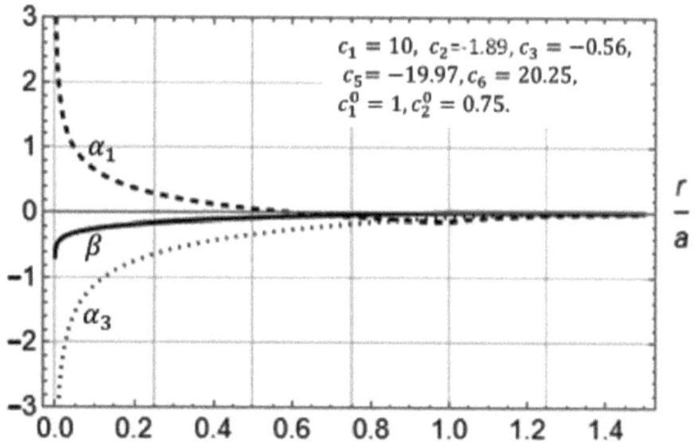

Fig. 9.16 The same as in Fig. 9.14 for a weak neutral inclusion with the elastic coefficient $c_1 = 10$

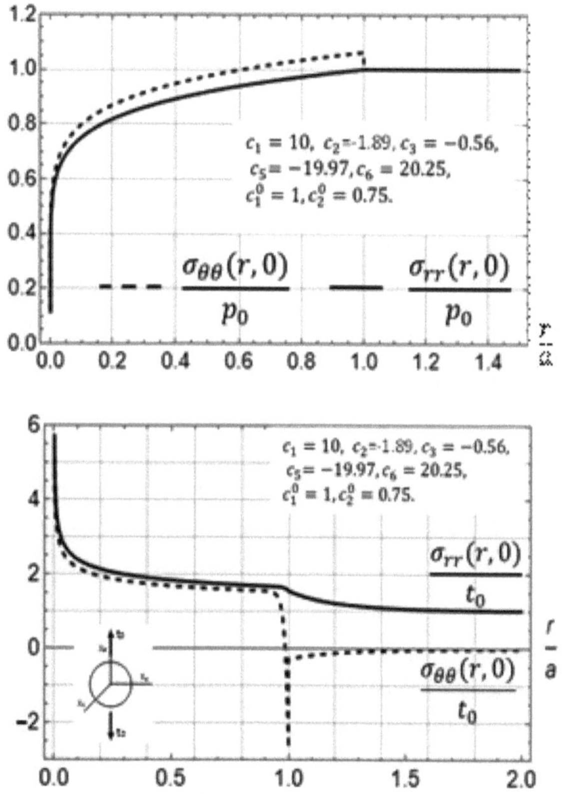

Fig. 9.17 Distributions of the radial σ_{rr} and hoop $\sigma_{\theta\theta}$ stresses in a weak neutral inclusion with the elastic coefficient $c_1 = 10$ subjected to hydrostatic pressure p_0 (above) and uniaxial tension t_0 in the direction of the x_3-axis (below)

(for strong neutral inclusions) or F^w (for weak neutral inclusions) with respect to the layer elastic coefficients should be found. The strong (weak) neutral inclusion exists if this minimum is equal to zero. For $h = a - k\delta a$ with a small step δa, $k = 1, 2, \ldots$, the function F^w has clearly defined zero-minima for h changing from

$h = a$ to $h = 0.1a$. The corresponding elastic coefficients of the layer define a weak neutral inclusion for the given value of the inclusion thickness, elastic constants of the isotropic core, and the constants of the host medium. The function F^s has minima in the vicinity of this branch of the elastic coefficients, but these minima are not equal to zero.

Let the elastic constants of the isotropic core be $c_1^{(1)} = 0.1, c_2^{(1)} = 0.45$ and the elastic constants of the host medium $c_1^0 = 1, c_2^0 = 0.75$. In Fig. 9.18, dependencies of the elastic constants of the weak neutral inclusions on the layer thickness h/a are shown $(0.1 < h/a < 1)$. The same dependencies for the core with the elastic constants $c_1^{(1)} = 10, c_1^{(1)} = 45$ are shown in Fig. 9.19.

The functions α_1, α_3 and β for a weak neutral inclusion with the elastic constants of the core $c_1^{(1)} = 0.1, c_2^{(1)} = 0.45$ and the layer thickness $h = 0.2a$ are shown in Fig. 9.20 (above). In Fig. 9.20 (below), the same functions for the inclusion without a layer ($h = 0$) are presented.

Stress distributions in a weak neutral inclusion with the layer thickness $h = 0.2a$ and the elastic constants of the isotropic core $c_1^{(1)} = 0.1, c_2^{(1)} = 0.45$ are shown in

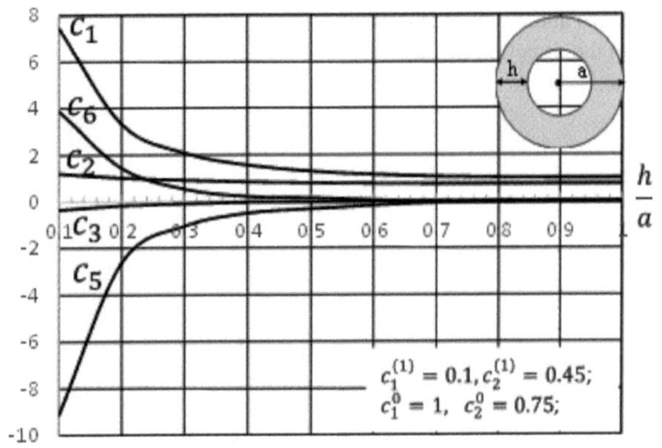

Fig. 9.18 Dependencies of the elastic coefficients of transverse isotropic layers on the layer thickness h/a for weakly neutral inclusions; the elastic coefficients of the isotropic core and the host medium are $c_1^{(1)} = 0.1$, $c_2^{(1)} = 0.45$, $c_1^0 = 1$, $c_2^0 = 0.75$

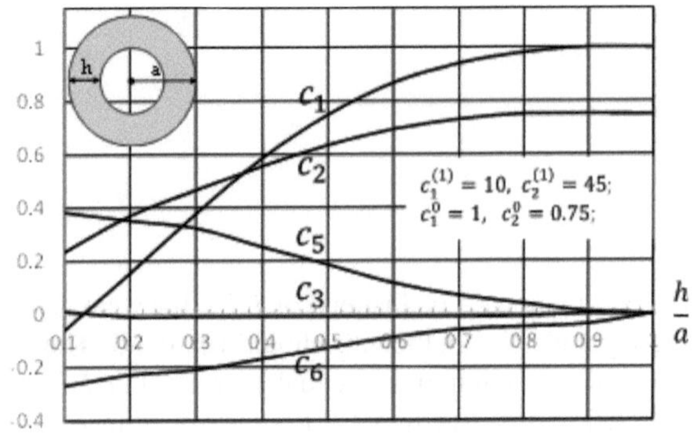

Fig. 9.19 The same as in Fig. 9.18 for the elastic coefficients of an isotropic core $c_1^{(1)} = 10, c_2^{(1)} = 45$

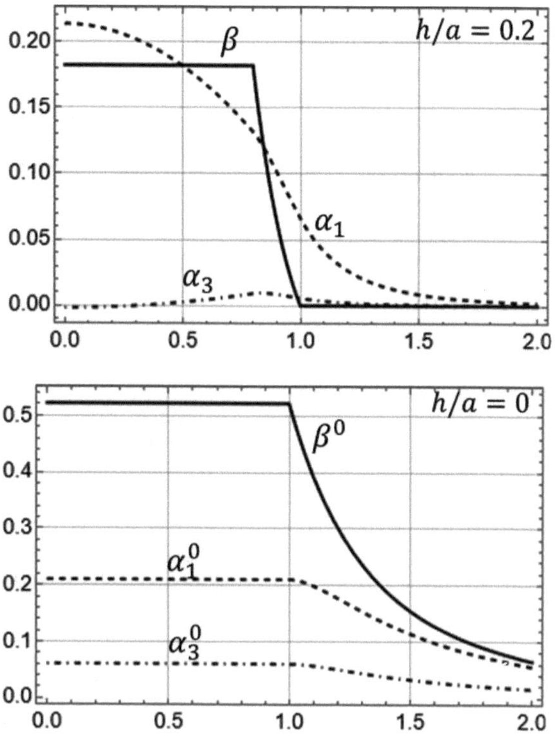

Fig. 9.20 Functions α_1, α_3 and β for a weak neutral inclusion of radius a with a transverse isotropic layer of thickness $h = 0.2a$; the elastic coefficients of the core and the host medium are $c_1^{(1)} = 0.1$, $c_2^{(1)} = 0.45$, $c_1^0 = 1$, $c_2^0 = 0.75$ (above); the same functions for the inclusion without a layer ($h = 0$) (below)

Fig. 9.21 (hydrostatic pressure) and Fig. 9.22 (1D-tension in the direction of the x_3-axis). Dashed lines in these figures are stress distributions in the inclusion without a layer.

The functions α_1, α_3 and β for a weak neutral inclusion with the elastic constants of the core $c_1^{(1)} = 10$, $c_2^{(1)} = 45$ and a layer of the thickness $h = 0.2a$ are shown in Fig. 9.23 (above), and for inclusion without a layer in Fig. 9.23 (below). The corresponding stresses in the medium with the inclusion are shown in Figs. 9.24 and 9.25.

9.3 Equivalent Inclusions

According to the definition of Norris and Parnell [1], a homogeneous inclusion is strongly equivalent to an original heterogeneous inclusion if the external field disturbances in the host medium caused by both inclusions coincide. For a weak equivalent inclusion, the principal terms of the external field disturbances due to the inclusions coincide only. We consider the problem of determining the elastic properties of a homogeneous isotropic inclusion that is weak equivalent to a radially transverse isotropic heterogeneous inclusion. Both inclusions have the same radius.

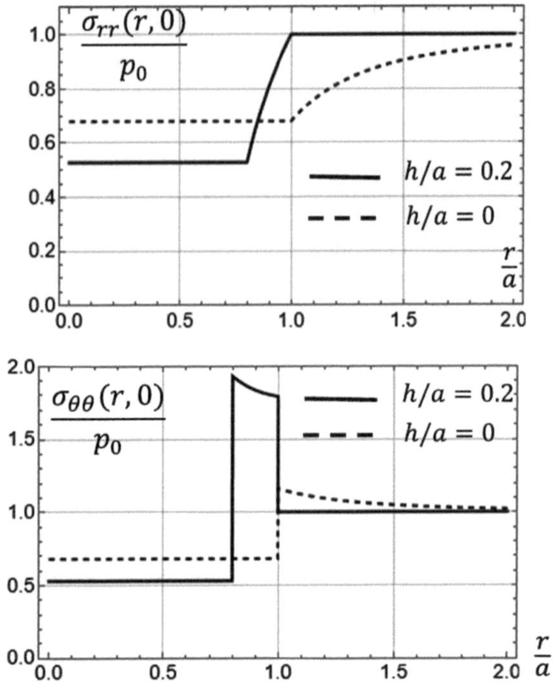

Fig. 9.21 Distributions of the radial σ_{rr} and hoop $\sigma_{\theta\theta}$ stresses in the weak neutral inclusion of radius a with an isotropic core and a transverse isotropic external layer of the thickness $h = 0.2a$ subjected to hydrostatic pressure p_0, the elastic coefficients of the core and the host medium are $c_1^{(1)} = 0.1, c_2^{(1)} = 0.45$, $c_1^0 = 1; c_2^0 = 0.75$; dashed lines are the same stresses for the inclusion without a layer

The strain field disturbance $\varepsilon^1(x)$ in the host medium caused by an inclusion follows from Eq. (4.12) in the form

$$\varepsilon^1(x) = \varepsilon(x) - \varepsilon^0 = -\int_V K(x - x')C^1(x')\varepsilon(x')dx'. \tag{9.58}$$

The principal term of the field $\varepsilon^1(x)$ far from the inclusion is the first term of the multipole expansion of the integral in this equation (Sect. 3.6)

$$\varepsilon^1(x) = -K(x)\int_V C^1(x')\varepsilon(x')dx' + o(r^{-3}), \tag{9.59}$$

$$K(x) = \frac{1}{r^3}K(n), n_\alpha = \frac{x_\alpha}{r}. \tag{9.60}$$

Here, it is taken into account that $K(x)$ is a homogeneous function of the order of −3. For a radially transverse isotropic inclusion subjected to a constant external strain field, the integral in Eq. (9.59) has the form

$$\int_V C^1(x)\varepsilon(x)dx = \int_V C^1(x)(E^1 + A(x))dx\varepsilon^0, \tag{9.61}$$

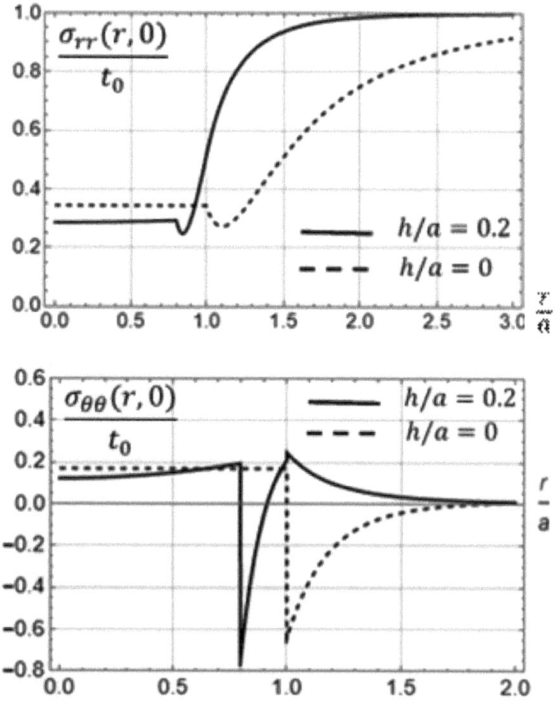

Fig. 9.22 The same as in Fig. 9.21 for the external 1D-tension t_0 in the direction of the x_3-axis

where tensor $A(x)$ is defined in Eqs. (9.36), (9.37). Thus the principal asymptotic term of the far strain field in the medium with an inclusion is defined by the integral in Eq. (9.59); this integral normalized by the inclusion volume v is an isotropic fourth rank tensor

$$H = \frac{1}{v}\int\limits_V (C(x) - C^0)(E^1 + A(x))dx = h_1 E^1 + h_2 E^2, \tag{9.62}$$

$$h_1 = \frac{3}{a^3}\int\limits_0^a \mathbb{I}_1(r)r^2 dr,\ h_2 = \frac{3}{a^3}\int\limits_0^a \mathbb{I}_2(r)r^2 dr. \tag{9.63}$$

Here, E^1, E^2 are the elements of the E-basis, a is the inclusion radius, the functions $\mathbb{I}_1(r)$, $\mathbb{I}_2(r)$ are defined in terms of the functions $\alpha_1(r)$, $\alpha_3(r)$ and $\beta(r)$ in Eq. (9.37)

$$\mathbb{I}_1(r) = c_1 - c_1^0 + \frac{1}{3}c_5 + \frac{2}{c_6}c_6 + \frac{1}{10}(10c_5 + 4c_6(1 + D) +$$
$$+ 2(c_1 - c_1^0)(3 + D)(5 + D) + c_5 D(7 + D))\alpha_1 + \frac{2}{15}(5c_5 + 2c_6 +$$
$$+ (c_1 - c_1^0)(3 + D)(5 + D) + (5c_3 + 6c_5 + 3c_6)D + (c_3 + c_5 + c_6)D^2))\alpha_3, \tag{9.64}$$

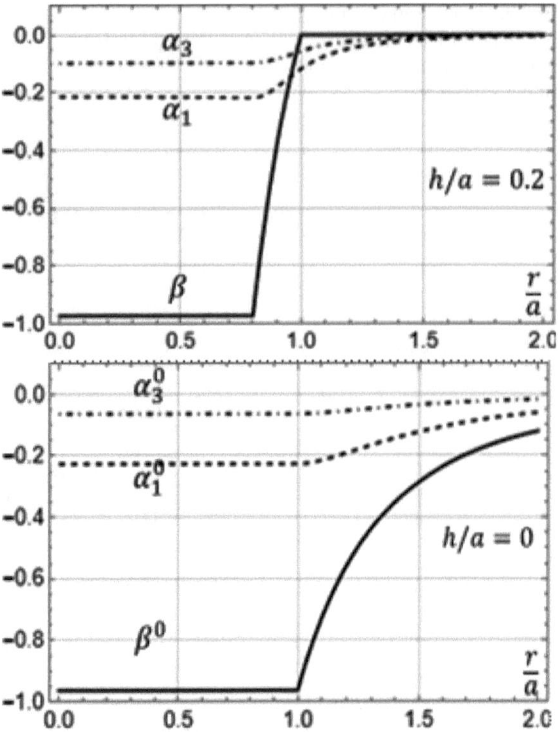

Fig. 9.23 The same as in Fig. 9.21 for the elastic coefficients of the core $c_1^{(1)} = 10, c_2^{(1)} = 45$

$$\mathbb{I}_2(r) = c_2 - c_2^0 + \frac{2}{3}c_3 + \frac{1}{15}c_6 + \frac{1}{30}(-10c_5 - 4c_6(1+D) -$$

$$- 2(c_1 - c_1^0)(3+D)(5+D) - c_5 D(7+D))\alpha_1 +$$

$$+ \frac{2}{45}(-5c_5 - 2c_6 - (c_1 - c_1^0)(3+D)(5+D) -$$

$$- ((5c_3 + 6c_5 + 3c_6)D + (c_3 + c_5 + c_6)D^2)\alpha_3 +$$

$$+ \frac{1}{9}(9(c_2 - c_2^0) + 6c_3 + c_5 + c_6 + (3(c_2 - c_2^0) +$$

$$+ 4c3 + c5 + c6)D + (c_1 - c_1^0)(3+D)\beta. \tag{9.65}$$

Here, $c_m = c_m(r)$, $m = 1, 2, 3, 5, 6$.

Then, we consider an isotropic spherical inclusion of radius a with the elastic stiffness tensor $\widehat{C} = \widehat{c}_1 E^1 + \widehat{c}_2 E^2$, where $\widehat{c}, \widehat{c}_2$ are constant coefficients. The field $\widehat{\varepsilon}(x)$ inside a homogeneous isotropic inclusion subjected to a constant external strain field has the form (Sect. 4.4)

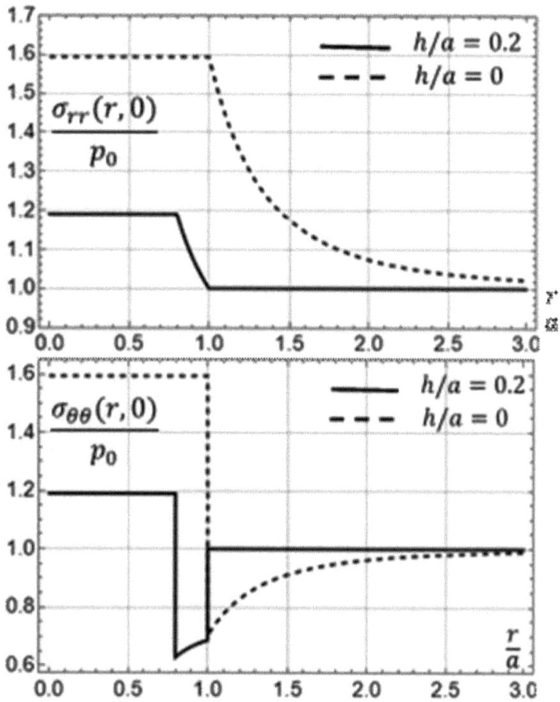

Fig. 9.24 The same as in Fig. 9.22 for the elastic coefficients of the core $c_1^{(1)} = 10$, $c_2^{(1)} = 45$

$$\widehat{\varepsilon} = (I + \widehat{A})\varepsilon^0,\ \widehat{A} = A_1 E^1 + A_2 E^2, \tag{9.66}$$

$$A_1 = \frac{15c_1^0(c_1^0 + c_2^0)}{2\widehat{c}_1(4c_1^0 + 3c_2^0) + 7\left(c_1^0\right)^2 + 9c_1^0 c_2^0} - 1; \tag{9.67}$$

$$A_2 = \frac{3(c_1^0 + c_2^0)((\widehat{c}_1 - c_1^0)(c_1^0 + 2c_2^0) - 5c_1^0(\widehat{c}_2 - c_2^0))}{(\widehat{c}_1 + 3\widehat{c}_2 + 2c_1^0)(8\widehat{c}_1 c_1^0 + 7\left(c_1^0\right)^2 + 6\widehat{c}_1 c_2^0 + 9c_{10}^0 c_2^0)}. \tag{9.68}$$

For a homogeneous isotropic inclusion, the integral (9.61) that defines the principal terms of the far-field in the host medium as follows

$$\widehat{H} = \frac{1}{v}\int\limits_V \left(\widehat{C} - C^0\right)\left(I + \widehat{A}\right)dx = \widehat{h}_1 E^1 + \widehat{h}_2 E^2, \tag{9.69}$$

where

$$\widehat{h}_1 = \frac{15c_1^0(c_1^0 + c_2^0)\,[\widehat{c}_1]}{15c_1^0(c_1^0 + c_2^0) + 2(4c_1^0 + 3c_2^0)\,[\widehat{c}_1]}, \tag{9.70}$$

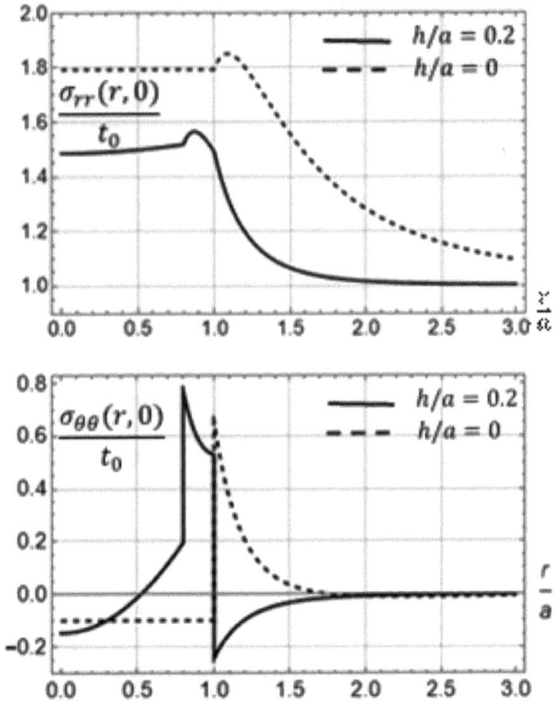

Fig. 9.25 The same as in Fig. 9.23 for the elastic coefficients of the core $c_1^{(1)} = 10, c_2^{(1)} = 45$

$$\widehat{h}_2 =$$

$$= \frac{3(c_1^0 + c_2^0)((c_1^0 + 2c_2^0) + 3(5c_1^0(c_1^0 + c_2^0) + (c_1^0 + 2c_2^0)\,[\widehat{c}_1])\,[\widehat{c}_2])}{(15c_1^0(c_1^0 + c_2^0) + 8c_1^0\,[\widehat{c}_1] + 6c_2^0\,[\widehat{c}_1])(3c_1^0 + 3c_2^0 + [\widehat{c}_1] + 3\,[\widehat{c}_2]))}, \tag{9.71}$$

$$[\widehat{c}_1] = \widehat{c}_1 - c_1^0, \;\; [\widehat{c}_2] = \widehat{c}_2 - c_2^0. \tag{9.72}$$

If the homogeneous isotropic inclusion is weak equivalent to the radially transverse isotropic inclusion, the tensors H and $\widehat{H}$ in Eqs. (9.62) and (9.69) should coincide, and therefore,

$$h_1 = \widehat{h}_1, h_2 = \widehat{h}_2. \tag{9.73}$$

These are two equations for the coefficients $\widehat{c}_1, \widehat{c}_2$ of the homogeneous isotropic inclusion that is weak equivalent to the radially transverse isotropic inclusion of the same radius.

We consider the equivalent inclusion problem for a homogeneous inclusion of arbitrary shape. The elastic stiffness tensor of the original and equivalent inclusions is C, but the equivalent inclusion is an ellipsoid with the given volume. For a constant external strain field ε^0, the field inside the original inclusion has in the form

$$\varepsilon(x) = \Lambda^{\varepsilon}(x)\varepsilon^{0}, x \in V, \tag{9.74}$$

where the function $\Lambda^{\varepsilon}(x)$ is assumed to be known, V is the region occupied by the inclusion. The principal term of the strain field far from the inclusion is defined by the integral in Eq. (9.61)

$$H = C^{1} \int_{V} \Lambda^{\varepsilon}(x)dx, C^{1} = C - C^{0}. \tag{9.75}$$

For an ellipsoidal inclusion, the function Λ^{ε} is constant inside the inclusion, and it has the form in Eq. (4.82)

$$\Lambda^{\varepsilon} = \left(I + A^{0}(a)C^{1}\right)^{-1}, \tag{9.76}$$

where the fourth rank tensor $A^{0}(a)$ depends on the aspect ratios of the ellipsoid and its orientation in space, the second rank tensor a defines the linear transformation of the x-space $y_{\alpha} = \left(a^{-1}\right)_{\alpha\beta} x_{\beta}$ that converts the ellipsoid in a unit ball. The principal terms of the far-field in the medium with the ellipsoidal inclusion is defined to the tensor

$$H^{0} = C^{1}\left(I + A^{0}(a)C^{1}\right)^{-1} v, \tag{9.77}$$

where v is the ellipsoid volume. Because the principal terms of the far fields of the original and weak equivalent inclusions coincide, the equation $H = H^{0}$ holds, and we come to the equation

$$\left(I + A^{0}(a)C^{1}\right)^{-1} = \overline{\Lambda}^{\varepsilon}, \ \overline{\Lambda}^{\varepsilon} = \frac{1}{v}\int_{V} \Lambda^{\varepsilon}(x)dx. \tag{9.78}$$

The expression for the tensor $A^{0}(a)$ follows from this equation in the form

$$A^{0}(a) = \left(\left(\overline{\Lambda}^{\varepsilon}\right)^{-1} - I\right)\left(C^{1}\right)^{-1}. \tag{9.79}$$

On the other hand, the tensor $A^{0}(a)$ is the following integral (Eq. (4.89))

$$A^{0}(a) = \frac{1}{4\pi}\int_{\Omega_{1}} \widetilde{K}(a^{-1}k)d\Omega_{k}, \tag{9.80}$$

where the function $\widetilde{K}(k)$ is in Eq. (3.16), and it depends on the elastic stiffness tensor C^{0} of the host medium only, Ω_{1} is the surface of a unit sphere in the k-space. Eq. (9.79) is an equation for the tensor a that determines the aspect ratios and the spatial orientation of the ellipsoidal inclusion that is weak equivalent to the original inclusion of a non-ellipsoidal shape. Both inclusions are homogeneous with the same elastic stiffness tensor and volume.

Remark. The tensor $A^0(a)$ depends on 6 components of the second rank tensor a. There is no guarantee that Eq. (9.79) can be satisfied with this number of constants for all possible forms of the original inclusions. The tensor a for an equivalent inclusion can be chosen to minimize the norm of the difference between the left- and right-hand sides of Eq. (9.79).

9.4 Notes

The problems of neutral inclusions in static elasticity and stationary conductivity are considered in [1], where an extensive survey of the works dedicated to these problems is presented. In the literature, most results concern the problem of neutral inclusions in conductivity and planar elasticity. The more challenging neutral inclusion problem in 3D-elasticity is considered in [4, 5].

References

1. Morris A, Parnell W (2020) Static elastic cloaking, low-frequency elastic wave transparency and neutral inclusions. Proc R Soc A 476:20190725
2. Brent R (1973) Algorithms for minimization without derivatives. Prentice-Hall
3. Wright S (2015) Coordinate descent algorithms. Math Program 151(1):3–34
4. Kanaun S (2024) Radially transverse isotropic inclusions in isotropic conductive media: local fields, effective properties, neutral inclusions. Int J Eng Sci 194:103965
5. Kanaun S (2024) Radially transverse isotropic inclusions in isotropic elastic media: local fields, neutral inclusions, effective elastic properties. Int J Eng Sci 200:104078

Chapter 10
The Homogenization Problem

Abstract Specific features of the homogenization problems for matrix composite materials and polycrystals are considered. Relations between the ensemble and spatial averages of the fields in heterogeneous media are discussed. The representative volume element of the medium is introduced. Integral equations for the fields in heterogeneous media are derived. Cross-property relations between effective characteristics of different physical natures are indicated.

Keywords Ensemble and spatial averages · Ergodic property · Representative volume element · Reference medium · Integral equations for the fields in heterogeneous media · Cross-property relations

The structure of heterogeneous materials considered in this part of the book has several specific features. First, the characteristic scales of fluctuations of physical properties of the materials are assumed to be much smaller than the sizes of the macroscopic specimen. Random functions that describe these properties are homogeneous in the broad sense [1]. Therefore, neglecting specific boundary effects, the principal information about physical fields in the composites can be obtained by consideration of an infinite heterogeneous medium. Second, it is assumed that the heterogeneous materials consist of an agglomeration of structural elements randomly distributed over sizes and orientations. In many cases, the shapes of these elements can be approximated by ellipsoids.

The functions that describe physical properties of this class of materials are statistically homogeneous random functions. It is known [1, 2] that complete information about a random function is contained in its characteristic functional, or in an infinite set of multipoint statistical moments. For instance, the complete solution of the problem of elasticity for a random (stochastic) composite is an infinite number of multipoint statistical moments of the random strain $\varepsilon(x)$ and stress $\sigma(x)$ tensors expressed in terms of the statistical moments of the elastic stiffness tensor $C(x)$ for non-random external loading (surface forces, body forces $q(x)$, temperature field $T(x)$, etc.). The principal difficulty in deriving the complete solutions is that in spite of linearity of the initial system of differential equations for the detailed physical fields, the problem is stochastically non-linear. For elasticity, the governing equation for the displacement vector $u(x)$ in heterogeneous media has the form

S. Kanaun, *Self-Consistent Methods for Composites*, Solid Mechanics and Its Applications 282, https://doi.org/10.1007/978-3-031-96739-9_10

$$\nabla_\lambda C_{\lambda\alpha\beta\mu}(x)\nabla_\beta u_\mu(x) = -q_\alpha(x). \tag{10.1}$$

The left-hand side of this equation contains the product of the statistically dependent random functions $C(x)$ and $\nabla u(x)$. Due to this nonlinearity, each statistical moment of a finite order of the random function $\nabla u(x)$ is expressed via an infinite set of moments of higher orders. Therefore, to make the problem solvable, one has to accept some simplifying assumptions that depend on the peculiarities of the composite microstructure. Note that for engineering applications, the complete information about random physical fields in a heterogeneous medium is unnecessary. As a rule, only the first and second statistical moments of these fields have interest for the applications.

The most important characteristic of physical fields in random heterogeneous media is the first statistical moment or the mathematical expectations of the fields. The first moment characterizes the average response of the micro heterogeneous material to external loading. If this response is known, the actual composite material can be replaced by some homogeneous material with the equivalent reaction to external loading. Determination of physical properties of a homogeneous medium that is equivalent to the actual heterogeneous material is the content of the homogenization problem. The solution of this problem provides relations that define macroscopic properties of the composite material in terms of the physical properties of the phases and details of the phase distributions in the material. The solution of the homogenization problem is the basis for the directed synthesis of composite materials with prescribed macro-properties.

Let each realization of the random set of structural elements of a composite be an element ω of a functional space with the known probabilistic measure $\mu(\omega)$ [2]. Hence, the physical field $u(x)$ (e.g., displacement vector) depends on ω, i.e., on a specific realization of the random set (ensemble) of the inclusions: $u(x) = u(x, \omega)$. The ensemble averaging of the field $u(x, \omega)$ at point x of the medium is the mean value of the random function $u(x, \omega)$ over the ensemble realizations ω, i.e.,

$$\langle u(x)\rangle = \int u(x, \omega)\mathrm{d}\mu(\omega), \ \langle \nabla u(x)\rangle = \int \nabla u(x, \omega)\mathrm{d}\mu(\omega). \tag{10.2}$$

Here, the integrals are understood as continual integrals over the functional space of the realizations of the medium microstructures ([1, 2]). In Fig. 10.1, the solid line is the detailed function $u(x)$, the dashed line is the mean value of this function.

Let L_* be a linear operator that allows calculating the mean $\langle u(x)\rangle$ of a random field $u(x)$ from the equation

$$\left(L_* \langle \nabla u\rangle\right)(x) = -q(x), \tag{10.3}$$

If such an operator exists, it is called the effective operator of the micro heterogeneous material. For statistically homogeneous composites, this operator describes a physical field in some homogeneous medium that is equivalent to the original

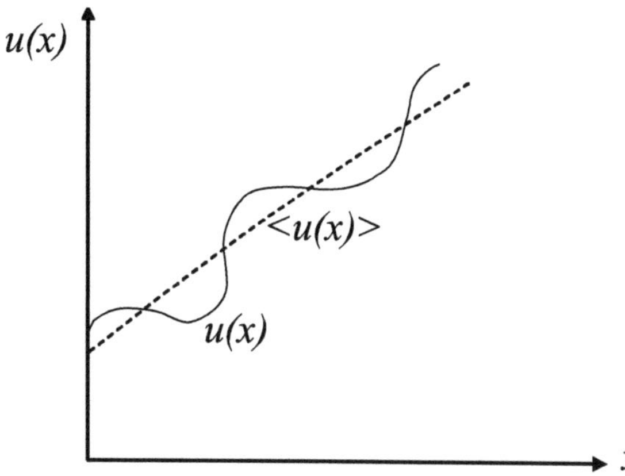

Fig. 10.1 Averaging of a random function $u(x)$ over the ensemble of realizations (dashed line), one of the realizations (solid line)

composite material. In the broad sense, the homogenization problem consists in the construction of the operator L_*.

In some cases, the ensemble averaging (10.2) can be replaced by the spatial averaging for a fixed typical realization of a random field . Let the property tensor of the medium be a homogeneous in space random function $C(x)$. If the medium is subjected to a constant external field at infinity, the field intensity (deformation) $\nabla u(x)$ is also a homogeneous in space random function. In this case, the ensemble averaging (10.2) can be changed with the spatial averaging over a fixed typical realization of the function $\nabla u(x)$

$$\langle \nabla u(x) \rangle = \lim_{W \to \infty} \frac{1}{W} \int_W \nabla u(x) \mathrm{d}x. \tag{10.4}$$

Here, W is a region that occupies the entire in 3D(2D)-space in the limit $W \to \infty$.

In practice, in place of the infinite region, the so-called representative volume V_0 of the heterogeneous material is considered . According to Hill [3], the volume V_0 should contain a sufficiently large number of structural elements (inclusions, grains) to be macroscopically homogeneous. The linear sizes of this volume should be much larger than the correlation radius of the random function $C(x)$. Then, the mean value $\langle \nabla u(x) \rangle$ of a random function $\nabla u(x)$ can be calculated on the basis of a typical fixed realization of this function in the characteristic volume V_0

$$\langle \nabla u(x) \rangle = \frac{1}{V_0} \int_{V_0} \nabla u(x) dx\ . \tag{10.5}$$

It is known [1] that the volume averaging (10.4) coincides with the averaging over the ensemble realizations (10.2) if the function $\nabla u(x)$ is ergodic. The stress and strain fields in a statistically homogeneous material are ergodic only for a constant external fields. If the external field changes in space, e.g., by wave propagation in composites, only the ensemble averaging can be used for the calculation of the mean $\langle \nabla u(x) \rangle$. For a non-homogeneous external field, the result of spatial averaging (10.5) does not coincide with ensemble averaging, and the concept of the representative volume element can be applied for the calculation of averaging specific field characteristics like energy density only.

The relation between the ensemble and spatial averages of physical fields in heterogeneous media has been discussed in the literature (see, e.g., the book [4]). For instance, it was shown that the systems described by Schrödinger equation have the so-called self-averaging properties. The latter means that spatial averaging of extensive physical fields similar to the energy density coincides with the ensemble averaging. The necessary conditions for the self-averaging are spatial homogeneity of the system and rapid decay of the correlations between physical parameters at different points of the system. Such conditions are fulfilled, e.g., for elastic fields in statistically homogeneous composite materials with a finite correlation radius of the physical property tensors.

10.1 Integral Equations for the Fields in Heterogeneous Media. The Reference Medium

In this section, we consider a heterogeneous medium with the property tensors $C(x)$ that is a homogeneous in space random function of oscillating type. For conductive media, $C(x)$ is the second rank conductivity tensor, and for elastic media, $C(x)$ is the fourth rank elastic stiffness tensor. The tensor $C(x)$ relates the field intensity $\varepsilon(x)$ and the flux density $\sigma(x)$ by the equation

$$\sigma(x) = C(x)\varepsilon(x). \tag{10.6}$$

For conductive media, $\varepsilon(x)$ and $\sigma(x)$ are vectors, and the right-hand side in this equation is the scalar product of the second rank tensor $C(x)$ and the vector $\varepsilon(x)$. For elasticity, $C(x)\varepsilon(x)$ is the convolution of the fourth rank tensor $C(x)$ and the second rank strain tensor $\varepsilon(x)$ over two indices. In the absence of body forces, the flux density satisfies the equation

$$\mathrm{div}\sigma(x) = 0, \tag{10.7}$$

where $\mathrm{div}\sigma(x) = \nabla_\alpha \sigma_\alpha(x)$. In the case of conductivity, the field intensity $\varepsilon(x)$ is gradient of a scalar potential $u(x)$

$$\varepsilon(x) = \nabla u(x). \tag{10.8}$$

In the case of elasticity, $u(x)$ is a vector and ∇ is changed with the symmetrized gradient def. The potential $u(x)$ satisfies the equation

$$\nabla C(x)\nabla u(x) = 0 \tag{10.9}$$

that follows from Eqs. (10.6)–(10.8).

For the analysis of fields in a heterogeneous medium, we introduce the reference homogeneous medium with a definitely positive property tensor $\widehat{C}$ and present the tensor $C(x)$ in the form

$$C(x) = \widehat{C} + C^1(x). \tag{10.10}$$

Thus Eq. (10.9) for the potential $u(x)$ can be written in the form

$$\nabla\widehat{C}\nabla u(x) = -\nabla C^1(x)\nabla u(x). \tag{10.11}$$

Application to Eq. (10.11) the integral operator with the kernel $\widehat{G}(x)$ that is Green's function of the operator $\nabla\widehat{C}\nabla$ results in the equation

$$u(x) = \widehat{u}(x) + \int \widehat{G}(x - x')\nabla' C^1(x')\nabla' u(x')dx'. \tag{10.12}$$

Here, $\widehat{u}(x)$ is the potential of the external field that would have existed in the reference medium for the same conditions at infinity. This potential satisfy the equation $\nabla\widehat{C}\nabla\widehat{u}(x) = 0$. For existence of the integral in Eq. (10.12), we have to accept that the function $C^1(x)$ has a finite support ($C^1(x)$ is equal to zero outside of a finite region in 3D-space). Later on, this restriction will be canceled. Using Gauss' theorem (Appendix C.1), the gradient ∇' in Eq. (10.12) can be overthrown on the kernel, and then, applying gradient operator ∇ to Eq. (10.12), we yield the equation for the field intensity $\varepsilon = \nabla u$ in the heterogeneous medium

$$\varepsilon(x) + \int \widehat{K}(x - x')C^1(x')\varepsilon(x')dx' = \widehat{\varepsilon}, \tag{10.13}$$

$$\widehat{K}(x) = -\nabla\nabla\widehat{G}(x), \qquad \widehat{\varepsilon} = \nabla\widehat{u}(x). \tag{10.14}$$

Here, the identity $\nabla'\widehat{G}(x - x') = -\nabla\widehat{G}(x - x')$ is used. A consequence of this equation is the equation for the flux density $\sigma(x)$ in Eq. (10.6) (see details in Sect. 4.1)

$$\sigma(x) - \int \widehat{S}(x - x')B^1(x')\sigma(x')dx' = \widehat{\sigma}, \tag{10.15}$$

$$\widehat{S}(x) = \widehat{C}\widehat{K}(x)\widehat{C} - \widehat{C}\delta(x), \;\; \widehat{\sigma} = \widehat{C}\widehat{\varepsilon}, \tag{10.16}$$

$$B^1(x) = B(x) - \widehat{B}, B(x) = C(x)^{-1}, \widehat{B} = \widehat{C}^{-1}. \tag{10.17}$$

For homogeneous in space random functions $C(x)$ and a constant external fields $\widehat{\varepsilon}, \widehat{\sigma}$, the functions

$$q(x) = C^1(x)\varepsilon(x), m(x) = B^1(x)\sigma(x) \tag{10.18}$$

under the integrals in Eqs. (10.13), (10.15) are homogeneous in space random functions of oscillating type. For extension of Eqs. (10.13), (10.15) on infinite heterogeneous media, we have to define actions of the integral operators in these equations on the class of oscillating functions in the entire space.

Homogeneous random functions of the oscillating type can be presented in the forms of exponential series with, generally speaking, incommensurable wave vectors k^j:

$$q\,(x) = q^0 + \sum_j q^j \exp\left(\mathrm{i}\,k^j \cdot x\right), \; m(x) = m^0 + \sum_j m^j \exp\left(\mathrm{i}\,k^j \cdot x\right). \tag{10.19}$$

Here, m^0, q^0 are constant components of $q(x)$ and $m(x)$. The coefficients m^j and q^j are such as these series converge, maybe in the generalized sense. Using the convolution property, the actions of the operators $\widehat{K}$ and $\widehat{S}$ on the function $\exp\left(\mathrm{i}\,k^j \cdot x\right)$ are presented in the forms:

$$\int \widehat{K}(x - x')e^{\mathrm{i}k^j \cdot x'}dx' =$$

$$= \int \widehat{K}(x - x')e^{\mathrm{i}k^j \cdot (x'-x)}dx'e^{\mathrm{i}k^j \cdot x} = \widetilde{K}(k^j)e^{\mathrm{i}k^j \cdot x}, \tag{10.20}$$

$$\int \widehat{S}(x - x')^{\mathrm{i}k^j \cdot x'}dx' = \widetilde{S}(k^j)e^{\mathrm{i}k^j \cdot x}, \tag{10.21}$$

where $\widetilde{K}(k)$ and $\widetilde{S}(k)$ are the Fourier transforms of the functions $\widehat{K}(x)$ and $\widehat{S}(x)$

$$\widetilde{K}(k) = k\widetilde{G}(k)k, \widetilde{G}(k) = \left(k\widehat{C}k\right)^{-1}, \tag{10.22}$$

$$\widetilde{S}(k) = \widehat{C}\widetilde{K}(k)\widehat{C} - \widehat{C}. \tag{10.23}$$

Since the functions $\widetilde{K}(k)$ and $\widetilde{S}(k)$ are homogeneous functions of the order of zero with respect to k, they are bounded for each $k^j\,(k^j \neq 0)$. Hence, the action of the operators $\widehat{K}$ and $\widehat{S}$ on the series in Eqs. (10.19) is transformed into analogous series with the coefficients $\widetilde{K}(k^j)q^j$, $\widetilde{S}(k^j)m^j$. These series converge if the original series (10.19) converges.

It remains to define the action of the operators $\widehat{K}$ and $\widehat{S}$ on the constants q^0 and m^0 in Eq. (10.19). For this end, we consider a model problem in which the "heterogeneity" with a constant property tensor C occupies the entire space. As a result, the medium is homogeneous with the property tensor C. The fields in the medium are induced by conditions at infinity, and we consider two types of such conditions. The first one is a fixed external flux density $\widehat{\sigma}$, and this condition is defined by the following procedure. First, we consider a finite region V with the boundary Ω and the following boundary conditions

$$t(x) = n(x)\widehat{\sigma}, x \in \Omega. \tag{10.24}$$

Here, $\widehat{\sigma}$ is a constant tensor, $n(x)$ is external normal to Ω. For such a condition, the field intensity and flux density in V are constant, and

$$\varepsilon = C^{-1}\widehat{\sigma}, \quad \sigma = \widehat{\sigma}. \tag{10.25}$$

This result does not depend on the size of the region V, and it holds when V tends to infinity and occupies the entire space. For this model problem, the integral equations Eqs. (10.13), (10.15) takes forms

$$\varepsilon + \int \widehat{K}(x - x')dx'(C - \widehat{C})\varepsilon = \widehat{C}^{-1}\widehat{\sigma}, \tag{10.26}$$

$$\sigma - \int \widehat{S}(x - x')dx' \left(C^{-1} - \widehat{C}^{-1}\right)\sigma = \widehat{\sigma}, \tag{10.27}$$

The solution of these equations coincide with ε and σ in Eq. (10.25) if the action of the operators $\widehat{K}$ and $\widehat{S}$ on constants are defined by the equations

$$\int \widehat{K}(x - x')dx' = \widehat{C}^{-1}, \quad \int \widehat{S}(x - x')dx' = 0. \tag{10.28}$$

For the second type of conditions (fixed external field intensity $\widehat{\varepsilon}$), we consider again a finite region V with the following potential at the boundary Ω of V

$$u(x) = \widehat{\varepsilon} \cdot x, \ x \in \Omega, \tag{10.29}$$

where $\widehat{\varepsilon}$ is a constant tensor. For the region V with a constant property tensor C, the field intensity and flux inside V take the forms

$$\varepsilon = \widehat{\varepsilon}, \quad \sigma = C\widehat{\varepsilon}. \tag{10.30}$$

This result does not depend on the size of the region V, and it remains when V tends to infinity and occupies the entire space. In this case, the solutions of Eqs. (10.26),

(10.27) coincide with ε and σ in Eq. (10.30) if the actions of the operators $\widehat{K}$ and $\widehat{S}$ on constants are defined by the equations

$$\int \widehat{K}(x - x')dx' = 0\,, \quad \int \widehat{S}(x - x')dx' = -\widehat{C}. \tag{10.31}$$

Thus, in what follows, we have to define what kind of external field (field intensity or flux density) is fixed in the problem.

Another way of deriving Eqs. (10.28) and (10.31) was considered in Sect. 3.4 (Eqs. (3.77), (3.78)).

10.2 The Effective Property Tensor

Using the function $q(x)$ in Eq. (10.18), we can present Eqs. (10.13), (10.15) for the detailed field intensity and flux density in the heterogeneous media in the forms

$$\varepsilon(x) = \widehat{\varepsilon} - \int \widehat{K}(x - x')\, q(x')dx', \tag{10.32}$$

$$\sigma(x) = \widehat{\sigma} - \int \widehat{S}(x - x')\widehat{B}q(x')dx'. \tag{10.33}$$

Here, the external fields $\widehat{\varepsilon}$ and $\widehat{\sigma}$ are assumed to be constant. Because the kernels $\widehat{K}(x)$, $\widehat{S}(x)$ and tensor $\widehat{B}$ are non-random, the fields $\varepsilon(x)$ and $\sigma(x)$ averaged over ensemble realizations of the random function $C(x)$ are

$$\langle\varepsilon(x)\rangle = \widehat{\varepsilon} - \int \widehat{K}(x - x')\, \langle q(x')\rangle d\, x', \tag{10.34}$$

$$\langle\sigma(x)\rangle = \widehat{\sigma} - \int \widehat{S}(x - x')\widehat{B}\langle q(x')\rangle d\, x'. \tag{10.35}$$

For a homogeneous in space random function $C(x)$ and a constant external fields $\widehat{\varepsilon}$ and $\widehat{\sigma}$, the field $q(x)$ is a homogeneous in space random function. Thus, the average $\langle q(x)\rangle$ of this function is a constant tensor. If ergodic hypothesis holds, $\langle q(x)\rangle$ can be found by spatial averaging of a fixed typical realization of $q(x)$

$$\langle q\rangle = \lim_{W\to\infty} \frac{1}{W}\int_W q(x)d\, x, \tag{10.36}$$

where W is a region that occupies the entire space in the limit $W \to \infty$. Taking into account the linearity of the problem, the field $\varepsilon(x)$ is a linear function of the external field $\widehat{\varepsilon}$ for each realization of $C(x)$

$$\varepsilon(x) = \Lambda^{\varepsilon}(x)\widehat{\varepsilon}. \tag{10.37}$$

Here, $\Lambda^{\varepsilon}(x)$ is a second rank (for conductivity) or fourth rank (for elasticity) tensor function. Substituting this equation in Eq. (10.36) for $q(x)$, we obtain for the mean $\langle q \rangle$

$$\langle q \rangle = P^{*}\widehat{\varepsilon}, \quad P^{*} = \lim_{W\to\infty} \frac{1}{W} \int\limits_{W} \left(C(x) - \widehat{C}\right) \Lambda^{\varepsilon}(x)dx. \tag{10.38}$$

Let the external field intensity $\widehat{\varepsilon}$ be fixed. Thus action of the operators K and S on constants is defined in Eq. (10.31), and from Eqs. (10.34), (10.35) we have

$$\langle \varepsilon \rangle = \widehat{\varepsilon}, \tag{10.39}$$

$$\langle \sigma \rangle = \widehat{\sigma} + \langle q \rangle = (\widehat{C} + P^{*})\widehat{\varepsilon} = C^{*}\langle \varepsilon \rangle. \tag{10.40}$$

Here, the tensor $C^{*} = \widehat{C} + P^{*}$ relates the averaged field intensity $\langle \varepsilon \rangle$ and flux density $\langle \sigma \rangle$ in the heterogeneous medium. This tensor is the effective property tensor of the heterogeneous medium.

Using the function $m(x)$ in Eq. (10.18), Eqs. (10.13), (10.15) are presented in the forms

$$\varepsilon(x) = \widehat{\varepsilon} + \int \widehat{K}\left(x - x'\right)\widehat{C}m(x')dx', \tag{10.41}$$

$$\sigma(x) = \widehat{\sigma} + \int \widehat{S}\left(x - x'\right) m(x')dx'. \tag{10.42}$$

Averaging $\varepsilon(x)$, $\sigma(x)$ over ensemble realizations of these random functions, we obtain

$$\langle \varepsilon(x) \rangle = \widehat{\varepsilon} + \int \widehat{K}\left(x - x'\right) dx' \widehat{C}\langle m(x') \rangle, \tag{10.43}$$

$$\langle \sigma(x) \rangle = \widehat{\sigma} + \int \widehat{S}\left(x - x'\right) dx' \langle m(x') \rangle. \tag{10.44}$$

Let the external flux density $\widehat{\sigma}$ is fixed in the problem. Because of the linearity of the problem, the detailed flux density $\sigma(x)$ is

$$\sigma(x) = \Lambda^{\sigma}(x)\widehat{\sigma}, \tag{10.45}$$

and the average $\langle m(x)\rangle$ is

$$\langle m(x)\rangle = Q^*\widehat{\sigma}, \quad Q^* = \lim_{W\to\infty}\frac{1}{W}\int_W \left(B(x)-\widehat{B}\right)\Lambda^{\sigma}(x)dx. \tag{10.46}$$

In this case, regularization of the operators $\widehat{K}$ and $\widehat{S}$ on constants is defined by Eqs. (10.28), and we obtain

$$\langle\sigma(x)\rangle = \widehat{\sigma},$$

$$\langle\varepsilon(x)\rangle=\widehat{B}\widehat{\sigma} + Q^*\widehat{\sigma} = B^*\langle\sigma(x)\rangle, \tag{10.47}$$

where $B^* = \widehat{B} + Q^*$ is the effective resistivity tensor. Because $B^* = (C^*)^{-1}$, the tensors P^* and Q^* relate by the equation

$$Q^* = (\widehat{C} + P^*)^{-1} - \widehat{B}. \tag{10.48}$$

The effective property tensors C^* and B^* should be definitely positive. It means that the scalars $\varepsilon C^*\varepsilon$ and $\sigma B^*\sigma$ are positive for any non-zero tensors ε and σ. For conductivity, the tensors C^* and B^* should be symmetric with respect to permutation of indices. This symmetry is a consequence of the existence of the potential w that is a single-valued function of the electric field e, and the current vector j is expressed as in terms of w as follows

$$j_\alpha = \frac{\partial w}{\partial e_\alpha}. \tag{10.49}$$

The complete differential of the function w is

$$dw = \frac{\partial w}{\partial e_\alpha}de_\alpha = j_\alpha de_\alpha. \tag{10.50}$$

On the other hand, for the media with linear conductive properties, we have

$$j_\alpha = C^*_{\alpha\beta}e_\beta, \; w = \frac{1}{2}e_\alpha C^*_{\alpha\beta}e_\beta. \tag{10.51}$$

and therefore, from Eqs. (10.49) and (10.51) we yield

$$C^*_{\alpha\beta}e_\beta = \frac{1}{2}\left(C^*_{\alpha\beta} + C^*_{\beta\alpha}\right)e_\beta. \tag{10.52}$$

Because e_β is an arbitrary vector, the tensor $C^*_{\alpha\beta}$ should be symmetric with respect to index permutations $C^*_{\alpha\beta} = C^*_{\beta\alpha}$.

For elasticity, the fourth rank tensors C^* and B^* should be symmetric with respect to the first and second pairs of indices, as well as about permutations of the index pairs. These symmetries are consequences of existence of a single-valued definitely positive function W of the strain tensor $\varepsilon_{\alpha\beta}$ such as the stress tensor $\sigma_{\alpha\beta}$ is

$$\sigma_{\alpha\beta} = \frac{\partial W}{\partial \varepsilon_{\alpha\beta}}, \quad W = \frac{1}{2}\varepsilon_{\alpha\beta} C^*_{\alpha\beta\gamma\delta} \varepsilon_{\gamma\delta}. \tag{10.53}$$

Similar to the case of conductivity, considering the complete differential of W, one can show that the tensor C^* satisfies the equalities

$$C^*_{\alpha\beta\gamma\delta} = C^*_{\beta\alpha\gamma\delta} = C^*_{\alpha\beta\delta\gamma} = C^*_{\gamma\delta\alpha\beta}. \tag{10.54}$$

Note that existence of the potential functions w and W in Eqs. (10.49) and (10.53) follows from the fundamental principles of thermodynamics (Onzager's principal).

The tensors P^* and Q^* in Eqs. (10.38) and (10.46) are results of averaging the functions $(C(x) - \widehat{C})\Lambda^{\varepsilon}(x)$ and $(B(x) - \widehat{B})\Lambda^{\sigma}(x)$ over ensemble realizations of the random function $C(x)$ and $B(x)$. The functions $\Lambda^{\varepsilon}(x)$ and $\Lambda^{\sigma}(x)$ define detailed fields in the heterogeneous medium subjected to a constant external field. These averages are the principal unknowns of the homogenization problem. Self-consistent methods allow the construction of reliable approximations for these averages.

10.3 Cross-Properties Relations

We consider a characteristic volume V of a two-phase micro heterogeneous medium, and let V_1 and V_2 $(V_1 + V_2 = V)$ be subregions occupied by different materials. The homogenization problem consists of the calculation of the effective property tensor of the composite. Behind these problems, there is a hidden suggestion that the effective properties, no matter how different in context and interpretation, should be interconnected. Such interconnections are called the cross-property relations.

The simplest cross-property relations are due to a mathematical analogy between the equations that govern different phenomena: dielectric and magnetic permittivity, electric conductivity, heat transfer, etc. In these cases, it is sufficient to obtain results for one of these properties, and for the rest, only the interpretations are changed. In some cases, however, the mathematical analogy is not so obvious, and some efforts are needed to clarify it. Many non-trivial cross-property relations are those where there is no direct mathematical analogy, and the relations result from the underlying internal structure of the medium. Here, we present one of the relations of this kind.

We consider a heterogeneous material that consists of the host medium (matrix) and a random set of inclusions with physical properties that differ from the properties of the host medium. The representative volume V of this material is subjected to surface forces and a constant temperature field T. The constitutive equation for the medium with incorporated thermal effects can be written in the form

$$\varepsilon_{\alpha\beta}(x) = B_{\alpha\beta\lambda\mu}(x)\sigma_{\lambda\mu}(x) + \alpha_{\alpha\beta}(x)T, \tag{10.55}$$

$$s(x) = \alpha_{\alpha\beta}(x)\sigma_{\alpha\beta}(x) + c_p(x)T. \tag{10.56}$$

Here, $B(x)$ is the elastic compliances tensor, $\alpha(x)$ is the tensor of thermal expansion coefficients, $c_p(x)$ is the heat capacity at constant stresses, $\varepsilon(x)$ and $\sigma(x)$ are the strain and stress tensors, $s(x)$ is the entropy density. $B(x)$, $\alpha(x)$ and $c_p(x)$ are random functions that are B^0, α^0, c_p^0 when point x is in the matrix and B, α, c_p if x is in inclusions. Let external forces be applied to the border S of the volume V_0

$$t_\alpha^0(x) = \sigma_{\alpha\beta}^0 n_\beta(x). \tag{10.57}$$

Here, σ^0 is a constant tensor, and $n(x)$ is the external normal to S. In this case, the mean stress field in V coincides with σ^0, as follows from the equation

$$\left\langle \sigma_{\alpha\beta}(x) \right\rangle = \frac{1}{V}\int\limits_V \sigma_{\alpha\beta}(x)\mathrm{d}x = \frac{1}{2V}\int\limits_\Omega \left(t_\alpha^0 x_\beta + t_\beta^0 x_\alpha\right) d\Omega = \sigma_{\alpha\beta}^0. \tag{10.58}$$

Here, Gauss' theorem is used.

The functions $B(x)$, $\alpha(x)$ are presented as the following sums

$$B(x) = B^0 + B^1 V_1(x), \quad \alpha(x) = \alpha^0 + \alpha^1 V_1(x), \tag{10.59}$$

$$c_p(x) = c^0 + c^1 V_1(x), \tag{10.60}$$

$$B^1 = B - B^0, \quad \alpha^1 = \alpha - \alpha^0, \; c^1 = c - c^0, \tag{10.61}$$

where $V_1(x)$ is the characteristic function of the region V_1 occupied by the inclusions: $V_1(x) = 1$, $x \in V_1$; $V_1(x) = 0$, $x \notin V_1$. After averaging Eqs. (10.55), (10.56) over the volume V we obtain

$$\langle\varepsilon\rangle = B^0 \langle\sigma\rangle + pB^1\overline{\sigma} + \langle\alpha\rangle T, \tag{10.62}$$

$$\langle s\rangle = \alpha^0 \langle\sigma\rangle + p\alpha^1\overline{\sigma} + \left\langle c_p\right\rangle T, \tag{10.63}$$

where p is the volume fraction of the inclusions, $\overline{\sigma}$ is the stress field averaged over the region occupied by the inclusions,

$$\langle\alpha\rangle = (1-p)\alpha^0 + p\alpha, \left\langle c_p\right\rangle = c^0 + pc^1.$$

Because of the linearity of the problem, the tensor $\overline{\sigma}$ is a linear function of the mean stress field $\langle\sigma\rangle = \sigma^0$ and of the temperature T

$$\overline{\sigma} = \Lambda \langle\sigma\rangle + \lambda T, \tag{10.64}$$

where Λ and λ are unknown fourth and second rank tensors. Substituting this equation into Eqs. (10.62) and (10.63) we obtain

$$\langle\varepsilon\rangle = B^* \langle\sigma\rangle + \alpha^* T, \tag{10.65}$$

$$\langle s\rangle = \alpha^* \langle\sigma\rangle + c_p^* T. \tag{10.66}$$

Here, B^*, α^* and c_p^* are the effective thermo-elastic characteristics of the heterogeneous medium. It follows from Eqs. (10.62)–(10.64) that these tensors are defined by the equations

$$B^* = B^0 + pB^1\Lambda = \langle B\rangle + pB^1 (\Lambda - I), \tag{10.67}$$

$$\alpha^* = \langle\alpha\rangle + pB^1\lambda = \langle\alpha\rangle + p\alpha^1 (\Lambda - I), \tag{10.68}$$

$$c_p^* = \langle c_p\rangle + p\alpha^1\lambda, \tag{10.69}$$

where I is a unit fourth rank tensor, and

$$\langle B\rangle = B^0 + pB^1. \tag{10.70}$$

It is seen from these equations that the effective tensors α^* and B^* are not independent, but they are expressed via the tensor Λ. Indeed, excluding this tensor from Eqs. (10.67) and (10.68), we have

$$\alpha^* - \langle\alpha\rangle = \alpha^1 \left(B^1\right)^{-1} \left(B^* - \langle B\rangle\right), \tag{10.71}$$

It follows from Eqs. (10.66) and (10.69) that

$$c_p^* - \langle c_p\rangle = \alpha^1 \left(B^1\right)^{-1} \left(\alpha^* - \langle\alpha\rangle\right). \tag{10.72}$$

Suppose that both components in the heterogeneous medium are isotropic with bulk elastic moduli K_0 and K and thermal expansion coefficients α_0, α. Then, Eqs. (10.71), (10.72) are reduced to the following ones

$$\alpha_* - \langle\alpha\rangle = \frac{\alpha - \alpha_0}{K^{-1} - K_0^{-1}} \left(\frac{1}{K_*} - \left\langle\frac{1}{K}\right\rangle\right), \tag{10.73}$$

$$c_p^* - \langle c_p\rangle = \frac{9(\alpha - \alpha_0)}{K^{-1} - K_0^{-1}} \left(\alpha_* - \langle\alpha\rangle\right). \tag{10.74}$$

Thus, if the effective bulk modulus of the composite K_* is known, one can calculate its effective coefficient of thermo-expansion α_* from (10.73), and *vice versa*, if the coefficient α_* is known, the effective bulk modulus K_* can be calculated from the same equation. This is an exact result that does not depend on the microstructure of the composite. In particular, it is correct for polycrystalline microstructures.

Equation (10.73) was firstly obtained in [5]. Almost at the same time, this equation was derived and generalized in [6, 7]. Equation (10.74) was obtained in [7]. The background of various cross-property relations was discussed and implemented in [8, 9]. The approximate theory of cross-property relations was developed in [10–12].

References

1. Prokhorov Y, Rosanov Y (1969) The probability theory: basic concepts, limit theorems. Random Processes. Springer, Berlin
2. Gichman I, Skorochod A (1977) Introduction to the theory of random processes. Nauka, Moscow, p 567
3. Hill R (1963) Elastic properties of reinforced solids: some theoretical principles. J Mech Phys Solids 11:357–372
4. Lifshitz I, Gradeskul S, Pastur L (1982) Introduction to the theory of disordered systems. Nauka, Moscow, p 358
5. Levin V (1967) On the thermal expansion coefficients of heterogeneous materials. Proc Acad Sci USSR Mech Solids 1:88–93
6. Shapery N (1968) Thermal expansion coefficients of composite materials, based on energy principles. J Composite Mater 2:380–404
7. Rosen B, Hashin Z (1970) Effective thermal expansion coefficient and specific heats of composite materials. Int J Eng Sci 8:157–173
8. Grabovsky Y, Milton G (1998) Exact relations for composites. Towards a complete solution. Documenta Math, Extra Volume ICM 1–10
9. Milton G (1997) Composites: a myriad of microstructure independent relations. In: Tatsumi T (ed) Theoretical and applied mechanics. Elsevier, pp 443–459
10. Kachanov M, Sevostianov I, Shafiro B (2001) Explicit cross-property correlations for porous materials with anisotropic microstructures. J Mech Phys Solids 49:1–25
11. Sevostianov I, Kachanov M (2002) Explicit cross-property correlations for anisotropic two-phase composite materials. J Mech Phys Solids 50:253–282
12. Sevostianov I, Kachanov M (2004) Connection between elastic and conductive properties of microstructures with Hertzian contacts. Proc Royal Soc Lond A 460:1529–1534

Chapter 11
The Effective Medium Method

Abstract The hypotheses of the effective medium method (EMM) and its modifications (Kerner's method, differential effective medium method) are formulated. The method and its modifications are applied to the solution of the homogenization problem for conductive and elastic matrix composite materials with ellipsoidal inclusions. Predictions of the method are compared with experimental data and results of numerical simulations. The results of the application of the method to the media with elliptical cracks are presented.

Keywords Effective medium method · Kerner's method · Differential effective medium method · Matrix composites · Media with multiple cracks

The effective medium method (EMM) can be used for the solution of the homogenization problem for heterogeneous media that consist of a set of homogeneous in space similar structural elements. Examples of such elements are inclusions in matrix composite materials or grains of polycrystalline aggregates. The method is based on the following hypotheses.

H_1. *Each structural element of the heterogeneous medium can be considered as an isolated one embedded into the homogeneous medium with the overall (effective) properties of the medium.*

H_2. *The field acting on each structural element coincides with the averaged field over the heterogeneous medium.*

The first hypothesis reduces the problem of interaction between the structural elements to the problem for an isolated element embedded into a homogeneous host medium with the effective properties of the composite (the one-particle problem). The hypotheses of the EMM allow for revealing the dependencies of the tensors P^* and Q^* in Eqs. (10.38) and (10.46) on the shapes and properties of the structural elements.

Further, two types of conditions at infinity are considered: a fixed field intensity ε^0 or a fixed flux density σ^0. For a fixed field intensity, the mean field in the composite coincides with the external field: $\langle\varepsilon\rangle = \varepsilon^0$, and for a fixed flux density, $\langle\sigma\rangle = \sigma^0$.

S. Kanaun, *Self-Consistent Methods for Composites*, Solid Mechanics and Its Applications 282, https://doi.org/10.1007/978-3-031-96739-9_11

For this reason, the second hypothesis can be formulated as follows: the field acting on each structural element coincides with the external field applied to the heterogeneous medium.

11.1 The Effective Medium Method for Matrix Composites

We start with the application of the EMM to the solution of the homogenization problem for matrix composite material. Such materials consist of a homogeneous host medium with the property tensor C^0 and a set of isolated homogeneous inclusions with the property tensor C homogeneously distributed in the host medium. According to the hypotheses of the EMM, the field inside each inclusion coincides with the field in an isolated inclusion embedded into the medium with the effective property of the composite and subjected to a constant external field applied to the medium (Fig. 11.1).

For matrix composites, it is reasonable to choose the properties of the reference medium coincided with the property of the host medium (the matrix material), e.g., $\widehat{C} = C^0$. For such a choice, the integral presentations (10.13), (10.15) of the field intensity and the flux density in the heterogeneous medium take the forms

$$\varepsilon(x) = \varepsilon^0 - \int K(x - x')C^1(x')\varepsilon(x')dx', \tag{11.1}$$

$$\sigma(x) = \sigma^0 + \int S(x - x')B^1(x')\sigma(x')dx', \tag{11.2}$$

where

$$C^1(x) = \left(C - C^0\right) V(x), \; B^1(x) = \left(B - B^0\right) V(x), \tag{11.3}$$

and $V(x)$ is the characteristic function of the region occupied by the inclusions, the kernels $K(x)$, $S(x)$ are defined in Eqs. (10.14), (10.16) with $\widehat{C} = C^0$. Therefore, the

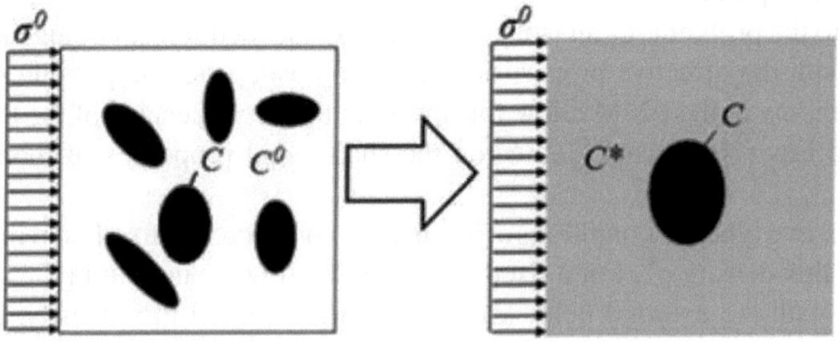

Fig. 11.1 The one-particle problem of the effective medium method

functions $C^1(x)$, $B^1(x)$ are equal to zero outside the inclusions, and the integrals in Eqs. (11.2), (11.3) are defined by the fields $\varepsilon(x)$ and $\sigma(x)$ inside inclusions only.

The fields ε and σ inside each inclusion can be found from the solution of the one-particle problem that is a consequence of the hypothesis H_1. The integral equations of this problem for the kth inclusion are

$$\varepsilon(x) + \int\limits_{v_k} K^*(x-x')C^{*1(k)}(x')\varepsilon(x')dx' = \varepsilon^0, \; x \in v_k, \tag{11.4}$$

$$\sigma(x) - \int\limits_{v_k} S^*(x-x')B^{*1(k)}(x')\sigma(x')dx' = \sigma^0, \; x \in v_k, \tag{11.5}$$

$$C^{*1(k)}(x) = C^{(k)}(x) - C^*, \; B^{*1(k)}(x) = B^{(k)}(x) - B^*. \tag{11.6}$$

Here, the kernels $K^*(x)$, $S^*(x)$ are defined in Eqs. (10.14), (10.16) with $\widehat{C} = C^*$, and $G^*(x)$ is Green's function of the homogeneous effective medium. For the linearity of the problem, the solutions of Eqs. (11.4), (11.5) can be presented in the forms

$$\varepsilon = \Lambda^{\varepsilon(k)}(x)\varepsilon^0, \; \sigma = \Lambda^{\sigma(k)}(x)\sigma^0, x \in v_k, k = 1, 2, ... \tag{11.7}$$

Here, the functions $\Lambda^{\varepsilon(k)}(x)$ and $\Lambda^{\sigma(k)}(x)$ depend on the properties of the inclusion, effective medium, and the inclusion shape. If the one-particle problem can be solved, the tensors P^* and Q^* in Eqs. (10.38), (10.46) take the forms

$$P^* = \lim_{W\to\infty} \frac{1}{W} \sum_{k=1}^{N} \int\limits_{v_k} \left(C^{(k)}(x) - C^0\right) \Lambda^{\varepsilon(k)}(x)dx, \tag{11.8}$$

$$Q^* = \lim_{W\to\infty} \frac{1}{W} \sum_{k=1}^{N} \int\limits_{v_k} \left(B^{(k)}(x) - B^0\right) \Lambda^{\sigma(k)}(x)dx, \tag{11.9}$$

where N is the number of inclusions in the region V. Then, the expressions for the effective property tensors C^* and B^* follow from Eqs. (10.40), (10.47)

$$C^* = C^0 + P^*, \; B^* = B^0 + Q^*. \tag{11.10}$$

Because P^* and Q^* depend on the effective property tensors C^*, B^* via the solution of the one-particle problems, Eqs. (11.10) are the implicit equations for the tensors C^* and B^*.

11.2 The Effective Conductivity Tensor of Composites with Ellipsoidal Inclusions

In the case of homogeneous ellipsoidal inclusions, the solutions of the integral equations (11.4), (11.5) are considered in Sect. 4.4. The tensor $\Lambda^{\varepsilon(k)}$ and $\Lambda^{\sigma(k)}$ in Eq. (11.7) are constant, and they have the forms

$$\Lambda^{\varepsilon(k)} = \left(I + A^*(a)(C^{(k)} - C^*)\right)^{-1}, \tag{11.11}$$

$$\Lambda^{\sigma(k)} = \left(I - D^*(a)(B^{(k)} - B^*)\right)^{-1}, \tag{11.12}$$

where

$$D^*(a) = C^*A^*(a)C^* - C^*. \tag{11.13}$$

Here, the tensor $A^*(a)$ is in Eq. (4.200), and the second rank tensor $a = a^{(k)}$ defines the aspect ratios and orientation of the kth inclusion, C^* is the effective property tensor of the composite. For a medium with identical ellipsoidal inclusions, the tensors P^* and Q^* in Eqs. (11.8), (11.9) are

$$P^* = p(C - C^0)\left(I + A^*(a)(C - C^*)\right)^{-1}, \tag{11.14}$$

$$Q^* = p(B - B^0)(I - D^*(a)(B - B^*))^{-1}, \tag{11.15}$$

where p is the inclusion volume fraction. As a result, the effective property tensors C^* and B^* in Eqs. (11.10) take the forms

$$C^* = C^0 + p(C - C^0)\left(I + A^*(a)(C - C^*)\right)^{-1}, \tag{11.16}$$

$$B^* = B^0 + p(B - B^0)(I - D^*(a)(B - B^*))^{-1}. \tag{11.17}$$

The tensor C^* presents in the both sides of Eq. (11.16). Therefore, Eq. (11.16) is an implicit equation for the effective property tensor C^*. Similarly, Eq. (11.17) is an implicit equation for the effective tensor $B^{*.}$.

Note that for anisotropic host media and anisotropic inclusions, the tensors C^* and B^* in Eqs. (11.16), (11.17) can be not symmetric with respect to the permutation of indices. To show this, we consider formal expansion of the second term in the right-hand side of Eq. (11.14) into the series in terms of the tensor $A^*(C - C^*)$

$$\begin{aligned} &(C - C^0)(I + A^*(C - C^*))^{-1} = \\ &= (C - C^0) - (C - C^0)A^*(C - C^*) + \\ &+(C - C^0)A^*(C - C^*)A^*(C - C^*) - \dots \end{aligned} \tag{11.18}$$

The second term in the right-hand side of this equation is symmetric with respect to indices if either $C^* = C^0$ or the tensors C^*, C^0,C and A^* have the same principal axes. In other cases, the tensor C^* in Eq. (11.16) is not symmetric. The same holds for the tensor B^* in Eq. (11.17). Because symmetry of the tensors C^* and B^* follows from fundamental principles of thermodynamics, the lack of symmetry is an inevitable defect of the effective medium method in application to matrix composite materials.

If the principal axes of the tensors C^*, C^0,C and A^* coincide, the solutions of Eqs. (11.16) and (11.17) result in the effective property tensors that relates by the equation $B^* = (C^*)^{-1}$.

Let an isotropic conductive medium with the conductivity coefficient c_0 contain a set of isotropic spherical inclusions with conductivity coefficients c. In this case, the tensor A^* in Eq. (11.11) is

$$A^*_{\alpha\beta} = \frac{1}{3c_*}\delta_{\alpha\beta}. \tag{11.19}$$

For homogeneous and isotropic distributions of inclusions in space, the effective medium is isotropic, and the equation for the effective conductivity coefficient c_* follows from Eq. (11.16) in the form

$$c_* = c_0 + 3p\frac{(c - c_0)c_*}{c + 2c_*}. \tag{11.20}$$

The physically correct solution of this equation is

$$c_* = \frac{1}{4}\left(2c_0 - c + 3p(c - c_0) + \sqrt{8cc_0 + (c - 2c_0 - 3p(c - c_0))^2}\right). \tag{11.21}$$

Equation (11.17) results in the same equation for the effective conductivity coefficient c_*.

For the medium with non-conductive inclusions ($c = 0$) we have from Eq. (11.21)

$$c_* = \left(1 - \frac{3}{2}p\right)c_0,\ p < \frac{2}{3};\ c_* = 0,\ p > \frac{2}{3}. \tag{11.22}$$

If $c \to \infty$ (highly conductive inclusions), the coefficient c_* takes the form

$$c_* = \frac{c_0}{1 - 3p}. \tag{11.23}$$

Therefore, c_* tends to infinity as $p \to 1/3$.

Let the host medium be isotropic and the inclusions be identical spheroids with the rotation axes parallel to the x_3-axis. If the conductivity tensor C of the inclusion material is isotropic, the effective conductivity tensor C^* is transverse isotropic, and in the Cartesian coordinate system x_1, x_2, x_3, the matrix of this tensor is diagonal with the components $C^*_{11} = C^*_{22} = c^*_1$, $C^*_{33} = c^*_3$. The solution of the one-particle problem for an isotropic spheroidal inclusion embedded into an infinite transversely isotropic medium is presented in Eqs. (4.211)–(4.213). Using this solution, Eq. (11.16) results in the system of equations for the effective conductivity coefficients c_1^* and c_3^*

$$c_1^* = c_0 + p(c - c_0)\left(1 + A_1^*(c - c_1^*)\right)^{-1}, \tag{11.24}$$

$$c_3^* = c_0 + p(c - c_0)\left(1 + A_3^*(c - c_3^*)\right)^{-1}, \tag{11.25}$$

where the coefficients A_1^* and A_3^* are

$$A_1^* = \frac{\gamma^2}{2}\int_0^\pi \frac{(\sin\theta)^3\, d\theta}{c_3^* + \gamma^2 c_1^* + (c_3^* - \gamma^2 c_1^*)\cos 2\theta}, \tag{11.26}$$

$$A_3^* = \int_0^\pi \frac{(\cos\theta)^2 \sin\theta d\theta}{c_3^* + \gamma^2 c_1^* + (c_3^* - \gamma^2 c_1^*)\cos 2\theta}, \quad \gamma = \frac{a_3}{a}, \tag{11.27}$$

Here, $a = a_1 = a_2$ and a_3 are semiaxes of the spheroidal inclusions.

The results of solution of Eqs. (11.24), (11.25) are presented in Figs. 11.2 and 11.3 for the inclusion conductivity $c = 20c_0$, $c = 0.01c$ and the inclusion aspect ratios $\gamma = 0, 1, \infty$.

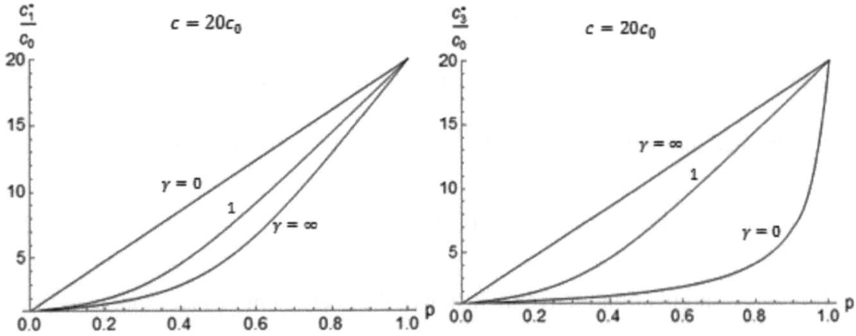

Fig. 11.2 Effective conductivity coefficients of the composite with an isotropic host medium with conductivity c_0 and isotropic spheroidal inclusions c of the same orientations and the conductivity coefficient $c = 10c_0$, γ is the aspect ratios of the inclusions

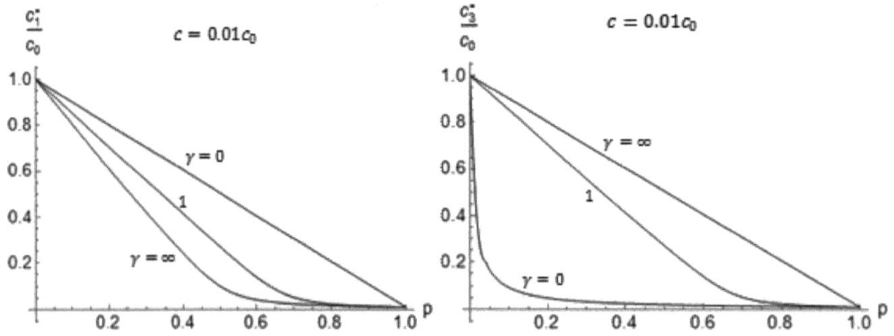

Fig. 11.3 The same as in Fig. 11.2 for $c = 0.01c_0$

11.3 Kerner's Method

Another version of the effective medium method in application to the matrix composite materials was proposed in [1]. In this version, the hypotheses of the method are formulated as follows.

- The field inside each inclusion in the composite coincides with the field inside the same inclusion with a layer of the matrix material at the inclusion interface embedded into the homogeneous medium with the effective properties of the composite. The thickness h of the layer depends on the volume fraction p of the inclusions. The volume v of the inclusion and the volume V^K of the inclusion with the layer (Kerner's cell) are related by the equation: $v = pV^K$.
- The layered inclusion is subjected to the external field applied to the composite (Fig. 11.4).

We consider an isotropic conductive host medium with isotropic spherical inclusions. For homogeneous and isotropic distributions of the inclusions in space, the effective conductivity tensor is isotropic. The solution of the one-particle problem for inclusion with a layer is obtained in Sect. 9.1.1. It follows from this solution that

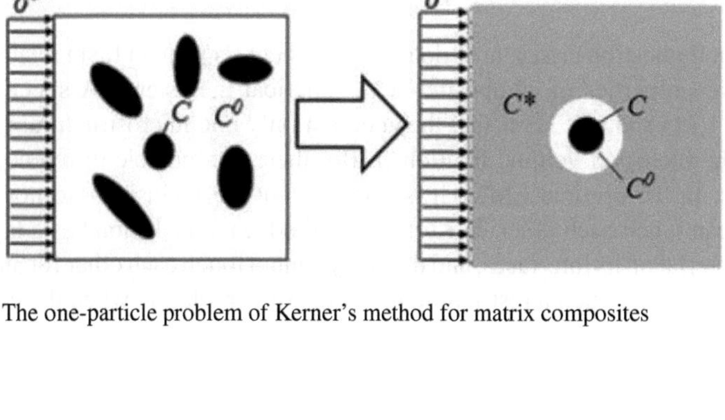

Fig. 11.4 The one-particle problem of Kerner's method for matrix composites

the tensor Λ^{ε} in Eq. (11.7) is constant inside the inclusion (in the core of the layered inclusion) and it has the form

$$\Lambda^{\varepsilon}_{\alpha\beta} = \lambda_* \delta_{\alpha\beta}, \tag{11.28}$$

$$\lambda_* = \frac{9c_0c_*}{2(1-p)(c_0^2 + cc_*) + 2(2+p)c_0c_* + (1+2p)c_0c}. \tag{11.29}$$

The equation for the effective conductivity coefficient c_* follows from Eqs. (11.8) and (11.10)

$$c_* = c_0 + p(c - c_0)\lambda_*. \tag{11.30}$$

This algebraic equation for c_* has an explicit solution

$$c_* = c_0 \frac{2(1-p)c_0 + (1+2p)c}{(2+p)c_0 + (1-p)c}. \tag{11.31}$$

For non-conductive inclusions ($c = 0$), we yield

$$c_* = c_0 \frac{2(1-p)}{(2+p)}, \tag{11.32}$$

and for inclusions with the infinite conductivity ($c \to \infty$),

$$c_* = c_0 \frac{(1+2p)}{(1-p)}. \tag{11.33}$$

Comparison of the predictions of the original effective medium method and Kerner's method for highly conductive and non-conductive spherical inclusions are presented in Figs. 11.5 and 11.6. In these figures, solid lines correspond to the original EMM and dashed lines to Kerner's method. The methods provide close values for the effective conductivity in the region of small volume fractions, but they deviate substantially for $p > 0.2$. Note that for composites with spherical inclusions, the volume fraction p close to 1 can be achieved for a wide distribution of the inclusion over the radii only.

A natural question arises: how do predictions (11.21) and (11.31) relate to actual conductive properties of composites with spherical inclusions? A specific feature of Eqs. (11.21) and (11.31) is that they contain only one microstructural parameter, namely the inclusion volume fraction p. But there is a notable difference in these equations. In the original EMM, it is implied that the inclusions do not intersect, but they can touch each other. In Kerner's method, each inclusion has a layer of the matrix material on its interfaces, and thus they cannot touch each other for any volume fraction $p < 1$. By such a restriction, the effect of percolation when the current can

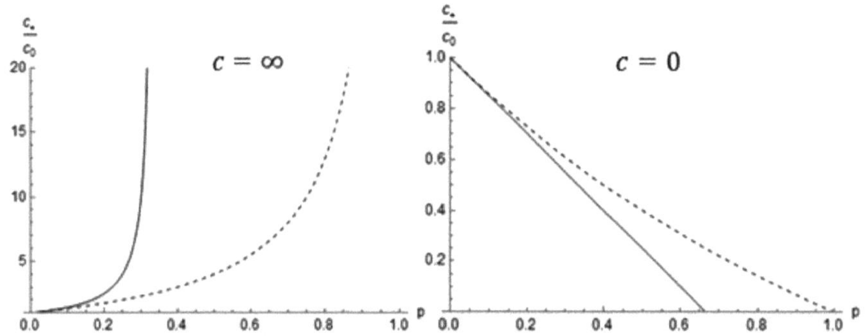

Fig. 11.5 Predictions of the effective field method (sold line) and Kerner's method (dashed line) for the effective conductivity c^* of an isotropic medium with isotropic spherical inclusions of infinite conductivity (left) and with non-conductive inclusions (right), c_0 is the conductivity of the host medium

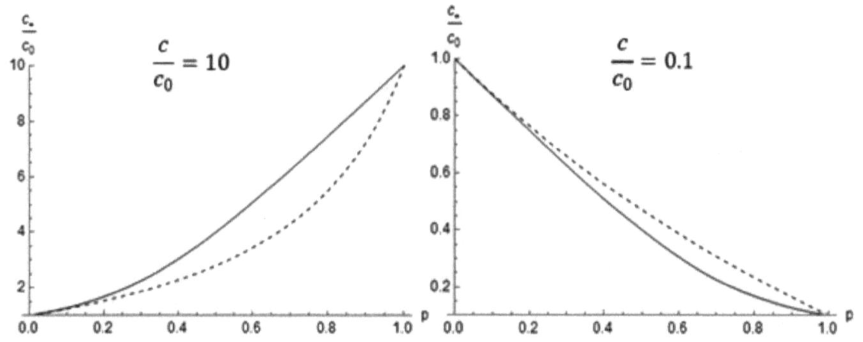

Fig. 11.6 Predictions of the effective medium method (sold line) and Kerner's method (dashed line) for the effective conductivity c^* of an isotropic medium with isotropic spherical inclusions of the conductivity $c = 10c_0$ (left) and of the conductive $c = 0.1c_0$ (right), c_0 is the conductivity of the host medium

pass through the inclusion phase only cannot occur. The original EMM predicts for composites with the inclusions of infinite conductivity that the effective conductivity c_* turns to infinity for the volume fraction of inclusions $p = 1/3$. For the composite with non-conductive inclusions, the effective conductivity turns to zero for $p = 2/3$. These volume fractions could be interpreted as the percolation thresholds for the composites with spherical inclusions. But it is known that percolation thresholds depend strongly on the statistical model of the inclusion distribution in the host medium, e.g., the size distribution of particles affects essentially the values of the percolation thresholds. In the original EMM, these statistics are not accounted for. In Kerner's method, there is an additional restriction of the absence of contacts between inclusions. The existence of the percolation thresholds in the original EMM can hardly be reasonably explained. But in numerical simulation of the effective conductive properties of the composites with highly conductive spherical particles of the same radius, the percolation threshold near the inclusion volume fraction $p = 0.33$ was indicated [2].

Application of Kerner's method to the inclusions of non-spherical shapes encounters difficulties in the solution of the one-particle problem. Even for an ellipsoidal inclusion with a layer of constant thickness, the one-particle problem can be solved only numerically.

11.4 The Effective Medium Method for Elastic Matrix Composites

In the case of elasticity, the equation for the effective stiffness tensor of matrix-inclusion composites keeps the form (11.16), where C^0, C and C^* are the fourth rank tensors of the elastic stiffness of the host medium, the inclusions, and the effective medium. Similar to the effective conductivity tensors, for composites with anisotropic inclusions, the effective medium method can result in a non-symmetric effective stiffness tensor with respect to the permutation of the index pairs.

Let isotropic spherical inclusions be homogeneously distributed in an isotropic matrix, and λ_0, μ_0 and λ, μ be Lame's parameters of the matrix and inclusion materials. In this case, the effective medium is isotropic. The solution of the one-particle problem of the EMM for a homogeneous spherical inclusion embedded in an isotropic host medium with Lame's parameters λ_*, μ_* of the effective medium is given by Eqs. (4.82) and (4.134). For these equations, the tensor P^* in Eq. (11.14) takes the form

$$P^* = p\left(C - C^0\right)\left(I + A(C^*)(C - C^*)\right)^{-1}, \tag{11.34}$$

$$A(C^*) = \frac{1}{3(3K_* + 4\mu_*)}E^2 + \frac{3(K_* + 2\mu_*)}{5\mu_*(3K_* + 4\mu_*)}\left(E^1 - \frac{1}{3}E^2\right). \tag{11.35}$$

Here, E^1, E^2 are the elements of the E-basis (Appendix A.1), K_*, μ_* are the bulk and shear moduli of the effective medium. From Eq. (11.16) follows the system of algebraic equations for the effective bulk K_* and shear μ_* moduli of the composite

$$K_* = K_0 + p\,(K - K_0)\left(1 + \frac{3(K - K_*)}{3K_* + 4\mu_*}\right)^{-1}, \tag{11.36}$$

$$\mu_* = \mu_0 + p\,(\mu - \mu_0)\left(1 + \frac{6\,(\mu - \mu_*)\,(K_* + 2\mu_*)}{5\mu_*\,(4\mu_* + 3K_*)}\right)^{-1}. \tag{11.37}$$

This is the system of algebraic equations of the second order with respect to K_* and μ_*. The solution of this system can be obtained numerically by a simple iteration process

$$K_*^{(n+1)} = F_K\left(K_*^{(n)}, \mu_*^{(n)}\right), \mu_*^{(n+1)} = F_\mu\left(K_*^{(n)}, \mu_*^{(n)}\right). \tag{11.38}$$

Here, $F_K(K_*, \mu_*)$, $F_\mu(K_*, \mu_*)$ are the right-hand sides of Eqs. (11.36), (11.37), $K_*^{(n)}, \mu_*^{(n)}$ are the effective bulk and shear moduli at the nth step of the iteration process.

To visualize the solution of the system (11.36), (11.37), we consider the case of a random set of spherical pores with $K = 0$, $\mu = 0$ in a homogeneous host medium. From Eq. (11.36) we obtain

$$K_* = (1-p)K_0\left(1 + \frac{3K_*}{4\mu_*}\right). \tag{11.39}$$

Substituting K_* from this equation into Eq. (11.37), we yield the quadratic equation for μ_*

$$8\mu_*^2 + (3K_0(1-p) - 4\mu_0(2-5p))\,\mu_* - 3\mu_0 K_0(1-p) = 0. \tag{11.40}$$

For large bulk modulus K_0, the effective bulk and shear moduli of the composite are assessed as follows

$$K_* = \frac{1}{3}K_0(3-p) - \frac{40p^2\mu_0}{27(3-p)} + O\left(K_0^{-1}\right), \tag{11.41}$$

$$\mu_* = \frac{3\mu_0(1-2p)}{(3-p)} + O\left(K_0^{-1}\right). \tag{11.42}$$

For absolutely rigid spheres ($K, \mu \to \infty$), Eqs. (11.36), (11.37) result in the effective bulk and shear moduli of the composite

$$K_* = \frac{K_0}{1-p} + \frac{8p\mu_0}{3(1-p)(2-5p)} + O\left(K_0^{-1}\right), \tag{11.43}$$

$$\mu_* = \frac{2\mu_0}{2-5p} + O\left(K_0^{-1}\right). \tag{11.44}$$

In this case, shear modulus μ_* becomes infinite when $p = 2/5$.

The shortcomings of the effective medium method in the case of matrix composites were noted in a number of works (see, e.g., [3]). The effective elastic moduli of the medium with rigid spherical inclusions ($C^0C^{-1} = O(\delta)$, $\delta << 1$) or pores ($C = 0$) calculated from equations Eqs. (11.36), (11.37) differ substantially from those measured in experiments if the volume fraction of the inclusions (pores) p exceeds 0.3 (see Fig. 11.8).

Better correspondence with experimental data can be obtained for the analogy of Kerner's version of the EMM [1, 4–6]. In this version, a layer of matrix material is introduced in the interface between the effective medium and the inclusion (three-phase model). For a matrix composite with spherical inclusion, the one-particle problem of Kerner's method is formulated as follows. Let a spherical inclusion with elastic stiffness tensor C have a spherical layer at the interface with the properties of the matrix material C^0. The layered inclusion is embedded into the homogeneous effective medium with the elastic stiffness tensor C^*. The one-particle problem consists in the calculation of the field inside the core of the layered inclusion by action of a constant external field at infinity. For spherical inclusions of the radius a, the thickness h of the layer relates to the inclusion volume fraction p by the equation [1, 4]

$$\left(\frac{a}{a+h}\right)^3 = p \,, \tag{11.45}$$

In the Cartesian coordinate system x_1, x_2, x_3 with the origin at the inclusion center, the strain field in the layered spherical inclusion is presented in the E-basis as follows (Sect. 8.4)

$$\varepsilon(r,n) = \Lambda^{\varepsilon}(r,n)\varepsilon^0, \tag{11.46}$$

$$\Lambda^{\varepsilon}(r,n) = \sum_{j=1}^{6} \epsilon_j(r)E^j(n),\, r = |x|,\, n_i = \frac{x_i}{r}, \tag{11.47}$$

where the scalar coefficients $\epsilon_j(r)$ are defined in Eqs. (8.60)–(8.64)

$$\epsilon_1 = 1 + (3 + D)\alpha_1,\ \epsilon_2 = \frac{1}{3}\left(\beta - 3\alpha_1 + (2 + D)\alpha_3\right), \tag{11.48}$$

$$\epsilon_3 = D(\alpha_1 + \alpha_3), \quad \epsilon_4 = \frac{1}{3}D(\beta - 3\alpha_1 - (2 + D)\alpha_3), \tag{11.49}$$

$$\epsilon_5 = D(4\alpha_3 + (1 + D)\alpha_1),\ \epsilon_6 = D(2 - D)(\alpha_1 - \alpha_3). \tag{11.50}$$

Here, $D = rd/dr$, α_1, α_3 and β are three scalar functions of the radial coordinate r. For isotropic inclusion and layer materials, the functions $\alpha_1, \alpha_2, \beta$ in the inclusion have forms

$$\alpha_1(r) = Y_1 + Y_2 r^2,\ \alpha_3 = Y_5 + Y_6 r^2 - \frac{1}{2},\ \beta = y_1 - 1, \tag{11.51}$$

where Y_1, Y_2, Y_5, Y_6 and y_1 are constants. The algorithm of construction of these constants is presented in Sect. 8.4. The values of these constants depend on the tensors C^0, C, C^* and the thickness h of the layer.

The tensor P^* in Eq. (11.8) is

$$P^* = p(C - C^0)\overline{\Lambda}^{\varepsilon}, \tag{11.52}$$

where $\overline{\Lambda}^{\varepsilon}$ is the following integral

$$\overline{\Lambda}^{\varepsilon} = \frac{1}{v}\int_0^a r^2 dr \int_{\Omega_1} \Lambda^{\varepsilon}(r,n)d\Omega_n \tag{11.53}$$

Here, Ω_1 is the surface of a unit sphere in the x-space, $v = 4\pi a^3/3$ is the inclusion volume. After integrating over Ω_1, we obtain

$$\overline{\Lambda}^{\varepsilon} = \mathbb{I}_1\left(E^1 - \frac{1}{3}E^2\right) + \mathbb{I}_2 E^2, \tag{11.54}$$

$$\mathbb{I}_1 = \frac{3}{a^3}\int_0^a \left(\epsilon_1(r) + \frac{1}{3}\epsilon_5(r) + \frac{2}{15}\epsilon_6(r)\right) r^2 dr = y_1, \tag{11.55}$$

$$\mathbb{I}_2 = \frac{1}{a^3}\int_0^a \left(\epsilon_1(r) + 3\epsilon_2(r) + \epsilon_3(r) + \epsilon_4(r) + \frac{1}{3}\epsilon_5(r) + \frac{1}{5}\epsilon_6(r)\right) r^2 dr =$$

$$= \frac{3}{15}\left(15Y_1 + 21Y_2 + 10Y_5 + 14Y_{6.}\right). \tag{11.56}$$

Here, Y_i, y_1 are the constants in Eq. (11.51) that define the solution inside the inclusion. Then, the equations for the effective bulk K_* and shear μ_* moduli of the composite follow from the equation $C^* = C^0 + P^*$

$$K_* = K_0 + p(K - K_0)\mathbb{I}_2(K_*, \mu_*), \tag{11.57}$$

$$\mu_* = \mu_0 + p(\mu - \mu_0)\mathbb{I}_1(K_*, \mu_*). \tag{11.58}$$

This is the system of non-linear equations for K_*, μ_*.

The solution of Eqs. (11.57), (11.58) can be found by the simple iteration procedure in Eq. (11.38), where $K_*^{(n)}$, $\mu_*^{(n)}$ are the effective bulk and shear moduli at the nth step of the iteration process. In some cases, the simple iteration process can diverge (e.g., for porous media with $p > 0.3$). The iteration scheme based on the equations

$$K_*^{(n+1)} = \alpha\left(F_K\left(K_*^{(n)}, \mu_*^{(n)}\right) - K_*^{(n)}\right) + K_*^{(n)}, \tag{11.59}$$

$$\mu_*^{(n+1)} = \alpha\left(F_\mu\left(K_*^{(n)}, \mu_*^{(n)}\right) - \mu_*^{(n)}\right) + \mu_*^{(n)}. \tag{11.60}$$

is more efficient in this case. Here, α is a numerical parameter ($|\alpha| < 1$) that should be taken for convergence of the iteration process.

Figure 11.7 presents the comparison of the theoretical and experimental values of the effective Young modulus $E_* = 9K_*\mu_*/(3K_* + \mu_*)$ and Poisson's ratio $\nu_* = (3K_* - 2\mu_*)/(6K_* + 3\mu_*)$ for composites with stiff spherical inclusions. In the experiments, a polymer matrix with spherical glass particles of the volume fraction p from $p = 0$ till $p = 0.5$ was studied. The Young moduli of the particle and the matrix phases relate as $E = 25.249E_0$, Poisson's ratio of the particles is $\nu = 0.23$ and of the matrix phase is $\nu_0 = 0.394$. In these figures, lines 1 correspond to the original EMM, lines 2 to Kerener's version of the EMM, black circles are experimental data presented in [7]. Lines 3 in this figure are predictions of the effective field method considered in the next chapter.

Figure 11.8 shows dependencies of the effective Young modulus and Poisson ratio of a homogeneous isotropic medium with spherical pores. In the literature, there are a number of experimental data on the elastic properties of porous materials. The

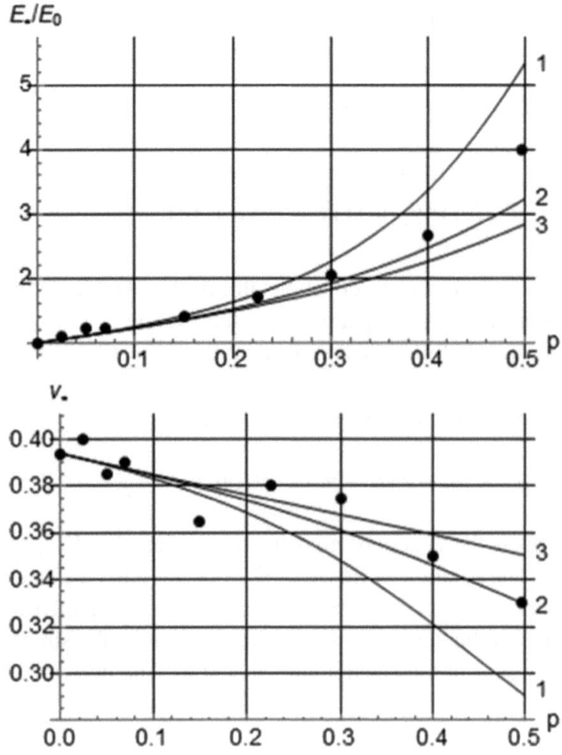

Fig. 11.7 Dependencies of the effective Young moduli E^* and Poisson ratio ν^* of the composite with the polymer matrix and spherical stiff inclusions on inclusion volume fraction p, Young's moduli of the matrix E_0 and the inclusions E relate as $E = 25.249E_0$, Poisson's ratio of the particles is $\nu = 0.23$ and the matrix is $\nu_0 = 0.394$; line 1 corresponds to the EMM, line 2 to Kerener's version of the EMM, circles are experimental data in [7]. Lines 3 corresponds to the effective field method considered in the next chapter

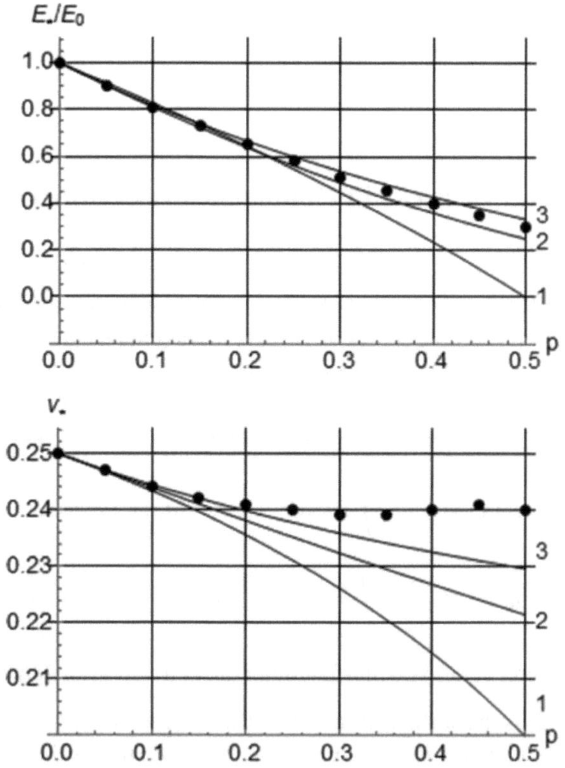

Fig. 11.8 Dependencies of the effective Young modulus E^* and Poisson ratio ν^* of the homogeneous material with a random set of spherical pores, Poisson's ratio of the host medium is $\nu_0 = 0.25$. Lines 1 correspond to the original EMM, lines 2 to Kerner's version of the EMM, and line 3 to the effective field method. Black points are results of numerical simulations [8]

dispersion of these data is substantial, and a possible reason is the difference in pore shapes from ideal spheres in the experimental specimens. The pore shapes are difficult to control, and usually pores are not spheres but spheroids, and the deviation from ideal spheres affects the effective elastic properties of the medium. The black points in Fig. 11.8 are results of numerical simulation of the effective elastic properties of porous material by the finite element method [8]. The Poisson ratio of the host medium is $\nu_0 = 0.25$. Predictions of the original EMM are lines 1, predictions of Kerner's method are lines 2, and of the effective field method—lines 3.

The three methods result in close predictions of the effective elastic moduli when the inclusion volume fraction p is less than 0.2. In general, Kerener's method describes the experimental data better than the original EMM. For volume fractions p of stiff inclusions more than $p = 0.3$, the predictions of the methods deviate, but Kerner's method corresponds better to experimental data and numerical simulations.

For matrix composites with anisotropic ellipsoidal inclusions of the same orientations, the equations of the EMM, as well as of Kerner's method, can result in a non-symmetric effective stiffness tensor of the composite. Application of Kerner's method for composites with ellipsoidal inclusions encounters technical difficulties in the solution of the one-particle problem. For an ellipsoid with a layer of constant thickness on the inclusion interface, the one-particle problem can be solved only numerically.

11.5 Cracks in Homogeneous Conductive Media

In his section, the effective medium method is applied to the calculation of the effective conductivity tensor of a homogeneous medium containing a random set of elliptical cracks. For an infinite homogeneous medium with the conductivity tensor C^0 containing cracks and subjected to a constant current j^0 at infinity, the electric field $e(x)$ and current $j(x)$ are presented in the integral form that is similar to the presentations (6.94), (6.95) of these fields in the case of an isolated crack

$$e_\alpha(x) = e^0_\alpha + \int K_{\alpha\beta}(x - x')C^0_{\beta\mu}n_\mu(x')b(x')\Omega(x')dx', \tag{11.61}$$

$$j_\alpha(x) = j^0_\alpha + \int S_{\alpha\beta}(x - x')n_\beta(x')b(x')\Omega(x')dx'. \tag{11.62}$$

Here, $\Omega(x)$ is delta-function concentrated at the crack surfaces $\Omega = \bigcup_i \Omega_i$, where Ω_i is the surface of the ith crack, the function $n(x)$ coincides with the normal to the crack plane in each crack. The function $b(x)$ is the jump of the electric potential on the crack surfaces, $e^0 = \left(C^0\right)^{-1} j^0$. The kernels of these equations are expressed in terms of Green's function $g(x)$ of the host medium

$$K_{\alpha\beta}(x) = -\nabla_\alpha\nabla_\beta g(x), \tag{11.63}$$

$$S_{\alpha\beta}(x) = -C^0_{\alpha\gamma}\nabla_\gamma\nabla_\delta g(x)C^0_{\delta\beta} - C^0_{\alpha\beta}\delta(x). \tag{11.64}$$

In the case of cracked media, the principal hypotheses of the effective medium method is formulated as follows.

- Each crack can be considered as an isolated one in the effective medium with the conductivity tensor C^*.
- The external field acting on the crack coincides with the external current j_0 applied to the medium.

The integral equation of the one-particle problem for a crack follows from the boundary conditions on the crack surfaces

$$n(x) \cdot j(x) = 0, x \in \Omega, \tag{11.65}$$

and this equation has the form

$$\int_{\Omega} T_*(x - x')b(x')d\Omega' = n \cdot j^0, x \in \Omega. \tag{11.66}$$

Here, the kernel $T_*(x)$ is defined by the equation

$$T_*(x) = -n_\alpha S^*_{\alpha\beta}(x)n_\beta, \tag{11.67}$$

$$S^*_{\alpha\beta}(x) = -C^*_{\alpha\gamma}\nabla_\gamma\nabla_\delta g^*(x)C^*_{\delta\beta} - C^*_{\alpha\beta}\delta(x), \tag{11.68}$$

where C^* is the conductivity tensor of the effective medium, $g^*(x)$ is Green's function of the effective medium. It is shown in Sect. 6.6 that for an elliptical crack, the function $b(x)$ is presented in the form

$$b(x_1, x_2) = T_*^{-1}(n \cdot j^0)z(x_1, x_2), \tag{11.69}$$

$$z(x_1, x_2) = \sqrt{1 - \left(\frac{x_1}{a_1}\right)^2 - \left(\frac{x_2}{a_2}\right)^2}. \tag{11.70}$$

Here, (x_1, x_2, x_3) are the Cartesian coordinate system with the axes x_1, x_2 directed along the ellipse semiaxes a_1, a_2, and x_3 is orthogonal to the crack plane. It follows from Eqs. (6.108)–(6.110) that the scalar T_* is the following integral

$$T_* = -\frac{1}{8\pi}\int_0^{2\pi} d\varphi \int_{-\infty}^{\infty} \widetilde{S}\left(\frac{\cos\varphi}{a_1}, \frac{\sin\varphi}{a_2}k_3\right) dk_3, \tag{11.71}$$

$$\widetilde{S}(k_1, k_2, k_3) = \frac{n_\alpha C^*_{\alpha\gamma}k_\gamma k_\delta C^*_{\delta\beta}n_\beta}{(kC^*k)} - n_\alpha C^*_{\alpha\beta}n_\beta, \tag{11.72}$$

where $(kC^*k) = k_\alpha C^*_{\alpha\beta}k_\beta$. Therefore, the integral $T_* = T_*(C^*, n)$ depends on the orientation of the crack ellipse in relation to the principal axes of the effective conductivity tensor C^*.

Using the solution of the one-particle problem, the electric field and current vectors in the cracked medium are presented in the forms

$$e_\alpha(x) = e^0_\alpha + \int K_{\alpha\beta}(x - x')C^0_{\beta\mu}q_{\mu\nu}(x')dx' j^0_\nu, \tag{11.73}$$

$$j_\alpha(x) = j^0_\alpha + \int S_{\alpha\beta}(x - x')q_{\beta\nu}(x')dx' j^0_\nu, \tag{11.74}$$

where

$$q_{\alpha\beta}(x) = n_\alpha(x)n_\beta(x)T_*(x)^{-1}Z(x)\Omega(x). \tag{11.75}$$

Here, the function $T_*(x)$ coincides with the constant T_* in Eq. (11.71) and $Z(x)$ coincides with the function $z(x_1, x_2)$ in Eq. (11.70) on the surface of each crack. The electric field and current averaged over realizations of the random crack sat are

$$\langle e(x)\rangle = \left(C^0\right)^{-1} j_0 + \int K(x-x')dx'C^0\Lambda(C^*)j^0, \tag{11.76}$$

$$\langle j(x)\rangle = j^0 + \int S(x-x')dx'\Lambda(C^*)j^0, \tag{11.77}$$

where

$$\Lambda_{\alpha\beta}(C^*) = \left\langle n_\alpha(x)n_\beta(x)T_*(x)^{-1}Z(x)\Omega(x)\right\rangle. \tag{11.78}$$

For homogeneous in space random crack sets and a constant external field j^0, the mean $\Lambda(C^*)$ is constant, and it can be moved from the integral sign. Using the ergodic hypothesis, the tensor Λ can be calculated by spatial averaging of a typical realization of the crack set, and hence we have

$$\Lambda_{\alpha\beta}(C^*) = \lim_{W\to\infty}\frac{1}{W}\int_W n_\alpha(x)n_\beta(x')\left(T_*(x)\right)^{-1}Z(x)\Omega(x) =$$

$$= \lim_{W\to\infty}\frac{2\pi}{3W}\sum_{i=1}^{N} a_1^{(i)}a_2^{(i)}\left(T_*^{(i)}\right)^{-1} n_\alpha^{(i)}n_\beta^{(i)}. \tag{11.79}$$

Here, W is the region in 3D-space that occupies the entire space in the limit $W\to\infty$, N is the number of cracks inside W, $a_1^{(i)}a_2^{(i)}$ are semiaxes of the ith crack, $n^{(i)}$ are the normal to the ith crack surface and $T_*^{(i)}$ is the scalar in Eq. (11.71) for the ith crack. Thus for the tensor $\Lambda(C^*)$, we yield the equation that follows from Eq. (11.79)

$$\Lambda(C^*) = \frac{2\pi}{3v_0}\left\langle a_1a_2T_*^{-1}n\otimes n\right\rangle, \tag{11.80}$$

where v_0 is the volume that falls at one crack, and $\langle\cdot\rangle$ is the average over ensemble realizations of the crack set. Because for a fixed external current j^0 Eqs. (10.28) hold

$$\int K(x-x')dx' = \left(C^0\right)^{-1} = B^0,\ \int S(x-x')dx' = 0, \tag{11.81}$$

and therefore,

$$\langle j(x)\rangle = j^0, \ \ \langle e(x)\rangle = \left(B^0 + \Lambda(C^*)\right)\langle j(x)\rangle . \tag{11.82}$$

It follows from these equations that the effective resistivity tensor B^* of the cracked medium have the form

$$B^* = B^0 + \Lambda(C^*). \tag{11.83}$$

The solution of this equation can be found by a simple iteration procedure

$$C^{*(n)} = \ \left(B^0 + \Lambda(C^{*(n-1)})\right)^{-1}, n = 1, 2, ..., \tag{11.84}$$

where $C^{*(n)}$ is the nth iteration, and $C^{*(0)} = C^0$. The tensor $T_* = T_*(C^*, n)$ should be calculating for $C^* = C^{*(n-1)}$ at each step of the iteration process and for possible crack orientations n in order to perform averaging in Eq. (11.80).

Equations (11.83) defines 6 independent components of the effective conductivity tensor C^*. The volume of calculations can be reduced if the symmetry of the effective medium is known in advance. For instance, for an isotropic medium with penny shape cracks of the same orientation, the effective medium is transverse isotropic with the isotropy axis orthogonal to the crack planes. In this case, tensor C^* has the form

$$C^*_{\alpha\beta} = c_*\theta_{\alpha\beta} + c_{n*}n_\alpha n_\beta, \theta_{\alpha\beta} = \delta_{\alpha\beta} - n_\alpha n_\beta, \tag{11.85}$$

where c_*, c_{n*} are scalar coefficients, n is the normal to the crack planes. In this case, calculating the integral in Eq. (11.71), we yield

$$T_* = \frac{\pi}{2a}\sqrt{c_*c_{n*}}, \tag{11.86}$$

where a is the crack radius. The tensor $\Lambda(C^*)$ in Eq. (11.80) is

$$\Lambda_{\alpha\beta}(C^*) = \frac{\tau}{\sqrt{c_*c_{n*}}}n_\alpha n_\beta, \ \tau = \frac{4a^3}{3v_0}. \tag{11.87}$$

As a result, Eq. (11.83) for C^* takes the form

$$\frac{1}{c_*}\theta_{\alpha\beta} + \frac{1}{c_{n*}}n_\alpha n_\beta = \frac{1}{c_0}\theta_{\alpha\beta} + \left(\frac{1}{c_0} + \frac{\tau}{\sqrt{c_*c_{n*}}}\right)n_\alpha n_\beta, \tag{11.88}$$

where c_0 is the conductivity of the host medium. Equating the coefficients of the linearly independent tensors $\theta_{\alpha\beta}$ and $n_\alpha n_\beta$, we obtain

$$c_* = c_0, \ \ \frac{1}{c_{n*}} = \frac{1}{c_0} + \frac{\tau}{\sqrt{c_0c_{n*}}}, \tag{11.89}$$

and for c_{n*} we yield

$$\frac{c_{n*}}{c_0} = 1 - \frac{\tau}{2}\left(\sqrt{4+\tau^2} - \tau\right). \tag{11.90}$$

For a homogeneous distribution of cracks over the orientations, the effective medium is isotropic $C^*_{\alpha\beta} = c_*\delta_{\alpha\beta}$, and for T_* we obtain

$$T_* = \frac{\pi}{2a} c_*. \tag{11.91}$$

Averaging over the orientations in Eq. (11.80) results in the expression for the tensor $\Lambda(c_*)$

$$\Lambda_{\alpha\beta} = \frac{\tau}{3c_*}\delta_{\alpha\beta}, \tag{11.92}$$

and from Eq. (11.83) we obtain the conductivity of the effective medium in the form

$$\frac{c_*}{c_0} = 1 - \frac{\tau}{3}. \tag{11.93}$$

Note that the EMM predicts that the effective conductivity of the medium with the homogeneous distribution of cracks over orientations turns to zero at the crack density $\tau = 3$, but for the cracks of the same orientations in Eq. (11.90), the coefficient c_{n*} is finite for any value of the crack density.

11.6 Elastic Medium with Random Sets of Elliptical Cracks

The strain and stress fields in the cracked elastic medium subjected to a constant external field are presented in the integral forms that follow from Eqs. (6.52), (6.53)

$$\varepsilon_{\alpha\beta}(x) = \varepsilon^0_{\alpha\beta} + \int K_{\alpha\beta\lambda\mu}(x-x')C^0_{\lambda\mu\nu\rho}n_\nu(x')b_\rho(x')\Omega(x')dx', \tag{11.94}$$

$$\sigma_{\alpha\beta}(x) = \sigma^0_{\alpha\beta} + \int S_{\alpha\beta\lambda\mu}(x-x')n_\lambda(x')b_\mu(x')\Omega(x')dx'. \tag{11.95}$$

Here, $\Omega(x)$ is delta-function concentrated at the crack surfaces, the function $n(x)$ coincides with the normal to the crack plane in each crack, and $b(x)$ coincides with the crack opening vector in the surface of each crack. Repeating derivation of the previous section, we yield the effective elastic compliance B^* of the cracked medium in the form

$$B^* = B^0 + \Lambda(C^*), \tag{11.96}$$

where B^0 are the elastic compliance tensor of the host media,

$$\Lambda_{\alpha\beta\gamma\delta}(C^*) = \frac{2\pi}{3v_0} \left\langle a_1 a_2 n_\alpha \left(T_*^{-1}\right)_{\beta\gamma} n_\delta \right\rangle \Big|_{(\alpha\beta)(\gamma\delta)}, \tag{11.97}$$

For elliptical cracks, the expression for the tensor T_* in this equation follows from Eq. (5.34)

$$(T_*)_{\alpha\beta} = -\frac{1}{8\pi} \int_0^{2\pi} d\varphi \int_{-\infty}^{\infty} \widetilde{S}_{3\alpha\beta 3} \left(\frac{\cos\varphi}{a_1}, \frac{\sin\varphi}{a_2} k_3 \right) dk_3, \tag{11.98}$$

$$\widetilde{S}_{\alpha\beta\gamma\delta}(k_1, k_2, k_3) = C^*_{\alpha\beta\mu\nu} \widetilde{K}_{\mu\nu\rho\tau}(k_1, k_2, k_3) C^*_{\rho\tau\gamma\delta} - C^*_{\alpha\beta\gamma\delta}. \tag{11.99}$$

Here, the components of the tensors are taken in the basis of the Cartesian system with the origin at the crack center, the axes x_1, x_2 directed along the ellipse semaxes, and the x_3-axis is orthogonal to the crack plane.

For an isotropic host medium with penny shape cracks of the same radii a homogeneously distributed over the orientations, the effective medium is isotropic, and its stiffness tensor C^* in the E-basis has the form

$$C^* = 2\mu_* E_1 + \frac{2\mu_* \nu_*}{1 - 2\nu_*} E^2, \tag{11.100}$$

In this case, the components of the tensor T_* follow from Eqs. (5.52)

$$(T_*)_{\alpha\beta} = T^*_{11} \theta_{\alpha\beta} + T^*_{33} n_\alpha n_\beta, \tag{11.101}$$

$$T^*_{11} = \frac{\pi\mu_*}{8a} \frac{(2 - \nu_*)}{(1 - \nu_*)}, \quad T^*_{33} = \frac{\pi\mu_*}{4a\,(1 - \nu_*)}, \tag{11.102}$$

where μ_*, ν_* are the shear modulus and Poisson's ratio of the effective medium. As a result, the tensor $\Lambda(C^*)$ in Eq. (11.97) is

$$\Lambda_{\alpha\beta\gamma\delta}(C^*) = \frac{2\pi a^2}{45 v_0} \left(\left(\frac{3}{T^*_{11}} + \frac{2}{T^*_{33}} \right) E_1 + \left(\frac{1}{T^*_{33}} - \frac{1}{T^*_{11}} \right) E_2 \right). \tag{11.103}$$

Dependencies of the effective shear modulus μ_* and Poisson's ratio ν_* of the cracked medium on the crack density parameter

$$\tau = \frac{4a^3}{3v_0} \tag{11.104}$$

are shown in Fig. 11.9 for Poisson's ratio of the host medium $\nu_0 = 0.3$.

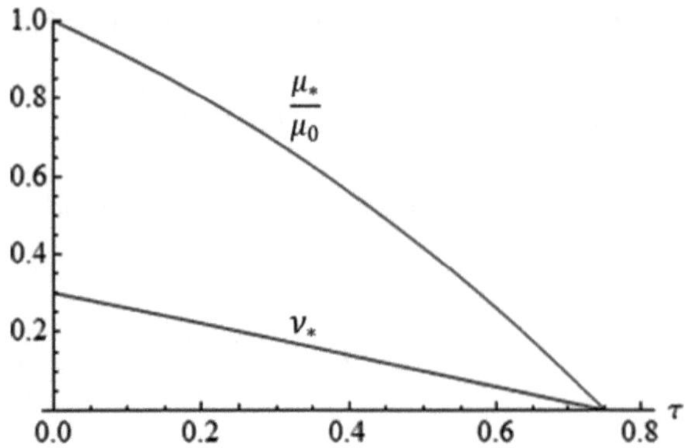

Fig. 11.9 Dependencies of the effective shear modulus μ^* and Poisson's ratio ν^* on the crack density parameter τ for an isotropic medium containing penny shape cracks homogeneously distributed over orientations, Poisson's ratio of the host medium is $\nu_0 = 0.3$

For penny shape cracks of the same orientation, the effective medium is transverse isotropic. In the P-bais, the elastic stiffness tensor C^0 of an isotropic host medium is presented in the forms

$$C^0 = k_0 P^2(n) + 2m_0 \left(P^1(n) - \frac{1}{2} P^2(n) \right) + l_0 \left(P^3(n) + P^4(n) \right) +$$

$$+ 4\mu_0 P^5(n) + n_0 P^6(n), \tag{11.105}$$

$$k_0 = \lambda_0 + \mu_0,\ m_0 = \mu_0,\ l_0 = \lambda_0,\ n_0 = \lambda_0 + 2\mu_0,$$

where λ_0, μ_0 are Lame's parameters of the host medium, n is the normal to the crack planes. On the same basis, the effective stiffness tensor C^* of the medium with cracks has the form

$$C^* = k_* P^2(n) + 2m_* \left(P^1(n) - \frac{1}{2} P^2(n) \right) + l_* \left(P^3(n) + P^4(n) \right) +$$

$$+ 4\mu_* P^5(n) + n_* P^6(n), \tag{11.106}$$

The equation for the effective compliance tensor $B^* = (C^*)^{-1}$ of the cracked medium follows from Eqs. (11.96), (11.97)

$$B^* = B^0 + \tau \left(\Lambda_5^* P^5(n) + \Lambda_6^* P^6(n) \right), \tau = \frac{4a^3}{3v_0}, \tag{11.107}$$

where $B^0 = \left(C^0 \right)^{-1}$, a is the crack radii, v_0 is the mean volume that falls at one crack, P^5 and P^6 are the elements of P-basis (Appendix A.2). The scalar coefficients Λ_5^*, Λ_6^* are expressed in terms of the effective elastic coefficients in Eq. (11.106)

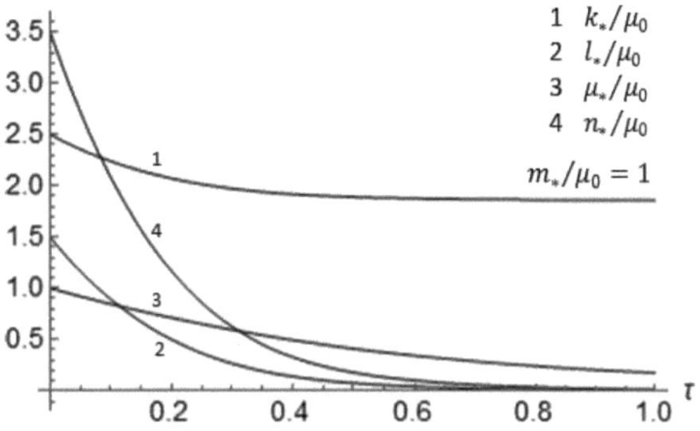

Fig. 11.10 Dependencies of the effective elastic coefficients in Eq. (11.106) on the crack density parameter τ for the isotropic medium with shear modulus μ_0 and Poisson's ratio $\nu_0 = 0.3$ containing a set of penny shape cracks of the same orientation

$$\Lambda_5^* = \frac{1}{4}\left(\sqrt{m_*\mu_*} + \frac{(k_* + m_*)n_* - l_*^2}{n_0\left(\sqrt{\xi_1} + \sqrt{\xi_2}\right)}\right), \tag{11.108}$$

$$\Lambda_6^* = \frac{1}{2}\frac{(k_* + m_*)n_* - l_*^2}{n_*\left(\sqrt{\xi_1} + \sqrt{\xi_2}\right)\sqrt{\xi_1\xi_2}}. \tag{11.109}$$

Here, ξ_1, ξ_2 are the roots of the square equation

$$\xi^2 - a_*\xi + b_* = 0, \tag{11.110}$$

where

$$a_* = \frac{1}{\mu_* n_*}\left(n_*(k_* + m_*) - l_*\left(2\mu_* + l_*\right)\right), \quad b_* = \frac{k_* + m_*}{n*}. \tag{11.111}$$

Equations (11.108)–(11.111) is consequences of the solution of the one-particle problem for a penny shape crack in a transverse isotropic host medium in Eqs. (5.110)–(5.111).

For the isotropic host medium with Poisson's ratio $\nu_0 = 0.3$, the results of numerical solutions of Eq. (11.107) are presented in Fig. 11.10 for the effective elastic coefficients of the tensor C^* in Eq. (11.106).

11.7 The Differential Effective Medium Method

The differential effective medium method (DEMM) is based on the concept that the matrix composite with a finite volume fraction of inclusions can be simulated by sequential addition to the host material (matrix), a small volume fraction of particles of the second phase. After each addition, the homogenization procedure is performed,

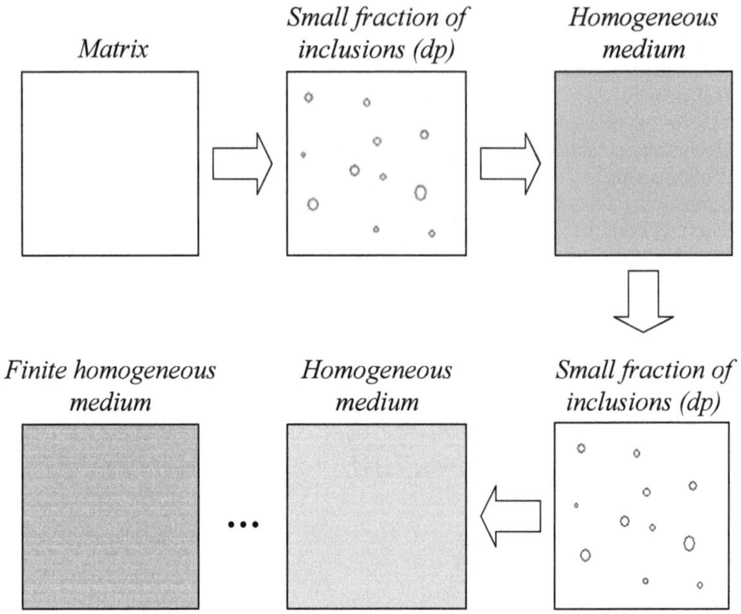

Fig. 11.11 The differential effective medium method

and for a new portion of the inclusions, the new composite is considered as a homogeneous host material. The process is continued until the required volume fraction $\overline{p}$ of the inclusions is achieved (Fig. 11.11). Note that at each step of the process, a small volume fraction δp of the particles is added, and therefore, the asymptotic equations for the elastic stiffness tensor of the composites with a small volume fraction of inclusions can be used (the dilute approximation). For ellipsoidal inclusions in a homogeneous matrix with the property tensor C^0, the dilute approximation has the form

$$C^* = C^0 + \delta p \left\langle (C - C^0)\Lambda^{\varepsilon}(C^0) \right\rangle, \tag{11.112}$$

$$\Lambda^{\varepsilon}\left(C^0\right) = \left(I + A(C^0)(C - C^0)\right)^{-1}. \tag{11.113}$$

Here, the tensor $A(C^0)$ for ellipsoidal inclusions is defined in Eq. (11.11), and averaging in Eq. (11.112) is performed over orientations and the size distribution of the inclusions. Hence for the described process, we can write

$$C^*(p + \delta p) = C^*(p) + \delta p \left\langle (C - C^*(p))\Lambda^{\varepsilon}(C^*(p)) \right\rangle, \tag{11.114}$$

where the tensor $\Lambda^{\varepsilon}(C^*)$ is determined by the same Eq. (11.113). In the limit $\delta p \to 0$, Eq. (11.114) is transformed in the differential equation for the unknown function $C^*(C, C^0, p)$

$$\frac{dC^*}{dp} = \left\langle (C - C^*(p))\Lambda^{\varepsilon}(C^*(p)) \right\rangle . \tag{11.115}$$

The initial condition for this equation is

$$C^*(p)|_{p=0} = C^0. \tag{11.116}$$

It was noted in [9] that the total volume fraction $\overline{p}$ of particles introduced into the matrix does not coincide with the volume fraction p of the second phase. The reason is that by adding a new portion δp of particles, some of them will "fall" into the region occupied by the previously deposited ones. Only the part $(1 - p)\delta p$, that "falls" within the matrix, will increase the second phase concentration p. Thus in the limit $\delta p \to 0$, we have

$$dp = (1 - p)\delta p. \tag{11.117}$$

and Eq. (11.115) is transformed into the following one

$$(1 - p)\frac{dC^*}{dp} = \left\langle \left(C - C^*(p)\right) \Lambda^*(C^*(p)) \right\rangle . \tag{11.118}$$

We consider an isotropic conductive medium with spherical isotropic conductive inclusions. The conductivity tensor of the host medium is $C^0_{\alpha\beta} = c_0\delta_{\alpha\beta}$, and the same for the inclusions is $C_{\alpha\beta} = c\delta_{\alpha\beta}$, the inclusions are homogeneously distributed in the matrix. In this case $C^*_{\alpha\beta} = c_*\delta_{\alpha\beta}$, and the equation for the scalar function $c_*(p)$ follows from Eq. (11.118) in the form

$$\frac{dc_*}{dp} = \frac{3c_*(c - c_*)}{(1 - p)(c + 2c_*)} \tag{11.119}$$

with the initial condition

$$c_*|_{p=0} = c_0. \tag{11.120}$$

The solution of this equation can be presented in the implicit form

$$\left(\frac{c - c_*}{c - c_0}\right)\left(\frac{c_0}{c_*}\right)^{1/3} = 1 - p. \tag{11.121}$$

For highly conductive inclusions ($c \to \infty$), we have

$$c_* = \frac{c_0}{(1 - p)^3}, \tag{11.122}$$

and for non-conductive inclusions ($c = 0$)

$$c_* = c_0(1 - p)^{3/2}. \tag{11.123}$$

Hence, in contrast to the original effective medium method, the DEMM predicts finite values of the effective conductivity for the complete region of the inclusion volume fractions $0 \leq p < 1$.

Let us consider an elastic composite material with an isotropic matrix containing a homogeneous in space random set of isotropic spherical inclusions. In this case, the composite material is isotropic, and its effective elastic stiffness tensor C^* satisfies the differential equation that follows from Eq. (11.118)

$$(1 - p)\frac{dC^*}{dp} = \left((C - C^*)^{-1} + A^*\right)^{-1}, \tag{11.124}$$

$$A^* = \frac{1}{3(3K_* + 4\mu_*)}E^2 + \frac{3(K_* + 2\mu_*)}{5\mu_*(3K_* + 4\mu_*)}\left(E^1 - \frac{1}{3}E^2\right). \tag{11.125}$$

Here, K_* and μ_* are the bulk and shear effective elastic moduli of the composite, K, μ the same moduli for the inclusion material. This equation is equivalent to the system of two differential equations for K_* and μ_*

$$(1 - p)\frac{dK_*}{dp} = \frac{(K - K_*)(3K_* + 4\mu_*)}{3K + 4\mu_*}, \tag{11.126}$$

$$(1 - p)\frac{d\mu_*}{dp} = \frac{5\mu_*(\mu - \mu_*)(3K_* + 4\mu_*)}{4\mu_*(3\mu + 2\mu_*) + 3K_*(2\mu + 3\mu_*)} \tag{11.127}$$

with the initial conditions: $K_*(0) = K_0$, $\mu_*(0) = \mu_0$. Numerical solutions of these equations are presented in Fig. 11.12 (for spherical rigid inclusions) and in Fig. 11.13 (for spherical pores). Poison's ratio of the matrix phase is $\nu_0 = 0.25$. In Figs. 11.12, 11.13, lines 1 are predictions of the original EMM, lines 2 are for the DEMM, and lines 3 are the prediction of the effective field method considered in Chap. 13. Lines with black points in their figures are results of numerical (finite element) simulations of the effective elastic properties of the composite presented in [8].

Note that for multiphase composites that contain several families of inclusions of different shapes and elastic properties (hybrid composites), the result of the DEMM depends on the order of implantation of the portions of different inclusion families into the host material. The detailed discussion of this problem is presented in [10].

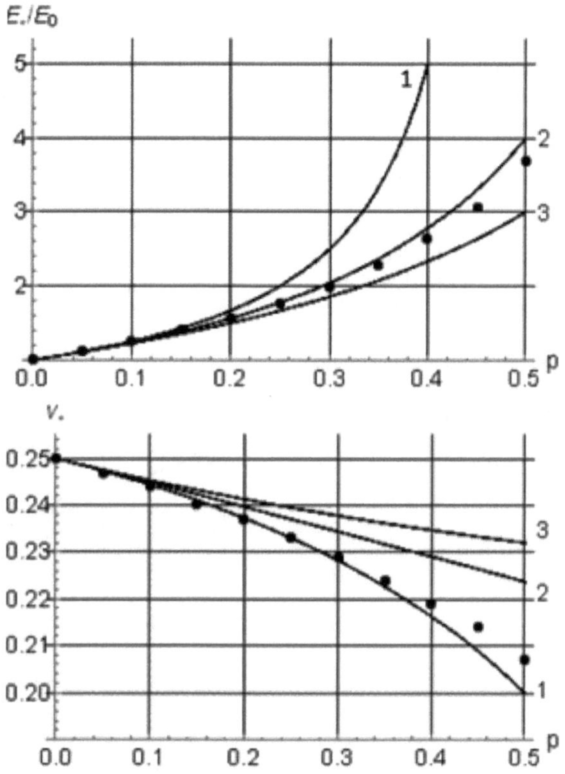

Fig. 11.12 Dependencies of the effective Young modulus E^* and Poisson ratio ν^* of the homogeneous material with random set of absolutely rigid spherical inclusions, Poisson's ratio of the host medium is $\nu_0 = 0.25$. Lines 1 correspond to the original EMM, lines 2 to the differential EMM, and line 3 to the effective field method. Black points are results of numerical simulation of the effective elastic properties by finite element method [8]

11.8 Notes

For calculation of the effective elastic moduli of matrix composite materials, the effective medium method was used in [11–14]. Kerner's version of the EMM was proposed in [1]. Some mistakes of this work were corrected in [4, 5]. A survey of application of Kerner's method for the solution of the homogenization problem for matrix-inclusion composites can be found in [15].

A survey of publications devoted to the solution of the homogenization problems for conductive matrix composite is presented in [16]. Relations of the effective medium method with variational bounds of estimation of effective properties of matrix composites are considered in [17]. Conductive properties of composite materials with random sets of thin inclusions (Sect. 9.3) were considered in [18].

Effective elastic moduli of a medium with cracks were calculated by the EMM in [19] and by the DEMM in [20].

The DEMM was first proposed in [21, 22] and used for a medium containing rigid particles in viscous fluid [23] (effective viscosity of fluid suspensions) and

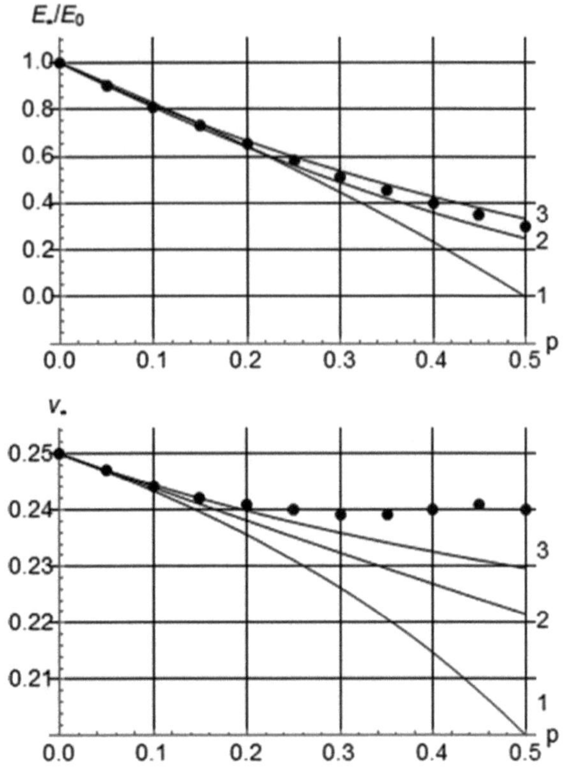

Fig. 11.13 Dependencies of the effective Young modulus E^* and Poisson ratio ν^* of the homogeneous material with a random set of spherical pores, Poisson's ratio of the host medium is $\nu_0 = 0.25$. Lines 1 correspond to the original EMM, lines 2 to the differential EMM, and line 3 to the effective field method. Black points are results of numerical simulation of the effective elastic properties by the finite element method [8]

in [24] (effective elastic moduli of matrix composites). Later on, the DEMM was used for the solution of the homogenization problems of various types of matrix composites in [9, 10, 25, 26]. The detailed discussion of the DEMM is presented in [27]. Versions of the effective medium method were also discussed in [28–30].

References

1. Kerner E (1956) The elastic and thermoelastic properties of composite media. Proc Phys Soc 69(B):808–813
2. Kim D, Lim J, Yu J (2019) Efficient prediction of the electrical conductivity and percolation threshold of nanocomposite containing spherical particles with three-dimensional random representative volume elements by random filler removal. Compos B Eng 168:387–397
3. Christensen R (1979) Mechanics of composite materials. Wiley-Interscience, New York, p 384
4. Christensen R, Loo L (1979) Solutions for effective shear properties in three phase and cylinder models. J Mech Phys Solids 27:315–330
5. Smith J (1974) Correction and extension of Van der Pole's method for calculating the shear modulus of particulate composite. J Res Natl Bar Stand 78A:355–361
6. Van der Poel C (1958) On the rheology of concentrated dispersions. Rheol Acta B1:198–205

7. Smith J (1976) The elastic constants of a particulate filled glassy polymer. Polymer Eng Sci 16:394–399
8. Segurado J, Llorca J (2002) A numerical approximation to the elastic properties of sphere-reinforced composites. J Mech Phys Solids 50:2107–2121
9. McLaughlin R (1977) A study of the differential scheme for composite materials. Int J Eng Sci 15:237–244
10. Norris A (1985) A differential scheme for effective moduli of composites. Mech Mater 4:1–16
11. Skorohod V (1961) Calculation of the effective isotropic moduli of disperse solid system. Powder Metall 1:50–55
12. Budiansky B (1965) On the elastic moduli of some heterogeneous materials. J Mech Phys Solids 13:223–234
13. Hill R (1965) Self-consistent mechanics of composite materials. J Mech Phys Solids 13:213–222
14. Levin V, Markov M (2005) Elastic properties of inhomogeneous transversely isotropic rocks. Int J Solids Struct 42:393–408
15. Kowalczyk-Gajewska K, Mazdziarz M (2020) Elastic properties of nanocrystalline materials of hexagonal symmetry: the core-shell model and atomistic estimates. Int J Eng Sci 157:103393
16. Markov K (2000) Elementary micromechanics of heterogeneous solids. In: Markov K, Preziosi L (eds) Heterogeneous media: micromechanics modeling methods and simulations. Birkhauser, Boston, pp 1–162
17. Willis J (1977) Bounds and self-consistent estimates for the overall properties of anisotropic composites. J Mech Phys Solids 25:185–202
18. Levin V, Markov M (2004) Electroconductivity of a medium with thin low-resistivity inclusions. J Electrostatics 61:129–145
19. Budiansky B, O'Connell R (1976) Elastic module of a cracked solid. Int J Solids Struct 12:81–91
20. Hashin Z (1988) The differential scheme and its application to cracked materials. J Mech Phys Solids 36:719–734
21. Bruggeman D (1935) Berechnung verschiedener physikalischer Konstanten von heterogenen Substanzen. Ann Phys 24:636–679
22. Bruggeman D (1937) Berechnung verschiedener physikalischer Konstanten von heterogenen Substanzen. III. Die Elastische Konstanten der Quasiisotropen Mischkoerper aus isotropen Substanzen. Ann Phys 29:160–178
23. Brinkman H (1952) The viscosity in concentrated suspension and solutions. J Chem Phys 20:571–585
24. Roscoe R (1952) The viscosity of suspension and solution. J Appl Phys 3:267–269
25. Zimmerman R (1991) Elastic moduli of a solid containing spherical inclusions. Mech Mater 12:17–24
26. Markov M, Levin V, Mousatov A, Kazatchenko E (2005) Elastic properties of double-porosity rocks using the differential effective medium model. Geophys Prospecting 53:733–754
27. Cleary M, Chen I, Lee S (1987) Self-consistent techniques for heterogeneous solids. ASCE J Eng Mech 106:861–887
28. Berryman J (1980) Long-wavelength propagation in composite elastic media. II: ellipsoidal inclusions. J Acoustic Soc Am 68:1820–1831
29. Berryman J, Berge P (1996) Critique of two explicit schemes for estimating elastic properties of multiphase composites. Mech Mater 22:149–164
30. Mukerji T, Berryman J, Mavko G, Berge P (1995) Differential effective medium modeling of rock elastic moduli with critical porosity constraints. Geophys Res Lett 22:555–558

Chapter 12
Effective Medium Method for Polycrystalline Materials

Abstract Application of the effective medium method to the solution of the homogenization problem for polycrystalline materials is performed. Conductive and elastic properties of polycrystals with spheroidal transverse isotropic grains are considered in detail. Effective thermo-expansion coefficients of elastic polycrystalline materials are obtained. Examples of clay polycrystalline materials and polycrystalline zinc are considered in detail.

Keywords Polycrystals with ellipsoidal grains · Grain orientations · Effective conductive properties · Effective elastic properties · Effective thermo-expansion coefficients

Polycrystalline microstructures are common for many metals, metallic blends, rock materials, plants, etc. Such materials consist of agglomerations of particles (grains), and the host medium (matrix phase) is absent. For polycrystalline metals, the grains are monocrystals with a random distribution of the crystallographic axes over the orientations in the grains. The microstructure of many rock materials consists of several types of grains with different physical properties. A sketch of such a microstructure is shown in Fig. 12.1. In this chapter, the effective medium method is applied to the calculation of the effective conductive and elastic properties of the polycrystals.

12.1 Effective Conductivity of Polycrystalline Materials

We consider the medium that is an agglomeration of grains of the same anisotropic material with a random distribution of the principal axes of the conductivity tensor C inside grains over orientations. The hypotheses of the effective medium method in application to the homogenization problem for polycrystals are formulated as follows.

H_1. *All grains of the polycrystal have ellipsoidal shapes. The conductivity tensor* C *inside each grain is constant, and the principal axes of this tensor coincide with the principal axes of the ellipsoidal grain.*

S. Kanaun, *Self-Consistent Methods for Composites*, Solid Mechanics and Its Applications 282, https://doi.org/10.1007/978-3-031-96739-9_12

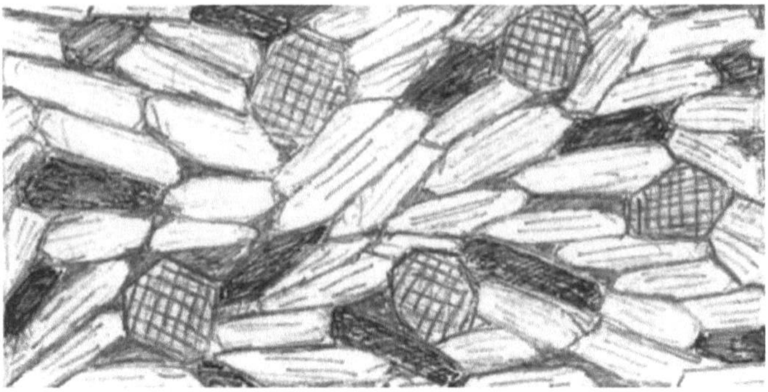

Fig. 12.1 A sketch of the micro structure of the polycrystalline rock material

H_2. For constant external field applied to the polycrystalline medium, each grain can be considered as an isolated one embedded into the homogeneous medium with the effective property tensor C^* of the polycrystal.

H_3. The external field that acts on each grain coincides with the averaged field over the polycrystalline medium.

The field intensity $\varepsilon(x)$ and flux density $\sigma(x)$ vectors in a polycrystalline medium satisfy the integral equations (10.13), (10.15). The property tensor $\widehat{C}$ of the reference medium in these equations is reasonable to choose coincided with the effective property tensor C^* of the polycrystal. For hypothesis H_3, the external fields $\widehat{\varepsilon}$, $\widehat{\sigma}$ are the averaged field intensity and flux density over the polycrystal ($\widehat{\varepsilon} = \langle\varepsilon(x)\rangle$, $\widehat{\sigma} = \langle\sigma(x)\rangle$). As a result, Eqs. (10.13), (10.15) take the forms

$$\varepsilon(x) = \langle\varepsilon(x)\rangle - \int K^*(x - x')C^{*1}(x')\varepsilon(x')dx', \tag{12.1}$$

$$\sigma(x) = \langle\sigma(x)\rangle + \int S^*(x - x')B^{*1}(x')\sigma(x')dx', \tag{12.2}$$

$$C^{*1}(x) = C(x) - C^*,\ B^{*1} = B(x) - B^*. \tag{12.3}$$

Here, the kernels $K^*(x)$ and $S^*(x)$ are defined in Eqs. (10.14), (10.16) with $\widehat{C} = C^*$.

According to the hypotheses H_1 and H_2, the field intensity and flux density inside the ith grain v^i ($i = 1, 2, \ldots$) are defined from the solution of the conductivity problem for an isolated ellipsoidal grain embedded into the homogeneous medium with the effective properties of the polycrystal and subjected to a constant external field $\langle\varepsilon\rangle$ ($\langle\sigma\rangle$). This problem is considered in Sect. 4.4, and its solution is presented in the forms (Eqs. (11.11), (11.12))

$$\varepsilon = \Lambda^{\varepsilon(i)} \langle \varepsilon(x) \rangle, \sigma = \Lambda^{\sigma(i)} \langle \sigma(x) \rangle, x \in v^i, \tag{12.4}$$

$$\Lambda^{\varepsilon(i)} = \left(I + A^{*(i)} C^{*1(i)}\right)^{-1}, \Lambda^{\sigma(i)} = \left(I - D^{*(i)} B^{*1(i)}\right)^{-1}, \tag{12.5}$$

$$C^{*1(i)} = C^{(i)} - C^*, B^{*1(i)} = B^{(i)} - B^*, \; i = 1, 2, ... \tag{12.6}$$

Here, $C^{(i)}$ is the conductivity tensor of the ith grain, $B^{(i)} = (C^{(i)})^{-1}$. The tensors $A^{*(i)}$and $D^{*(i)}$ are defined in Eqs. (4.200) and (11.13) with $\widehat{C} = C^*$. Tensors $A^{*(i)}$and $D^{*(i)}$ depend on the aspect ratios and orientation of the grain v^i. Taking into account Eqs. (12.4) for the fields inside the grains and Eqs. (12.1), (12.2), and (12.4), the field intensity and the flux density in the polycrystal are approximated by the following integrals

$$\varepsilon(x) = \langle \varepsilon(x) \rangle - \int K^*(x - x^{'}) C^{*1}(x^{'}) \Lambda^{\varepsilon}(x^{'}) dx^{'} \langle \varepsilon(x) \rangle, \tag{12.7}$$

$$\sigma(x) = \langle \sigma(x) \rangle + \int S^*(x - x^{'}) B^{*1}(x^{'}) \Lambda^{\sigma}(x^{'}) dx^{'} \langle \sigma(x) \rangle. \tag{12.8}$$

Here, the functions $C^{*1}(x)\Lambda^{\varepsilon}(x)$ and $B^{*1}(x)\Lambda^{\sigma}(x)$ coincide with $C^{*1(i)}\Lambda^{\varepsilon(i)}$ and $B^{*1(i)}\Lambda^{\sigma(i)}$ inside the ith grain ($i = 1, 2, ...$). After averaging Eqs. (12.7), (12.8) over the ensemble of realizations of the polycrystalline structure, and taking into account that the means of the stationary random functions $C^{*1}(x)\Lambda^{\varepsilon}(x)$ and $B^{*1}(x)\Lambda^{\sigma}(x)$ are constant, we obtain

$$\langle \varepsilon(x) \rangle = \langle \varepsilon(x) \rangle - \int K^*(x - x^{'}) dx^{'} \left\langle C^{*1}(x) \Lambda^{\varepsilon}(x) \right\rangle \langle \varepsilon(x) \rangle, \tag{12.9}$$

$$\langle \sigma(x) \rangle = \langle \sigma(x) \rangle + \int S^*(x - x^{'}) dx^{'} \left\langle B^{*1}(x) \Lambda^{\sigma}(x) \right\rangle \langle \sigma(x) \rangle. \tag{12.10}$$

If the conditions at infinity correspond to a fixed flux density, the integrals in this equation are defined in Eq. (10.28)

$$\int K^*(x - x^{'}) dx^{'} = B^*, \int S^*(x - x^{'}) dx^{'} = 0, \tag{12.11}$$

and for the averaged field intensity $\langle \varepsilon(x) \rangle$, we yield

$$\langle \varepsilon(x) \rangle = \langle \varepsilon(x) \rangle - B^* \left\langle C^{*1}(x) \Lambda^{\varepsilon}(x) \right\rangle \langle \varepsilon(x) \rangle. \tag{12.12}$$

It follows from this equation that the second term in the right-hand side should be equal to zero for any external field $\langle \varepsilon(x) \rangle$. As a result, we yield the identity

$$\left\langle C^{*1}(x) \Lambda^{\varepsilon}(x) \right\rangle = \left\langle \left(C(x) - C^*\right) \Lambda^{\varepsilon}(x) \right\rangle = 0. \tag{12.13}$$

Equation (12.13) is called the condition of self-consistency, and it can be interpreted as follows. The integral term in Eq. (12.7) is perturbations of the averaged field $\langle\varepsilon(x)\rangle$ caused by the heterogeneities (grains of the polycrystal). Equation (12.13) is the requirement that for the reference medium with the effective conductivity tensor C^* of the polycrystal, the averaged perturbations of the mean field $\langle\varepsilon(x)\rangle$ should be equal to zero.

Equation (12.13) can be written in the form

$$C^* = \langle C(x)\Lambda^{\varepsilon}(x)\rangle \langle\Lambda^{\varepsilon}(x)\rangle^{-1}. \tag{12.14}$$

Tensor $\Lambda^{\varepsilon}(x)$ in the right-hand side of this equation depends on the effective conductivity tensor C^*. Therefore, Eq. (12.14) is an implicit equation for the tensor C^*.

A similar equation for the effective resistivity tensor B^* can be derived using Eq. (12.10) if we accept that the field intensity is fixed in the problem. In this case,

$$\int K^*(x - x^{'})dx^{'} = 0, \int S^*(x - x^{'})dx^{'} = -C^*, \tag{12.15}$$

and from Eqs. (12.10), we yield the equality

$$\langle\sigma(x)\rangle = \langle\sigma(x)\rangle - C^* \left\langle B^{*1}(x)\Lambda^{\sigma}(x)\right\rangle \langle\sigma(x)\rangle. \tag{12.16}$$

It follows from this equation that the second term in the right-hand side is equal to zero for any $\langle\sigma(x)\rangle$, and therefore,

$$\left\langle B^{*1}(x)\Lambda^{\sigma}(x)\right\rangle = \left\langle(B(x) - B^*)\Lambda^{\sigma}(x)\right\rangle = 0. \tag{12.17}$$

This equation can be written in the form

$$B^* = \langle B(x)\Lambda^{\sigma}(x)\rangle \langle\Lambda^{\sigma}(x)\rangle^{-1}. \tag{12.18}$$

Similar to Eqs. (12.14), (12.18) is an implicit equation for the effective resistivity tensor B^* of the polycrystalline medium. Equations (12.14) and(12.18) are the reciprocal equations of the effective medium method in application to the homogenization problem for polycrystals.

Note important features of Eqs. (12.14), (12.18).

1. In Eqs. (12.14), (12.18), the averages are performed over sizes, properties, and orientations of grains. Shapes of the grains affect the tensors $A^{*(i)}$and $D^{*(i)}$ in Eqs. (12.5) via the grain aspect ratios. Because these tensors don't depend on the absolute grain sizes, the distribution of grains over sizes does not affect the solutions of Eqs. (12.14), (12.18).

2. Because the principal axes of the tensor C inside each grain coincide with the symmetry axes of the ellipsoidal grain (hypothesis H_1), the orientation

distribution function of the principal axes of the grain and the grain conductivity tensor C coincide. This function should be known for numerical solution of Eqs. (12.14), (12.18).

3. For grains with the same aspect ratios, the averages in Eqs. (12.14), (12.18) are performed over the grain orientations only.

4. If grains have the same aspect ratios and orientations, the average signs in Eqs. (12.14), (12.18) can be omitted, and we obtain

$$C^* = C, B^* = B. \tag{12.19}$$

Thus, for ellipsoidal grains of the same aspect ratios and orientations, the effective conductive properties of the polycrystal coincide with the properties of the grain material.

5. Relation between tensors C^* and B^* in Eqs. (12.14) and (12.18) has the form

$$\left(C^*\right)^{-1} = B^*. \tag{12.20}$$

To prove this equation, we consider a consequence of the definitions (12.5) of the tensors Λ^ε and Λ^σ and the relation between the tensors A^* and D^*

$$D^* = C^*A^*C^* - C^*. \tag{12.21}$$

From these equations follow the identities

$$C\Lambda^\varepsilon = C(I + A^*C^{*1})^{-1} = (I - D^*B^{*1})^{-1}C^* = \Lambda^\sigma C^*, \tag{12.22}$$

$$B\Lambda^\sigma = B(I - D^*B^{*1})^{-1} = (I + A^*C^{*1})^{-1}B^* = \Lambda^\varepsilon B^*. \tag{12.23}$$

After averaging these equations, we yield

$$\langle C\Lambda^\varepsilon\rangle = \langle\Lambda^\sigma\rangle\, C^*, \langle B\Lambda^\sigma\rangle = \langle\Lambda^\sigma\rangle\, B^*. \tag{12.24}$$

As a result, we have

$$\left(C^*\right)^{-1} = \langle\Lambda^\varepsilon\rangle\,\langle C\Lambda^\varepsilon\rangle^{-1} = \langle B\Lambda^\sigma\rangle\left(B^*\right)^{-1}B^*\,\langle\Lambda^\sigma\rangle^{-1} =$$

$$= \langle B\Lambda^\sigma\rangle\,\langle\Lambda^\sigma\rangle^{-1} = B^*, \tag{12.25}$$

and therefore, Eq. (12.20) holds.

6. The detailed Eq. (12.14) for the effective conductivity tensor has the form

$$\left\langle\left(C(x) - C^*\right)\left(I + A^*(C(x) - C^*)\right)^{-1}\right\rangle = 0. \tag{12.26}$$

Formal expansion of the tensor under the averaging sign results in the equation

$$\left(C - C^*\right)\left(I + A^*(C - C^*)\right)^{-1} = \left(C - C^*\right) - \left(C - C^*\right) A^*(C - C^*) +$$

$$+ \left(C - C^*\right) A^*(C - C^*)A^* \left(C - C^*\right) A^*(C - C^*) + ... \qquad (12.27)$$

Because C and A^* are symmetric tensors, all the terms in the right-hand side of this equation are symmetric if C^* is a symmetric tensor. It means that Eq. (12.26) can be satisfied by a symmetric tensor C^*. Therefore, unlike the case of matrix composites, application of the EMM to the solution of the homogenization problem for polycrystals results in symmetric tensors of the effective conductive properties.

12.2 Effective Conductivity Tensor of Polycrystals with Spheroidal Transverse Isotropic Grains

In this section, we consider polycrystals that consist of spheroidal grains with the same aspect ratios $\gamma = a_3/a$, where $a_1 = a_2 = a$, a_3 are grain semiaxes. We assume that the grain material is transverse isotropic with the isotropy axis directed along the symmetry axis of each grain. For the description of the grain orientations, we introduce a global Cartesian coordinate system (X_1, X_2, X_3) with the basis vectors of the axes G^1, G^2, G^3. The origin of this system is moved to the center of each grain by parallel transferring. We also introduce the local Cartesian systems (x_1, x_2, x_3) with the origin at the grain centers and the basis vectors g^1, g^2, g^3. The x_3-axis of the local system is directed along the symmetry axis of the grain (vector n in Fig. 12.2).

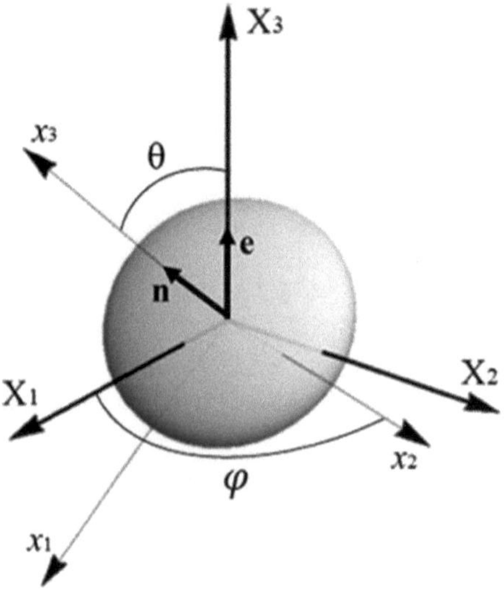

Fig. 12.2 Global and local coordinate systems for a spheroidal grain

Relations between the local and global bases are defined by the equation

$$g^i = Q^i_j G^i, \tag{12.28}$$

$$\left[Q^i_j\right] = \begin{bmatrix} \cos\varphi\cos\theta & \sin\varphi\cos\theta & -\sin\theta \\ -\sin\varphi & \cos\varphi & 0 \\ \cos\varphi\sin\theta & \sin\varphi\sin\theta & \cos\theta \end{bmatrix}. \tag{12.29}$$

Here, azimuthal θ and polar φ angles define the direction of the symmetry axis of the grain (vector n) in the global coordinate system. Note that for orthogonal matrix Q^i_j in Eq. (12.29), the components of vector G^3 in the local basis are $(-\sin\theta, 0, \cos\theta)$ and they don't depend on the polar angle φ. The axis x_2 and the basis vector g^2 of the local systems are always in the plane (X_1, X_2) of the global system.

For transverse isotropic grains, the conductivity tensor of the grain material has the form

$$C_{\alpha\beta}(n) = c_1\theta_{\alpha\beta}(n) + c_3 n_\alpha n_\beta,\ \theta_{\alpha\beta}(n) = \delta_{\alpha\beta} - n_\alpha n_\beta. \tag{12.30}$$

For a set of spheroidal grains with the same aspect ratios, the averages in Eqs. (12.14) and (12.18) are performed over the grain orientations only. If the orientation distribution function $\rho(\theta, \varphi)$ is known, the average of a tensor $T(\theta, \varphi)$ is the following integral

$$\langle T(\theta, \varphi)\rangle = \int_0^{2\pi} d\varphi \int_0^{\pi} T(\theta, \varphi)\rho(\theta, \varphi)\sin\theta d\theta. \tag{12.31}$$

Further, we assume that the orientation distribution function $\rho(\theta, \varphi)$ does not depend on the polar angle φ. For such a function, distribution over angle φ is homogeneous. An example of such a function is the orientation distribution function $\rho(\theta)$ of Owens-March's type [1, 2]

$$\rho(\theta) = \frac{b}{4\pi(b^2\cos\theta^2 + \sin\theta^2)^{3/2}},\quad 0 \le b \le 1. \tag{12.32}$$

This function satisfies the normalization condition

$$\int_0^{2\pi} d\varphi \int_0^{\pi} \rho(\theta)\sin\theta d\theta = 1. \tag{12.33}$$

For this function, the effective conductivity tensor C^* of the polycrystal is transverse isotropic, and the unit vector e of the isotropy axis of the effective medium coincides with the basis vector G^3 of the global system (Fig. 12.2). In the global basis, the tensor C^* is presented in the form

$$C^*_{\alpha\beta} = c^*_1\theta_{\alpha\beta} + c^*_3 e_\alpha e_\beta, \theta_{\alpha\beta} = \delta_{\alpha\beta} - e_\alpha e_\beta, \tag{12.34}$$

where $e = G^3$, c^*_1, c^*_3 are effective conductivity coefficients.

In detail, Eq. (12.14) for C^* has the form

$$C^* = \left\langle C(x)(I + A^*(C(x) - C^*))^{-1}\right\rangle \left\langle (I + A^*(C(x) - C^*))^{-1}\right\rangle^{-1}. \tag{12.35}$$

Here, the tensors $A^* = A^*(C^*, \varphi, \theta)$, $C(x) = C(\varphi, \theta)$ depend on the orientation of a grain with respect to the global coordinate system. The tensor A^* is the integral over unit sphere Ω_1 in Eq. (4.200)

$$A^* = \frac{1}{4\pi}\int_0^{2\pi} d\varphi' \int_0^{\pi} \widetilde{K}^*\left(a^{-1}k(\varphi', \theta')\right)\sin\theta' d\theta', \tag{12.36}$$

For transverse isotropic effective media with the conductivity tensor C^* in Eq. (12.34), the tensor $\widetilde{K}^*(k)$ has the form

$$\widetilde{K}^*_{\alpha\beta}(k) = \frac{m_\alpha m_\beta}{c^*_1(1-(m\cdot e)^2) + c^*_3(m\cdot e)^2}, m_\alpha = \frac{k_\alpha}{|k|}. \tag{12.37}$$

In the local basis g^i of each grain, the tensor a that determines the shape of a spheroidal grain is diagonal and

$$a_{\alpha\beta} = a(\delta_{1\alpha}\delta_{1\beta} + \delta_{2\alpha}\delta_{2\beta}) + a_3\delta_{3\alpha}\delta_{3\beta}, \tag{12.38}$$

where a, a_3 are spheroid's semiaxes. We introduce vector m'

$$m'(\theta', \varphi') = a^{-1}m(\theta', \varphi'), \tag{12.39}$$

where $m(\theta', \varphi')$ is a vector on the unit sphere with the components

$$m_1(\theta', \varphi') = \sin\theta' \cos\varphi', m_2(\theta', \varphi') = \sin\theta' \sin\varphi',$$
$$m_3(\theta', \varphi') = \cos\theta'. \tag{12.40}$$

The components of the vector $m'/|m'|$ in the local basis of each grain are

$$\frac{m'_1}{|m'|} = \frac{a_3\sin\theta' \cos\varphi'}{\Delta'(\theta', \varphi')}, \frac{m'_2}{|m'|} = \frac{a_3\sin\theta' \sin\varphi'}{\Delta'(\theta', \varphi')}, \frac{m'_3}{|m'|} = \frac{a\cos\theta'}{\Delta'(\theta', \varphi')},$$
$$\Delta'(\theta', \varphi') = \sqrt{a_3^2\left(\sin\theta'\right)^2 + a^2\left(\cos\theta'\right)^2}. \tag{12.41}$$

Then, the tensor A^* in Eq. (12.36) takes the form

$$A^* = \tag{12.42}$$

$$= \frac{1}{4\pi} \int_0^{2\pi} d\varphi' \int_0^{\pi} \widetilde{K}^* \left(\frac{a_3 \sin\theta' \cos\varphi'}{\Delta(\theta', \varphi')}, \frac{a_3 \sin\theta' \sin\varphi'}{\Delta(\theta', \varphi')}, \frac{a \cos\theta'}{\Delta(\theta', \varphi')} \right) \sin\theta' d\theta'.$$

We consider the tensor A^* for specific grain shapes.

1. If $a_3 \ll a$, the grain aspect ratio $\gamma = a_3/a$ is small, and the grains are oblate spheroids (disks). In the limit $\gamma \to 0$, we have

$$\lim_{a_3 \to 0} \widetilde{K}^* \left(\frac{a_3 \sin\theta' \cos\varphi'}{\Delta(\theta', \varphi')}, \frac{a_3 \sin\theta' \sin\varphi'}{\Delta(\theta', \varphi')}, \frac{a \cos\theta'}{\Delta(\theta', \varphi')} \right) =$$

$$= \widetilde{K}^*(0, 0, 1) = \widetilde{K}^*(n), \tag{12.43}$$

where n is the unit vector of the grain symmetry axis, and in the local basis of the grain, vector n has the components (0, 0, 1). It is taken into account that $\widetilde{K}^*(k)$ is a homogeneous function of the order of zero, and for any scalar parameter λ, $\widetilde{K}^*(\lambda k) = \widetilde{K}^*(k)$. Thus

$$A^* = \widetilde{K}^*(n), \tag{12.44}$$

and for very oblate spheroids, the integral in Eq. (12.42) is calculated explicitly.

2. If $a_3 \gg a$, $\gamma \gg 1$, the grains are prolate spheroids (fibers), and in the limit $\gamma \to \infty$, we have

$$\lim_{a_3 \to \infty} \widetilde{K} \left(\frac{a_3 \sin\theta' \cos\varphi'}{\Delta(\theta', \varphi')}, \frac{a_3 \sin\theta' \sin\varphi'}{\Delta(\theta', \varphi')}, \frac{a \cos\theta'}{\Delta(\theta', \varphi')} \right) =$$

$$= \widetilde{K}^*(\cos\varphi', \sin\varphi', 0). \tag{12.45}$$

As a result, we have

$$A^* = \frac{1}{2\pi} \int_0^{2\pi} \widetilde{K}^*(\cos\varphi', \sin\varphi', 0) d\varphi', \tag{12.46}$$

and in this case, the double integral in Eq. (12.42) is reduced to 1D integral.

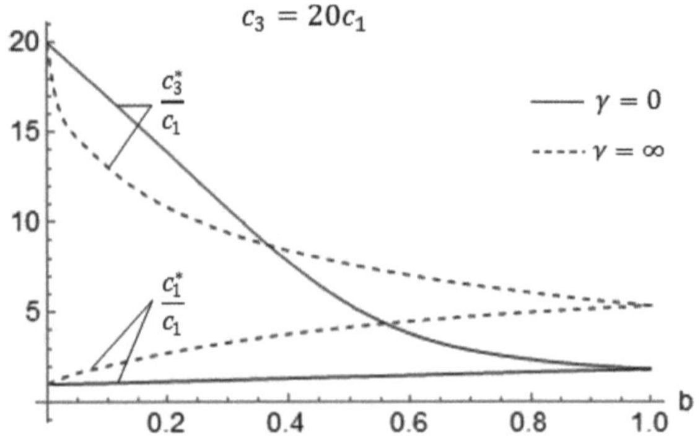

Fig. 12.3 Dependencies of the effective conductivity coefficients c_1^* and c_3^* of the polycrystal with spheroidal transverse isotropic grains on the parameter b of the orientation distribution function in Eq. (12.32); the conductivity coefficient of the grain are c_1, c_3 and $c_3 = 20c_1$

3. For spherical grains ($a = a_3, \gamma = 1$), tensor A^* is the double integral

$$A^* = \frac{1}{4\pi} \int_0^{2\pi} d\varphi' \int_0^{\pi} \widetilde{K}^*(\sin\theta' \cos\varphi', \sin\theta' \sin\varphi', \cos\theta') \sin\theta' d\theta'. \tag{12.47}$$

Equation (12.35) for the effective conductivity tensors C^* of the polycrystal can be solved iteratively. The nth iteration $C^{*(n)}$ and the previous iteration $C^{*(n-1)}$ relate by the equation

$$C^{*(n)} = \Phi(C^{*(n-1)}),\ \Phi(C^*) = \langle C(x)\Lambda^*(x)\rangle \langle \Lambda^*(x)\rangle^{-1}, \tag{12.48}$$

$$\Lambda^*(x) = (I + A^*(C(x) - C^*))^{-1} \tag{12.49}$$

with $C^{*(0)} = \langle C(x)\rangle$.

We consider the polycrystals with transverse isotropic grains and the conductivity coefficients in Eq. (12.30) $c_3 = 20c_1$ and $c_3 = 0.1c_1$ for the aspect ratios $\gamma = 0, \infty$. The distribution function $\rho(\theta)$ of the grains over orientations is defined in Eq. (12.32). Dependencies of the effective conductivity coefficients c_1^*, c_3^* on the parameter b in the function $\rho(\theta)$ are shown in Figs. 12.3 and 12.4.

12.3 The Effective Medium Method for Elastic Polycrystals

In the case of elastic polycrystals, the equations for the effective elastic stiffness C^* and compliance B^* tensors keep the forms in Eqs. (12.14), (12.18)

$$C^* = \langle C(x)(I + A^*(C(x) - C^*))^{-1}\rangle \langle (I + A^*(C(x) - C^*))^{-1}\rangle^{-1}, \tag{12.50}$$

$$B^* = \langle B(x)(I - D^*(B(x) - B^*))^{-1}\rangle \langle (I - D^*(B(x) - B^*))^{-1}\rangle^{-1}, \tag{12.51}$$

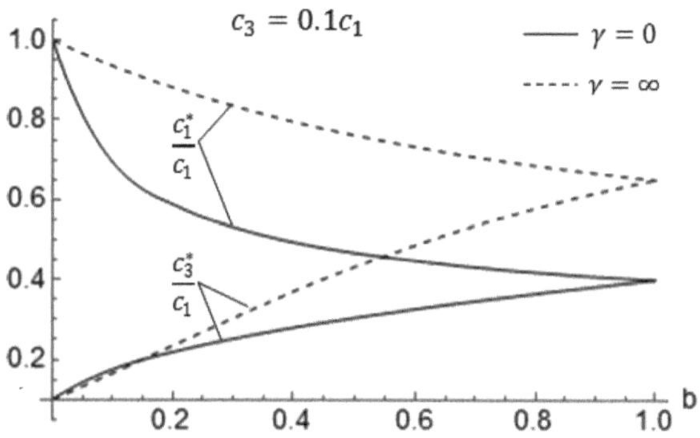

Fig. 12.4 The same as in Fig. 12.3 for $c_3 = 0.1c_1$

where $C(x)$, A^*, C^* and $B(x)$, B^*, D^* are fourth rank tensors, $C(x)$ is the elastic stiffness tensor of the grain material, $B(x) = (C(x))^{-1}$. The tensors A^* and D^* are defined in Eqs. (4.89), (4.92), where tensor C^0 should be changed with C^* and the function $K(x)$ the function $K^*(x)$ for the effective medium. The solutions of Eqs. (12.50), (12.51) have the properties 1-6 indicated in Sect. 12.1. In particular, the tensors C^* and B^* that satisfy Eqs. (12.50), (12.51) are symmetric with respect to the permutation of the index pairs as well as permutations of the indices inside the pairs.

12.3.1 Thermo-Expansion Coefficients of Elastic Polycrystals

Let the tensors of elastic compliance B and the second rank tensor of thermo-expansion coefficients α of the grain material be constant inside each grain. The symmetry axes of these tensors coincide with principal axes of the ellipsoidal grains randomly distributed over orientations. By heating the polycrystal from temperature $T = 0$ to $T \neq 0$, there appear temperature (internal) stresses in the polycrystal. It is shown in Sect. 4.8 that the temperature stresses $\sigma(x)$ in heterogeneous thermo-elastic media satisfy the integral equation

$$\sigma(x) = \int S^*(x - x')B^{*1}(x')\sigma(x')dx' + \int S^*(x - x')\alpha^{*1}(x')T dx', \qquad (12.52)$$

$$B^{1*}(x) = B(x) - B^*, \alpha^{*1}(x) = \alpha(x) - \alpha^*. \qquad (12.53)$$

Here, B^* and α^* are the effective elastic compliance tensor and the tensor of thermo-expansion coefficients of the polycrystalline medium. The kernel $S^*(x)$ of the integral operator in this equation is defined in Eq. (10.16) with $\widehat{C} = C^*$. It is assumed that the external stress field σ^0 is equal to zero, and for $T = 0$, the internal stresses are absent in the polycrystal. The strain tensor $\varepsilon(x)$ in the polycrystalline medium subjected to a constant temperature field T is defined in Eq. (4.32) by $\sigma^0 = 0$

$$\varepsilon(x) = \alpha^* T - \int K^*(x - x')C^* \left(B^{*1}(x')\sigma(x') + \alpha^{*1}(x')T\right) dx'. \tag{12.54}$$

Here, the kernel $K^*(x)$ is defined in Eqs. (10.14) with $\widehat{C} = C^*$. The tensor of effective thermo-expansion coefficients of the polycrystal is the averaged strain tensor $\varepsilon(x)$ by heating the polycrystal to the temperature $T = 1$. Thus, after averaging Eq. (12.54) over the ensemble of realizations of the random tensors C and α for $T = 1$ and using Eq. (10.28)

$$\int K^*\left(x - x'\right)dx' = \left(C^*\right)^{-1}, \tag{12.55}$$

we obtain

$$\alpha^* = \langle \varepsilon(x) \rangle = \langle \varepsilon(x) \rangle - \left\langle B^{*1}(x)\sigma(x) + \alpha^{*1}(x) \right\rangle. \tag{12.56}$$

It follows from this equation that the second term in its right-hand side should be equal to zero

$$\left\langle B^{*1}(x)\sigma(x) + \alpha^{*1}(x) \right\rangle = 0. \tag{12.57}$$

For the solution of the homogenization problem, we accept the hypotheses of the effective medium method H_1, H_2, H_3 (Sect. 12.1). Thus, each grain in the polycrystal has an ellipsoidal shape, and it can be considered as an isolated one in the homogeneous medium with the effective thermo-elastic properties of the polycrystal (hypothesis H_2). For $T = 1$, the integral equation (12.54) for the temperature stresses in the ith grain takes the form (the one-particle problem)

$$\sigma^i(x) = \int\limits_{v^i} S^*(x - x')B^{*1i}\sigma^i(x')dx' + \int\limits_{v^i} S^*(x - x')\alpha^{*1i}dx', x \in v^i, \tag{12.58}$$

$$B^{*1i} = B^i - B^*, \alpha^{*1i} = \alpha^i - \alpha^*, i = 1, 2, ... \tag{12.59}$$

Here, v^i is the volume of the ith grain. It follows from the property of polynomial conservation for ellipsoidal inclusions (Sect. 4.4) that the solution of Eq. (12.58) is constants inside each ellipsoidal region. Therefore, for the ith grain we have

$$D^{*i} = \int\limits_{v^i} S^*(x - x')dx' =$$

$$= \frac{1}{4\pi} \int\limits_{|k|=1} \widetilde{S}^*(a^{-1}k)dk, \quad a = a^i, x \in v^i, i = 1, 2, ..., \tag{12.60}$$

and the stress field inside the ith grain is defined by the equation

$$\sigma^i = \Lambda^{\sigma i} D^{*i} \alpha^{1*i},\ \Lambda^{\sigma i} = \left(I - D^{*i} B^{1*i}\right)^{-1},\ x \in v^i,\ i = 1, 2, ... \tag{12.61}$$

Substituting this equation for σ^i into Eq. (12.57), we yield

$$\left\langle B^{*1}(x)\Lambda^{\sigma}(x)D^{*}(x)\alpha^{*1}(x) + \alpha^{*1}(x)\right\rangle =$$

$$= \left\langle \left(I - B^{*1}(x)D^{*}(x)\right)^{-1}\left(\alpha(x) - \alpha^{*}\right)\right\rangle = 0. \tag{12.62}$$

Here, the tensors $B^{*1}(x)$, $D^{*}(x)$ and $\alpha^{*1}(x)$ coincide with the tensors B^{*1i}, D^{*i}, α^{*1i} in the ith grain v^i $(i = 1, 2, ...)$.The expression for the tensor of effective thermo-expansion coefficients of the polycrystal follows from Eq. (12.62)

$$\alpha^{*} = \left\langle \left(I - B^{*1}(x)D^{*}(x)\right)^{-1}\right\rangle^{-1} \left\langle \left(I - B^{*1}(x)D^{*}(x)\right)^{-1}\alpha(x)\right\rangle. \tag{12.63}$$

Here, the effective compliance tensor B^{*} should be found from Eq. (12.51). After solution of this equation, the right-hand side of Eq. (12.63) can be calculated.

12.3.2 Elastic Polycrystals with Spheroidal Transverse Isotropic Grains

In this section, we consider polycrystals with spheroidal grains of the same aspect ratios $\gamma = a_3/a$, where $a_1 = a_2 = a$, a_3 are ellipsoid semiaxes. The grain material is transverse isotropic with the isotropy axis directed along the symmetry axis of each grain. In the E-basis, the elastic stiffness tensor C of the grain material has the form

$$C(n) = c_1E^1 + c_2E^2 + c_3(E^3(n) + E^4(n)) + c_5E^5(n) + c_6E^6(n), \tag{12.64}$$

where c_i are independent elastic constants of the transverse isotropic material, n is the unit vector of the symmetry axis of the grain.

For spheroidal grains with the same aspect ratios, the averages in Eqs. (12.50), (12.51) are performed over the grain orientations only. For the definition of the spheroidal grains orientations, we introduce the global G^1, G^2, G^3 and local g^1, g^2, g^3 Cartesian bases indicated in Sect. 12.2. As a result, the grain orientation is defined by polar φ and azimuthal θ angles of the spherical coordinate system with the polar axis directed along the basis vector G^3. If the orientation distribution function does not depend on the angle φ, the effective stiffness tensor C^{*} of the polycrystal is transverse isotropic. For the unit vector e of the isotropy axis of the effective medium, $e = G^3$ (Fig. 12.2), and the tensor C^* is presented in the form similar to $C(n)$ in Eq. (12.64)

$$C^* = c_1^* E^1 + c_2^* E^2 + c_3^*(E^3(e) + E^4(e)) + c_5^* E^5(e) + c_6^* E^6(e), \tag{12.65}$$

where $c_i^*(i = 1, 2, 3, 5, 6)$ are effective elastic constants of the polycrystal.
Tensor A^* in Eq. (12.50) is the double integral

$$A^* = \frac{1}{4\pi} \int_0^{2\pi} d\varphi' \int_0^{\pi} \widetilde{K}^* \left(a^{-1} m(\varphi', \theta')\right) \sin\theta' d\theta', \tag{12.66}$$

where vector m in the local g^i-basis is

$$m = \cos\varphi' \sin\theta' g^1 + \sin\varphi' \sin\theta' g^2 + \cos\theta g^3. \tag{12.67}$$

For transverse isotropic effective media, the tensor $\widetilde{K}^*(k)$ under the integral sign in Eq. (12.66) takes the form

$$\widetilde{K}^*_{\alpha\beta\gamma\delta}(k) = \big(b_1 m_\alpha \delta_{\beta\gamma} m_\delta + b_2 m_\alpha m_\beta m_\gamma m_\delta +$$

$$+ b_3 (m \cdot e) m_\alpha (m_\beta e_\gamma + m_\gamma e_\beta) m_\delta + b_4 m_\alpha e_\beta e_\gamma m_\delta\big) |_{(\alpha\beta)(\gamma\delta)}, \tag{12.68}$$

$$m_\alpha = \frac{k_\alpha}{|k|}, \; (m \cdot e) = m_\alpha e_\alpha, \tag{12.69}$$

$$b_1 = \frac{1}{d_1}, b_2 = \frac{1}{d_1 \Delta} \left(d_3^2 (m \cdot e)^2 - d_2(d_1 + d_4)\right), \tag{12.70}$$

$$b_3 = \frac{1}{d_1 \Delta} (d_2 d_4 - d_3(d_1 + d_3(m \cdot e))), \tag{12.71}$$

$$b_4 = \frac{1}{d_1 \Delta} \left(d_3^2 (m \cdot e)^2 - d_4(d_1 + d_2)\right), \tag{12.72}$$

$$\Delta = (d_1 + d_2)(d_1 + d_4) - (d_2 d_4 - 2 d_2 d_3 + d_3^2)(m \cdot e)^2 + d_3^2 (m \cdot e)^4. \tag{12.73}$$

Here, the coefficients d_i are expressed in terms of effective elastic constants c_i^* of the polycrystal as follows

$$d_1 = \frac{1}{4} \left(2c_1^* + c_5^*(m \cdot e)^2\right), d_2 = \frac{1}{2} c_1^* + c_2^*, \tag{12.74}$$

$$d_3 = \frac{1}{4} c_5^* + c_3^*, d_4 = \frac{1}{4} c_5^* + c_6^*(m \cdot e)^2. \tag{12.75}$$

Tensor D^* in Eq. (12.51) and tensor A^* in Eq. (12.66) relate by the equation

$$D^* = C^*(e)A^*C^*(e) - C^*(e). \tag{12.76}$$

By transferring from the local basis g^i to the global basis G^i, the tensors A^* and D^* become the functions of the angles θ and φ

$$A^* = A^*(\theta, \varphi),\ D^* = D^*(\theta, \varphi). \tag{12.77}$$

Similar to the case of conductive media Eqs. (12.44), (12.46), (12.47), tensor A^* is simplified for the grains of specific shapes.

1. If $a_3 \ll a$, the grains are oblate spheroids (disks), and in the limit $a_3, \gamma \to 0$, we have

$$A^* = \widetilde{K}^*(n). \tag{12.78}$$

Here, n is the unit vector in the direction of the grain symmetry axis (Fig. 12.2). In the local basis of the grain, vector n has the components $(0, 0, 1)$. Thus, for very oblate spheroidal grains, the integral in Eq. (12.66) is calculated explicitly.

2. If $a_3 \gg a$, the grains are prolate spheroids (fibers), and in the limit $a_3 \to \infty$, we have

$$A^* = \frac{1}{2\pi}\int_0^{2\pi} \widetilde{K}^*(\cos\varphi', \sin\varphi', 0)d\varphi'. \tag{12.79}$$

Thus, for very prolate spheroids, the double integral in Eq. (12.66) is reduced to a 1D-integral.

3. For spherical grains ($a = a_3, \gamma = 1$), tensor A^* is the 2D integral

$$A^* = \frac{1}{4\pi}\int_0^{2\pi} d\varphi' \int_0^{\pi} \widetilde{K}^*(\sin\theta' \cos\varphi', \sin\theta' \sin\varphi', \cos\theta') \sin\theta' d\theta'. \tag{12.80}$$

For a transverse isotropic effective medium with the isotropy axis in the direction of vector e, tensor A^* in this equation is also transverse isotropic with the same isotropy axis

$$A^* = A_1^* E^1 + A_2^* E^2 + A_3^*(E^3(e) + E^4(e)) + A_5^* E^5(e) + A_6^* E^6(e). \tag{12.81}$$

Here, A_i^* are constant coefficients.

Equations (12.50) and (12.51) for the effective stiffness and compliance tensors of the polycrystal can be solved by iterations. In the simplest iterative process, the iteration $C^{*(n)}$ is calculated using the previous iteration $C^{*(n-1)}$ from the equations

$$C^{*(n)} = \Phi(C^{*(n-1)}),\ \Phi(C^*) = \langle C(x)\Lambda^*(x)\rangle \langle \Lambda^*(x)\rangle^{-1}. \quad (12.82)$$

$$\Lambda^*(x) = (I + A^*(C(x) - C^*))^{-1}. \quad (12.83)$$

For a large anisotropy of the elastic stiffness tensor of grains, the simplest iteration process can diverge, and the process based on the equation

$$C^{*(n)} = C^{*(n-1)} - \alpha \left(C^{*(n-1)} - \Phi(C^{*(n-1)})\right), 0 < \alpha < 1. \quad (12.84)$$

is more efficient. Here, parameter $\alpha < 1$ should be taken for convergence of the iteration process. The process is stopped if the tolerance

$$\delta = \frac{||C^{*(n)} - C^{*(n-1)}||}{||C^{*(n)}||} \quad (12.85)$$

become smaller than a prescribed number. Here, $||C||$ is norm of a tensor, and in the calculations, $\delta = 10^{-4}$ was accepted. More sophisticated methods of solution of Eqs. (12.50) and (12.51) were discussed in [3].

For carrying out algebraic operations with fourth rank tensors in Eqs. (12.50) and (12.51), it is convenient to transfer from fourth rank tensors symmetric with respect to the first and second pairs of indices to 6×6 non-symmetric matrices.

$$T_{\alpha\beta\gamma\delta} \Longleftrightarrow \overline{T}_{pq},\ p, q = 1, 2, 3, 4, 5, 6. \quad (12.86)$$

Algebraic operations with the matrices that correspond to the operations with fourth rank tensors are considered in Appendix A.6.

12.3.3 *Effective Elastic Properties of Clay Polycrystalline Aggregates*

As an example of application of the EMM, we consider the homogenization problem for the illite-dominated clay polycrystalline aggregates. Clay grains of such materials are transverse isotropic, and their independent elastic constants are presented in [4, 5]. In Voigt's notations [6], the dimensionless independent elastic constants $\overline{c}_{ij} = c_{ij}/c_{33}$ and constants $\overline{c}_i = c_i/c_{33}$ in Eq. (12.64) for clay grains are in Table 12.1.

Relations between Voigt's constants c_{ij} and constants c_i in Eq. (12.64) for the elastic stiffness tensor of transverse isotropic materials are

$$c_1 = c_{11} - c_{12},\ c_2 = c_{12},\ c_3 = c_{13} - c_{12}, \quad (12.87)$$

$$c_5 = 2(2c_{44} + c_{12} - c_{11}),\ c_6 = c_{11} - 2c_{13} + c_{33} - 4c_{44}. \quad (12.88)$$

Table 12.1 Elastic constants of the clay grain material

c_{ij}/c_{33}	$\bar{c}_{11}$	$\bar{c}_{12}$	$\bar{c}_{13}$	$\bar{c}_{33}$	$\bar{c}_{44}$
	3.2	0.83	0.66	1	0.7
c_i/c_{33}	$\bar{c}_1$	$\bar{c}_2$	$\bar{c}_3$	$\bar{c}_5$	$\bar{c}_6$
	2.37	0.83	−0.17	−1.94	0.88

The microstructure of illite-dominated clay polycrystals is described in [5]. Clay grains are spheroids with aspect ratios γ of the order of 0.05 (clay platelets). For such small values of the aspect ratios of grains, tensors A^* and D^* in Eqs. (12.50), (12.51) can be changed with their limit values for $\gamma \to 0$. Orientations of grains are described by the orientation distribution function of Owens-March type [2]

$$\rho(\theta) = \frac{b^{3/2}}{4\pi(b^3 \cos\theta^2 + \sin\theta^2)^{3/2}}, \quad 0 \leq b \leq 1. \tag{12.89}$$

It is assumed that the distribution with respect to polar angle φ is homogeneous. For illite-dominated clay polycrystals, the parameter b in this equation is about $0.4 \div 0.5$. Dependences of this function on the parameter b is shown in Fig. 12.5.

Numerical calculation of the effective stiffness tensor of the polycrystal using Eqs. (12.50) and (12.51) provides identical results, but the iteration process based on Eq. (12.51) for the effective compliance tensor B^* converges better. The results of calculations are presented in Fig. 12.6 for the dimensionless effective elastic constants $\left(\bar{c}_{ij}\right)^* = \left(c_{ij}\right)^* / c_{33}$. The dotted lines in these figures are the results of averaging the

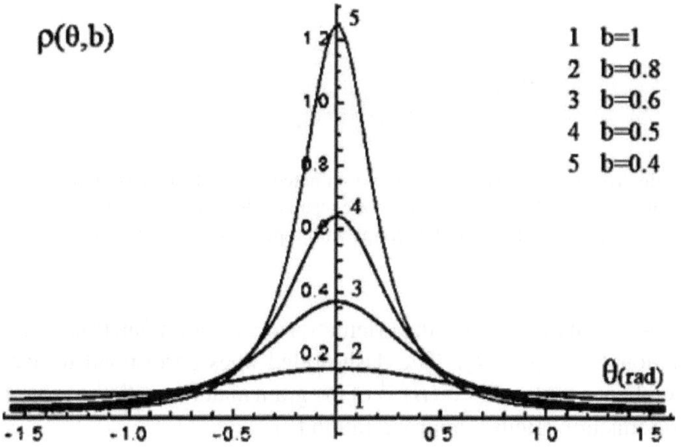

Fig. 12.5 The orientation distribution function in Eq. (12.89) for various values of parameter b

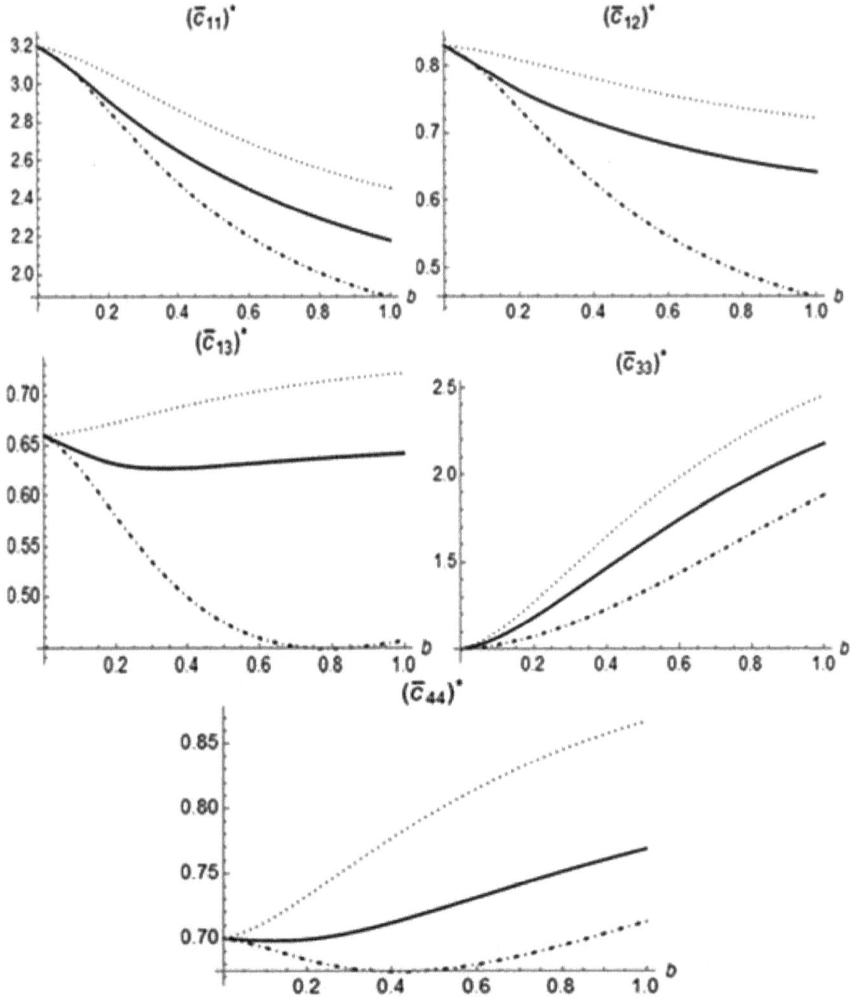

Fig. 12.6 Effective elastic constants of illite-dominated clay polycrystals; dotted lines are Voigt's averaging of grain material, dash-dotted lines are results of Reuss' averaging, black lines are results of the effective medium method for clay grains with aspect ratios $\gamma = 0$ (disks)

stiffness tensor C of the grains using orientation distribution function (12.89) (Voigt's approximation $C^* = \langle C \rangle$ [7]). The dash-dotted lines correspond to averaging the elastic compliance tensor $C^* = \langle B \rangle^{-1}$ of the grain material with the same orientation distribution function (Reuss' approximation [8]).

12.3.4 Effective Elastic and Thermo-Elastic Properties of Polycrystalline Zinc

In the next example, we consider the elastic and thermo-elastic properties of the polycrystals of zinc. Zinc is a transverse isotropic material which elastic constants are indicated in [6]. The independent elastic constants of zinc crystals in Eq. (12.64) and in Voigt's notations are in Table 12.2.

Calculations of the effective elastic constants of zinc polycrystals were performed for three types of grains: very oblate spheroids (disks), very prolate spheroids (fibers), and spherical grains. The orientation distribution function of the symmetry axis of grain materials is in Eq. (12.89). The dependences of the effective elastic constants on the parameter b in this function are presented in Fig. 12.7. The dotted lines in these figures are the results of Voigt's averaging, and the dash-dotted lines are Reuss' averaging.

In the local basis with the axis of isotropy directed along the x_3-axis, the second rank tensor of thermo-expansion coefficients of zinc crystals is diagonal [6]

$$\alpha_{\alpha\beta} = \alpha_{11}(\delta_{1\alpha}\delta_{1\beta} + \delta_{2\alpha}\delta_{2\beta}) + \alpha_{33}\delta_{3\alpha}\delta_{3\beta}, \quad \frac{\alpha_{11}}{\alpha_{33}} = 0.203. \tag{12.90}$$

Dimensionless effective coefficients of thermo-expansion of zinc polycrystals ($\overline{\alpha}^*_{ii} = \alpha^*_{ii}/\alpha_{33}$, $i = 1, 2, 3$) with three types of grains (disks, fibers, and spheres) for orientation distribution function of grains $\rho(\theta)$ in Eq. (12.89) are presented in Fig. 12.8. For the construction of the tensor of the effective thermo-expansion coefficients of the polycrystal, Eq. (12.63) of the effective medium method is used. Dotted lines in these figures are the results of averaging the tensor α over orientations with the orientation distribution function (12.89) (Voigt's averaging), and dash-dotted lines are the results of a similar averaging of the inverse tensor α^{-1}(Reuss' averaging). Dashed lines in Fig.5 are effective thermo-expansion coefficients of the polycrystals with spherical grains.

It is seen from Fig. 12.7 that the influence of the grain aspect ratios on the effective elastic constants of zinc polycrystals is not very strong: difference in the effective elastic constants of the polycrystals with disk-like, fiber-like, and spherical grains is about 10-15%. The influence of the grain aspect ratios on the effective thermo-expansion coefficients is even less. Such influence can be stronger for a higher anisotropy of the elastic stiffness tensor and the tensor of thermo-expansion coefficients of the grain material.

Table 12.2 Independent elastic constants of zinc crystals

$\overline{c}_{ij} = c_{ij}/c_{33}$	$\overline{c}_{11}$	$\overline{c}_{12}$	$\overline{c}_{13}$	$\overline{c}_{33}$	$\overline{c}_{44}$
	2.7	0.5	0.8	1	0.7
$\overline{c}_i = c_i/c_{33}$	$\overline{c}_1$	$\overline{c}_2$	$\overline{c}_3$	$\overline{c}_5$	$\overline{c}_6$
	2.2	0.5	0.3	−1.6	−0.7

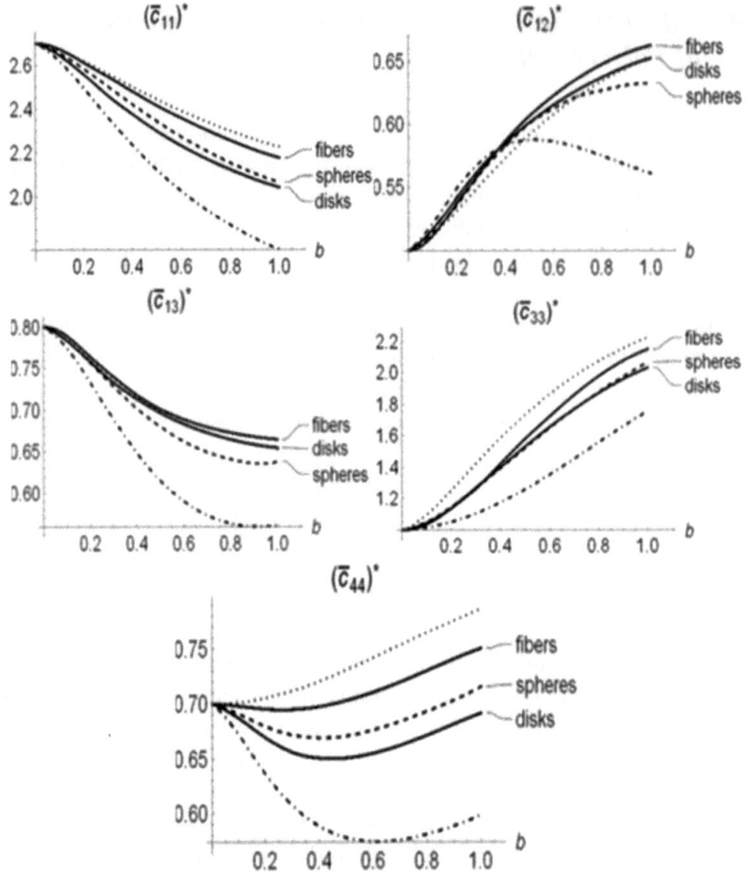

Fig. 12.7 Effective elastic constants of zinc polycrystals with various grain aspect ratios: $\gamma = 0$ (disks), $\gamma \to \infty$ (fibers) and $\gamma = 1$ (spherical grains); dotted lines are Voigt's averaging of grain material, dash-dotted lines are Reuss' averaging

12.4 The Effective Medium Method for Polycrystals with Different Families of Grains

Let the polycrystal consist of several families of grains with different shapes and elastic stiffness tensors. The orientation distribution functions can be different for different grain families. For this type of polycrystals, the hypotheses of the EMM are formulated as follows. Each grain can be considered as an isolated one embedded in a homogeneous medium with the effective elastic stiffness tensor C^* of the composite and subjected to the average strain field over the polycrystal. Equation (12.26) of the effective medium method for the tensor C^* takes the form

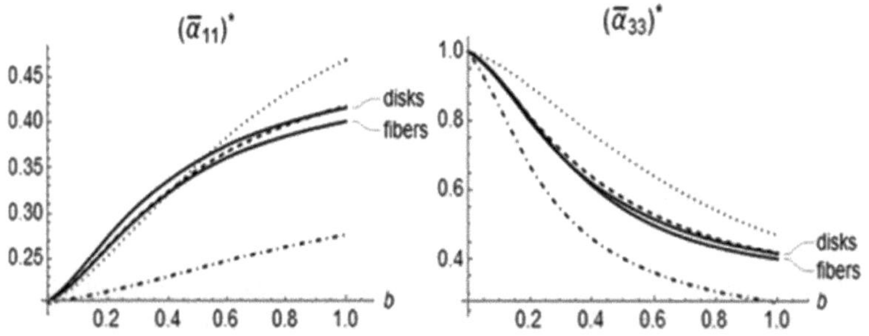

Fig. 12.8 Effective thermo-expansion coefficients of zinc polycrystals with various aspect ratios of spheroidal grains, dashed black lines correspond to spherical grains; dotted lines are the averages of the tensor of thermo-expansion coefficients α of grain material with orientation distribution function (12.89), dash-dotted line are the same averages of the inverse tensor α^{-1}

$$\left\langle (C^r - C^*)\left(I + A^*(C^r - C^*)\right)^{-1} \right\rangle =$$

$$= \sum_{r=1}^{n} p_r \left\langle (C^r - C^*)\left(I + A^{*r}(C^r - C^*)\right)^{-1} \right\rangle_r = 0. \tag{12.91}$$

Here, p_r is the volume fraction of the rth family of grains, the tensors A^{*r} depend on the tensor C^* and the aspect ratios of the grains of the rth family. The average under the sum sign is performed over the grain orientations in the rth family. The effective elastic stiffness tensor obtained from this equation is symmetric with respect to permutations of index pairs as well as to permutations of indices inside the pares. Because the symmetry of the effective medium is not known in advance, all 21 constants of the effective stiffness tensor have to be calculated from Eq. (12.91).

It was proposed in [9, 10] to use Eq. (12.91) for estimation of the effective stiffness tensor of matrix composite materials. In this version of the EMM, the matrix phase is simulated by a polycrystal with ellipsoidal grains of the same properties and orientations. Then, the effective elastic stiffness tensor of the matrix-inclusion composite is calculated from the equation similar to Eq. (12.91)

$$p\left\langle (C - C^*)\left(I + A^{*i}(C - C^*)\right)^{-1} \right\rangle +$$

$$+ (1 - p)(C^0 - C^*)\left(I + A^0(C^0 - C^*)\right)^{-1} = 0. \tag{12.92}$$

Here, C^0 is the stiffness tensor of the matrix material, C is the same for the inclusions, p is the inclusion volume fraction. This version of the EMM in application to matrix-inclusion composites is called the symmetric effective medium method (SEMM). Note that in this method, the shape of the ellipsoidal grains of the matrix phase cannot be uniquely defined because the matrix material is homogeneous. The tensor A^0 in Eq. (12.92) depends on the aspect ratios of the grains of the matrix phase. For different aspect ratios, the effective stiffness tensors C^* found from Eq. (12.92) will be different also, and such an ambiguity can hardly be accepted.

12.5 Conclusion

In the previous chapters, the effective medium method (EMM) was applied to solution of the homogenization problem for matrix composites and polycrystals. In application to matrix-inclusion composites, the EMM has the following drawbacks:

The effective elastic stiffness tensor obtained by the method cannot be symmetric with respect to the permutation of index pairs;

The reciprocal equations for the effective elastic stiffness tensor C^* and the elastic compliance tensor B^* can have different solutions;

Comparison with experimental data and numerical simulations of the effective elastic properties of matrix composites show that the method overestimates interactions between the inclusions.

Kerner's version of the EMM corresponds better to experimental data and numerical simulations. But this version requires numerical solution of the one-particle problems, except for the simplest case of spherical isotropic inclusions in an isotropic matrix. It results in additional technical difficulties by implementation of the EMM algorithm.

In application to the homogenization problem for polycrystals, the effective medium does not have the mentioned drawbacks. The effective elastic stiffness tensors of the polycrystals obtained by the EMM are symmetric with respect to the permutation of index pairs as well as the permutation of indices inside pairs. The solutions of the reciprocal equations for the effective elastic stiffness and elastic compliance tensors are related by the correct equation $(C^*)^{-1} = B^*$. Predictions of the EMM are close to experimental data.

12.6 Notes

The simplest approximate solutions of the homogenization problem for polycrystals were proposed in the first half of twentieth century by Voigt [7] and Reuss [8]. The authors suggested to find polycrystal effective elastic stiffness tensors by averaging the elastic stiffness tensor of the grain material $\langle C(x) \rangle$ (Voigt) or the elastic compliance tensor of the grain material $\langle B(x) \rangle$ (Reuss) over possible orientations of the crystal symmetry axes inside grains. For the same polycrystal, these two averages provide different results, and Hill [11] employed the arithmetical or geometrical means of Voigt's and Reuss' averages in the attempt to describe better experimental data. Later on, there were proposed more sophisticated average schemes of Voigt's and Reuss' types for calculation of the polycrystal effective elastic constants [12, 13].

Voigt's and Reuss' averages ignore the influence of grain shapes on the effective elastic properties of polycrystals as well as the equations that govern elastic fields in polycrystalline media. The homogenization methods that take into account these factors were intensively developed in the second half of the twentieth century. First solution of the homogenization problem based on the integral equations for the strain tensor in polycrystalline media was obtained in [14]. In this work, the solution of the integral equation was expanded into Neumann's series, and the first two terms of this series were kept only. For this approximation, the homogenization problem is reduced to constructing the correlation function $R(x - x^{'}) = \langle C(x) \otimes C(x^{'}) \rangle$ of the random elastic stiffness tensor $C(x)$ in the polycrystal and then, calculating integrals containing this function. This approach serves for small deviations of the elastic stiffness tensor $C(x)$ from its mean value $\langle C(x) \rangle$ only. If the fluctuations of $C(x)$ are not small, more terms in Neumann's series should be kept. Methods of calculation of partial sums of these series were considered in [15, 16]. The authors developed the diagram technique for calculation the average strain and stress tensors in polycrystalline media. In these works, the EMM was also applied to the calculation of the effective elastic stiffness tensor C^* for polycrystals with ellipsoidal grains. Numerical realizations of the method were performed for polycrystals with spherical grains, cubic symmetry of the grain material, and homogeneous distributions of grain crystallographic axes over orientations.

First, the effective medium method was applied to the calculation of the effective elastic stiffness tensor of polycrystals in [17, 18]. In these works, the equation of the EMM for polycrystals with ellipsoidal grains and arbitrary symmetry of the grain material was derived. A simpler way of deriving the same equations was pointed out in [16].

Numerical realization of the EMM for the polycrystals with spherical grains and non-isotropic macro-properties was performed in [3, 19, 20]. In [19, 20], the orientation distribution functions of the symmetry axes in the grains were expanded in a series of generalized spherical harmonics, and two terms of this series were kept only. For actual polycrystals, the coefficients in this series can be found experimentally by the X-ray method. In [19, 20], the results of calculations were presented for various values of these coefficients. The orientation distribution function of grains in polycrystalline rock materials can also be revealed by X-ray method [2]. For spheroidal transverse isotropic grains, the orientation distribution function of many rock polycrystals can be approximated by Owens-March's distribution function in Eq. (12.89) [1, 2].

This chapter is based on the paper [21].

References

1. Johansen T, Ruud B, Jakobsen M (2004) Effect of grain scale alignment on seismic anisotropy and reflectivity of shales. Geophys Prospect 52:133–149
2. Louis L, Day-Stirrat R, Hofmann R, Saxena N, Schleicher A (2018) Computation of effective elastic properties of clay from X-ray texture goniometry data. Geophysics 83(5):MR245–MR261
3. Morris P (1971) Iterative scheme for calculating polycrystal elastic constants. Int J Eng Sci 9:917–920
4. Sayers C, Boer L (2020) Effect of variations in microstucture on clay elastic anisotropy. Geophysics 85/2:MR73–MR82
5. Sevostianov I, Vernik L (2021) Micromechanics-based rock-physics model for inorganic shale. Geophysics 86/2:MR105–MR116
6. Nye J (1985) Physical properties of crystals, their representation by tensors and matrices. Clarendon Press, Oxford, p 329
7. Voigt W (1928) Lehrbuch der Kristallphysik. Teubner, Leipzig, Berlin
8. Reuss A (1929) Berechnung der Fliessgrenze von Mischkristallen auf Grund der Plastizitätsbedingung fűr Einkristalle. Z Angew Math Mech 9:49–58
9. Willis J (1977) Bounds and self-consistent estimates for the overall properties of anisotropic composites. J Mech Phys Solids 25:185–202
10. Berryman J, Berge P (1996) Critique of two explicit schemes for estimating elastic properties of multiphase composites. Mech Mater 22:149–164
11. Hill R (1952) The elastic behaviour of a crystalline aggregate. Proc Phys Soc 65(B):349–354
12. Aleksandrov K, Aisenberg L (1966) A method for the computation of the physical constants of polycrystalline materials (in Russian). Dokl Akad Nauk SSSR 167:1028–1031
13. Matthies S, Humbert M (1993) The realization of the concept of a geometric mean for calculating physical constants of polycrystalline materials. Physica Status Solidi (b) 177:K47–K50
14. Lifshits I, Rosentsveig L (1946) Zur Theorie der elastischen Eigenschaften von Polykristallen. Zh Eksp Teor Fiz 16:967–975
15. Zeller R, Dederichs P (1973) Elastic connstants of polycrystals. Physica Status Solidi (b) 65:831–842
16. Gubernatis J, Krumhansl J (1975) Macroscopic engineering properties of polycrystalline materials: elastic properties. J Appl Phys 46(5):1875–1883
17. Hershey A (1954) The elasticity of an isotropic aggregate of anisotropic cubic crystals. J Appl Mech 21:236–240
18. Kröner E (1958) Berechnung der elastischen Konstanten des Vielkristalls aus den Konstanten des Einkristalls. Z Phys 151:504–518
19. Kneer G (1965) Uber die Berechnung der Elastizitatsmoduln vielkristalliner Aggregate mit Textur. Phys Status Solidi 9:825–837
20. Morris P (1970) Elastic constants of polycrystals. Int J Eng Sci 8:49–61
21. Kanaun S (2023) Effective medium method in solution of the elastic and thermo-elastic homogenization problems for polycrystals with spheroidal transverse isotropic grains. Int J Eng Sci 190:103886

Chapter 13
The Effective Field Method

Abstract The self-consistent effective field method of solution of the homogenization problem for matrix composite materials is considered. The principal hypothesis of the method concerns the local exciting fields acting on the inclusions. This field is assumed to be constant and the same for all the inclusions (quasi-crystalline approximation). In addition to the shapes, volume fraction and properties of the inclusions, the method requires information about the shape of correlation hole for a typical inclusion in the composite. It is shown that the effective conductive and elastic properties of the composites depends strongly on the correlation hole aspect ratios. The method is applied to solutions of the homogenization problem for composites with random and regular set of ellipsoidal inclusions. The predictions of the method are compared with experimental data and results of numerical simulations. Well-known Mori-Tanaka method coincides with the effective field method by the additional assumption that the aspects of the correlation hole coincide with the aspect ratios of the ellipsoidal inclusions. It is shown that the considered version of the method can violate symmetry of effective conductivity and elastic stiffness tensors of the composite when it is applied to composites with several different families of the inclusions.

Keywords Matrix composites · Effective exciting field · Correlation hole · Mori-Tanaka method · Symmetry of the effective property tensors

This chapter is devoted to the effective field self-consistent method of solution of the homogenization problem for heterogeneous materials. This method originates from the similar method of quantum mechanics used for description of particle interactions in multiparticle systems. It can be applied to composites with homogeneous matrix phases containing sets of isolated inclusions. In this chapter, the algorithm of the method is developed, its predictions are compared with experimental data and the results of numerical simulations. The method is applied to solutions of the homogenization problem for composites with various types of fillers used in engineering applications.

S. Kanaun, *Self-Consistent Methods for Composites*, Solid Mechanics and Its Applications 282, https://doi.org/10.1007/978-3-031-96739-9_13

13.1 The Basic Version of the Effective Field Method

We consider a homogeneous host medium with the property tensor C^0 containing a random set of isolated inclusions with the property tensor C. The medium is subjected to a constant external field intensity ε^0 or flux density σ^0 at infinity. In the effective field method (EFM), the local exciting fields acting on the inclusions in the composite are the principal objects of consideration. The local exciting field ε^* (σ^*) is the sum of the external field applied to the medium and the fields induced in the region of the inclusion by all surrounding heterogeneities (Fig. 13.1).

The method is based on the following hypotheses.

H_1. Each inclusion in the composite can be considered as an isolated one in the original host medium subjected to a local exiting field ε^* (σ^*).

H_2. The local exiting field (effective field) is the same for all the inclusions (quasi-crystallne approximation).

The effective field is not an actual field that exists physically in the composite material. This field would have existed in the region occupied by the inclusion if the states of all the surrounding inclusions are fixed (frozen) and the considered inclusion is changed with the host medium. The effective field for the ith inclusion can be also reconstructed by fixing the actual field inside the inclusion and putting it into a homogeneous host medium. The external field that induced inside the isolated inclusion the same field as it has been inside this inclusion in the composite is the effective field.

In the simplest version of the method, the effective field is assumed to be constant and the same for all the inclusions (quasi-crystalline approximation). This term originates from the fields that act on each particle in a regular lattice of identical particles subjected to a constant external field (Fig. 13.2). By randomization of a regular lattice, this hypothesis implies that actual local exiting fields acting on the particles can be changed with the fields averaged over possible particle configurations Fig. 13.3.

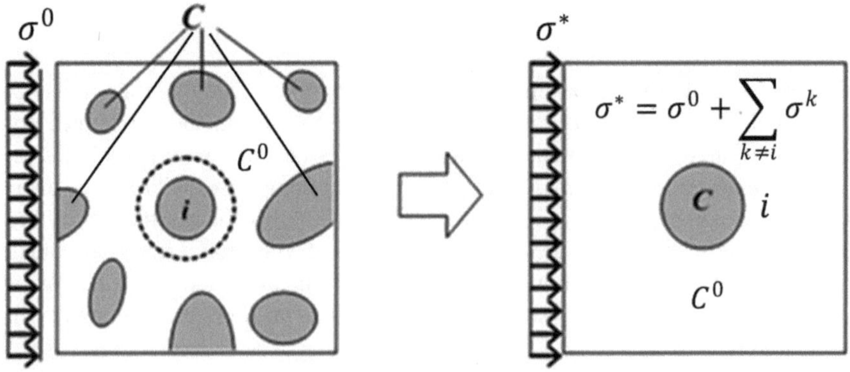

Fig. 13.1 The one-particle problem of the effective field method

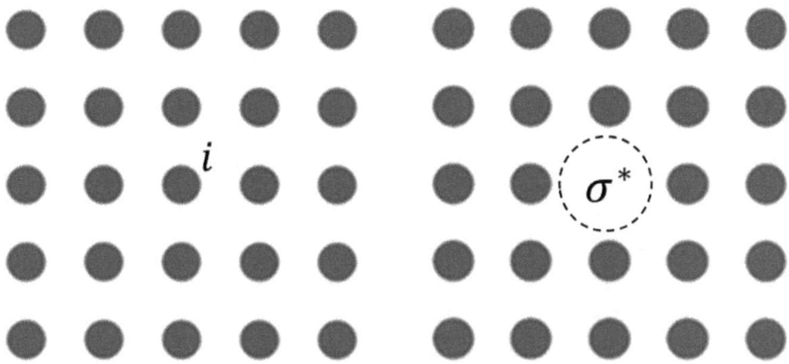

Fig. 13.2 The effective field σ^* is the field in the host medium in the region occupied by the ith inclusion if this inclusion is changed with the host medium but the fields inside other inclusions are fixed (frozen)

Fig. 13.3 The same as in Fig. 13.2 for a randomized regular inclusion lattice

Let an infinite homogeneous medium with the property tensor C^0 contains a homogeneous in space random set of isolated inclusions. The inclusions occupy regions v_i with the property tensors $C^{(i)}$, $i = 1, 2, ...$ The medium is subjected to a constant external field intensity ε^0 at infinity. The field intensity $\varepsilon(x)$ in the composite satisfies Eqs. (10.13), (10.14). For matrix composites, the properly tensor of the reference medium $\widehat{C}$ is reasonable to take equal to the property tensor of the matrix phase C^0. For such a choice, the integral in Eq. (10.13) is spread on the regions occupied by the inclusions only, the equation for the field intensity $\varepsilon(x)$ takes the form

$$\varepsilon(x) = \varepsilon^0 - \sum_i \int_{v_i} K(x - x')C^{1(i)}\varepsilon(x')dx', \tag{13.1}$$

$$C^{1(i)} = C^{(i)} - C^0, \; K(x) = -\nabla\nabla g(x). \tag{13.2}$$

Here, $g(x)$ is Green's function of the host medium. It follows from Eq. (13.1) that the field $\varepsilon(x)$ inside the ith inclusion is presented in the form

$$\varepsilon(x) = \varepsilon^{*i}(x) - \int_{v_i} K(x - x')C^{1(i)}\varepsilon(x')dx', \ x \in v_i, \tag{13.3}$$

$$\varepsilon^{*i}(x) = \varepsilon^0 - \sum_{k \neq i} \int_{v_k} K(x - x')C^{1(k)}\varepsilon(x')dx'. \tag{13.4}$$

It follows from these equations that the ith inclusion can be considered as an isolated one in the host medium (matrix) subjected to the external field $\varepsilon^{*i}(x)$ that is the sum of the external field ε^0 and the field induced in the region v_i by all surrounding inclusions.

13.1.1 Equation for the Effective Field

According to hypothesis H_2 of the effective field method, the effective field ε^* is assumed to be constant and the same for all the inclusions. Thus in Eq. (13.3), $\varepsilon^{*i} \approx \varepsilon^*$ for all i, and the field $\varepsilon(x)$ in the i the inclusion can be presented in the form

$$\varepsilon(x) = \Lambda^{\varepsilon(i)}(x)\varepsilon^*, \ x \in v_i, \ i = 1, 2, ..., \tag{13.5}$$

where the tensor $\Lambda^{\varepsilon i}(x)$ depends on the properties and shape of the ith inclusion. For construction of this tensor, the one-particle problem for the ith inclusion should be solved. We introduce the function

$$q(x) = C^1(x)\Lambda^\varepsilon(x), \tag{13.6}$$

where $\Lambda^\varepsilon(x)$ and $C^1(x)$ coincide with tensors $\Lambda^{\varepsilon(i)}(x)$ and $C^{1(i)}$ inside the ith inclusion. For the hypotheses of the EFM and Eqs. (13.1), (13.5), the field $\varepsilon(x)$ in the composite medium is approximated by the equation

$$\varepsilon(x) = \varepsilon^0 - \int K(x - x')q(x')V(x')dx'\varepsilon^*, \tag{13.7}$$

where $V(x)$ is the characteristic function of the region $V = \bigcup_i v_i$ occupied by the inclusions

$$V(x) = 1, x \in V, \ V(x) = 0, \ x \notin V. \tag{13.8}$$

We introduce the function

$$V\left(x';x\right)=\sum_{k\neq i}v_k\left(x'\right)\quad\text{if}\quad x\in v_i, \tag{13.9}$$

where $v_i(x)$ is the characteristic function of the region v_i, occupied by the ith inclusion. Using this function, the local exiting field acting on the inclusion at point $x\in V$ is presented in the form

$$\varepsilon^*(x)=\varepsilon^0-\int K\left(x-x'\right)q\left(x'\right)V\left(x';x\right)dx'\varepsilon^*,\ x\in V. \tag{13.10}$$

The average of the field $\varepsilon^*(x)$ by the condition that $x\in V$ results in the equation

$$\left\langle\varepsilon^*(x)|x\right\rangle=\varepsilon^0-\int K\left(x-x'\right)\left\langle q\left(x'\right)V\left(x';x\right)|x\right\rangle dx'\varepsilon^*. \tag{13.11}$$

Here $\langle\cdot\,|x\rangle$ is the averaging over ensemble realizations of the random set of inclusions by the condition that $x\in V$. It is reasonable to identify the mean $\langle\varepsilon^*(x)|x\rangle$ with the effective field ε^*

$$\left\langle\varepsilon^*(x)|x\right\rangle=\varepsilon^*. \tag{13.12}$$

As a result, Eq. (13.11) becomes a closed equation for the effective field ε^*.

$$\varepsilon^*=\varepsilon^0-\int K\left(x-x'\right)\left\langle q\left(x'\right)V\left(x';x\right)|x\right\rangle dx'\varepsilon^*. \tag{13.13}$$

The conditional mean under the integral in Eq. (13.11) is presented in the form

$$\left\langle q\left(x'\right)V\left(x';x\right)|x\right\rangle=p\left\langle q(x')|x'\right\rangle\Psi\left(x',x\right), \tag{13.14}$$

where

$$p=\langle V(x)\rangle,\ \Psi\left(x',x\right)=\frac{\left\langle V\left(x';x\right)|x\right\rangle}{\langle V(x)\rangle}. \tag{13.15}$$

Here, it is assumed that the inclusion shapes and properties are statistically independent on inclusion spatial locations, p is the inclusion volume fraction. For a homogeneous in space random set of inclusions, the function $\Psi\left(x,x'\right)$ depends on the difference $x-x'$

$$\Psi\left(x',x\right)=\Psi\left(x-x'\right). \tag{13.16}$$

The conditional mean $\langle q(x)|x\rangle$ is defined by the equation

$$\langle q(x)|x\rangle=\frac{1}{p}\langle q(x)V(x)\rangle \tag{13.17}$$

Assuming the ergodic property of the random function $q(x)$, the ensemble average $\langle q(x)V(x)\rangle$ can be changed with the spacial average for a fixed typical realization of the inclusion set, and we have

$$\langle q(x)V(x)\rangle = \lim_{W\to\infty}\frac{1}{W}\int_W C^1(x)\Lambda^\varepsilon(x)V(x)dx =$$

$$= \lim_{W\to\infty}\frac{1}{W}\sum_{i=1}^{N}\int_{v_i} C^{1(i)}(x)\Lambda^{\varepsilon i}(x)\mathrm{d}x\,. \tag{13.18}$$

Here, N is the number of inclusions in the region W that occupies the entire space in the limit $W\to\infty$. After averaging both sides of this equation once more over ensemble realizations of the random set of inclusions we obtain

$$\langle q(x)V(x)\rangle = \lim_{W\to\infty}\frac{\langle v\rangle\,N}{W}P^\varepsilon = pP^\varepsilon, \tag{13.19}$$

where $\langle v\rangle$ is the mean inclusion volume, and

$$p = \lim_{W\to\infty}\frac{\langle v\rangle\,N}{W} \tag{13.20}$$

The tensor P^ε is the following mean

$$P^\varepsilon = \frac{1}{\langle v\rangle}\left\langle\int_v C^1\Lambda^\varepsilon(x)dx\right\rangle. \tag{13.21}$$

As a result, the solution of Eq. (13.13) for the effective field ε^* takes the form

$$\varepsilon^* = \left(I - pA^*P^\varepsilon\right)^{-1}\varepsilon^0, \tag{13.22}$$

where

$$A^* = -\int K(x)\,\Psi(x)dx. \tag{13.23}$$

13.1.2 Correlation Function $\Psi(x)$

In this subsection, we consider the function $\Psi(x)$ in Eqs. (13.15), (13.23) in detail. The mean of a random function $f(x)$ by the condition that $x\in V$ is defined by the equation [1]

$$\langle f(x)|x\rangle = \frac{\langle f(x)V(x)\rangle}{\langle V(x)\rangle}. \tag{13.24}$$

For the conditional mean in Eq. (13.15), we have

$$\Psi(x) = \frac{\left\langle V\left(x'; x'+x\right)|x'+x\right\rangle}{\langle V(x)\rangle} = \frac{\left\langle V\left(x'; x'+x\right) V\left(x'+x\right)\right\rangle}{\langle V(x)\rangle^2}. \tag{13.25}$$

The numerator in this equation relates to the covariance $\left\langle V(x')V(x'+x)\right\rangle$ of the random function $V(x)$, and for $V(x) = \sum\limits_i v_i(x)$, we have

$$\left\langle V\left(x'\right) V\left(x'+x\right)\right\rangle = \tag{13.26}$$

$$= \lim_{W\to\infty}\frac{1}{W}\int\limits_W \left(\sum_i v_i\left(x'\right) v_i\left(x'+x\right) + \sum_{i,j\,(i\neq j)} v_i\left(x'\right) v_j\left(x'+x\right)\right) dx'.$$

Here, the integral $\int v_i\left(x'\right) v_i\left(x'+x\right)dx'$ is the volume of intersection of two identical regions v_i moved on vector x with respect to each other. The numerator $\left\langle V(x'; x'+x)V(x'+x)\right\rangle$ in Eq. (13.25) coincides with the covariance in Eq. (13.26) if the contributions of the realizations with points x and x' inside the same inclusion are omitted. The latter follows from the definition (13.9) of the function $V(x'; x)$. Therefore, taking away the first sum in Eq. (13.26), we yield for $\Psi(x)$ the equation

$$\Psi(x) = \frac{1}{p^2}\lim_{W\to\infty}\frac{1}{W}\int\limits_W \sum_{(i\neq j)} v_i\left(x'\right) v_j\left(x'+x\right)dx', \tag{13.27}$$

It follows from this equation that $\Psi(x)$ is a continuous function with the mean value equal to 1

$$\lim_{W\to\infty}\frac{1}{W}\int\limits_W \Psi(x)\mathrm{d}x = 1. \tag{13.28}$$

Because the inclusions do not intersect, point x cannot be covered by two different inclusions and therefore,

$$\Psi(x) = 0 \quad \text{when } x = 0. \tag{13.29}$$

The function $\Psi(x)$ consists of an oscillating function with the mean value equal to 1, and a function with a finite support. A sketch of $\Psi(x)$ for a homogeneous in space random set of spherical inclusions of radius a is shown in Fig. 13.4. Further, the function $\Psi(x)$ is called the correlation hole for a typical inclusion.

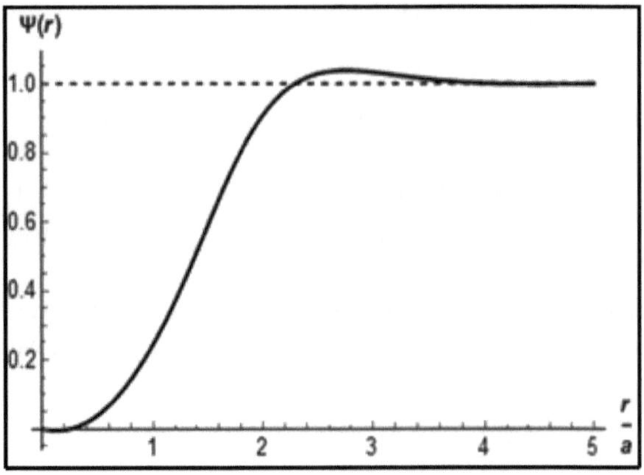

Fig. 13.4 A sketch of the correlation function $\Psi(x)$ for a homogeneous and isotropic in space random set of spherical inclusions of the radius a, $r = |x|$

The function $\Psi(x)$ depends on the geometrical properties of the inclusion distribution in the host medium. Symmetry (in the statistical sense) of the random set of inclusions defines symmetry of the function $\Psi(x)$. In particular, if the random set of inclusions is isotropic in space, $\Psi(x)$ is spherically symmetric, i.e., $\Psi(x) = \Psi(|x|)$. Deviation of the random set of inclusions from statistical isotropy results in anisotropy of the macro properties of the composite material in spite of isotropy of the matrix and inclusion phases (texture). In Fig. 13.5, a realization of a random set of inclusions is shown, where dashed lines indicate the regions around each inclusion where the probability to meet other inclusions is small. The shapes of these region determine symmetry of the correlation hole $\Psi(x)$.

In many cases, the texture symmetry can be described by a second rank tensor α. This tensor determines linear transformation $y = \alpha^{-1}x$ of the x-space that converts the function $\Psi(x)$ into spherically symmetric one

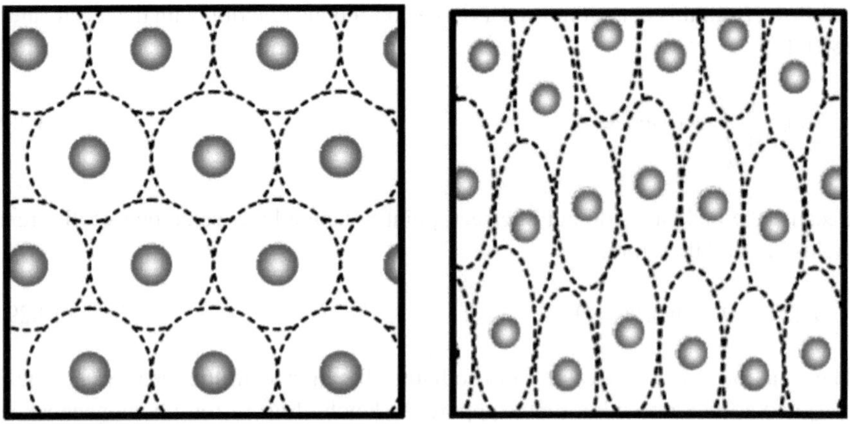

Fig. 13.5 Spatial distribution of inclusions in the case of the spherical and ellipsoidal symmetry of the function $\Psi(x)$. The dashed lines indicate the regions around each inclusion where the probability to meet other inclusions is small

$$\Psi(\alpha y) = \Psi'(|y|). \tag{13.30}$$

For calculation of the integral (13.23) we accept that the external field intensity ε^0 is fixed, and regularization of the integral operator with the kernel $K(x)$ on constants is defined by Eq. (10.28)

$$\int K(x)dx = 0. \tag{13.31}$$

Then, for the function $\Psi(x)$ with the properties (13.28)–(13.30), the integral A^* in Eq. (13.23) is calculated as follows (Appendix B.2)

$$A^* = -\int K(x)\,\Psi\,(x)\,\mathrm{d}x = \int K(x)\,(1 - \Psi(x))dx =$$

$$= A(\alpha) + (\det \alpha)\; pv \int K(\alpha\, y)\left(1 - \Psi'(|\, y|)\right) dy\,. \tag{13.32}$$

Here, the tensor $A(\alpha)$ is defined in Eq. (4.89), the last integral is understood in the sense of Cauchy principal value. This integral exists because the function $K(\alpha\, y)$ satisfies the equation [2, 3]

$$\int_{\Omega_1} K(\alpha\, y)d\Omega = 0, \tag{13.33}$$

where Ω_1 is the surface of unit sphere in the y-space. For Eq. (13.33), the last integral in Eq. (13.32) vanishes for any function $1 - \Psi'(|\, y|)$ that tends to zero at infinity faster than $|y|^{-1}$. Thus if Eq. (13.30) holds, the integral (13.23) takes the form

$$A^* = -\int K(x)\,\Psi\,(x)\,\mathrm{d}x = A(\alpha), \tag{13.34}$$

$$A(\alpha) = \frac{1}{4\pi}\int_{\Omega_1} \widetilde{K}(\alpha^{-1}k)\mathrm{d}\Omega. \tag{13.35}$$

Here, $\widetilde{K}(k)$ is Fourier transform of the function $K(x)$. Note that integral (13.34) depends neither on the detailed behavior of the function $\Psi(x)$ nor on the size of the correlation hole because $\widetilde{K}(k)$ is a homogeneous function of the order of zero. Only the shape of the correlation hole affects the value of the integral A^*. Substituting Eq. (13.34) in Eq. (13.22), we obtain the final equation for the effective field ε^* acting on each inclusion in the composite

$$\varepsilon^* = \Lambda^{\varepsilon *}\varepsilon^0,\; \Lambda^{\varepsilon *} = (I - pA(\alpha)P^{\varepsilon})^{-1}. \tag{13.36}$$

13.1.3 The Homogenization Problem

Let a homogeneous host medium with the property tensor C^0 contains a set of identical inclusions of the same orientations with the property tensor C. In the framework of the effective field method, the field $\varepsilon(x)$ inside each inclusion is defined in Eq. (13.5). Taking into account the equations

$$q(x) = C^1\varepsilon(x) = C^1\Lambda^{\varepsilon}(x)\varepsilon^*, \tag{13.37}$$

$$B^1\sigma(x) = (C^{-1} - (C^0)^{-1})C\varepsilon(x) = -B^0 q(x), \tag{13.38}$$

and Eqs. (10.13), (10.15) for $\widehat{C} = C^0, \widehat{\varepsilon} = \varepsilon^0, \widehat{\sigma} = C^0\varepsilon^0$, the fields $\varepsilon(x)$ and $\sigma(x)$ in the composite are approximated by the integrals

$$\varepsilon(x) = \varepsilon^0 - \int K(x - x')q(x')V(x')dx', \tag{13.39}$$

$$\sigma(x) = \sigma^0 - \int S(x - x')B^0 q(x')V(x')dx'. \tag{13.40}$$

After averaging these equations over ensemble realizations of the random set of inclusions, we yield

$$\langle\varepsilon(x)\rangle = \varepsilon^0 - \int K(x - x')dx' \langle q(x)V(x)\rangle, \tag{13.41}$$

$$\langle\sigma(x)\rangle = C^0\varepsilon^0 - \int S(x - x')dx' B^0 \langle q(x)V(x)\rangle. \tag{13.42}$$

It is taken into account that for homogeneous in space random sets of inclusions and a constant external field, the mean $\langle q(x)V(x)\rangle$ is constant. For Eq. (13.19), we have

$$\langle q(x)V(x)\rangle = pP^{\varepsilon}\varepsilon^* = pP^{\varepsilon}\left(I - pA^*P^{\varepsilon}\right)^{-1}\varepsilon^0. \tag{13.43}$$

$$P^{\varepsilon} = \frac{1}{v}C^1\int_v \Lambda^{\varepsilon}(x)dx. \tag{13.44}$$

If the external field intensity ε^0 is fixed, the integrals in Eqs. (13.41), (13.42) are defined in Eqs. (10.31)

$$\int K(x - x')dx' = 0, \quad \int S(x - x')dx' = -C^0. \tag{13.45}$$

Taking into account Eq. (13.43), we have

$$\langle\varepsilon(x)\rangle = \varepsilon^0,\ \langle\sigma(x)\rangle = \left(C^0 + pP^\varepsilon\left(I - pA^*P^\varepsilon\right)^{-1}\right)\langle\varepsilon(x)\rangle\,. \tag{13.46}$$

where the tensor P^ε is in Eq. (13.44), tensor A^* is the integral in Eq. (13.23). It follows from this equation that tensor C^*

$$C^* = C^0 + pP^\varepsilon\left(I - pA^*P^\varepsilon\right)^{-1}. \tag{13.47}$$

relates the averages of the field intensity $\langle\varepsilon\rangle$ and flux density $\langle\sigma\rangle$ over ensemble realizations of the random set of inclusions. Therefore, C^* is the tensor of the effective conductivity of the composite.

The equation for the effective property tensor can be derived on the basis of equation for the effective flux density $\sigma(x)$. Similar to the effective field intensity ε^*, we introduce the effective flux density σ^* that acts on each inclusion in the composite. Then, the flux density $\sigma(x)$ inside the ith inclusion takes the form

$$\sigma(x) = \Lambda^{\sigma(i)}(x)\sigma^*,\ x \in v_i,\ i = 1, 2, ..., \tag{13.48}$$

were tensor $\Lambda^{\sigma(i)}(x)$ is to be constructed from the solution of the one-particle problem for the ith inclusion. Field intensity $\varepsilon(x)$ and flux density $\sigma(x)$ in the medium are approximated by the equations

$$\varepsilon(x) = \varepsilon^0 + \int K(x - x')C^0P^\sigma(x')V(x')dx'\sigma^*, \tag{13.49}$$

$$\sigma(x) = \sigma^0 + \int S(x - x')P^\sigma(x')V(x')dx'\sigma^*, \tag{13.50}$$

$$P^\sigma(x) = B^1\Lambda^\sigma(x),\ B^1 = B - B^0,\ B = C^{-1},\ B^0 = \left(C^0\right)^{-1}. \tag{13.51}$$

Here, the function $\Lambda^\sigma(x)$ coincides with $\Lambda^{\sigma(i)}(x)$ inside ith inclusion. The local exciting flux density $\sigma^*(x)$ acting at point $x \in V$ has the form

$$\sigma^*(x) = \sigma^0 + \int S(x - x')P^\sigma(x')V(x; x')dx'\sigma^*,\ x \in V, \tag{13.52}$$

where the function $V(x; x')$ is defined in Eq. (13.9). After averaging this equation by the condition $x \in V$ and accepting that

$$\left\langle\sigma^*(x)|x\right\rangle = \sigma^*, \tag{13.53}$$

we obtain a closed equation for the effective flux density σ^*

$$\sigma^* = \sigma^0 - pD^* P^\sigma \sigma^*, \tag{13.54}$$

$$P^\sigma = \frac{1}{v}\int_v P^\sigma(x)dx, \quad D^* = -\int S(x)\Psi(x)dx. \tag{13.55}$$

Therefore, the effective field σ^* is

$$\sigma^* = \Lambda^{\sigma*}\sigma^0, \Lambda^{\sigma*} = \left(I + pD^* P^\sigma\right)^{-1}, \tag{13.56}$$

If the function $\Psi(x)$ (correlation hole) has ellipsoidal symmetry, the tensor D^* in this equation is

$$D^* = D(\alpha) = C^0 A(\alpha) C^0 - C^0, \tag{13.57}$$

where $A(\alpha)$ is defined in Eq. (13.35), and the tensor α determines the shape of the correlation hole.

If we introduce the function

$$m(x) = B^1\sigma(x), \tag{13.58}$$

and take into account the equation

$$C^1\varepsilon(x)V(x) = (C - C^0)C^{-1}\sigma(x)V(x) = -C^0 m(x)V(x), \tag{13.59}$$

the equations for the field intensity $\varepsilon(x)$ and flux density $\sigma(x)$ in the composite are approximated by the equations

$$\varepsilon(x) = \varepsilon^0 + \int K(x - x')C^0 m(x')V(x')dx', \tag{13.60}$$

$$\sigma(x) = \sigma^0 + \int S(x - x')m(x')V(x')dx', \tag{13.61}$$

where

$$m(x) = P^\sigma \Lambda^{\sigma*}\sigma^0, \ \Lambda^{\sigma*} = \left(I + pD^* P^\sigma\right)^{-1}. \tag{13.62}$$

After averaging Eqs. (13.60), (13.61) over ensemble realizations of the random set of inclusions, we obtain

$$\langle\varepsilon(x)\rangle = \varepsilon^0 + \int K(x - x')dx' C^0 \langle m(x)V(x)\rangle\,, \tag{13.63}$$

$$\langle\sigma(x)\rangle = \sigma^0 + \int S(x - x')dx' \langle m(x)V(x)\rangle\,. \tag{13.64}$$

where the mean $\langle m(x)V(x)\rangle$ is constant. Then, for a fixed external field flux σ^0, Eqs. (10.28) hold

$$\int S(x)dx = 0, \int K(x)dx = B^0, \tag{13.65}$$

and Eqs. (13.60), (13.61) take the forms

$$\langle\sigma(x)\rangle = \sigma^0, \langle\varepsilon(x)\rangle = \left(B^0 + pP^{\sigma}\Lambda^{\sigma *}\right)\langle\sigma(x)\rangle\,. \tag{13.66}$$

It follows from this equation that the effective resistivity tensor B^* of the composite is

$$B^* = B^0 + pP^{\sigma}\left(I + pD^* P^{\sigma}\right)^{-1}. \tag{13.67}$$

For the correlation hole of ellipsoidal shape, $D^* = D(\alpha)$ and

$$B^* = B^0 + pP^{\sigma}(I + pD(\alpha)P^{\sigma})^{-1}. \tag{13.68}$$

Equations (13.47) and (13.68) are the reciprocal equations of the effective field method for the effective property tensors of matrix composites.

13.2 Effective Property Tensors of Composites with Ellipsoidal Inclusions

13.2.1 Conductive Medium with Identical Ellipsoidal Inclusions

Let a homogeneous host medium with the property tensor C^0 contains a set of identical inclusions of the same orientations with the property tensor C. In the case of ellipsoidal inclusions, the solution of the one-particle problem is defined in Eq. (4.82), and tensor P^{ε} in Eq. (13.47) takes the form

$$P^{\varepsilon} = C^1\left(I + A(a)C^1\right)^{-1}, C^1 = C - C^0. \tag{13.69}$$

where the tensor $A(a)$ in Eq. (4.89) depends on the aspect ratio of the ellipsoids. The effective conductivity tensor of the medium with inclusions is

$$C^* = C^0 + pP^\varepsilon(I - pA(\alpha)P^\varepsilon)^{-1}. \tag{13.70}$$

Equations (13.62), (13.66) result in the equation for the effective resistivity tensor B^* of the composite

$$B^* = B^0 + pP^\sigma(I + pD(\alpha)P^\sigma)^{-1}, \tag{13.71}$$

where the tensor P^σ is

$$P^\sigma = B^1\left(I - D(a)B^1\right)^{-1},\ D(a) = C^0A(a)C^0 - C^0. \tag{13.72}$$

Note important properties of Eqs. (13.70) and (13.71)

1. In Eqs. (13.70) and (13.71), the symmetry of the texture affects the tensor $A(\alpha)$ and $D(\alpha)$, meanwhile shapes and properties of the inclusions determine the tensor P^ε and P^σ.

2. Because $A(\alpha)$, $D(\alpha)$, P^ε and P^σ are symmetric tensors, C^* and B^* in Eqs. (13.70) and (13.71) are also symmetric. It becomes clear after expansion of these tensors in terms of the inclusion volume fraction p. For instance,

$$C^0 + pP^\varepsilon(I - pA(\alpha)P^\varepsilon)^{-1} =$$

$$= C^0 + pP^\varepsilon + p^2P^\varepsilon A(\alpha)P^\varepsilon + p^3P^\varepsilon A(\alpha)P^\varepsilon A(\alpha)P^\varepsilon + \ldots \tag{13.73}$$

Because P^ε and $A(\alpha)$ are symmetric tensors, all terms in this series are symmetric.

3. The tensors C^* and B^* in Eqs. (13.70) and (13.71) relate by the equation

$$\left(C^*\right)^{-1} = B^*. \tag{13.74}$$

4. If the tensors $A(a)$ and $A(\alpha)$ coincide, the equations for the effective property tensors C^* and B^* are simplified

$$C^* = C^0 + pC^1\left(I + (1-p)A(a)C^1\right)^{-1}. \tag{13.75}$$

$$B^* = B^0 + pB^1\left(I - (1-p)D(a)B^1\right)^{-1}. \tag{13.76}$$

The condition $A(a) = A(\alpha)$ means that the symmetry of the correlation hole $\Psi(x)$ coincides with the symmetry of the inclusions, and therefore, the tensors α and a coincide.

5. For $p = 1$, Eqs. (13.75), (13.76) results in the equation $C^* = C$ that is formally correct.

Let the matrix material be isotropic ($C^0_{\alpha\beta} = c_0\delta_{\alpha\beta}$), and the inclusions be balls with the isotropic conductivity tensor $C_{\alpha\beta} = c\delta_{\alpha\beta}$. If the inclusion distribution in space is homogeneous and isotropic, the function $\Psi(x)$ is spherically symmetric, and the tensor $A(a)$ and $A^*(\alpha)$ are

$$A_{\alpha\beta}(a) = A^*_{\alpha\beta}(\alpha) = \frac{1}{3c_0}\delta_{\alpha\beta}. \tag{13.77}$$

As a result, the composite is isotropic ($C^*_{\alpha\beta} = c_*\delta_{\alpha\beta}$), and the effective conductivity coefficient c_* is

$$c_* = c_0 + p(c - c_0)\left(1 + (1-p)\frac{c - c_0}{3c_0}\right)^{-1}. \tag{13.78}$$

For highly conductive inclusions ($c \to \infty$),

$$c_* = \frac{1+2p}{1-p}c_0, \tag{13.79}$$

and for non-conductive inclusions, $c = 0$ and we have

$$c_* = \frac{2(1-p)}{2+p}c_0. \tag{13.80}$$

Equations (13.78)–(13.80) coincide with Eqs. (11.31)–(11.33) of Kerner's version of the effective medium method in application to isotropic media with spherical isotropic inclusions. Comparisons of the predictions of the effective field and effective medium methods are presented in Figs. 11.5 and 11.6. Equations (13.78)–(13.80) don't predict existence of percolation threshold in the effective conductivity of matrix-inclusion composites that are usually observed in experiments with matrix composites containing inclusions of large (small) conductivities.

Unlike Kerner's method, the effective field method can be extended to the case of ellipsoidal inclusions without difficulties. For a composite with spheroidal inclusions of the same orientations and aspect ratios γ, the tensor $A(a)$ in Eq. (13.70) takes the form in Eqs. (4.207), (4.208)

$$A(a) = \frac{1}{c_0}\left(A_1(\gamma)\theta_{\alpha\beta} + A_3(\gamma)n_\alpha n_\beta\right), \theta_{\alpha\beta} = \delta_{\alpha\beta} - n_\alpha n_\beta, \tag{13.81}$$

$$A_1(\gamma) = \beta(\gamma), A_3(\gamma) = 1 - 2\beta(\gamma), \tag{13.82}$$

$$\beta(\gamma) = \begin{cases} \dfrac{\gamma\left(\arccos(\gamma) - \gamma\sqrt{1-\gamma^2}\right)}{2(1-\gamma^2)^{3/2}}, & \gamma < 1 \\ \dfrac{\gamma\left(\gamma\sqrt{\gamma^2-1} - \operatorname{arccos\,h}(\gamma)\right)}{2(\gamma^2-1)^{3/2}}, & \gamma > 1 \end{cases}. \tag{13.83}$$

where n is the unit vector of inclusion rotation axes. Note that for $\gamma = 0$, $A_1 = 0$, $A_3 = 1$, and in the limit $\gamma \to \infty$, $A_1 \to 1/2$, $A_3 \to 0$.

The composite material with aligned spheroids is transversely isotropic, and its tensor of the effective conductivity $C^*_{\alpha\beta}$ is

$$C^*_{\alpha\beta} = c^*_1 \theta_{\alpha\beta} + c^*_3 n_\alpha n_\beta, \tag{13.84}$$

where the conductivity coefficients c^*_1 and c^*_3 are

$$c^*_1 = c_0 + pP_1 \left(1 - pA_1(\gamma_*)P_1\right)^{-1}, \tag{13.85}$$

$$c^*_3 = c_0 + pP_3 \left(1 - pA_3(\gamma_*)P_3\right)^{-1}, \tag{13.86}$$

$$P_1 = \left(\frac{1}{c - c_0} + \frac{1}{c_0} A_1(\gamma)\right)^{-1}, \quad P_3 = \left(\frac{1}{c - c_0} + \frac{1}{c_0} A_3(\gamma)\right)^{-1}. \tag{13.87}$$

Here, the coefficients $A_1(\gamma_*)$, $A_3(\gamma_*)$ are determined in Eqs. (13.82), (13.83) with the aspect ratio γ_* of the correlation hole $\Psi(x)$. If $\gamma = \gamma_*$, Eqs. (13.85) and (13.86) are simplified

$$c^*_1 = c_0 + p\left(\frac{1}{c - c_0} + \frac{(1-p)}{c_0} A_1\right)^{-1}, \tag{13.88}$$

$$c^*_3 = c_0 + p\left(\frac{1}{c - c_0} + \frac{(1-p)}{c_0} A_3\right)^{-1}. \tag{13.89}$$

For $\gamma = 0$, the composite contains parallel disk-like inclusions and the conductivity coefficients of such a material are

$$c^*_1 = c_0 + p(c - c_0), \tag{13.90}$$

$$c^*_3 = c_0\left(1 + \frac{2p(c - c_0)}{2c_0 + (1-p)(c - c_0)}\right). \tag{13.91}$$

The limit $\gamma \to \infty$ corresponds to the composite with a set of infinite circular fibers, and in this case,

$$c^*_1 = c_0\left(1 + p\frac{2p(c - c_0)}{c + c_0 - p(c - c_0)}\right), \tag{13.92}$$

$$c^*_3 = c_0 + p(c - c_0). \tag{13.93}$$

13.2.2 Conductive Medium with Ellipsoidal Inclusions of Random Orientations

In this subsection, we consider a homogeneous conductive medium with identical ellipsoidal inclusion randomly distributed over orientations. In the basis of the principal axes of each ellipsoid, the conductivity tensor C of the inclusions is the same. But in the global basis, components of the tensor C depends on the inclusion orientation. Tensor $A(a)$ in the solution of the one-particle problem (13.69) also depends on the inclusion orientation.

Let a host medium with conductivity tensor C^0 contain spheroidal transverse isotropic inclusions. The isotropy axis of tensor C coincides with the rotation axis of each inclusion. In this case, the orientations of the inclusions are defined by unit vectors n of their rotation axes. For composite with spheroidal inclusions of random orientations, a modified version of the effective field method can be proposed. This version reflects better specific features of the local exciting fields acting on the inclusions of different orientations. We assume that the effective field that acts on each inclusion depends on the inclusion orientation $\varepsilon^* = \varepsilon^*(n)$, and all the inclusions of the orientation n are subjected to the same constant effective field $\varepsilon^*(n)$. As a result, the field inside the ith inclusions v_i of the orientation $n^{(i)}$ is presented in the form

$$\varepsilon^{(i)} = \Lambda^{\varepsilon(i)}(n^{(i)})\varepsilon^*(n^{(i)}), x \in v_i. \tag{13.94}$$

Here the tensor $\Lambda^{\varepsilon(i)}(n^{(i)})$ depends on the inclusion orientation. We introduce the function

$$P^{\varepsilon}(x) = C^1(n)\Lambda^{\varepsilon}(n), \tag{13.95}$$

Here, $n = n(x)$, and $P^{\varepsilon}(x)$ coincides with the tensor $C^{1(i)}\Lambda^{\varepsilon(i)}(n^{(i)})$ inside the ith inclusion, $C^{1(i)} = C^{(i)} - C^0$. Then, the field $\varepsilon(x)$ in the composite is approximated by the following integral

$$\varepsilon(x) = \varepsilon^0 - \int K(x - x')P^{\varepsilon}(x')\varepsilon^*(n')V(x')dx', \tag{13.96}$$

where unit vector $n' = n(x')$ defines the orientation of the inclusion at the point x'.The local exiting field $\varepsilon^*(x)$ acting on the inclusion at point $x \in V$ is approximated by the equation

$$\varepsilon^*(x) = \varepsilon^0 - \int K\left(x - x'\right) P^{\varepsilon}\left(x'\right) \varepsilon^*(n')V\left(x'; x\right)dx', x \in V, \tag{13.97}$$

where the function $V\left(x'; x\right)$ is defined in Eq. (13.9). The average of the field $\varepsilon^*(x)$ by the condition that point x belongs to inclusions of the orientation n results in the equation

$$\left\langle \varepsilon^*(x)|x,n\right\rangle = \varepsilon^0 - \int K\left(x-x'\right)\left\langle P^\varepsilon\left(x'\right)\varepsilon^*(n')V\left(x';x\right)|x,n\right\rangle dx'. \tag{13.98}$$

Here $\langle\cdot\,|x,n\rangle$ is averaging by the condition that points x is in the inclusions of the orientation n. After identifying the mean $\langle\varepsilon^*(x)|x,n\rangle$ with the effective field $\varepsilon^*(n)$

$$\left\langle \varepsilon^*(x)|x,n\right\rangle = \varepsilon^*(n), \tag{13.99}$$

Equation (13.98) becomes a closed equation for the effective field $\varepsilon^*(n)$

$$\varepsilon^*(n) = \varepsilon^0 - \int K\left(x-x'\right)\left\langle P^\varepsilon\left(x'\right)\varepsilon^*(n')V\left(x';x\right)|x,n\right\rangle dx'. \tag{13.100}$$

For statistical independence of the properties and shapes of the inclusions on their spacial positions, the conditional mean under the integral in this equation is presented in the form

$$\left\langle P^\varepsilon\left(x'\right)\varepsilon^*(n')V\left(x';x\right)|x,n\right\rangle = p\left\langle P^\varepsilon(x')\varepsilon^*(n')|x'\right\rangle \Psi(x-x',n), \tag{13.101}$$

where

$$\Psi(x-x',n) = \frac{1}{p}\left\langle V\left(x';x\right)|x,n\right\rangle \tag{13.102}$$

is the function in Eq. (13.27) with the properties indicated in Eqs. (13.28), (13.29). If the correlation hole $\Psi(x,n)$ has the shape of an ellipsoid of the orientation n, the integral in Eq. (13.100) is

$$\int K\left(x-x'\right)\Psi(x-x',n)dx' = -A^*(n), \tag{13.103}$$

and the equation for $\varepsilon^*(n)$ take the form

$$\varepsilon^*(n) = \varepsilon^0 + pA^*(n)\left\langle P^\varepsilon(n)\varepsilon^*(n)\right\rangle. \tag{13.104}$$

Similar to the case of inclusions of the same orientates, we assume that there exists an ellipsoidal regions around each inclusion with small probability to meet other inclusions (Fig. 13.6). For an isotropic host medium, tensor $A^*(n) = A(n,\gamma_*)$, and $A(n,\gamma)$ is defined by the right-hand side of Eq. (13.81) with $\gamma = \gamma_*$, and γ_* is the aspect ratio of the correlation hole that corresponds to the inclusion of the orientation n. If the inclusion can be arbitrary close to each other without mutual intersections, $\gamma_* = \gamma$, where γ is the inclusion aspect ratio.

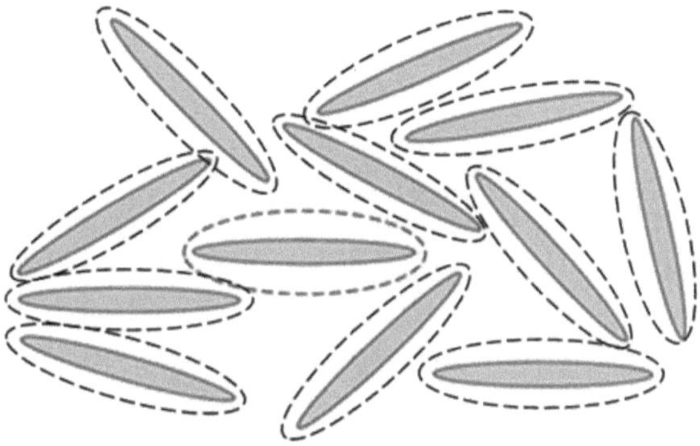

Fig. 13.6 The host medium with a set of inclusions of random orientations, the dashed lines show regions around the inclusions where the probability to meet other inclusions is small

After multiplication of Eq. (3.98) with the tensor $P^{\varepsilon}(n)$ and averaging its both parts over the ensemble of realization of the inclusion set, we yield a closed equation for the mean $\langle P^{\varepsilon}(n)\varepsilon^*(n)\rangle$

$$\left\langle P^{\varepsilon}(n)\varepsilon^*(n)\right\rangle = \langle P^{\varepsilon}(n)\rangle\, \varepsilon^0 + p\,\langle P^{\varepsilon}(n)A(n,\gamma_*)\rangle \left\langle P^{\varepsilon}(n)\varepsilon^*(n)\right\rangle. \tag{13.105}$$

The solution of this equation is

$$\left\langle P^{\varepsilon}(n)\varepsilon^*(n)\right\rangle = (I - p\,\langle P^{\varepsilon}(n)A(n,\gamma_*)\rangle)^{-1} \langle P^{\varepsilon}(n)\rangle\, \varepsilon^0. \tag{13.106}$$

Then, the approximate equations for the fields $\varepsilon(x)$ and $\sigma(x)$ take the forms

$$\varepsilon(x) = \varepsilon^0 - \int K(x-x')P^{\varepsilon}(n')\varepsilon^*(n')V(x')dx', \tag{13.107}$$

$$\sigma(x) = \sigma^0 - \int S(x-x')B^0 P^{\varepsilon}(n')\varepsilon^*(n')V(x')dx'. \tag{13.108}$$

After averaging these equations over ensemble realizations of the inclusion set, we obtain

$$\langle\varepsilon(x)\rangle = \varepsilon^0,\ \langle\sigma(x)\rangle = C^0\varepsilon^0 + p\left\langle P^{\varepsilon}(n)\varepsilon^*(n)\right\rangle. \tag{13.109}$$

Here, it is assumed that external field ε^0 is fixed in the problem, and Eqs. (13.45) hold. Substituting in Eq. (13.109) the mean $\left\langle C^1\Lambda^{\varepsilon}(n)\varepsilon^*(n)\right\rangle$ from Eq. (13.106), we yield

$$\langle\sigma(x)\rangle = C^* \langle\varepsilon(x)\rangle\,, \tag{13.110}$$

where

$$C^* = C^0 + p(I - p\,\langle P^{\varepsilon}(n)A(n,\gamma_*)\rangle)^{-1} \langle P^{\varepsilon}(n)\rangle \tag{13.111}$$

is the effective conductivity tensor of the composite. For ellipsoidal inclusions, the tensor $P^{\varepsilon}(n)$ in Eq. (13.111) has the form

$$P^{\varepsilon}(n) = C^1 \left(I + A(n,\gamma)C^1\right)^{-1}. \tag{13.112}$$

For an isotropic host medium with spheroidal inclusions, tensor $A(n,\gamma)$ in Eqs. (13.111), (13.112) is defined in Eqs. (13.81)–(13.83)

$$A(n,\gamma) = \frac{1}{c_0}\left(A_1(\gamma)\theta(n) + A_3(\gamma) n\otimes n\right). \tag{13.113}$$

Here, γ is the aspect ratio of the inclusions. For the dimensionless functions $A_1(\gamma)$, $A_3(\gamma)$ are in Eqs. (13.82), (13.83),the product $P^{\varepsilon}(n)A(n,\gamma_*)$ has the form

$$P^{\varepsilon}(n)A^*(n) = \frac{A_1(\gamma_*)(c-c_0)}{c_0 + A_1(\gamma)(c-c_0)}\theta(n) + \frac{A_3(\gamma_*)(c-c_0)}{c_0 + A_3(\gamma)(c-c_0)} n\otimes n. \tag{13.114}$$

If the distribution of inclusions over orientation is homogeneous, then

$$\langle n\otimes n\rangle = \frac{1}{3}I,\ I_{\alpha\beta} = \delta_{\alpha\beta}, \tag{13.115}$$

and the means $\langle P^{\varepsilon}(n)\rangle$ and $\langle P^{\varepsilon}(n)A(n,\gamma_*)\rangle$ are isotropic

$$\langle P^{\varepsilon}(n)\rangle = \left(\frac{2c_0(c-c_0)}{3\left(c_0 + A_1(\gamma)(c-c_0)\right)} + \frac{c_0(c-c_0)}{3\left(c_0 + A_3(\gamma)(c-c_0)\right)}\right) I, \tag{13.116}$$

$$\langle P^{\varepsilon}(n)A(n,\gamma_*)\rangle =$$

$$= \left(\frac{2A_1(\gamma_*)(c-c_0)}{3\left(c_0 + A_1(\gamma)(c-c_0)\right)} + \frac{A_3(\gamma_*)(c-c_0)}{3\left(c_0 + A_3(\gamma)(c-c_0)\right)}\right) I. \tag{13.117}$$

For homogeneous distribution of inclusions over orientations, the effective medium is isotropic. The effective coefficient of conductivity c_* can be found from Eqs. (13.111), (13.114) and (13.117). The tensor C^* depends on the aspect ratios of the inclusions and the correlation hole $\Psi(x,n)$.

Remark. Expansion of Eq. (13.111) for C^* in terms of the inclusion volume fraction p results in the series

$$C^* = C^0 + p\langle P^{\varepsilon}(n)\rangle + p^2 \langle P^{\varepsilon}(n)A(n,\gamma_*)\rangle \langle P^{\varepsilon}(n)\rangle + ... \tag{13.118}$$

For the tensor $P^{\varepsilon}(n)$ in Eq. (13.112), the term proportional to p^2 in this equation can be not symmetric with respect to index permutations.

13.2.3 Elastic Medium with a Set of Ellipsoidal Inclusions

In this subsection, we consider a homogeneous elastic medium with the stiffness tensor C^0 containing a set of identical ellipsoidal inclusions of the same orientation with the stiffness tensor C. Application of the effective field method for solution of the homogenization problem is similar to the case of conductivity. The hypothesis on a constant effective strain field ε^* acting on each inclusion results in the equation for ε^* similar to (13.36)

$$\varepsilon^* = (I - pA(\alpha)P^{\varepsilon})^{-1}\varepsilon^0. \tag{13.119}$$

$$P^{\varepsilon} = C^1\left(I + A(a)C^1\right)^{-1}, C^1 = C - C^0, \tag{13.120}$$

and $A(\alpha)$ and $A(a)$ are the forth rank tensor defined in Eq. (4.89), the second rank tensor a converts an ellipsoidal inclusion into a unit ball by the coordinate transformation $y = a^{-1}x$, the second rank tensor α transforms the ellipsoidal correlation hole $\Psi(x)$ into a spherically symmetric function. The tensor P^{ε} origins to the solution of the one-particle problem of elasticity for an inclusion in a homogeneous elastic medium subjected to a constant external field in Eq. (4.82). The equation for the effective elastic stiffness tensor C^* of the composite coincides with Eqs. (13.70)

$$C^* = C^0 + pP^{\varepsilon}(I - pA(\alpha)P^{\varepsilon})^{-1}. \tag{13.121}$$

Let the shape of the inclusion and the correlation hole $\Psi(x)$ coincide. Then,

$$A(\alpha) = A(a), \tag{13.122}$$

and Eq. (13.121) for C^* is simplified

$$C^* = C^0 + pC^1\left(I + (1-p)A(a)C^1\right)^{-1}. \tag{13.123}$$

In the case of elasticity, the reciprocal equations for the effective stress field σ^* and the effective compliance tensor B^* are similar to Eqs. (13.56), (13.68)

$$\sigma^* = (I + pD(\alpha)P^{\sigma})^{-1}\sigma^0, \tag{13.124}$$

$$B^* = B^0 + pP^{\sigma}(I + pD(\alpha)P^{\sigma})^{-1}. \tag{13.125}$$

where

$$B^0 = \left(C^0\right)^{-1}, B^1 = B - B^0, B = C^{-1}, \tag{13.126}$$

and forth rank tensors P^{σ} and $D(a)$ are

$$P^{\sigma} = B^1 \left(I - D(a)B^1\right)^{-1}, \tag{13.127}$$

$$D(a) = C^0 A(a) C^0 - C^0. \tag{13.128}$$

If the shape of the inclusions concede with the shape of the correlation hole $\Psi(x)$, $D(a) = D(\alpha)$, and

$$B^* = B^0 + pB^1\left(I - (1-p)D(a)B^1\right)^{-1} \tag{13.129}$$

The tensors C^* and B^* in Eqs. (13.123), (13.129) relate by the equation

$$\left(C^*\right)^{-1} = B^*. \tag{13.130}$$

The EFM allows us to estimate concentration of elastic fields on inclusions in matrix composite materials. According to the hypotheses of the method, each inclusion is subjected to the constant external effective field ε^*. It follows from the solution of the solution of the one-particle problem (Sect. 3.3) that the strain and stress tensors inside the inclusions are

$$\varepsilon^+ = \left(I + A(a)C^1\right)^{-1}\varepsilon^*, \sigma^+ = \left(I - D(a)B^1\right)^{-1}\sigma^*. \tag{13.131}$$

Interactions between inclusions are taken into account via the effective external fields ε^*, σ^* acting on the inclusions. The strain and stress concentrations in the matrix phase at the points of inclusion boundaries Ω are defined in Eqs. (4.66) of Chap. 3.

$$\varepsilon^-(n) = F^{\varepsilon}(n)\varepsilon^+, \sigma^-(n) = F^{\sigma}(n)\varepsilon^+. \tag{13.132}$$

Here, n is the external normal to the inclusion surface at point $x \in \Omega$, $F^{\varepsilon}(n)$, $F^{\sigma}(n)$ are stress concentration factors in Eqs. (4.66).

Let a composite material consist of an isotropic matrix and a set of isotropic identical spheroidal inclusions homogeneously distributed in the matrix. If there is no restriction for the inclusion locations (except their intersections), the shape of the correlation hole $\Psi(x)$ and the shape of the inclusions coincide. Hence $A(a) = A(\alpha) = A(\gamma)$, where γ is the aspect ratio of the spheroids. In the E-basis, the tensor $A(\gamma)$ is defined in Eqs. (4.140), (4.141). The same tensor in the P-basis is

$$A(\gamma) = A_1' P^2 + A_2'\left(P^1 - \frac{1}{2}P^2\right) + A_3'(P^3 + P^4) + A_5' P^5 + A_6' P^6, \tag{13.133}$$

$$A_1' = \frac{1}{2\mu_0}(\beta_1 - 4\kappa_0\beta_2), \ A_2' = \frac{1}{\mu_0}(\beta_1 - 2\kappa_0\beta_2), \tag{13.134}$$

$$A_3' = -\frac{\kappa_0}{\mu_0}(\beta_1 - 4\beta_2),\ A_5' = \frac{1}{\mu_0}\left(1 - \beta_1 - 4\kappa_0\left(\beta_2 + \beta_3\right)\right), \tag{13.135}$$

$$A_6' = \frac{1}{\mu_0}(1 - 2\beta_1 - \kappa_0\left(3\beta_2 + 6\beta_3 + \beta_4\right)). \tag{13.136}$$

Here, $P^i = P^i(n)$ and n is unit vector of the rotation axes of the spheroids, the functions $\beta_i = \beta_i(\gamma)$ are defined in Eqs. (4.140)–(4.142).

In the P-basis, the effective stiffness tensor C^* of a transverse isotropic medium with a set of identical spheroidal inclusions is presented in the form

$$C^* = k_* P^2 + 2m_*\left(P^1 - \frac{1}{2}P^2\right) + \lambda_*\left(P^3 + P^4\right) + 4\mu_* P^5 + n_* P^6, \tag{13.137}$$

In this notation, k_* is the plane strain bulk modulus for lateral dilatation without longitudinal extension, m_* is the shear modulus for any transversal to the n direction, μ_* is the longitudinal shear modulus, n_* is the modulus of longitudinal uniaxial extension, λ_* is the associated cross-modulus. These five independent elastic moduli of a transversely isotropic medium relate to five components c_{ij}^* of the matrix presentation of this tensor in Voigt's notations [4]

$$c_{11}^* = k_* + m_*,\ c_{12}^* = k_* - m_*,\ c_{13}^* = \lambda_*,\ c_{33}^* = n_*, \tag{13.138}$$

$$c_{23}^* = c_{13}^*,\ c_{44}^* = c_{55}^* = \mu_*,\ c_{66}^* = m_*. \tag{13.139}$$

The elastic constants $k_*, m_*, \lambda_*, \mu_*, n_*$ depend on the volume fraction p and the aspect ratio γ of the inclusions by the equations

$$k_* = k_0 + \frac{p}{\Delta}(k_1 + \mu_1(3\lambda_1 + 2\mu_1)A_6'(\gamma)(1-p)), \tag{13.140}$$

$$m_* = \mu_0 + \frac{p\mu_1}{1 + 2(1-p)\mu_1 A_2'(\gamma)}, \tag{13.141}$$

$$\lambda_* = \lambda_0 + \frac{p}{\Delta}(\lambda_1 + 2\mu_1(3\lambda_1 + 2\mu_1)A_3'(\gamma)(1-p)), \tag{13.142}$$

$$\mu_* = \mu_0 + \frac{p\mu_1}{1 + (1-p)\mu_1 A_5'(\gamma)}, \tag{13.143}$$

$$n_* = n_0 + \frac{p}{\Delta}(n_1 + 4\mu_1(3\lambda_1 + 2\mu_1)A_1'(\gamma)(1-p)), \tag{13.144}$$

$$\Delta = (1 - (-4\lambda_1 A_3'(\gamma) - 4k_1 A_1'(\gamma) - n_1 A_6'(\gamma))(1-p) -$$
$$- 4\mu_1(3\lambda_1 + 2\mu_1)\left(A_3'(\gamma)^2 - A_1'(\gamma)\,A_6'(\gamma)\right)(1-p)^2. \tag{13.145}$$

Here, λ_0, μ_0 are Lame's parameters of the matrix material, λ, μ are the same for the inclusions,

$$\lambda_1 = \lambda - \lambda_0, \mu_1 = \mu - \mu_0, m_1 = m - m_0, n_1 = n - n_0. \tag{13.146}$$

For the isotropic material, the constant k_0, m_0, n_0 are

$$k_0 = \lambda_0 + \mu_0, m_0 = \mu_0, n_0 = \lambda_0 + 2\mu_0. \tag{13.147}$$

Similar relations hold for the elastic constants of the inclusion material.

For absolutely rigid inclusions ($\lambda, \mu \to \infty$), dependencies of the effective elastic constants on the volume fraction and inclusion aspect ratios are shown in Figs. 13.7, 13.8 and 13.9.

The same dependencies of the effective elastic constants for spheroidal pores ($\lambda = \mu = 0$) are shown in Figs. 13.10, 13.11 and 13.12.

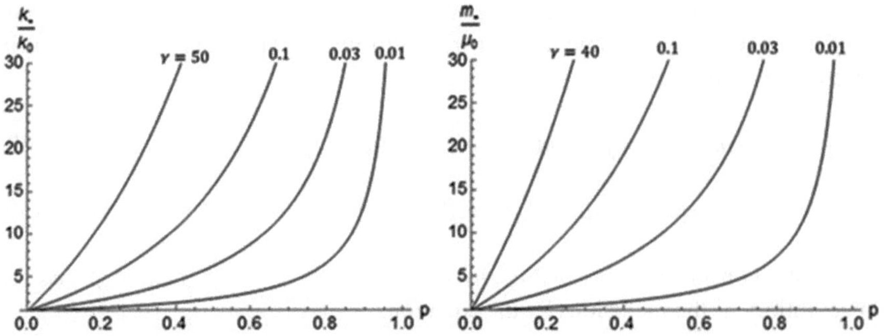

Fig. 13.7 Dependencies of the effective elastic coefficients K_* and m_* of the transverse isotropic composite with an isotropic matrix and absolutely rigid spheroidal inclusions of the same orientations on the inclusion aspect ratios γ and volume fractions p; k_0, μ_0 are bulk and shear moduli of the host medium

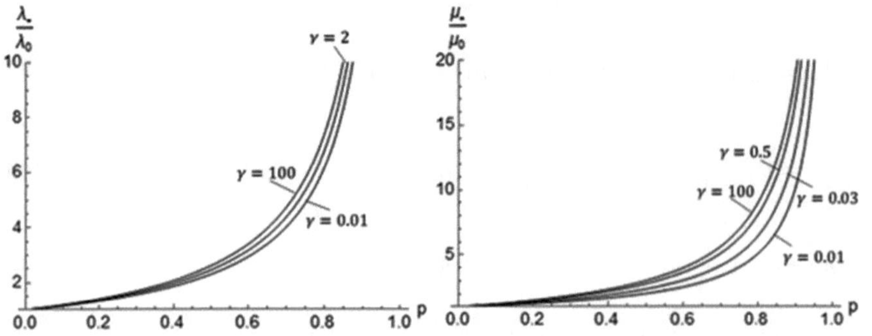

Fig. 13.8 The same as in Fig. 13.7 for the coefficients λ_*, μ_*; λ_0, μ_0 are Lame's parameters of the host medium

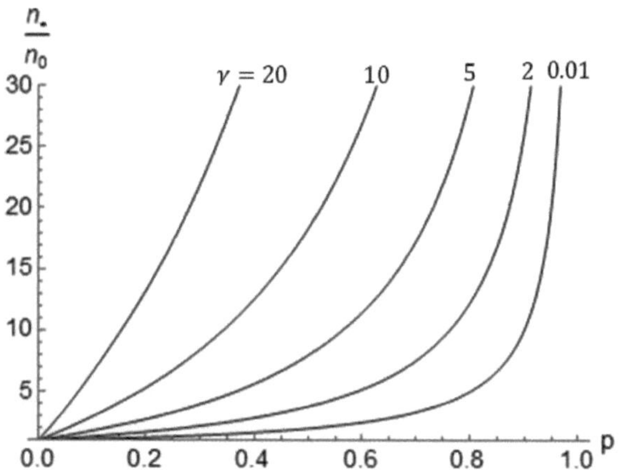

Fig. 13.9 The same as in Figs. 13.7 and 13.8 for the coefficient n_*, $n_0 = \lambda_0 + 2\mu_0$

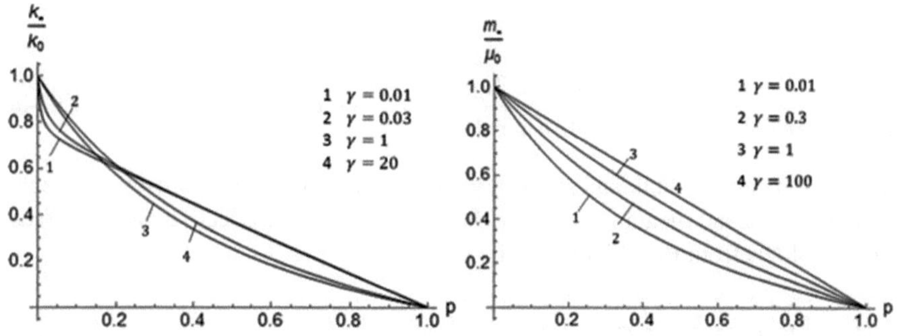

Fig. 13.10 Dependencies of the effective elastic coefficients K_* and m_* of the transverse isotropic composite with an isotropic matrix and spheroidal pores of the same orientations on the inclusion aspect ratios γ and volume fractions p; k_0,μ_0 are bulk and shear moduli of the host medium

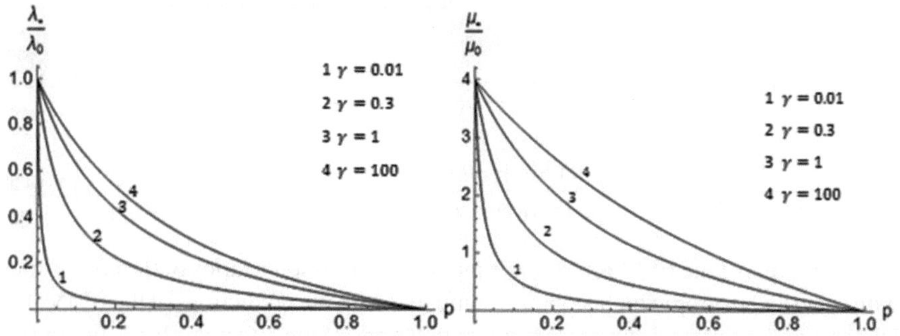

Fig. 13.11 The same as in Fig. 13.10 for the coefficients λ_*, μ_*; λ_0, μ_0 are Lame's parameters of the host medium

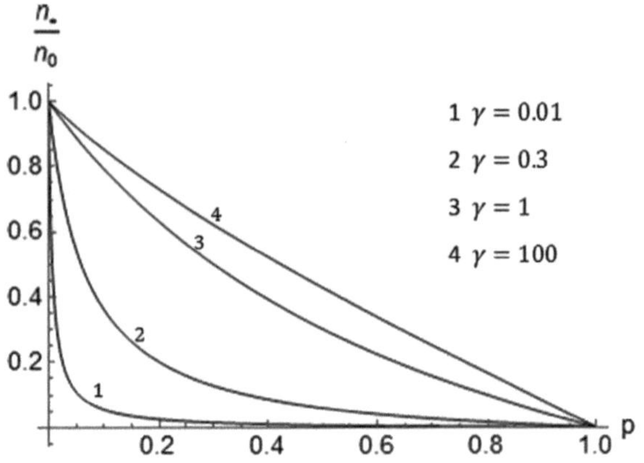

Fig. 13.12 The same as in Figs. 13.10 and 13.11 for the coefficient n_*, $n_0 = \lambda_0 + 2\mu_0$

For spherical inclusions ($\gamma = 1$), the effective stiffness tensor of the composite is isotropic, and tensor A takes the form

$$A = \frac{3}{3K_0 + 4\mu_0} E^2 + \frac{6(K_0 + 2\mu_0)}{5\mu_0(3K_0 + 4\mu_0)} \left(E^1 - \frac{1}{3} E^2 \right). \tag{13.148}$$

Here, K_0, μ_0 are the bulk and shear moduli of the host medium. For isotropic inclusions with bulk and shear moduli K, μ, the expression for the effective elastic stiffness tensor follows from Eq. (13.123) in the form

$$C^* = K_* E^2 + 2\mu_* \left(E^1 - \frac{1}{3} E^2 \right), \tag{13.149}$$

$$K_* = K_0 + p \frac{K_0 (K - K_0)}{K_0 + (1 - p) s_1 (K - K_0)}, \tag{13.150}$$

$$\mu_* = \mu_0 + p \frac{\mu_0 (\mu - \mu_0)}{\mu_0 + (1 - p) s_2 (\mu - \mu_0)}, \tag{13.151}$$

$$s_1 = \frac{3K_0}{3K_0 + 4\mu_0}, \quad s_2 = \frac{6 (K_0 + 2\mu_0)}{5 (3K_0 + 4\mu_0)}. \tag{13.152}$$

Predictions of the EFM for Young's moduli and Poisson's ratios of the composites with spherical isotropic inclusions are compared with the predictions of the EMM, the DEMM, and numerical (finite element) simulations of the effective elastic properties of the composites with absolutely rigid spherical inclusions and spherical pores in Figs. 11.12 and 11.13. Comparison of the EFM with experimental data for the composite with stiff inclusions are in Fig. 11.8. These comparisons show that the EFM provides correct predictions of the effective elastic constants of the composites for inclusion volume fractions $p < 0.3$. For larger volume fractions, the method underestimate inclusion interactions.

For elastic medium with random distribution of ellipsoidal inclusions over orientations, the EFM results in the elastic stiffness tensors of the composites that are similar to those in Eq. (13.111)

$$C^* = C^0 + p(I - p\langle P^\varepsilon(Q,a)A(Q,\alpha)\rangle)^{-1}\langle P^\varepsilon(Q,a)\rangle . \tag{13.153}$$

Here, tensor Q determines the orientation of an inclusion with respect to a global coordinate system. Averaging in Eq. (13.153) is performed over the distribution function of the tensor Q. Tensor $P^\varepsilon(Q,a)$ depends on the shape of the ellipsoid and its orientation

$$P^\varepsilon(Q,a) = C^1\left(I + A(Q,a)C^1\right)^{-1}, C^1 = C - C^0. \tag{13.154}$$

Here, the tensor $A(Q,a)$ is defined in Eq. (4.89), and tensor a defines the shape of the ellipsoids. Tensor $A(Q,\alpha)$ depends on the shape of the correlation hole for the inclusions of the orientation Q. The shape of the correlation hole can be different from the ellipsoid shapes, and tensors α and a cannot coincide.

13.2.4 Elastic Medium Reinforced with Unidirected Fibers

We consider a composite reinforced with a set of identical unidirectional infinite fibers with elliptical cross sections. Spatial distribution of the fibers is statistically homogeneous and isotropic. According to the effective field method, each fiber is subjected to a constant local effective field ε^* that is the same for all the fibers. The one-particle problem of the EFM for a fiber v_i has the form

$$\varepsilon = \Lambda^\varepsilon\varepsilon^*, \Lambda^\varepsilon = \left(I + A(\gamma)C^1\right)^{-1}, \tag{13.155}$$

where $C^1 = C - C^0$, C and C^0 are the elastic stiffness tensors of the fiber and the host medium, $\gamma = a_2/a_1$ is the fiber aspect ratio, a_1, a_2 are the ellipse semiaxes. For an isotropic host medium with Lame's parameters λ_0, μ_0, the tensor $A(\gamma)$ in the T-basis (Appendix A.3) has the form

$$A(\gamma) = A_1T^1 + A_2T^2 + A_4(T^4 + T^5) + A_{10}T^{10} + A_{12}(T^{11} + T^{12}), \tag{13.156}$$

$$A_1 = \frac{\gamma}{2\mu_0(1+\gamma)^2}\left(2(1+\gamma) - \kappa_0(1+2\gamma)\right), \tag{13.157}$$

$$A_2 = \frac{\gamma}{2\mu_0(1+\gamma)^2}\left(2(1+\gamma) - \kappa_0(2+\gamma)\right), A_4 = -\frac{\kappa_0\gamma}{2\mu_0(1+\gamma)^2}, \tag{13.158}$$

$$A_{10} = \frac{1}{4\mu_0}\left(1 + \frac{\kappa_0\gamma}{2(1+\gamma)^2}\right), A_{12} = \frac{1}{4\mu_0(1+\gamma)}, \kappa_0 = \frac{\lambda_0 + \mu_0}{\lambda_0 + 2\mu_0}. \tag{13.159}$$

The basic tensors T^i are defined in Eqs. (4.112) with n^3-vector directed along the fiber axis, and vectors n^1, n^2 along the ellipse semiaxes a_1, a_2. For $\gamma = 1$, the inclusions are circular cylinders, and the tensor A is transverse isotropic. In the P-basis with the vector $n^3 = n$ of the cylinder generatrix, this tensor is

$$A = \frac{1}{4\mu_0}\left((2-\kappa_0)P^1(n) - \frac{\kappa_0}{2}P^2(n) + 2P^5(n)\right). \tag{13.160}$$

The algorithm of application of the EFM to the solution of the homogenization problem is similar to the one developed in Sects. 13.1 and 13.2 for the medium with ellipsoidal inclusions. The effective strain field ε^* and the effective elastic stiffness tensor C^* are presented in a form similar to (13.36), (13.70)

$$\varepsilon^* = (I - pA(\gamma)P^\varepsilon)^{-1}\varepsilon^0, \quad P^\varepsilon = C^1\Lambda^\varepsilon, \tag{13.161}$$

$$C^* = C^0 + pP^\varepsilon\left(I - pA(\gamma)P^\varepsilon\right)^{-1}. \tag{13.162}$$

Here, p is the fiber volume fraction. In the P-basis, the tensors C^0 for isotropic host medium is presented in the forms

$$C^0 = k_0P^2 + 2m_0\left(P^1 - \frac{1}{2}P^2\right) + l_0\left(P^3 + P^4\right) + 4\mu_0P^5 + n_0P^6, \tag{13.163}$$

where

$$k_0 = \lambda_0 + \mu_0, m_0 = \mu_0, l_0 = \lambda_0, n_0 = \lambda_0 + 2\mu_0. \tag{13.164}$$

For transverse isotropic fibers, the elastic stiffness tensor C is

$$C = kP^2 + 2m\left(P^1 - \frac{1}{2}P^2\right) + l\left(P^3 + P^4\right) + 4\mu P^5 + nP^6. \tag{13.165}$$

In the P-basis, the effective stiffness tensor C^* in Eq. (13.162) takes the form

$$C^* = k_*P^2 + 2m_*\left(P^1 - \frac{1}{2}P^2\right) + l_*\left(P^3 + P^4\right) + 4\mu_*P^5 + n_*P^6, \tag{13.166}$$

where five effective elastic constants are

$$k_* = k_0 + p\frac{(k - k_0)\,\mu_0}{\Delta}, \tag{13.167}$$

$$m_* = m_0\left(1 + \frac{p\,(m - m_0)}{m_0 + (1 - p)\,\beta_0\,(m - m_0)}\right), \tag{13.168}$$

$$l_* = l_0 + p\frac{(l - l_0)\,\mu_0}{\Delta}, \tag{13.169}$$

$$\mu_* = \mu_0\left(1 + \frac{2p\,(\mu - \mu_0)}{2\mu_0 + (1 - p)\,(\mu - \mu_0)}\right), \tag{13.170}$$

$$n_* = n_0 + p\,(n - n_0) - p\,(1 - p)\,\frac{\mu_0(l - l_0)^2}{n_0\Delta}, \tag{13.171}$$

$$\Delta = \mu_0 + 2\,(1 - p)\,(k - k_0)\,\frac{\mu_0}{n_0}, \quad \beta_0 = \frac{n_0 + \mu_0}{2n_0}. \tag{13.172}$$

In Fig. 13.13, predictions of the EFM for the effective elastic moduli of the composite reinforced with unidirected cylindrical fibers are compared with the experimental data presented in [5]. In this work, the composites with the epoxy resin matrix ($E_0 = 5.27GPa$, $\nu_0 = 0.3$) and the transversely isotropic carbon fibers

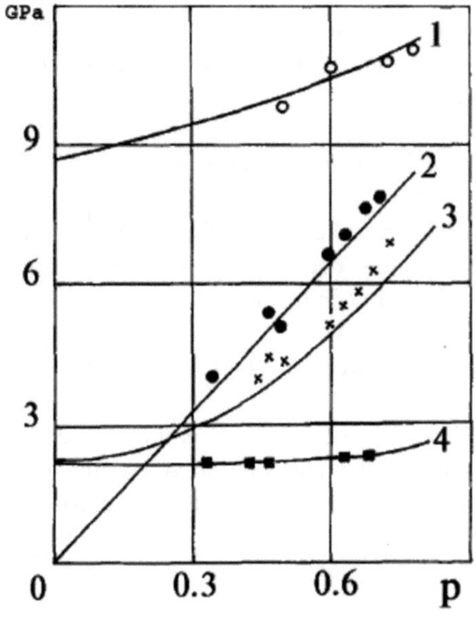

Fig. 13.13 Comparison of the EFM predictions for the effective elastic moduli of the composite reinforced with unidirectional cylindrical fibers with experimental data presented in [5]

($E_1 = 8GPa$, $E_3 = 410.6GPa$, $\nu_{12} = 0.568$, $\nu_{13} = 0.273$, $\mu_3 = 10.2GPa$) are studied for various volume fractions p of the fibers. Line 1 in Fig. 13.13 corresponds to the EFM predictions for the parameter $\lambda_* + 2m_*$, (light circles (∘) are the experimental data), $2 - m_*(\bullet)$, $3 - \mu_*(\times)$, $4 - n_*(\blacksquare)$.

13.3 Mori-Tanaka Method

An approach known in the literature as Mori-Tanaka method is a particular case of the EFM in application to the homogeneous host medium containing random sets of identical ellipsoidal inclusions. The equations of Mory-Tanaka method coincide with the equations of the EFM by an additional assumption that the symmetry of the correlation function $\Psi(x)$in Eq. (13.15) (correlation hole) coincides with the symmetry of ellipsoidal inclusions.

In the original derivation [6], the authors used the concepts of eigenstrains, equivalent inclusion, energy transformation, etc., and at the first glance, this derivation is totally different in spirit from the concept of the effective field method. Later on, it was shown in [7] that the method is based in fact on the following hypotheses.

- *All inclusions in the composite are ellipsoids with a constant elastic stiffness tensor inside,*
- *Each inclusion behaves as an isolated one in the original matrix subjected to a constant external strain (stress) field that is the mean strain (stress) field over the matrix phase.*

We consider a composite that consists of a homogeneous matrix and identical ellipsoidal homogeneous inclusions of the same orientations randomly distributed in the matrix. The mean stress and strain fields in the composite satisfy obvious relations

$$\langle\sigma\rangle = p\langle\sigma\rangle_{inc} + (1-p)\langle\sigma\rangle_m, \quad \langle\varepsilon\rangle = p\langle\varepsilon\rangle_{inc} + (1-p)\langle\varepsilon\rangle_m, \tag{13.173}$$

where p is the volume fraction of the inclusions, $\langle\cdot\rangle_{inc}$ and $\langle\cdot\rangle_m$ are averaging over the regions occupied by the inclusions and the matrix phase. Because Hooke's law holds for each phase, we have

$$\langle\sigma\rangle = pC\langle\varepsilon\rangle_{inc} + (1-p)C^0\langle\varepsilon\rangle_m = C^0\langle\varepsilon\rangle + pC^1\langle\varepsilon\rangle_{inc}. \tag{13.174}$$

Here, $C^1 = C - C^0$, C is the elastic stiffness tensor of the inclusions, C^0 is the same for the matrix phase. According to the hypotheses of Mori-Tanaka method, the average strain fields inside the inclusions $\langle\varepsilon\rangle_{inc}$ has the form

$$\langle\varepsilon\rangle_{inc} = \Lambda\langle\varepsilon\rangle_m, \quad \Lambda = (I + AC^1)^{-1}, \quad x \in V \tag{13.175}$$

where the tensor A is defined in Eq. (4.89). Substitution of this equation into the second Eq. (13.173) results in the equation for the mean strain field $\langle\varepsilon\rangle$ in the composite

$$\langle\varepsilon\rangle = (p\Lambda + (1-p)I)\langle\varepsilon\rangle_m, \tag{13.176}$$

and for the mean strain field over the matrix phase, we obtain

$$\langle\varepsilon\rangle_m = (I - p(I-\Lambda))^{-1}\langle\varepsilon\rangle = (I - pAC^1\Lambda)^{-1}\langle\varepsilon\rangle. \tag{13.177}$$

Here, Eq. (13.175) for Λ is taken into account. Equation for $\langle\varepsilon\rangle_{inc}$ follows from Eqs. (13.175), (13.177)

$$\langle\varepsilon\rangle_{inc} = \Lambda\left(I - pAC^1\right)^{-1}\langle\varepsilon\rangle. \tag{13.178}$$

As a result, Eq. (13.174) for the average stress field $\langle\sigma\rangle$ in the composite takes the form

$$\langle\sigma\rangle = \left(C^0 + pC^1\Lambda\left(I - pAC^1\Lambda\right)^{-1}\right)\langle\varepsilon\rangle. \tag{13.179}$$

The effective elastic stiffness tensor C^* relates the averages $\langle\sigma\rangle$ and $\langle\varepsilon\rangle$ and therefore,

$$C^* = C^0 + pP\left(I - pAP\right)^{-1}, P = C^1\Lambda. \tag{13.180}$$

Equation (13.180) coincides with Eq. (13.70) of the effective field method by an additional assumption that the symmetry of the correlation function $\Psi(x)$ (correlation hole) coincides with the inclusion symmetry, and therefore, $A(\alpha) = A(a) = A$.

Simplicity of Eq. (13.180) makes Mori-Tanaka method attractive for many researches. But it was noted in [8–10] that application of Eq. (13.180) to the composites with several different families of inclusions (hybrid composites) results in non-symmetric effective elastic stiffness tensors.

Let a homogeneous host medium contains two families of different ellipsoidal inclusions. The elastic properties, aspect ratios and orientations of the ellipsoids coincide in each family but they are different for different families. In this case, Eqs. (13.173) for the mean strain and stress tensors in the composite take the forms

$$\langle\sigma\rangle = p_1\langle\sigma\rangle^1_{inc} + p_2\langle\sigma\rangle^2_{inc} + (1-p_1-p_2)\langle\sigma\rangle_m, \tag{13.181}$$

$$\langle\varepsilon\rangle = p_1\langle\varepsilon\rangle^1_{inc} + p_2\langle\varepsilon\rangle^2_{inc} + (1-p_1-p_2)\langle\varepsilon\rangle_m, \tag{13.182}$$

where p_1, p_2 are the volume fractions of inclusions of the first and second family, $\langle\sigma\rangle^i_{inc}$, $\langle\varepsilon\rangle^i_{inc}$ are stress and strain fields averaged over the region occupied by the inclusions of the ith family. Because Hooke's law holds for each phase, the average stress field in the composite is

$$\langle\sigma\rangle = p_1 C_1 \langle\varepsilon\rangle_{inc}^1 + p_2 C_2 \langle\varepsilon\rangle_{inc}^2 + (1 - p_1 - p_2) C^0 \langle\varepsilon\rangle_m =$$

$$= C^0 \langle\varepsilon\rangle + p_1 C_1^1 \langle\varepsilon\rangle_{inc}^1 + p_2 C_2^1 \langle\varepsilon\rangle_{inc}^2, \tag{13.183}$$

$$C_1^1 = C_1 - C^0, C_2^1 = C_2 - C^0. \tag{13.184}$$

The hypothesis of Mori-Tanaka method relates the strain field inside inclusions with the mean strain over the matrix phase. Thus the equation for $\langle\varepsilon\rangle_{inc}^i$ takes the form

$$\langle\varepsilon\rangle_{inc}^1 = \Lambda^1 \langle\varepsilon\rangle_m, \quad \langle\varepsilon\rangle_{inc}^2 = \Lambda^2 \langle\varepsilon\rangle_m, \tag{13.185}$$

$$\Lambda^1 = (I + A^1 C_1^1)^{-1}, \quad \Lambda^2 = (I + A^2 C_2^1)^{-1}. \tag{13.186}$$

Substitution of these equations into Eq. (13.182) results in the expression for the mean strain field in the composite

$$\langle\varepsilon\rangle = \left(p_1 \Lambda^1 + p_2 \Lambda^2 + (1 - p_1 - p_2) I\right) \langle\varepsilon\rangle_m, , \tag{13.187}$$

and for the mean strain field over the matrix phase, we obtain

$$\langle\varepsilon\rangle_m = \left(I - p_1 A^1 C_1^1 \Lambda^1 - p_2 A^2 C_2^1 \Lambda^2\right)^{-1} \langle\varepsilon\rangle. \tag{13.188}$$

Equation for $\langle\varepsilon\rangle_{inc}^i$ follows from Eqs. (13.185), (13.188)

$$\langle\varepsilon\rangle_{inc}^1 = \Lambda^1 \left(I - p_1 A^1 C_1^1 \Lambda^1 - p_2 A^2 C_2^1 \Lambda^2\right)^{-1} \langle\varepsilon\rangle, \tag{13.189}$$

$$\langle\varepsilon\rangle_{inc}^2 = \Lambda^2 \left(I - p_1 A^1 C_1^1 \Lambda^1 - p_2 A^2 C_2^1 \Lambda^2\right)^{-1} \langle\varepsilon\rangle, \tag{13.190}$$

and Eq. (13.187) for the average stress field $\langle\sigma\rangle$ in the composite takes the form

$$\langle\sigma\rangle = C^* \langle\varepsilon\rangle\,. \tag{13.191}$$

where the effective elastic stiffness tensor C^* is

$$C^* = C^0 + p_1 P^1 \left(I - p_1 A^1 P^1 - p_2 A^2 P^2\right)^{-1} +$$

$$+ p_2 P^2 \left(I - p_1 A^1 P^1 - p_2 A^2 P^2\right)^{-1}, \tag{13.192}$$

$$P^1 = C_1^1 \Lambda, P^2 = C_2^1 \Lambda^2. \tag{13.193}$$

Let us expand the right-hand side of Eq. (13.192) in terms of the inclusion volume fractions p_1, p_2

$$C^* = C^0 + p_1 P^1 + p_2 P^2 + p_1^2 P^1 A^1 P^1 + p_2^2 P^2 A^2 P^2 +$$
$$+ p_1 p_2 (P^1 A^2 P^2 + P^2 A^1 P^1) + ... \tag{13.194}$$

For the conductivity problems, the second rank tensors A^1, A^2, P^1, P^2 in this equations are symmetric as well as the tensors $P^1 A^1 P^1$ and $P^2 A^2 P^2$. But the tensors $P^1 A^2 P^2$ and $P^2 A^1 P^1$ are not symmetric because P^1, P^2 are different tensors as well as A^1, A^2. For elasticity, A^1, A^2, P^1, P^2 are fourth rank tensor symmetric with respect to first and second pairs of indices as well as to the index pair permutations. Meanwhile the tensors $P^1 A^2 P^2$ and $P^2 A^1 P^1$ are not symmetric with respect to the index pairs for P^1, P^2 and A^1, A^2 are different tensors.

The drawback of Eq. (13.192) originates from the assumption that all the inclusions in the composite are subjected to the same external field $\langle \varepsilon \rangle_m$. In application to hybrid composites, the version of the effective field method considered in the previous sections has the same defect: the effective conductivity tensor can be not symmetric, and the effective elastic stiffness tensor can be not symmetric with respect to index pair permutations. The reason is the assumption that the effective field $\varepsilon^* (\sigma^*)$ is the same for all the inclusions. The version of the EFM that allows improving this drawback is considered in Chap. 17.

13.4 The Effective Field Method for 2D-Elastic Composites

In the 2D-case, the expression for the effective elastic stiffness tensor C^* of composites has the same form (13.70), where the tensors are to be replaced by their 2D-analogies

$$C^* = C^0 + p P^\varepsilon \left(I - p A^* P^\varepsilon\right)^{-1}. \tag{13.195}$$

For identical elliptical inclusions of the same orientations, the tensor P^ε in this equation is defined by the equation

$$P^\varepsilon = C^1 \left(I + A(a) C^1\right)^{-1}. \tag{13.196}$$

Here, $C^1 = C - C^0$, and tensor $A(a)$ depends on the inclusion semiaxes and its spatial orientation. For an anisotropic host medium this tensor is defined in terms of 1D-integral

$$A(a) = \frac{1}{2\pi}\int_0^{2\pi} \widetilde{K}\left(a^{-1}k(\varphi)\right) d\varphi,\ k(\varphi) = \cos(\varphi)e^1 + \sin(\varphi)e^2, \tag{13.197}$$

where e^1, e^2 is Cartesian basis in 2D-space with the origin in the ellipse center, linear transformation of x-space $y = a^{-1}x$ converts the ellipse in unit circle. Tensor $\widetilde{K}(k)$ is Fourier transform of tensor $K(x)$ that is second derivatives of Green's function for 2D homogeneous elastic media.

In the case of isotropic medium, $\widetilde{K}(k)$ have the forms

$$\widetilde{K}(k) = \frac{1}{\mu_0}\left(E^5(m) - \kappa_0 E^6(m)\right), m_\alpha = \frac{k_\alpha}{|k|}, \tag{13.198}$$

where $\kappa_0 = (2(1-\nu_0))^{-1}$ for the plane strain case, and $\kappa_0 = (1+\nu_0)/2$ for the plane stress case, ν_0 is Poisson's ratio of the host medium. In the E-basis for the plane case, tensor $K(x)$ is defined by the equation (Appendix B.3)

$$K(x) = \frac{1}{2\pi\mu_0|x|^2}\left(E^1 - 2E^5 - \right.$$

$$\left. - \frac{\kappa_0}{2}\left(4E^1 - E^2 - 12E^5(n) + 8E^6(n)\right)\right), \tag{13.199}$$

Tensor A^* in Eq. (13.195) is the 2D-integral

$$A^* = \int K(x)\left(1 - \Psi(x)\right) dx, \tag{13.200}$$

where the correlation function $\Psi(x)$ in the 2D-case has the same form as in the 3D-case (Eq. (13.27))

$$\Psi(x) = \frac{1}{p^2}\lim_{W\to\infty}\frac{1}{W}\int_W \sum_{(i\neq j)} v_i\left(x'\right) v_j\left(x' + x\right)dx', \tag{13.201}$$

Here, $v_i(x)$ is the characteristic function of the 2D-region v_i occupied by the $i$the inclusion, W is the region that occupied the entire space when $W \to \infty$. In the 2D-case, the function $\Psi(x)$ has the same properties as in the 3D case

$$\Psi(0) = 0,\ \lim_{W\to\infty}\frac{1}{W}\int_W \Psi(x)dx = 1, \tag{13.202}$$

If symmetry of the function $\Psi(x)$ is defined by tensor α such as $\Psi(\alpha y) = \Psi'(|y|)$,

$$A^* = A(\alpha), \tag{13.203}$$

where $A(\alpha)$ is defined in Eq. (13.197) with $a = \alpha$. If symmetry of the correlation hole $\Psi(x)$ coincides with the symmetry of the elliptical inclusions, then $A(\alpha) = A(a)$, and the equation for the effective stiffness tensor is

$$C^* = C^0 + pC^1 \left(I + (1-p)A(a)C^1\right)^{-1}. \tag{13.204}$$

13.4.1 The Effective Field Method for Composites with Regular Sets of Inclusions

The EFM can be applied to the composites when the correlations functions of a random set of inclusions $\Psi(x)$ does not have ellipsoidal shapes. In particular, the method can be used for solution of the homogenization problem for composites with regular lattice of isolated inclusions in a homogeneous matrix.

Let identical inclusions form a regular lattice in a homogeneous host medium. The effective elastic stiffness tensor C^* of a regular composite is defined by the equation

$$\langle\sigma(x)\rangle = C^*\langle\varepsilon(x)\rangle. \tag{13.205}$$

Here, $\varepsilon(x)$, $\sigma(x)$ are detailed strain and stress fields in the composite by action of a constant external field $\varepsilon^0(\sigma^0)$ at infinity. The means $\langle\varepsilon(x)\rangle$ and $\langle\sigma(x)\rangle$ can be understood as the strain and stress fields at a fixed point x of the composite averaged over all possible spatial translations of the inclusion lattice. Thus the averaging in (13.205) can be performed over ensemble realizations of the random function $V(x)$ defined by the equation

$$V(x) = V_0(x + r), \tag{13.206}$$

where $V_0(x)$ is the characteristic function of a fixed inclusion lattice, and r is a stochastic vector homogeneously distributed in space. Note that for a constant external strain (stress) field, the ergodic properties of the fields $\varepsilon(x)$ and $\sigma(x)$ are fulfilled automatically, and the general scheme of the EFM can be applied to regular composite structures without changes.

We consider the correlation function $\Psi(x)$ in Eq. (13.25) for regular lattices of circular inclusions in 2D-space. If $v_i(x)$ is the characteristic function of the region occupied by ith inclusion, the integral $\int v_i(x' + x)v_i(x')dx'$ is the area of intersection of two identical circles moved at the vector x with respect to each other and therefore,

$$\int v_i\left(x'\right) v_i\left(x' + x\right) dx' = \pi a^2 j(x), \tag{13.207}$$

$$j(x) = \begin{cases} \frac{2}{\pi}\left(\arccos\left(\frac{|x|}{2a}\right) - \frac{|x|}{2a}\sqrt{1-\left(\frac{|x|}{2a}\right)^2}\right), & x < 2a \\ 0, & x > 2a \end{cases}, \tag{13.208}$$

where a is the inclusion radius. For a regular lattice of identical inclusions, the covariance $\langle V(x')V(x'+x)\rangle$ is a periodic function of the variable x. In the vicinity of the point $x = 0$, this function is defined in Eq. (13.207). Therefore, the covariance $\langle V(x')V(x'+x)\rangle$ has the form

$$\langle V(x')V(x'+x)\rangle = p\sum_m j(x-m), \tag{13.209}$$

where p is the volume fraction of inclusions, m is the vector of the lattice of the inclusion centers. The function $\Psi(x)$ in Eq. (13.201) takes the form

$$\Psi(x) = \frac{1}{p}\sum_{m\neq 0} j\,(x-m)\,, \tag{13.210}$$

where the function $j(x)$ is defined in Eq. (13.208). It follows from this equation that the integral in Eq. (13.200) for the effective strain field ε^* is presented in the form of the following converging series

$$\int K(x)\,(1-\Psi\,(x))dx' = A^*\,, \tag{13.211}$$

$$A^* = A^0 + \sum_{m\neq 0} pv\int K\,(x)\left(1-\frac{1}{p}j\,(x-m)\right)dx\,. \tag{13.212}$$

Here, the tensor A^0 for an isotropic host medium and a circular inclusion has the form

$$A^0 = \frac{1}{8\mu_0}\left(2(2-\kappa_0)E^1 - \kappa_0 E^2\right). \tag{13.213}$$

Let circular inclusions form a lattice with an equilateral triangular cell. Because symmetry of the function $\Psi(x)$ coincides with the symmetry of the lattice, the integral in the right-hand side of Eq. (13.211) is the tensor with the symmetry of the function $\Psi(x)$. It is known [11] that the basis of fourth rank tensors with the symmetry of a equilateral triangular lattice consists of isotropic tensors only. Therefore, the integral in Eq. (13.212) has the following tensor structure

$$pv\int K\,(x)\,(1-\Psi\,(x))dx = \alpha_1 E^1 + \alpha_2 E^2, \tag{13.214}$$

where α_1 and α_2 are scalar coefficients. For the tensor $K(x)$ in Eq. (13.199), the following index convolutions are equal to zero

$$K_{\alpha\alpha\beta\beta}(x) = K_{\alpha\beta\beta\alpha}(x) = 0. \tag{13.215}$$

After performing the same convolutions to the both sides of Eq. (13.214), we yield

$$2\alpha_1 + 4\alpha_2 = 0, 3\alpha_1 + 2\alpha_2 = 0. \tag{13.216}$$

It follows from these equations that $\alpha_1 = \alpha_2 = 0$. Therefor, for the lattice of equilateral triangular cells of circular inclusions, the integral in the right-hand side of Eq. (13.212) vanishes. As a result, we obtain the effective elastic stiffness tensor C^* of the composite with the triangular lattice of circular inclusions in the form

$$C^* = C^0 + pC^1\left(I + (1-p)\, A^0 C^1\right)^{-1}, \tag{13.217}$$

where the tensor A^0 is in Eq. (13.213).

For a square lattice of circular inclusions, the integral in Eq. (13.212) is not equal to zero and it has the symmetry of the square lattice. The basis of the forth rank tensors of such symmetry consists of three tensors: E^1, E^2 and tensor Q that in the basis of the unit vectors $e^{(1)}$, $e^{(2)}$ of the square cell is

$$Q_{\alpha\beta\lambda\mu} = \sum_{i=1}^{2} \delta_{i\alpha}\delta_{i\beta}\delta_{i\lambda}\delta_{i\mu},\ \alpha, \beta, \lambda, \mu = 1, 2. \tag{13.218}$$

Therefore, the integral in Eq. (13.212) is

$$pv \int K(x)\,(1 - \Psi(x))\, dx = \alpha_1 E^1 + \alpha_2 E_2 + \alpha_3 Q. \tag{13.219}$$

For the properties (13.215) of the function $K(x)$, there are two liner relations for the coefficients α_i in this equation. As a result, the integral in Eq. (13.219) is defined with precision of a scalar function $\alpha(p)$ of the inclusion area fraction p

$$pv \int K(x)\,(1 - \Psi(x))\, dx = \alpha(p)\, \Gamma, \tag{13.220}$$

$$\Gamma_{\alpha\beta\lambda\mu} = \frac{2\kappa_0}{\mu_0}\left(2E^1_{\alpha\beta\lambda\mu} + E^2_{\alpha\beta\lambda\mu} - 4\sum_{i=1}^{2} \delta_{i\alpha}\delta_{i\beta}\delta_{i\lambda}\delta_{i\mu}\right). \tag{13.221}$$

Here, the scalar coefficient $\alpha(p)$ is the sum of the integrals

$$\alpha(p) = -\sum_m{}' \int k(x)\left(1 - \frac{1}{p} j(x-m)\right) dx. \tag{13.222}$$

where

$$k(x) = \frac{1}{8\pi|x|^2}\left(1 - 8\left(n \cdot e^{(1)}\right)^2\left(n \cdot e^{(2)}\right)^2\right), \quad n = \frac{x}{|x|}. \tag{13.223}$$

The plot of the function $\alpha(p)$ obtained by numerical summation of the series (13.222) is presented on Fig. 13.14.

The tensor of the effective elastic stiffness tensor of the composite with a square lattice of circular inclusions is defined by the equation

$$C^* = C^0 + pC^1\left(I + (1-p)\,A^0C^1 + p\alpha(p)\Gamma C^1\right)^{-1}, \tag{13.224}$$

where the tensors A^0 and Γ are in Eqs. (13.213) and (13.221).

In [12], the solution of plane problem for periodical lattices of circular inclusions in homogeneous isotropic host medium subjected to a constant external field was found in the form of series of doubly periodic functions. The values of the effective elastic moduli presented in [12] can be considered as exact ones. In Fig. 13.15, exact values of effective elastic moduli of triangular lattices of circular inclusions with an equilateral cell (Eq. (13.217)) are compared with predictions of the EFM. Here E_0, μ_0 and E, μ are Young's and shear elastic moduli of the matrix and inclusions; E_*, μ_* are the effective moduli of the composite. The solid lines correspond to the exact solutions, the dashed lines with light circles are predictions of the EFM.

In Figs. 13.16 and 13.17, the effective elastic moduli of an elastic plane with a square lattice of the circular inclusions calculated from Eq. (13.224) are presented. Exact values of these moduli given in [12] are solid lines in this figure, and predictions of the EFM are dashed lines with light circles.

These figures show that for the composites with regular structures, the EFM approaches are close to the exact effective elastic moduli when the ratio of the matrix and the inclusions Young moduli is inside the bounds

$$0.1 \leq \frac{E_0}{E} \leq 10. \tag{13.225}$$

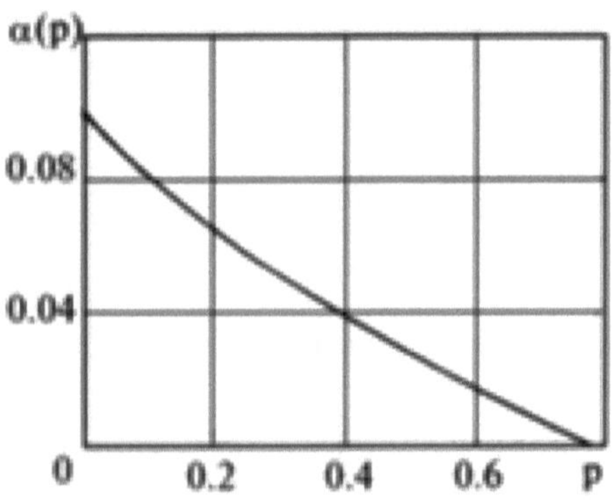

Fig. 13.14 Function $\alpha(p)$ in Eqs. (13.221) for a plane with a square lattice of circular inclusions

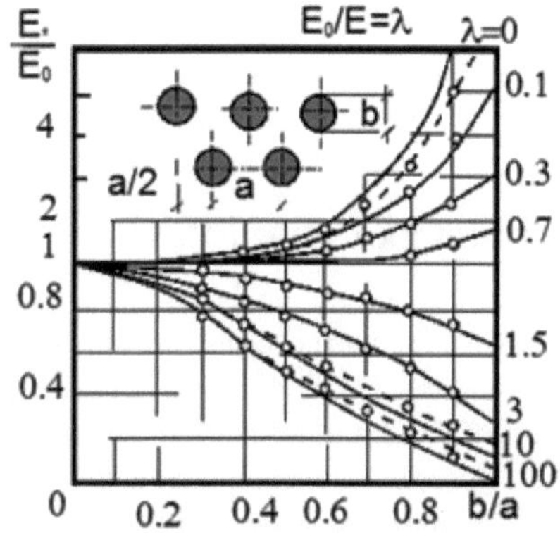
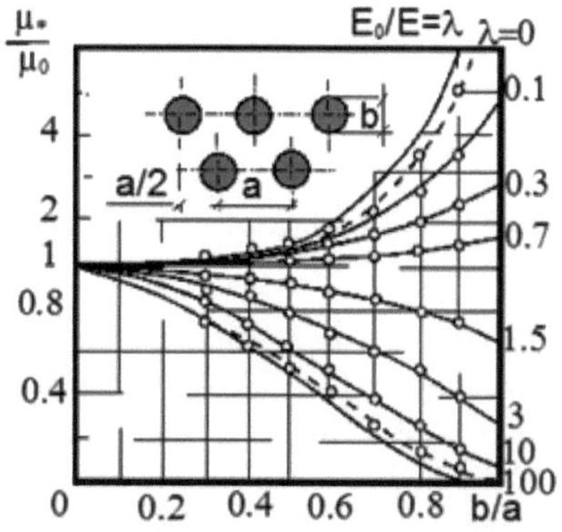

Fig. 13.15 Comparison of the exact and approximate values of the effective Young and shear moduli of composites with a triangle lattice of circular inclusions

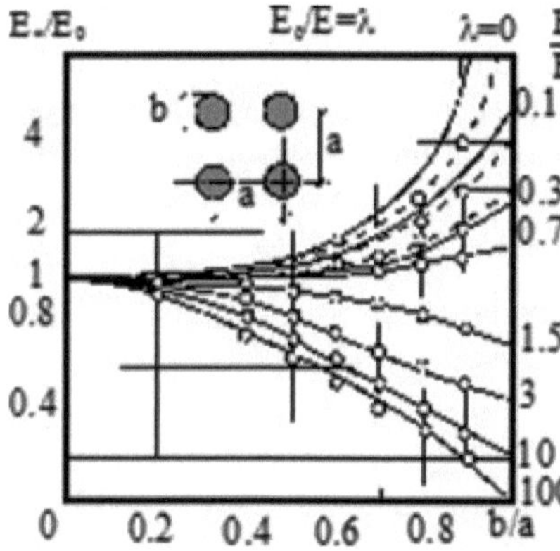
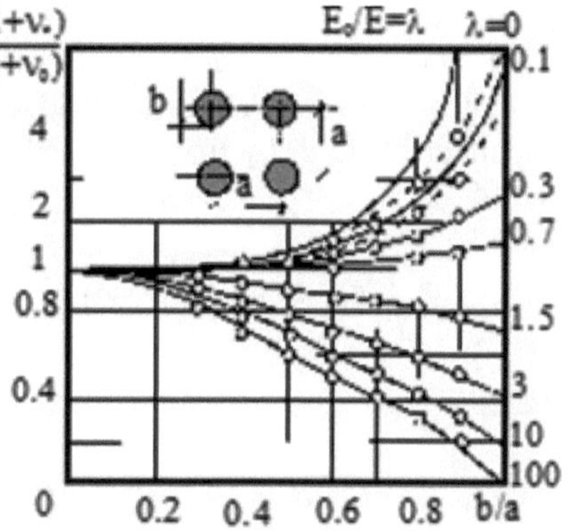

Fig. 13.16 Comparison of the exact and approximate effective Young and shear moduli of a plane with circular inclusions

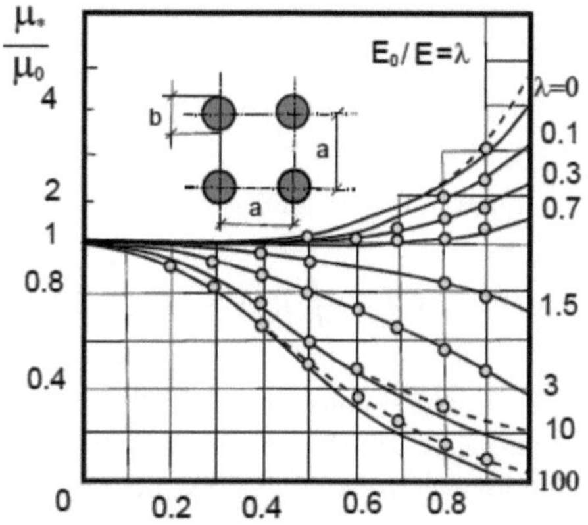

Fig. 13.17 The same as in Fig. 13.16 for the shear modulus of the composites with square lattices of circular inclusions

For $E_0/E > 10$ or $E_0/E < 0.1$ and the inclusion fractions $p \geq 0, 4$, the errors of the EFM are significant and these errors increases with the inclusion fraction p.

13.5 Thermo-elastic Deformation of Matrix Composites

In this Section, we consider thermal strains and stresses in matrix composite materials with ellipsoidal inclusions. Let the composite medium is subjected to a constant temperature T, and it is unconstrained at infinity. The hypotheses of the effective field method do not change in this case. Thus it is assumed that each inclusion in the composite behaves as an isolated one embedded into the homogeneous host medium (matrix), and the local external field that acts on the inclusion is constant and the same for all the inclusions. The elastic stiffness tensor of the host medium is C^0, the tensor of temperature expansion coefficients is α^0. The same tensors for the inclusion material are C and α.

Let $V(x)$ be the characteristic function of the region occupied by a random set of isolated inclusions. The stress field $\sigma(x)$ in the composite subjected to a constant temperature field T by unconstrained conditions at infinity satisfies the equation (Sect. 4.2):

$$\sigma(x) = \int S(x - x')\left(B^1\sigma(x') + \alpha^1 T\right)V(x')dx'. \tag{13.226}$$

where,

$$B^1 = B - B^0, B = C^{-1}, B^0 = \left(C^0\right)^{-1}, \alpha^1 = \alpha - \alpha^0. \tag{13.227}$$

Using equations

$$S(x) = C^0 K(x) C^0 - C^0\delta(x), \tag{13.228}$$

$$\sigma(x) = C(x)\left(\varepsilon(x) - \alpha(x)T\right), \int K(x - x')C^0\alpha^0 T dx' = \alpha^0 T, \tag{13.229}$$

the equation for the total strain tensor $\varepsilon(x)$ follows from Eq. (13.226) in the form

$$\varepsilon(x) = \alpha^0 T + \int K(x - x')C^0\left(B^1\sigma(x') + \alpha^1 T\right)V(x')dx'. \tag{13.230}$$

The integral equation of the one-particle problem for an isolated inclusion in a thermo-elastic homogeneous medium subjected to constant temperature field T and a homogeneous external stress field σ^* has the form:

$$\sigma(x) - \int_v S(x - x') B^1 \sigma(x') dx' =$$

$$= \sigma^* + \int_v S(x - x') \alpha^1 dx', x \in v, \tag{13.231}$$

where v is the region occupied be the inclusion. The solution of this problem is presented in the form

$$\sigma(x) = \sigma^T(x) + \sigma^s(x), \tag{13.232}$$

where the functions $\sigma^T(x)$ and $\sigma^s(x)$ are the solutions of the integral equations

$$\sigma^s(x) - \int_v S(x - x') B^1 \sigma^s(x') dx' = \sigma^*, \tag{13.233}$$

$$\sigma^T(x) - \int_v S(x - x') B^1 \sigma^T(x') dx' = \int_v S(x - x') \alpha^1 T dx'. \tag{13.234}$$

For an ellipsoidal inclusion, the solutions of these equations are constant inside region v and have the forms (Sect. 4.5)

$$\sigma^s = \Lambda^\sigma \sigma^*, \ \sigma^T = \Lambda^T T, \ x \in v. \tag{13.235}$$

$$\Lambda^\sigma = \left(I - DB^1\right)^{-1}, \ \Lambda^T = \Lambda^\sigma D \alpha^1. \tag{13.236}$$

where the tensor $D = D(a)$ is defined in Eq. (4.92), the second rank tensor a is determined by the aspect ratios and orientation of the ellipsoidal inclusion.

In the framework of the effective field method, the average effective stress field σ^* acting on each inclusion is constant and the same for all the inclusion. As a result, the stress tensor and the tensors of total deformation are approximated by the equations that follow from Eqs. (13.226), (13.230)

$$\sigma(x) = \int S(x - x') \left(P^s \sigma^* + P^T T\right) V(x') dx', \tag{13.237}$$

$$\varepsilon(x) = \alpha^0 T + \int K(x - x') C^0 \left(P^s \sigma^* + P^T T\right) V(x') dx', \tag{13.238}$$

where

$$P^s = B^1 \Lambda^\sigma, \ P^T = B^1 \Lambda^T + \alpha^1 = \left(I + P^s D\right) \alpha^1. \tag{13.239}$$

The local exciting field $\sigma^*(x)$ acting on point $x \in V$, is approximated as follows

$$\sigma^*(x) = \int S(x - x')\left(P^s\sigma^* + P^T T\right)V\left(x'; x\right)dx', \tag{13.240}$$

where function $V\left(x'; x\right)$ is defined in Eq. (13.9). Averaging this equation by the condition that $x \in V$ and accepting that $\langle \sigma^*(x)\,|x\rangle = \sigma^*$, we yield a closed equation for the tensor σ^*

$$\sigma^* = \int S(x - x')\left(P^s\sigma^* + P^T T\right)\left\langle V\left(x'; x\right)|x\right\rangle dx'. \tag{13.241}$$

Here, the conditional mean $\left\langle V\left(x'; x\right)|x\right\rangle$ is presented in the form

$$\left\langle V\left(x'; x\right)|x\right\rangle = p\Psi(x - x'),\ \Psi\left(x - x'\right) = \frac{\left\langle V\left(x'; x\right)|x\right\rangle}{\langle V(x)\rangle}, \tag{13.242}$$

where p is inclusion volume fraction, $\Psi(x)$ is the correlation hole for a typical inclusion in the composite. From Eq. (13.241) follows the equation for the effective field

$$\sigma^* = -p\left(I + pDP^s\right)^{-1}DP^T T. \tag{13.243}$$

Averaging Eq. (13.238) for the strain tensor over ensemble realizations of the random set of inclusions, we obtain the equation for the mean strain field in the composite

$$\begin{aligned}\langle \varepsilon\rangle &= \alpha^0 T + p\left(P^s\sigma^* + P^T T\right) = \\ &= \alpha^0 T + p(I + pDP^s)^{-1}P^T T.\end{aligned} \tag{13.244}$$

Because $\langle \varepsilon\rangle$ is the mean strain tensor of the composite subjected to temperature T, $\langle \varepsilon\rangle$ coincides with the effective tensor $\alpha^*$of the temperature expansion if $T = 1$. Thus, the tensor α^* is

$$\begin{aligned}\alpha^* &= \alpha^0 + p(I + pDP^s)^{-1}P^T = \\ &= \alpha^0 + p(I - (1 - p)DB^1)^{-1}\alpha^1.\end{aligned} \tag{13.245}$$

We consider an isotropic host medium with spherical isotropic inclusions. In this case, all the tensors in Eq. (13.245) are isotropic and

$$D = -\frac{4\mu_0 K_0}{3K_0 + 4\mu_0}E^2 - \frac{2\mu_0(9K_0+8\mu_0)}{5\,(3K_0 + 4\mu_0)}\left(E^1 - \frac{1}{3}E^2\right), \tag{13.246}$$

$$B^1 = \frac{K_0 - K}{9K_0K} E^2 + \frac{\mu_0 - \mu)}{2\mu_0\mu}\left(E^1 - \frac{1}{3}E^2\right), \tag{13.247}$$

$$\alpha^0_{\alpha\beta} = \alpha_0\delta_{\alpha\beta},\ \alpha^1_{\alpha\beta} = (\alpha - \alpha_0)\delta_{\alpha\beta}, \alpha^*_{\alpha\beta} = \alpha_*\delta_{\alpha\beta}. \tag{13.248}$$

Here, K_0, μ_0 are the bulk and shear moduli of the host medium, K, μ are the same for the inclusions. The effective coefficient of thermo expansion of the composite is

$$\alpha_* = \alpha_0 + p\,(\alpha - \alpha_0)\left(1 - 4\,(1 - p)\,\frac{\mu_0\,(K - K_0)}{K\,(3K_0 + 4\mu_0)}\right)^{-1}. \tag{13.249}$$

Temperature stresses σ^+ in the inclusions can be assessed from the equation

$$\sigma^+ = \left(I - DB^1\right)^{-1}\sigma^* = -p(I - (1 - p)DB^1)^{-1}D(I - DB^1)^{-1}T \tag{13.250}$$

The error in the obtained equations for α^* was discussed in [13, 14]. It was noted that for the fiber-glass plastics, the relative error in calculating α^* from Eq. (13.245) does not exceed 10% up to the fiber concentration corresponded to dense packing.

13.6 The Point Defect Model in the Theory of Composite Materials

In this section we consider the approximation based on the replacement of the inclusions of finite sizes by point defects. Let an isolated inclusion with the elastic stiffness tensor $C(x)$ be placed in a homogeneous elastic medium with the stiffness tensor C^0,and V be the region occupied by the inclusion. The medium is subjected to a constant strain field ε^0 (stress field σ^0) at infinity. The fields $\varepsilon(x)$ and $\sigma(x)$ in the medium with the inclusion are presented in the forms

$$\varepsilon\,(x) = \varepsilon^0 + \int K\left(x - x'\right) C^0 m\left(x'\right)d\,x', \tag{13.251}$$

$$\sigma\,(x) = \sigma^0 + \int S(x - x')m\left(x'\right)dx', \tag{13.252}$$

where $m(x)$ is the equivalent density of the dislocation moments induced in the region occupied by the inclusion (see Sect. 3.3), and

$$m(x) = B^1\,(x)\,\sigma\,(x)\,V\,(x)\,,\ B^1(x) = B\,(x) - B^0, \tag{13.253}$$

$$B\,(x) = C(x)^{-1},\ \ B^0 = \left(C^0\right)^{-1}. \tag{13.254}$$

For a constant external stress σ^0, the stress field $\sigma(x)$ inside the inclusion is presented in the form

$$\sigma(x) = \Lambda^{\sigma}(x)\,\sigma^0,\ x \in V, \tag{13.255}$$

where the tensor $\Lambda^{\sigma}(x)$ is determined from the solution of the one-particle problem for an isolated inhomogeneity subjected to a constant external field σ^0. Thus for $m(x)$ we have

$$m(x) = P^{\sigma}(x)\,\sigma^0,\ P^{\sigma}(x) = B^1(x)\,\Lambda^{\sigma}(x). \tag{13.256}$$

Transition to the point defect corresponds to replacement of the density $m(x)$ by the first term of the multipole expansion (Sect. 13.6)

$$m(x) = P^0\sigma^0\delta(x-\xi) + \dots,\quad P^0 = \int P^{\sigma}(x)dx, \tag{13.257}$$

where ξ is a point inside the inclusion, and further ξ is taken at the center of weight of the inclusion. In particular, for an ellipsoidal inclusion, the tensor P^0 has the following form

$$P^0 = vB^1(I - D(a)\,B^1)^{-1}, \tag{13.258}$$

where v is the inclusion volume, the tensor $D(a)$ depends on the ellipsoid shape and it is defined in Eq. (4.92). If the inhomogeneity is an elliptical crack, then

$$P^0_{\alpha\beta\gamma\delta} = \frac{2}{3}\pi a_1 a_2 n_\alpha \left(T^0\right)^{-1}_{\beta\gamma} n_\delta\,|_{(\alpha\beta)(\gamma\delta)}, \tag{13.259}$$

where n is the normal to the crack plane, a_1, a_2 are the crack semiaxes. Tensor T^0 is defined in Eq. (5.34).

Let the medium contain a set of point defects, and σ^{*i} be the local exciting field that acts on the ith defect. Then, the stress and strain fields in the medium are presented in the forms that follow from Eqs. (13.251)–(13.252)

$$\varepsilon(x) = \varepsilon^0(x) + \sum_i \int K\left(x-x'\right) C^0 P^{0(i)}\sigma^{*i}\delta\left(x'-\xi_i\right)dx', \tag{13.260}$$

$$\sigma(x) = \sigma^0(x) + \sum_i \int S\left(x-x'\right) P^{0(i)}\sigma^{*i}\delta\left(x'-\xi_i\right)dx'. \tag{13.261}$$

Here $P^{0(i)}\sigma^{*i}$ is the coefficient of $\delta(x-\xi_i)$ in the multipole series for the density $m_i(x)$ for the ith inclusion. The local exciting field σ^* for the kth inclusion has the form

$$\sigma^{*k} = \sigma^0(\xi_k) + \sum_{i \neq k} \int S(\xi_k - x')\, P^{0(i)} \sigma^{*i} \delta(x' - \xi_i) dx', \tag{13.262}$$

where $\xi_k (k = 1, 2, ...)$ are the position vector of the kth point defects.

Let the points ξ_i be a realization of a homogeneous in space random point set X. We introduce the generalized functions $X(x)$ and $X(x, x')$ by the equations

$$X(x) = \sum_i \delta(x - \xi_i), \quad X(x'; x) = \sum_{i \neq k} \delta(x' - \xi_i) \text{ if } x = \xi_k. \tag{13.263}$$

By transition to the model of point defects, Eqs. (13.251), (13.252) for the stress and strain fields take the forms

$$\sigma(x) = \sigma^0(x) + \int S(x - x')\, P^0(x')\, \sigma^*(x')\, X(x') dx', \tag{13.264}$$

$$\varepsilon(x) = \varepsilon^0(x) + \int K(x - x')\, C^0 P^0(x')\, \sigma^*(x')\, X(x') dx', \tag{13.265}$$

where the functions $P^0(x)$ and $\sigma^*(x)$ coincide with $P^{0(i)}$ and σ^{*i} at the points $x = \xi_i$, $i = 1, 2, ...$. Eq. (13.262) for $\sigma^*(x)$ can be presented in the form

$$\sigma^*(x) = \sigma^0(x) + \int S(x - x')\, P^0(x')\, \sigma^*(x')\, X(x'; x) dx', \quad x \in X. \tag{13.266}$$

If the external field σ^0 is constant, the tensor $\sigma^*(x)$ at the points $\xi_i \in X$ is a homogeneous random function. For hypothesis H_1 of the effective field method (Sect. 13.1), the field σ^* is assumed to be the same for all defects. After averaging both sides of Eq. (13.266) over ensemble realizations of the sets X and P^0 by the condition $x \in X$, the equation for the tensor σ^* takes the form

$$\sigma^* = \sigma^0 + p \int S(x - x') \overline{P}^0 \Psi(x - x') dx' \sigma^*, \tag{13.267}$$

$$\overline{P}^0 = \frac{1}{v} P^0, \quad \Psi(x - x') = \frac{1}{n_0} \langle X(x'; x) \,|x \rangle, \tag{13.268}$$

where n_0 is the number of defects in a unit volume of the medium, $p = n_0 v$ are the volume fractions of the inclusions. The correlation function $\Psi(x)$ in these equations has the properties

$$\Psi(0) = 0, \quad \lim_{W \to \infty} \frac{1}{W} \int_W \Psi(x) dx = 1. \tag{13.269}$$

Calculating the integral in Eq. (13.266) and solving this equation with respect to σ^*, we obtain

$$\sigma^* = \left(I + pD\overline{P}^0\right)^{-1} \sigma^0, \quad D = \int S(x)(1 - \Psi(x))dx. \tag{13.270}$$

Averaging the equations for the stress and strain tensors in the medium with point defects in Eqs. (13.264), (13.265), taking into account that

$$\langle X(x) \rangle = n_0, \tag{13.271}$$

and definition (13.65) for the action of the operators S and K on constants, we yield

$$\langle \sigma \rangle = \sigma^0, \quad \langle \varepsilon \rangle = \varepsilon^0 + p\overline{P}^0 \left(I + pD\overline{P}^0\right)^{-1} \sigma^0. \tag{13.272}$$

As a result, the effective compliance tensor B^* of the medium with point defects takes the form

$$\langle \varepsilon \rangle = B^* \langle \sigma \rangle \,, \quad B^* = B^0 + p\overline{P}^0 \left(I + pD\overline{P}^0\right)^{-1}. \tag{13.273}$$

Suppose that there exists a linear transformation α of the x-space that converts the function $\Psi(x)$ into a spherically symmetric function

$$y = \alpha^{-1} x, \; \Psi(\alpha y) = \Psi'(|y|). \tag{13.274}$$

In this case, the tensor D in Eq. (13.270) takes the form

$$D = D(\alpha) = C^0 A(\alpha) C^0 - C^0, \tag{13.275}$$

where the tensor $A(\alpha)$ is defined in Eq. (4.89).

If the point defects simulate ellipsoidal homogeneous inclusions, the tensor P^0 for has form (13.258), and Eq. (13.273) results in the following equation for the effective elastic stiffness tensor C^*

$$C^* = C^0 + p\overline{P}^0 \left(I - pA(\alpha)\overline{P}^0\right)^{-1}, \quad \overline{P}^0 = C^1 \left(I + A(\alpha) C^1\right)^{-1}. \tag{13.276}$$

This equation coincides with Eq. (13.70) for C^* in the case of inclusions of finite sizes. Hence the point defect model results in the same expressions for the effective elastic stiffness tensor of the composites as by application of the EFM to the composites with inclusions of finite sizes.

13.7 Conclusion

In this chapter, the effective field method is applied to solution of the homogenization problem for matrix composites with a homogeneous in space random set of inclusions. The method results in finite equations for the effective property tensors of composites with various types of fillers. This is an advantage of the EFM over the effective medium method that reduces the homogenization problem to the equations that can be solved only numerically except some trivial cases.

Another important difference of the EFM from the EMM is that the equations of the EFM contain an additional structural characteristic of the composites: the correlation function of the random set of inclusions. This function allows taking into account details of spatial distributions of the inclusions in the host medium. For correlation functions with ellipsoidal symmetry, the equations of the EFM contain aspect ratios of the correlation holes only. Thus for application of the EFM, besides the volume fraction, properties and shapes of inclusions, the aspect ratios of the correlation hole for a typical inclusion should be available.

Comparison with experimental data and numerical simulations show that the simplest version of the EFM underestimates interaction between the inclusions, meanwhile the EMM overestimates the interactions.

The version of the EFM considered in this chapter as well as its simplest Mory-Tanaka version result in the loss of symmetry of the effective property tensors of composites with different families of inclusions (hybrid composites). The versions of the EFM that allow correcting these defects are considered in Chaps. 17 and 18.

13.8 Notes

In implicit forms, the effective field method was used for the calculation of effective elastic properties of composites with ellipsoidal inclusions in [16, 17]. An explicit formulation of the EFM in application to the homogenization problem for matrix composites with ellipsoidal inclusions and its limit forms (elliptical cracks and long cylindrical fibers) was used in [14, 18, 19]. In [20], the EFM was used for evaluation of stress concentrations on the inclusions in the matrix composite materials.

Mori-Tanaka method was first proposed in [6] and then developed in [7, 21] for the calculation of thermal conductivities and elastic properties of multi-phase composites. Relation of Mori-Tanaka approximation with variational bounds for the effective elastic moduli of composites was analyzed in [22]. Specific features of Mori-Tanaka method were discussed in [23].

Applications of the EFM to the solution of the homogenization problem for composites with regular lattices of ellipsoidal inclusions were considered in [19, 24].

In [25], the EFM was used to calculate the effective elastic moduli of composites reinforced with thin stiff flakes or ribbons. The elastic moduli of media with cracks were calculated by the EFM [26–28]. Conductive properties of composite materials with random sets of thin inclusions were considered in [29]. The results of these works are presented in Chaps. 14 and 15.

In [30], the EFM was applied to the calculation of the effective elastic moduli of composites reinforced with short axi-symmetric fibers (Chap. 16).

The coefficients of temperature expansions of composites with ellipsoidal inclusions and concentration of temperature stresses on the inclusions were found by the EFM in [14, 31]. Temperature deformation of the medium containing spherical multi-layered inclusions was considered in [32], and the medium reinforced with infinite cylindrical multi-layered fibers in [33].

Section 13.6 is based on the work [15], where the problem of modeling of the finite inclusions by point defects was considered.

References

1. Prokhorov Y, Rosanov Y (1969) The probability theory: basic concepts, limit theorems, random processes. Springer, Berlin
2. Mikhlin S (1962) Multi-dimensional singular integrals and integral equations. Phys Math Lit, Moscow, p 254
3. Eskin G (1981) Boundary-value problems for elliptic pseudo-differential equations. Math Soc 232. (New York, USA)
4. Nye J (1985) Physical properties of crystals, their representation by tensors and matrices. Clarendon Press, Oxford, p 329
5. Dean G, Lockett F (1973) Determination of the mechanical properties of fiber composites by ultrasonic techniques. Analysis of the test methods for high modulus fibers and composites. Am Soc Test Mater STP N521:326–346
6. Mori S, Tanaka K (1973) Average stress in matrix and average elastic energy of materials with misfitting inclusions. Acta Metall 21:571–574
7. Benveniste Y (1987) A new approach to the application of Mori-Tanaka's theory in composite materials. Mech Mater 6:147–157
8. Qiu Y, Weng G (1990) On the application of the Mori-Tanaka theory involving transversely isotropic spheroidal inclusions. Int J Eng Sci 28:1121–1137
9. Benveniste Y (1991) On diagonal and elastic symmetry of the approximate effective stiffness tensor of heterogeneous media. J Mech Phys Solids 39:927–946
10. Ferrari M (1991) Asymmetry and the high concentration limit of the Mori-Tanaka effective medium theory. Mech Mater 11:251–256
11. Lochin V, Sedov L (1963) Nonlinear tensor functions of several tensor arguments. Appl Math Mech (PMM) 27:393–417
12. Grigoliuk E, Filshtinsky L (1970) Perforated plates and shells. Nauka, Moscow, p 556
13. Kochetkov E, Leont'ev N, Ugodchikov N (1985) Effective thermo-elastic properties of regular fiber reinforced composites. In: Ugodchikov N (ed) Applied problems of strength and plasticity of materials, vol 29. Gorky State University, Gorky, pp 66–72
14. Levin V (1976) On the determination of elastic and thermoelastic constants of composite materials. Proc Acad Sci USSR Mech Solids 6:136–145
15. Kanaun S (1982) On the model of point defects in the mechanics of elastic inhomogeneous media. Proc Acad Sci USSR Mech Solids 4:109–118
16. Walpole L (1966) On bounds for the overall elastic modules of inhomogeneous systems. I, II J Mech Phys Solids 14:151–301
17. Walpole L (1969) On the overall elastic modules of composite materials. J Mech Phys Solids 17:235–251

18. Kanaun S (1975) The method of self-consistent field in the problem of effective elastic properties of composites. J Appl Mech Techn Phys 4:194–203
19. Kanaun S (1977) The approximation of self-consistent field for composite elastic media. J Appl Mech Techn Phys 2:160–169
20. Levin V (1977) On the stress concentration on inclusions in composite materials. Appl Math Mech (PMM) 41:735–743
21. Benveniste Y (1986) On Mori-Tanaka's method in cracked bodies. Mech Res Commun 13:193–201
22. Norris A (1989) An examination of the Mori-Tanaka effective medium approximation for multiphase composites. Trans ASME 88(56):84–88
23. Berryman J, Berge P (1996) Critique of two explicit schemes for estimating elastic properties of multiphase composites. Mech Mater 22:149–164
24. Kanaun S (1980) Interaction of periodical crack systems in an elastic medium. Appl Mech 16:36–42
25. Kanaun S, Kudryavtseva L (1988) Elastic properties of composites reinforced with thin hard inclusions. Mech Composite Mater 1:129–136
26. Kanaun S (1980) Poisson's set of cracks in an elastic solid. Appl Math Mech (PMM) 44:1129–1139
27. Kanaun S (1993) Effective elastic properties of the medium with an array of thin inclusions. Rev Mex Fis 39:74–83
28. Levin V, Markov M, Kanaun S (2004) Effective field method for seismic properties of cracked rocks. J Geophys Res 109(B08202):1–13
29. Levin V, Markov M (2004) Electroconductivity of a medium with thin low-resistivity inclusions. J Electrostatics 61:129–145
30. Kanaun S (1990) Elastic properties of composites reinforced with short axi-symmetric fibers. Appl Mech 26:56–63
31. Levin V (1982) On thermoelastic stresses in composite media. Appl Math Mech (PMM) 46:502–506
32. Kanaun S, Kudryavtzeva L (1987) Thermal stresses in composites with multi-layered spherical inclusions. Proc Acad Sci USSR Mech Solids 4:113–121
33. Kanaun S, Kudrjavtzeva L (1989) Thermo elastic properties of composite materials reinforced with unidirectional multi-layered fibers. Appl Math Mech (PMM) 53:798–807

Chapter 14
Effective Field Method for Composites with Cracks and Thin Soft Inclusions

Abstract The effective field method is applied to solution of the homogenization problem for a media with random sets of thin soft ellipsoidal inclusions or elliptical cracks. Specific features of application of the method for cracked media are indicated and discussed. Elliptical cracks of various orientations in 3D and 2D elastic media are considered. The predictions of the method are compared with experimental data and exact solutions for regular crack systems in 2D case.

Keywords Cracked medium · Effective field method · Elastic and conductive media with cracks · Isotropic media with elliptical cracks · 2D-problems for cracked media

Cracked medium is an adequate model for simulation of effective elastic and conductive properties of materials with dispersed damage. In this Chapter, the effective field method is applied to solution of the homogenization problem for homogeneous media with random sets of thin soft ellipsoidal inclusions and elliptical cracks. Details of application of the EFM for solution of the homogenization problem for cracked media are considered, predictions of the method are compared with experimental data.

14.1 Conductive Medium with a Random Set of Elliptical Cracks

We consider application of the effective field method to solution of the homogenization problem for a homogeneous medium with the conductivity tensor C^0 containing a random set of elliptical cracks. The one-particle conductivity problem for a medium containing a thin inclusion of low conductivity that covers also the crack problem is considered in Sect. 6.6. Integral presentations (6.94), (6.95) for the electric field $e(x)$ and electric current $j(x)$ vectors in the homogeneous medium with a crack can be extended in the case of the medium with multiple cracks as follows

S. Kanaun, *Self-Consistent Methods for Composites*, Solid Mechanics and Its Applications 282, https://doi.org/10.1007/978-3-031-96739-9_14

$$e_\alpha(x) = e^0_\alpha + \int K_{\alpha\beta}(x - x')C^0_{\beta\gamma}n_\beta(x')b(x')\Omega(x')dx', \tag{14.1}$$

$$j_\alpha(x) = j^0_\alpha + \int S_{\alpha\beta}(x - x')n_\beta(x')b(x')\Omega(x')dx', \tag{14.2}$$

Here, $\Omega(x) = \sum_i \Omega_i(x)$ and $\Omega_i(x)$ the delta-function concentrated on the surface Ω_i of the ith crack, scalar $b(x)$ is the jump of the field potential on the crack, $n(x)$ is the normal to the crack surfaces and

$$b(x) = b^{(i)}(x), n(x) = n^{(i)}(x) \text{ if } x \in \Omega_i. \tag{14.3}$$

The equation for the potential jump $b(x)$ follows from the boundary condition on the crack surfaces

$$n(x) \cdot j(x) = 0, x \in \Omega. \tag{14.4}$$

For an isolated elliptical crack Ω with semiaxes a_1, a_2 subjected to a constant external current vector j^0, the function $b(x)$ takes the form in Eqs. (6.107), (6.106) for $\lambda_0 = 0$

$$b(x) = T_0^{-1}\left(n \cdot j^0\right) z(x), x \in \Omega, \tag{14.5}$$

where

$$z(x) = \sqrt{1 - \left(\frac{x_1}{a_1}\right)^2 - \left(\frac{x_2}{a_2}\right)^2} \tag{14.6}$$

and x_1, x_2 are the axes of the cartesian coordinate system directed along the crack semiaxes. For an isotropic medium with the conductivity coefficient c_0, the scalar T_0 is defined in Eq. (6.113)

$$T^0 = \frac{c_0}{\sqrt{a_1 a_2}} \frac{E(1 - \gamma^2)}{\sqrt{\gamma}}, \gamma = \frac{a_2}{a_1}, a_1 > a_2, \tag{14.7}$$

where $E(\varsigma)$ is the elliptical integral of the second kind.

We consider a set of identical cracks of the same orientation randomly distributed in the host medium. The medium is subjected to a constant current vector j^0 at infinity. According to the hypothesis of the effective field method, each crack can be considered as an isolated one subjected to a constant effective current j^* that is the same for all cracks, and $j^* \neq j^0$. For this hypothesis, the electric field and current vectors in the cracked medium are approximated by the equation

$$e_\alpha(x) = e^0_\alpha + \int K_{\alpha\beta}(x - x')C^0_{\beta\gamma}\Lambda_{\gamma\delta}Z(x')\Omega(x')dx' j^*_\delta, \tag{14.8}$$

$$j_\alpha(x) = j^0_\alpha + \int S_{\alpha\beta}(x - x')\Lambda_{\beta\gamma}Z(x')\Omega(x')dx' j^*_\gamma, \tag{14.9}$$

where

$$\Lambda_{\alpha\beta} = T_0^{-1}n_\alpha n_\beta, \tag{14.10}$$

and the function $Z(x)$ coincides with the function $z(x)$ in Eq. (14.6) at the surface of each crack. The function

$$\Omega(x'; x) = \sum_{i\neq j} \Omega_i(x'), \quad \text{when } x \in \Omega_j, \tag{14.11}$$

allows presenting the local exiting field $j^*(x)$ that acts at points x on the crack surfaces in the form

$$j^*(x) = j^0 + \int S(x - x')\Lambda Z(x')\Omega(x'; x)dx' j^*, \quad x \in \Omega. \tag{14.12}$$

Averaging this equation by the condition that $x \in \Omega$, we obtain the equation

$$\left\langle j^*(x)|x\right\rangle = j^0 + \int S(x - x')\Lambda\left\langle Z(x')\Omega(x'; x)|x\right\rangle dx' j^*, \tag{14.13}$$

Here, $\langle f(x)|x\rangle = \langle f(x)|x \in \Omega\rangle$ is the mean under condition that $x \in \Omega$. Identifying the mean $\langle j^*(x)|x\rangle$ with the effective field j^*, we yield a closed equation for the effective current j^*

$$j^* = j^0 + \int S(x - x')\Lambda\left\langle Z(x')\Omega(x'; x)|x\right\rangle \mathrm{d}x' j^* \tag{14.14}$$

For calculation of the conditional mean in Eq. (14.14), first we consider the averages $\langle\Omega(x)\rangle$ and $\langle Z(x)\Omega(x)\rangle$ of the generalized functions $\Omega(x)$ and $Z(x)\Omega(x)$ over the ensemble realizations of the random crack set. For ergodic property, the average over the ensemble realizations can be changed with the spatial average for a typical realization of the crack set, and we have

$$\langle\Omega(x)\rangle = \lim_{W\to\infty}\frac{1}{W}\int_W \Omega(x)dx = \lim_{W\to\infty}\frac{1}{W}\sum_{i=1}^{N}\int_{\Omega_i} d\Omega = \frac{\pi}{v_0}a_1a_2, \tag{14.15}$$

$$\langle Z(x)\Omega(x)\rangle = \lim_{W\to\infty}\frac{1}{W}\int_W Z(x)\Omega(x)dx =$$

$$= \lim_{W\to\infty}\frac{1}{W}\sum_{i=1}^{N}\int_{\Omega_i} z(x)d\Omega = \frac{2\pi}{3v_0}a_1a_2. \tag{14.16}$$

Here, W is a region that occupies the entire 3D-space in the limit $W\to\infty$, N is the number of cracks in W, v_0 is the volume that falls at one crack.

The conditional mean $\langle f(x)|x\rangle$ of a random function $f(x)$ is defined by the equation [1]

$$\langle f(x)|x\in\Omega\rangle = \frac{\langle f(x)\Omega(x)\rangle}{\langle\Omega(x)\rangle}, \tag{14.17}$$

Thus the conditional mean under the integral in Eq. (14.14) is presented in the form

$$\left\langle Z(x')\Omega(x';x)|x\right\rangle = \frac{1}{\langle\Omega(x)\rangle}\left\langle Z(x')\Omega(x';x)\Omega(x)\right\rangle =$$

$$= \frac{1}{\langle\Omega(x)\rangle}\lim_{W\to\infty}\frac{1}{W}\sum_{\substack{i,j\\(i\neq j)}}\int_W Z(x')\Omega_i(x')\Omega_j(x)dx' =$$

$$= \frac{2\pi}{3v_0}a_1a_2\Psi(x-x'). \tag{14.18}$$

Here, the function $\Omega(x';x)$ is defined in Eq. (14.11). The correlation function $\Psi(x)$ in this equation has the properties

$$\Psi(0) = 0,\ \lim_{W\to\infty}\frac{1}{W}\int_W \Psi(x)dx = 1. \tag{14.19}$$

The first equation is a consequence the assumption that cracks do not intersect, the second equation can be proved by integrating both parts of Eq. (14.18) over the region W with further transition to the limit $W\to\infty$. As a result, Eq. (14.11) takes the form

$$j_\alpha^* = j_\alpha^0 + \tau\int S_{\alpha\beta}(x-x')\Psi(x-x')dx'\overline{\Lambda}j_\beta^*, \tag{14.20}$$

$$\tau = \frac{4}{3v_0}(a_1a_2)^{3/2},\ \overline{\Lambda} = \frac{\pi}{2\sqrt{a_1a_2}}\Lambda, \tag{14.21}$$

where tensor Λ is defined in Eq. (14.10).

Further, we assume that there exists a linear transformation of x-space $y = \alpha^{-1}x$ that converts the function $\Psi(x)$ into a spherically symmetric one

$$\Psi(x) = \Psi(\alpha y) = \Psi'(|y|). \tag{14.22}$$

This assumption corresponds to the random crack configurations with ellipsoidal regions around each crack where the probability to meet other cracks is small (Fig. 14.1). The function $\Psi(x)$ is called the correlation hole for a typical crack in the cracked medium.

Using the properties (14.19), (14.22) of the function $\Psi(x)$ and the regularization (11.81) for a fixed external current,

$$\int S(x)dx = 0, \int K(x)dx = B^0, \tag{14.23}$$

the integral in Eq. (14.20) is calculated as follows (Appendix B.2)

$$\int S(x)\Psi(x)dx = -\int S(x)\,(1 - \Psi(x))\,dx =$$

$$= -\int S(x)\,(1 - \Psi(x))\,dx = -D^*,\ D^* = \frac{1}{4\pi}\int\limits_{\Omega_1} \widetilde{S}(\alpha^{-1}k)d\Omega_k. \tag{14.24}$$

Finally, the equation for the effective current j^* follows from Eq. (14.20) in the form

$$j^* = (I + \tau D^*\overline{\Lambda})^{-1} j^0. \tag{14.25}$$

After averaging Eqs. (14.8), (14.9) for $e(x)$ and $j(x)$ over the ensemble of realization of the crack set, we obtain

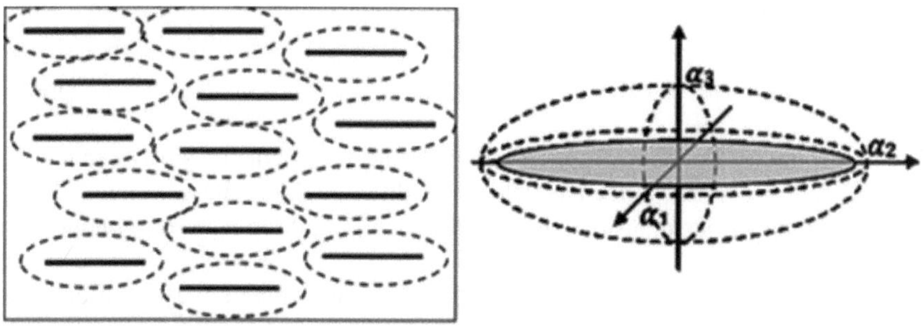

Fig. 14.1 A random set of cracks with regions around each crack with small probability to meet other cracks

$$\langle e(x)\rangle = e^0 + \int K(x - x')dx'C^0 \Lambda \left\langle Z(x')\Omega(x')\right\rangle j^*, \tag{14.26}$$

$$\langle j(x)\rangle = j^0 + \int S(x - x')dx'\Lambda \left\langle Z(x')\Omega(x')\right\rangle j^*, \tag{14.27}$$

For a fixed external current j^0, Eqs. (14.23) for the integrals in these equations hold, and using Eq. (14.16) for the mean $\langle Z(x)\Omega(x)\rangle$, we yield

$$\langle e(x)\rangle = e^0 + \tau\overline{\Lambda} j^*, \langle j(x)\rangle = j^0. \tag{14.28}$$

From these equations and Eq. (14.25), we obtain relation between the mean electric field and current in the cracked medium

$$\langle e(x)\rangle = B^* \langle j(x)\rangle, \tag{14.29}$$

$$B^* = B^0 + \tau\overline{\Lambda}(I + \tau D^*\overline{\Lambda})^{-1}. \tag{14.30}$$

Here, B^* is the effective resistivity tensor of the cracked medium.

The tensor D^* depends on the shape of the correlation hole $\Psi(x)$, and D^* relates to the tensor A^* in Eq. (4.200) for $A^* = A(\alpha)$ by the equation

$$D^* = C^0 A^* C^0 - C^0. \tag{14.31}$$

If the cracks can by arbitrary close to each other, the aspect ratios of the correlation hole $\gamma_1^* = \alpha_3/\alpha_1, \gamma_2^* = \alpha_3/\alpha_2$ tends to zero, and for an isotropic host medium we have

$$A^*_{\alpha\beta} = \frac{1}{c_0} n_\alpha n_\beta, D^*_{\alpha\beta} = c_0(\delta_{\alpha\beta} - n_\alpha n_\beta). \tag{14.32}$$

For such D^*, product of this tensor with the tensor $\overline{\Lambda}$ in Eq. (14.21) vanishes, and the field j^* and the effective resistivity tensor $B^*$are

$$j^* = j^0, B^* = B^0 + \tau\overline{\Lambda}. \tag{14.33}$$

The first equation can be interpreted as the absence of interactions between cracks. Only if $\gamma_1^*, \gamma_2^* \neq 0$, the effective field j^* is different from the external field j^0, and crack interactions contribute to the values of j^* and the effective conductivity tensor B^*.

For circular cracks in an isotropic medium with the conductivity coefficient c_0, $\gamma_1^* = \gamma_2^* = \gamma_*$, and tensor D^* is transverse isotropic

$$D^*_{\alpha_\beta} = c_0(A_1(\gamma_*) - 1)\theta_{\alpha\beta}(n) + c_0\left(A_3(\gamma_*) - 1\right) n_\alpha n_\beta, \tag{14.34}$$

where

$$A_1(\gamma_*) = \frac{\gamma_* \left(\arccos(\gamma_*) - \gamma_* \sqrt{1-\gamma_*^2}\right)}{2\left(1-\gamma_*^2\right)^{3/2}}, \tag{14.35}$$

$$A_3(\gamma_*) = 1 - 2A_1(\gamma_*). \tag{14.36}$$

It follows from Eq. (14.30) that the effective conductivity tensor $C^* = (B^*)^{-1}$ of an isotropic medium with a random set of circular cracks of the same orientations is transverse isotropic, and

$$C^*_{\alpha\beta} = c_0(\delta_{\alpha\beta} - n_\alpha n_\beta) + c_3^* n_\alpha n_\beta, \tag{14.37}$$

where the conductivity coefficient c_3^* is

$$c_3^* = c_0\left(1 - \frac{\tau}{1+\tau A_3(\gamma_*)}\right). \tag{14.38}$$

Dependencies the coefficient c_3^* on the crack density τ and the aspect ratio γ_* of the correlation hole are shown in Fig. 14.2. The dashed line in this figure is the prediction of the effective medium method for the isotropic medium with circular cracks of the same orientation indicated in Eq. (11.90).

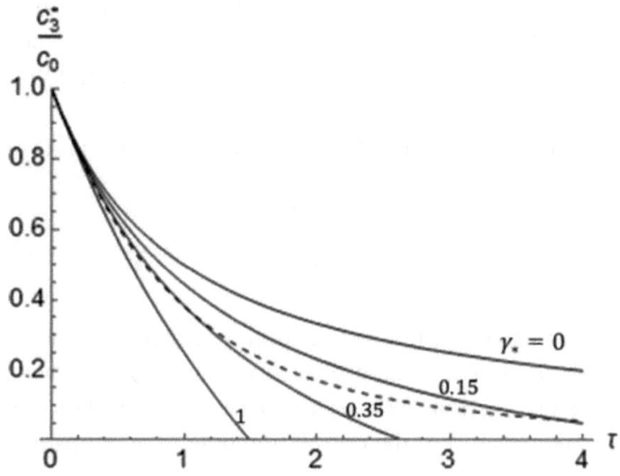

Fig. 14.2 The effective conductivity coefficient c^* for the medium with a random set of circular cracks of the same orientations; γ_* is the aspect ratio of the correlation hole around a typical crack; the dashed line is prediction of the effective medium method for the coefficient c_{eff}

14.2 Effective Conductivity of the Medium with Cracks of Various Orientations

The orientation of an elliptical crack in 3D-space is defined by three Euler angles with respect to a global cartesian coordinate system. For simplicity, we consider circular cracks which orientations are defined by unit vector n normal to the crack surface. For the host medium containing cracks of various orientations, it is reasonable to assume that the effective field acting on each crack depends on the crack orientation $j^* = j^*(n)$. The approximate Eq. (14.12) for the effective current acting at points on the crack surfaces is presented in the form

$$j^*(x) = j^0 +$$

$$+ \int S(x - x')\Lambda(n')Z(x')j^*(n')\Omega(x; x')dx', \quad x \in \Omega, \tag{14.39}$$

where $n' = n(x')$ is the normal to the crack plane at point x'. After averaging this equation by the condition that x is on the crack surface of the orientation n, we yield

$$\left\langle j^*(x)|x, n\right\rangle = j^0 +$$

$$+ \int S(x - x')\left\langle\Lambda(n')Z(x')j^*(n')\Omega(x; x')|x, n\right\rangle dx'. \tag{14.40}$$

Here, $\langle j^*(x)|x, n\rangle = j^*(n)$ is the effective field that acts on all the cracks of the orientation n. The mean under the integral in this equation is presented in the form similar to the case of cracks of the same orientations

$$\left\langle\Lambda(n')Z(x')j^*(n')\Omega(x; x')|x, n\right\rangle = \tau\left\langle\overline{\Lambda}(n)j^*(n)\right\rangle\Psi(x - x', n), \tag{14.41}$$

$$\Psi(x - x', n) = \frac{\left\langle\Omega(x; x')|x, n\right\rangle}{\langle\Omega(x)\rangle}, \tau = \frac{4}{3v_0}(a_1a_2)^{3/2}, \overline{\Lambda} = \frac{\pi}{2\sqrt{a_1a_2}}\Lambda. \tag{14.42}$$

For cracks of various orientation we accept that there is an ellipsoidal region around each crack where the probability to meet other cracks is small (Fig. 14.3).

The function $\Psi(x, n)$ describes the correlation hole around the cracks of the orientation n. For the correlation hole of an ellipsoidal shape

$$\int S(x - x')\Psi(x - x', n)dx' = -D^*(n), \tag{14.43}$$

where $D^*(n)$ is defined in Eq. (14.24). As a result, we have from Eq. (14.40)

$$j^*(n) = j^0 - \tau D^*(n)\left\langle\overline{\Lambda}(n)j^*(n)\right\rangle. \tag{14.44}$$

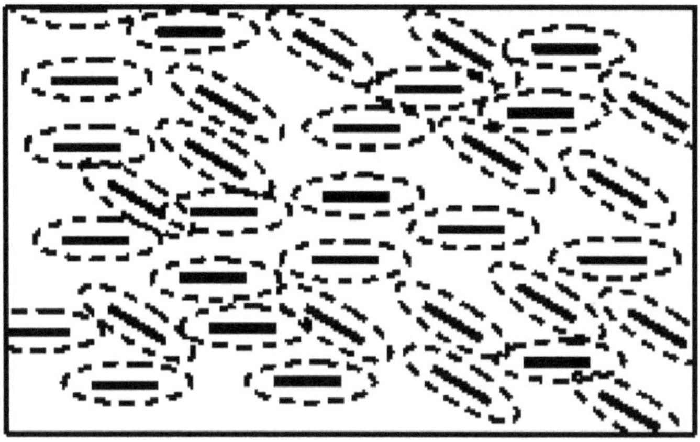

Fig. 14.3 The host medium with cracks of various orientation, the dashed lines indicate the regions around each crack with small probability to meet other cracks

Here, averaging is performed over the crack orientations n. Multiplying this equation with the tensor $\overline{\Lambda}(n)$ and averaging the result over the crack orientations, we obtain

$$\left\langle\overline{\Lambda}(n)j^*(n)\right\rangle = \left\langle\overline{\Lambda}(n)\right\rangle j^0 - \tau\left\langle\overline{\Lambda}(n)D^*(n)\right\rangle^{-1}\left\langle\overline{\Lambda}(n)j^*(n)\right\rangle. \tag{14.45}$$

Thus for the mean $\left\langle\overline{\Lambda}(n)j^*(n)\right\rangle$, we yield

$$\left\langle\overline{\Lambda}(n)j^*(n)\right\rangle = \left(1 + \tau\left\langle\overline{\Lambda}(n)D^*(n)\right\rangle\right)^{-1}\left\langle\overline{\Lambda}(n)\right\rangle j^0. \tag{14.46}$$

The approximate equations for the electric field and current in the cracked medium are presented in the forms

$$\langle e(x)\rangle = e^0 + \tau\int K(x-x')dx'C^0\left\langle\overline{\Lambda}(n)j^*(n)\right\rangle, \tag{14.47}$$

$$\langle j(x)\rangle = j^0 + \tau\int S(x-x')dx'\left\langle\overline{\Lambda}(n)j^*(n)\right\rangle. \tag{14.48}$$

For a fixed external electric current Eqs. (14.23) hold, and we have

$$\langle j(x)\rangle = j^0, \tag{14.49}$$

$$\langle e(x)\rangle = B^0 j^0 + \tau\left\langle\Lambda(n)j^*(n)\right\rangle =$$

$$= B^0 j^0 + \left(1 + \tau\left\langle\overline{\Lambda}(n)D^*(n)\right\rangle\right)^{-1}\left\langle\overline{\Lambda}(n)\right\rangle j^0. \tag{14.50}$$

As a result, the effective resistivity tensor of the cracked medium takes the form

$$B^* = B^0 + \tau\left(1 + \tau\left\langle\overline{\Lambda}(n)D^*(n)\right\rangle\right)^{-1}\langle\Lambda(n)\rangle\,. \tag{14.51}$$

Here, $\langle\cdot\rangle$ means averaging over crack orientations.

For an isotropic host medium with the conductivity coefficient c_0, the tensors $\overline{\Lambda}(n)$ is

$$\overline{\Lambda}_{\alpha\beta}(n) = \frac{1}{c_0}n_\alpha n_\beta, \tag{14.52}$$

and tensor $D^*(n)$ is defined in Eq. (14.34)

$$D^*_{\alpha\beta}(n) = D_1(\gamma^*)\theta_{\alpha\beta}(n) + D_3(\gamma_*)n_\alpha n_\beta, \tag{14.53}$$

$$D_1(\gamma_*) = c_0(A_1(\gamma_*) - 1),\; D_3(\gamma_*) = c_0\left(A_3(\gamma_*) - 1\right)n_\alpha n_\beta, \tag{14.54}$$

$$A_3(\gamma_*) = 1 - 2A_1(\gamma_*), \tag{14.55}$$

where the coefficients $A_1(\gamma_*)$ is in Eqs. (14.35), (14.36). For the homogeneous distribution of cracks over orientations, we have

$$\langle\Lambda(n)\rangle = \frac{1}{3c_0}I,\; \left\langle\Lambda(n)D^*(n)\right\rangle = \frac{D_3(\gamma_*)}{3c_0}I, \tag{14.56}$$

Here, I is unit second rank tensor. The effective medium is isotropic with the conductivity coefficient

$$\frac{c_*}{c_0} = \left(1 + \frac{\tau}{(3 + \tau(A_3(\gamma_*) - 1))}\right)^{-1}, \tag{14.57}$$

$$A_3(\gamma_*) - 1 = \frac{\gamma_*\left(\gamma_*\sqrt{1-\gamma_*^2} - \arccos(\gamma_*)\right)}{\left(1-\gamma_*^2\right)^{3/2}} \tag{14.58}$$

Dependencies of the effective conductivity on the crack density τ and the aspect ratio γ_* are shown in Fig. 14.4. The dashed line in this figure is the prediction of the effective medium method for the effective conductivity of the medium with penny shape cracks homogeneously distributed over the orientations (Eq. (11.93)).

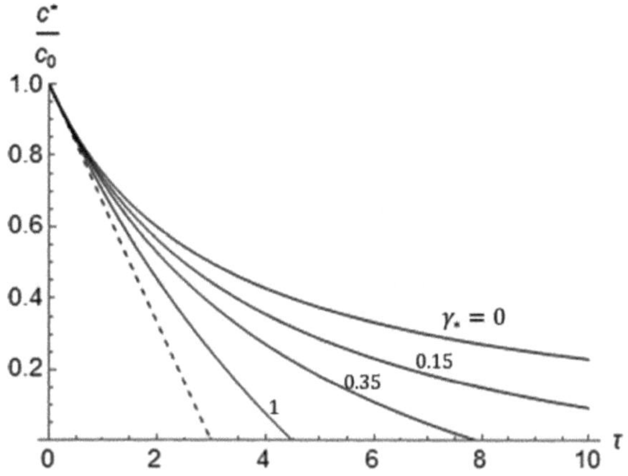

Fig. 14.4 Dependencies of the effective conductivity on the crack density τ and the aspect ratio γ_*, the dashed line is predictions of the effective medium method for the effective conductivity of the host medium with penny shape cracks homogeneously distributed over orientations

14.3 Elastic Medium with Cracks and Crack-Like Inclusions

In this Section, we consider a homogeneous elastic medium with thin inclusions which elastic moduli considerably smaller than the moduli of the host medium $(C(C^0)^{-1} = O(\delta_2),\ \delta_2 << 1)$. The principal terms of the asymptotic solution of the one-particle problem for such an inclusion with respect to the small parameters δ_2 and $\delta_1 = h/a$ (h is the transverse size of the inclusion, and a are the characteristic size of the inclusion middle surface) are considered in Chap. 6.

We start with a homogeneous elastic medium containing a random set of identical elliptical crack of the same orientation. Let Ω_i be the surface of the ith crack, and $\Omega_i(x)$ is delta-function concentrated on Ω_i. According to hypothesis H_2 of the effective field method, each crack in the cracked medium is subjected to the same constant local effective stress field σ^*. This assumption results in the following integral presentations for the strain and stress tensors in the cracked medium

$$\varepsilon(x) = \varepsilon^0 + \int K(x - x')C^0 \Lambda Z(x')\Omega\left(x'\right)dx'\sigma^*, \tag{14.59}$$

$$\sigma\left(x\right) = \sigma^0 + \int S\left(x - x'\right)\Lambda Z\left(x'\right)\Omega\left(x'\right)dx'\sigma^*. \tag{14.60}$$

Here, $\Omega\left(x\right)$ is delta function concentrated on the surfaces of all cracks

$$\Omega\left(x\right) = \sum_i \Omega_i\left(x\right). \tag{14.61}$$

The function $Z(x)$ coincides with the function

$$z(x)=\sqrt{1-\frac{x_1^2}{a_1^2}-\frac{x_2^2}{a_2^2}} \tag{14.62}$$

on the surface of each crack; the axes x_1, x_2 of the cartesian coordinate system with the origin at the crack center are directed along the crack semiaxes a_1, a_2. The forth rank tensor Λ is defined from the solution of the one-particle problem for an isolated crack subjected to a constant external stress field (Sect. 5.3).

$$\Lambda=\frac{2\sqrt{a_1a_2}}{\pi}\overline{\Lambda}$$

$$\overline{\Lambda}_{\alpha\beta\gamma\delta}=\frac{\pi}{2\sqrt{a_1a_2}}n_\alpha\left(T_0^{-1}\right)_{\beta\gamma}n_\delta\Big|_{(\alpha\beta)(\gamma\delta)}, \tag{14.63}$$

For an arbitrary anisotropic host medium, the tensor T_0 in this equation is the double integral in Eq. (5.34). For isotropic host medium, tensor $\overline{\Lambda}$ is presented in the T-basis (Appendix A.3) as follows

$$\overline{\Lambda}=\frac{1}{4}\left(\Lambda_1T^{11}+\Lambda_2T^{12}\right)+\Lambda_3T^3, \tag{14.64}$$

Here, the tensors T^3, T^{11}, T^{12} are

$$T^3_{\alpha\beta\gamma\delta}=n_\alpha n_\beta n_\gamma n_\delta,\ n=n^3,$$

$$T^{11}_{\alpha\beta\gamma\delta}=4n_\alpha n^1_\beta n^1_\gamma n_\delta\big|_{(\alpha\beta)(\gamma\delta)},\ T^{12}_{\alpha\beta\gamma\delta}=4n_\alpha n^2_\beta n^2_\gamma n_\delta\big|_{(\alpha\beta)(\gamma\delta)}, \tag{14.65}$$

and n^1, n^2 are unit vectors directed along the ellipse semiaxes a_1, a_2. The scalar coefficients Λ_i in Eq. (13.160) are expressed in terms of complete elliptical integrals

$$\Lambda_1=\frac{\pi(1-\nu_0)}{\mu_0\left(c_1+\nu_0c_2\right)},\ \Lambda_2=\frac{\pi(1-\nu_0)}{\mu_0\left(c_1+\nu_0c_3\right)}, \tag{14.66}$$

$$\Lambda_3=\frac{\pi(1-\nu_0)}{\mu_0c_1},\ c_1=\frac{E(\eta)}{\sqrt{\gamma}},\ c_2=\frac{\gamma^2K(\eta)-E(\eta)}{\sqrt{\gamma}\eta}, \tag{14.67}$$

$$\eta=1-\gamma^2,\ \gamma=\frac{a_2}{a_1},\ a_1\geq a_2. \tag{14.68}$$

Here, ν_0 is Poisson's ratio of the host medium, $K(\eta)$ and $E(\eta)$ are the complete elliptic integrals of the first and second kinds

$$K(\eta) = \int_0^{\pi/2} \frac{d\theta}{\sqrt{1-\eta\sin^2\theta}}, \; E(\eta) = \int_0^{\pi/2} \sqrt{1-\eta\sin^2\theta d\theta}. \tag{14.69}$$

For a circular crack in an isotropic medium, $a_1 = a_2, \gamma = 1$ and the coefficients Λ_i take the forms

$$\Lambda_1 = \Lambda_2 = \frac{4}{\mu_0}\frac{(1-\nu_0)}{(2-\nu_0)}, \; \Lambda_3 = \frac{2(1-\nu_0)}{\mu_0}, \tag{14.70}$$

and in this case, tensor $\overline{\Lambda}$ can be presented in the P-basis

$$\overline{\Lambda} = \frac{4}{\mu_0}\frac{(1-\nu_0)}{(2-\nu_0)} P^5(n) + \frac{2(1-\nu_0)}{\mu_0} P^6(n) \tag{14.71}$$

with unit vector n normal to the crack plane.

The effective compliance tensor of the elastic medium with identical elliptical cracks of the same orientation has form in Eq. (14.30)

$$B^* = B^0 + \tau\overline{\Lambda}(I + \tau D^*\overline{\Lambda})^{-1}, \quad \tau = \frac{4}{3}\frac{(a_1 a_2)^{3/2}}{v_0} \tag{14.72}$$

where $B^0 = \left(C^0\right)^{-1}$, tensor $\overline{\Lambda}$ is defined in Eq. (14.63). Tensor D^* depends on the shape of the correlation holes $\Psi(x)$. If this shape is ellipsoidal and a second rank tensor α^* defines the shape and orientation of the ellipsoid,

$$D^* = C^0 A(\alpha^*) C^0 - C^0, \tag{14.73}$$

where tensor $A(\alpha^*)$ in the T-basis is in Eq. (4.120). It is assumed that the principal axis of the correlation hole coincide with the principal axes of the crack, and $\gamma_1^* = \alpha_3^*/\alpha_1^*, \gamma_1^* = \alpha_3^*/\alpha_2^*$ are aspect ratios of the correlation hole. If there are no restrictions for crack mutual positions, α_3^* tends to zero, and tensor D^* takes the form

$$D^* = -\mu_0\left(P^1(n) + (2\kappa_0 - 1)P^2(n)\right). \tag{14.74}$$

In the T-basis, the tensors $P^1(n)$ and $P^2(n)$ are

$$P^1(n) = T^1 + T^2 + \frac{1}{2}T^{10}, \; P^2(n) = T^1 + T^2 + T^4 + T^5. \tag{14.75}$$

For D^* in Eq. (14.74), the product $D^*\Lambda$ in Eq. (14.72) vanishes, and the effective compliance tensor B^* is a linear function of the crack density parameter τ

$$B^* = B^0 + \tau\overline{\Lambda}, \tau = \frac{4}{3}\frac{(a_1a_2)^{3/2}}{v_0}. \tag{14.76}$$

In the T-basis, this tensor has the form

$$B^* = B^0 + \tau(\Lambda_1 T^{11} + \Lambda_2 T^{12} + \Lambda_3 T^3). \tag{14.77}$$

Here, $B^0 = \left(C^0\right)^{-1}$, and the cracked medium has 3 additional elastic constants.

For circular cracks, the medium is transverse isotropic, and the effective compliance tensor B^* is

$$B^* = \frac{(1-\nu_0)}{4\mu_0(1+\nu_0)}P^1 + \frac{1}{2\mu_0}P^2 - \frac{\nu_0}{2\mu_0(1+\nu_0)}\left(P^3+P^4\right) +$$

$$+\frac{1}{\mu_0}\left(1+4\tau\frac{(1-\nu_0)}{(2-\nu_0)}\right)P^5 + \frac{1}{2\mu_0(1+\nu_0)}\left(1+4\tau\left(1-\nu_0^2\right)\right)P^6. \tag{14.78}$$

The case $\gamma = 1$ corresponds to the crack distribution with a spherical region around each crack with small probability to meet other cracks in this region. In this case, the function $\Psi(x)$ has spherical symmetry $\Psi(x) = \Psi(|x|)$, and the tensor D^* is isotropic

$$D^* = \frac{2\mu_0}{15(1-\nu_0)}\left((1+5\nu_0)E^2 - (7-5\nu_0)E^1\right). \tag{14.79}$$

For spheroidal regions with aspect ratio γ^* around each circular crack, tensor D^* in the P-basis has the form

$$D^* = d_1 P^2 + d_2\left(P^1 - \frac{1}{2}P^2\right) + d_3\left(P^3+P^4\right) + d_5 P^5 + d_6 P^6, \tag{14.80}$$

where $P^i = P^i(n)$, and the coefficients d_i depends on the aspect ratio γ^* of the correlation hole

$$d_1 = -\mu_0\left((1-4\kappa_0)(2\beta_1-1) + 8\kappa_0\beta_2\right), \tag{14.81}$$

$$d_2 = -2\mu_0\big(1 - 2\beta_1 + 4\kappa_0\beta_2\big), \tag{14.82}$$

$$d_3 = 2\mu_0\big((1-4\kappa_0)\beta_1 + 8\kappa_0\beta_2\big), \tag{14.83}$$

$$d_5 = -4\mu_0\left((1+4\kappa_0)\beta_1 - 16\kappa_0\beta_2\right), \quad d_6 = -32\kappa_0\mu_0\beta_2, \tag{14.84}$$

Here $\beta_i = \beta_i(\gamma^*)$, $i = 1, 2$ and

$$\beta_1(\gamma) = \frac{\gamma^2}{4}\int_0^{\pi}\frac{(\sin\theta)^3\,d\theta}{\Delta(\theta)},\ \beta_2(\gamma) = \frac{\gamma^4}{16}\int_0^{\pi}\frac{(\sin\theta)^5}{\Delta(\theta)^2}d\theta, \tag{14.85}$$

$$\Delta(\theta) = (\gamma\sin\theta)^2 + (\cos\theta)^2. \tag{14.86}$$

The effective compliance tensor of the medium with circular cracks of the same orientation is

$$B^* = B^0 + \frac{4\tau\Lambda_1}{4+\tau\Lambda_1 d_5}P^5(n) + \frac{\tau\Lambda_3}{1+\tau\Lambda_3 d_6}P^6(n)\ , \tag{14.87}$$

In the P-basis, the effective stiffness tensor $C^* = (B^*)^{-1}$ is

$$C^* = k_* P^2 + 2m_*\left(P^1 - \frac{1}{2}P^2\right) + \lambda_*\left(P^3 + P^4\right) + 4\mu_* P^5 + n_* P^6. \tag{14.88}$$

The dependencies of the effective parameters k_*, λ_*, μ_*, n_* on the crack density parameter τ and the correlation hole aspect ratio γ^* are shown in Figs. 14.5 and 14.6, the host medium is isotropic with shear modulus μ_0 and Poisson ratio $\nu_0 = 0.3$, in Eq. (14.88), the constant m_* is equal to μ_0.

In the case of thin soft ellipsoidal inclusions, $C\left(C^0\right)^{-1} = O(\delta_2)$, $\delta_2 \ll 1$ and the inclusion aspect ratios $\gamma_1, \gamma_2 \ll 1$, the tensor T_0 in Eq. (14.63) should be changed with the tensor

$$T_0' = \frac{1}{2h}n\cdot C\cdot n + T_0, \tag{14.89}$$

where h is the smallest ellipsoid semiaxis, C is elastic stiffness tensor of the inclusion material.

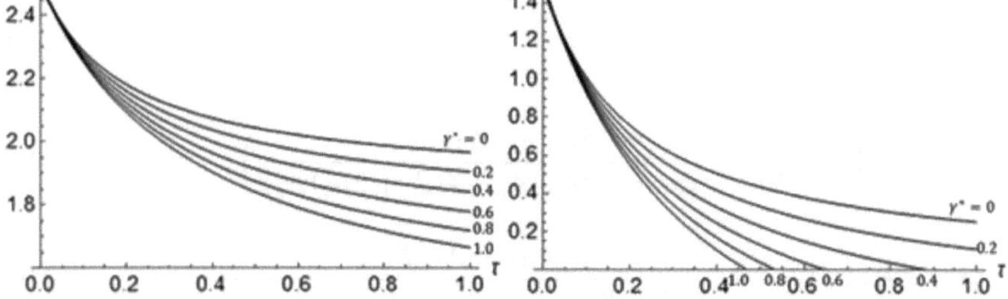

Fig. 14.5 Dependencies of the effective elastic constants E_1^* and E_3^* in Eq. (14.90) on the crack density τ and the correlation hole aspect ratio γ_* for an isotropic medium with a random set of circular cracks of the same orientations

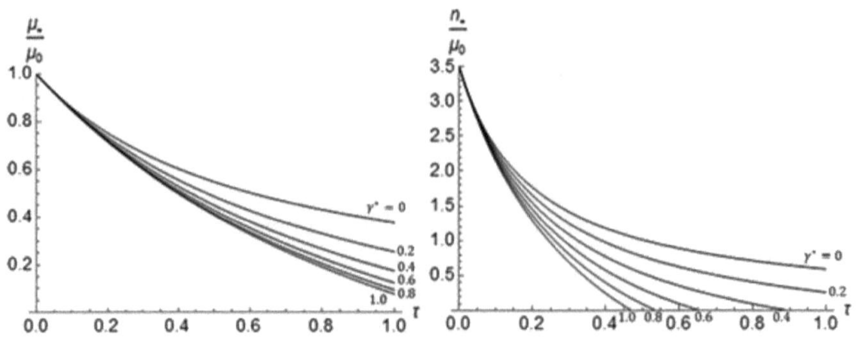

Fig. 14.6 The same as in Fig. 14.5 for the effective elastic constants ν^*_{12} and ν^*_{13} in Eq. (14.90)

For the medium with elliptical cracks of various orientations, the effective compliance tensor $B^*$of the medium has the form in Eq. (14.51) when the tensors $\overline{\Lambda}$ and D^* are changed with their analogs for elasticity

$$B^* = B^0 + \tau\left(1 + \tau\left\langle\overline{\Lambda}(n)D^*(n)\right\rangle\right)^{-1}\left\langle\overline{\Lambda}(n)\right\rangle. \tag{14.90}$$

Here, $\langle\cdot\rangle$ means averaging over crack orientations. For instants, in the case of homogeneous distribution of circular cracks over orientations, we have

$$\left\langle\overline{\Lambda}(n)\right\rangle = \Lambda_1\left\langle P^5\right\rangle + \Lambda_3\left\langle P^6\right\rangle, \tag{14.91}$$

$$\left\langle\overline{\Lambda}(n)D^*(n)\right\rangle = d_3\Lambda_3\left\langle P^4\right\rangle + \frac{1}{2}d_5\Lambda_1\left\langle P^5\right\rangle + d_6\Lambda_3\left\langle P^6\right\rangle, \tag{14.92}$$

$$\left\langle P^4\right\rangle = \frac{1}{3}E^2, \left\langle P^5\right\rangle = \frac{1}{3}E^1, \left\langle P^6\right\rangle = \frac{1}{15}\left(2E^1 + E^2\right), \tag{14.93}$$

where E^1, E^2 are the elements of the E-basis. The effective stiffness tensor of the cracked medium is isotropic and

$$C^* = K_* E_2 + 2\mu_*\left(E_1 - \frac{1}{3}E^2\right), \tag{14.94}$$

$$\mu_* = \mu_0\left(1 - \frac{4\tau\,(5\Lambda_1 + 2\Lambda_3)\,\mu_0}{30 + \tau\,(5(d_5 + 4\mu_0)\,\Lambda_1 + 4(d_6 + 2\mu_0)\Lambda_3)}\right), \tag{14.95}$$

$$K_* = K_0\left(1 - \frac{6\tau\,(\Lambda_1 + \Lambda_3)\,K_0}{6 + \tau\,((d_5 + 8k_0)\,\Lambda_1 + 2(3(d_3 + K_0) + d_6)\Lambda_3)}\right), \tag{14.96}$$

$$K_0 = \frac{2\mu_0(1 + \nu_0)}{3(1 - 2\nu_0)}, \; K_* = \frac{2\mu_*(1 + \nu_*)}{3(1 - 2\nu_*)}. \tag{14.97}$$

In Fig. 14.8 dependencies of the effective shear modulus μ_* and Poisson ratio ν_* of the isotropic medium with circular cracks homogeneously distributed over orientations on the crack density τ and aspect ratio γ^* of the correlation hole are shown.

It follows from Figs. 14.4, 14.5, 14.6 and 14.7 that the crack interaction effect depends strongly on the parameter γ^*. When $\gamma^* = 0$, the effective field σ^*coincides with the external field σ^0 applied to the medium, and crack interactions absent. Interactions become stronger when γ^* increases. This effect originates to a specific stress field induced by a crack in an elastic medium. The regions with large stresses are localized near the crack edges but the regions near the crack central parts are unloaded. If crack positions are not restricted, the regions of large and small stress levels are compensated, and the average crack opening coincides with the opening of an isolated crack. If crack statistics is such as the probability of crack appearance in the regions of low stresses is small, the average crack opening grows and the

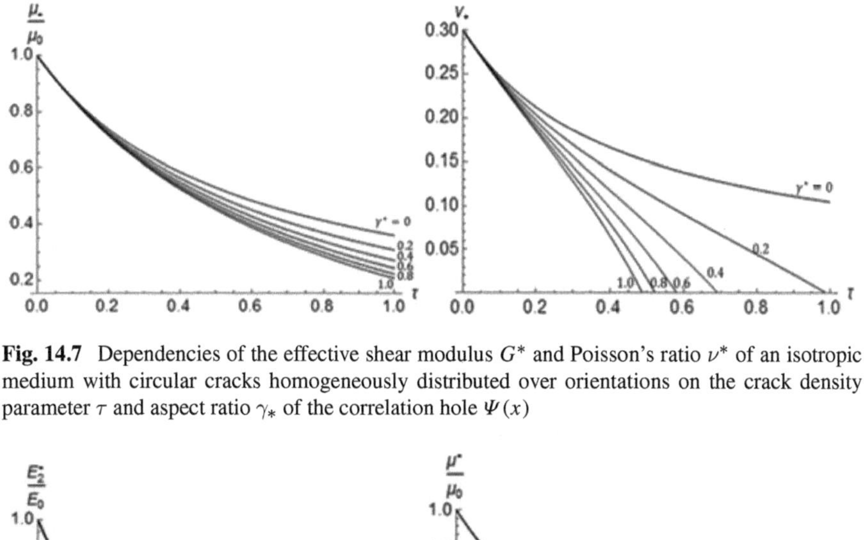

Fig. 14.7 Dependencies of the effective shear modulus G^* and Poisson's ratio ν^* of an isotropic medium with circular cracks homogeneously distributed over orientations on the crack density parameter τ and aspect ratio γ_* of the correlation hole $\Psi(x)$

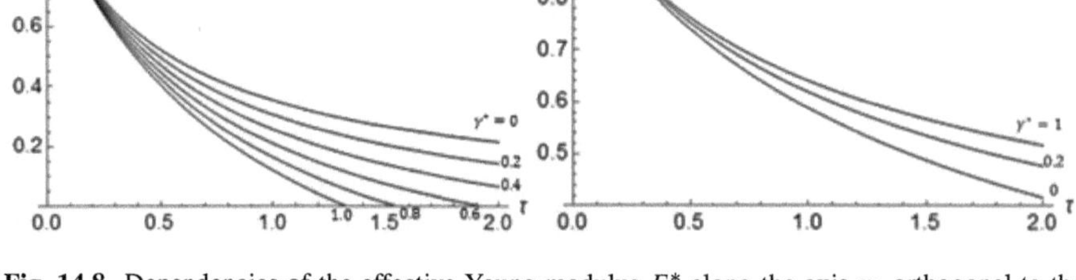

Fig. 14.8 Dependencies of the effective Young modulus E_2^* along the axis x_2 orthogonal to the crack lines and the effective shear modulus G_{12}^* of the medium with a random set of cracks of the same orientations are shown for the values of the parameter $0 < \gamma^* < 1$, $\nu_0 = 0.3$

interaction effect becomes stronger. The larger is parameter γ^* the smaller is the probability to meet a crack in the regions of low stresses.

14.4 2D Medium with Cracks and Thin Soft Inclusions

In this Section, the 2D problem of elasticity for a medium with a set of direct cracks is considered. The solution of the homogenization problem by the effective field method is compared with predictions of other self-consistent methods and experimental data.

14.4.1 *The One-Particle 2D-Problem for a Direct Crack*

Let a crack occupy the interval $(-l, l)$ of the x_1-axis in the elastic plane subjected to a constant external stress field σ^0 at infinity, x_1, x_2 are cartesian coordinate axes in the plane. Crack opening vector $b(x_1)$ is the solution of the integral equation (5.12) that in the 2D-case takes the form

$$\int_{-l}^{l} T_{\alpha\beta}(x_1 - x_1', 0)b_\beta(x_1')dx_1 = n_\beta\sigma^0_{\beta\alpha}, x_1 \in (-l, l). \tag{14.98}$$

where n is the normal to the crack line, the kernel $T(x_1, x_2)$ is expressed in terms of the function $S(x)$ for 2D-elastic media

$$T_{\alpha\beta}(x_1, x_2) = -n_\gamma S_{\gamma\alpha\beta\delta}(x_1, x_2)n_\delta. \tag{14.99}$$

For a direct crack, the solution of Eq. (14.98) has the form

$$b_\alpha(x_1) = B_\alpha\sqrt{1 - \frac{x_1^2}{l^2}}, \tag{14.100}$$

where B is a constant vector. From Eq. (14.98) we have

$$T^0_{\alpha\beta}B_\beta = n_\beta\sigma^0_{\beta\alpha}, \tag{14.101}$$

$$T^0_{\alpha\beta} = \int_{-l}^{l} T_{\alpha\beta}(x_1, 0)\sqrt{1 - \frac{x_1^2}{l^2}}dx_1 = \frac{1}{l}\int_{-1}^{1} T_{\alpha\beta}(\varsigma, 0)\sqrt{1 - \varsigma^2}d\varsigma. \tag{14.102}$$

Here, $\varsigma = x_1/l$, and it is taken into account that $T(x)$ is a homogeneous function of the order of -2. Transferring to Fourier transforms of the integrand functions, the constant tensor T^0 is presented in the form

$$T^0 = \frac{1}{2l}\int_{-\infty}^{\infty}\frac{1}{k}\widetilde{T}(k)J_1(k)dk, \tag{14.103}$$

where the equation

$$\int_{-l}^{l}\sqrt{1-\varsigma^2}e^{ik\varsigma}d\varsigma = \frac{\pi}{k}J_1(k), \tag{14.104}$$

is taking into account, $J_1(k)$ is Bessel's function of the first kind. The function $\widetilde{T}(k)$ is Fourier transform of the function $T(\varsigma, 0)$ in Eq. (14.102)

$$\widetilde{T}_{\alpha\beta}(k) = \int_{-\infty}^{\infty}T_{\alpha\beta}(\varsigma, 0)e^{i\varsigma k}d\varsigma = -\frac{1}{2\pi}\int_{-\infty}^{\infty}\widetilde{S}_{2\alpha\beta2}(k, k_2)dk_2, \tag{14.105}$$

Because $\widetilde{T}(k)$ is a homogeneous function of the order of 1, $\widetilde{T}(k) = |k|\widetilde{T}(1)$ and therefore,

$$T^0 = \frac{1}{l}\widetilde{T}(1)\int_0^{\infty}J_1(k)dk = \frac{1}{l}\widetilde{T}(1). \tag{14.106}$$

Finally, the equation for the tensor T^0 takes the form

$$T^0_{\alpha\beta} = -\frac{1}{2\pi l}\int_{-\infty}^{\infty}n_\gamma\widetilde{S}_{\gamma\alpha\beta\delta}(1, k_2)n_\delta dk_2, \tag{14.107}$$

and the solution of the one-particle 2D-problem for a direct crack is

$$b_\alpha(x_1) = (T^0)^{-1}_{\alpha\beta}n_\gamma\sigma^0_{\gamma\beta}z(x_1),\ z(x_1) = \sqrt{1-\frac{x_1^2}{l^2}}\,. \tag{14.108}$$

For a thin elliptical inclusion filled with a soft material, tensor T^0 in this equation should be changed with $\overline{T}^0$

$$\overline{T}^0 = \frac{n\cdot C\cdot n}{2h} + T^0, \tag{14.109}$$

where h is the smallest ellipse semiaxis, C is the elastic stiffness tensor of the inclusion material.

For an isotropic host medium, tensor $\widetilde{S}(k_1, k_2)$ has the form (Appendix B.3)

$$\widetilde{S}(k_1, k_2) = -4\mu_0\kappa_0 P^1(m), m_\alpha = \frac{k_\alpha}{|k|}, \tag{14.110}$$

and hence

$$n_\gamma \widetilde{S}_{\gamma\alpha\beta\delta}(1, k_2) n_\delta = -4\mu_0\kappa_0 \left(\frac{1}{(1+k_2^2)^2} n_\alpha n_\beta + \frac{k_2^2}{(1+k_2^2)^2} e_\alpha e_\beta \right), \tag{14.111}$$

where e is unit vector of the x_1-axis. After integration in Eq. (14.107), we yield the equation for tensor T^0

$$T^0_{\alpha\beta} = \frac{\mu_0\kappa_0}{l} \delta_{\alpha\beta}. \tag{14.112}$$

Thus crack opening of a direct crack in an isotropic elastic medium subjected to external stress field σ^0 (the solution of the 2D one-particle crack problem) is

$$b_\alpha(x_1) = \frac{l}{\mu_0\kappa_0} n_\beta \sigma^0_{\beta\alpha} z(x_1), -l < x_1 < l, \tag{14.113}$$

where function $z(x_1)$ is in Eq. (14.108).

14.4.2 The Effective Field Method for the Elastic Plane with Multiple Cracks

For integral presentations of the stress and strain fields in the 2D homogeneous elastic medium with multiple cracks, we introduce the delta-function $l_i(x)$, concentrated in the line l_i of the ith crack. The function $l_i(x)$ acts on any smooth function $\varphi(x)$ in 2D-space be the rule

$$\int l_i(x)\varphi(x)dx = \int_{l_i} \varphi(x)dl. \tag{14.114}$$

Here, the integral on the left is spread over the entire 2D-plane. If $L(x) = \sum_i l_i(x)$ is delta function concentrated on all the cracks, the strain and stress tensors in the cracked medium are presented in the forms

$$\varepsilon(x) = \varepsilon^0 + \int K(x - x')C^0 n(x')b(x')L\left(x'\right)dx', \tag{14.115}$$

$$\sigma\left(x\right) = \sigma^0 + \int S\left(x - x'\right) n(x')b(x')L\left(x'\right)dx'. \tag{14.116}$$

where $x = (x_1, x_2)$, and the kernels $K(x)$ and $S(x)$ are defined in Eqs. (11.63), (11.64) for the 2D-case, C^0 is the stiffness tensor of the host medium, vector $n(x)$ coincides with the normal n^i to the crack line l_i and vector $b(x)$ coincides with the crack opening $b^i(x)$ on the ith crack.

For the principal hypothesis of the effective field method, each crack is subjected to a constant effective stress field σ^* that is the same for all the cracks. It follows from Eq. (14.108) that vector $b(x)$ on L is defined by the equation

$$b(x) = \left(T^0\right)^{-1} n(x)Z(x)\sigma^* = \Lambda(x)Z(x)\sigma^*, \tag{14.117}$$

where the function $Z(x)$ coincides with the function $z(x)$ in Eq. (14.108) in each crack line, tensor $\Lambda(x)$ coincides with the tensor Λ^i when $x \in l_i$

$$\Lambda^i_{\alpha\beta\gamma\delta} = n^i_\alpha \left(T^{0i}\right)^{-1}_{\beta\gamma} n^i_\delta \Big|_{(\alpha\beta)(\gamma\delta)}. \tag{14.118}$$

Thus the strain and stress tensors in the cracked medium are approximated by the equations

$$\varepsilon(x) = \varepsilon^0 + \int K(x - x')C^0 \Lambda(x')Z(x')L\left(x'\right)dx'\sigma^*, \tag{14.119}$$

$$\sigma\left(x\right) = \sigma^0 + \int S\left(x - x'\right) \Lambda(x')Z\left(x'\right) L\left(x'\right)dx'\sigma^*. \tag{14.120}$$

Let cracks of the same length $2l$ and orientations n compose a homogeneous random set in the plane. In this case, the averages of the strain and stress tensors over the crack positions follow from Eqs. (14.119), (14.120) in the forms

$$\langle\varepsilon(x)\rangle = \varepsilon^0 + \int K(x - x')dx'C^0 \Lambda\left\langle Z(x')L\left(x'\right)\right\rangle\sigma^*, \tag{14.121}$$

$$\langle\sigma\left(x\right)\rangle = \sigma^0 + \int S\left(x - x'\right)dx'\Lambda\left\langle Z\left(x'\right) L\left(x'\right)\right\rangle\sigma^*. \tag{14.122}$$

Using the ergodic property of the crack set, the average $\left\langle Z\left(x'\right) L\left(x'\right)\right\rangle$ in these equations is

$$\langle Z(x)L(x)\rangle = \lim_{W\to\infty}\frac{1}{W}\int\limits_W Z(x)L(x)dx =$$

$$= \lim_{W\to\infty}\frac{1}{W}\sum_{i=1}^{N}\int\limits_{l_i} z(x)dl = \frac{\pi}{2s_0}l. \tag{14.123}$$

Here, W is the region that that occupied the entire 2D-plane in the limit $W\to\infty$, N is the number of cracks in W, s_0 is the region that falls on one crack. For a fixed external stress field σ^0, the integral in Eqs. (14.121), (14.122) are

$$\int K(x-x')dx' = \left(C^0\right)^{-1}, \quad \int S\left(x-x'\right)dx' = 0, \tag{14.124}$$

and for the averages $\langle\sigma(x)\rangle$ and $\langle\varepsilon(x)\rangle$, we yield

$$\langle\sigma(x)\rangle = \sigma^0,\ \langle\varepsilon(x)\rangle = \varepsilon^0 + \tau\overline{\Lambda}\sigma^*, \tag{14.125}$$

$$\tau = \frac{\pi l^2}{s_0},\ \overline{\Lambda} = \frac{1}{2l}\Lambda. \tag{14.126}$$

Here, the tensor Λ is defined in Eq. (14.118).

The local exiting field that acts on a point $x\in$ follows from Eq. (14.120)

$$\sigma^*(x) = \sigma^0 + \int S\left(x-x'\right)\Lambda Z\left(x'\right)L\left(x';x\right)dx'\sigma^*, \tag{14.127}$$

where

$$L\left(x';x\right) = \sum_{i\neq j} l_i(x') \text{ if } x\in l_j. \tag{14.128}$$

Averaging Eq. (14.127) by the condition that $x\in L$ and accepting that $\langle\sigma^*(x)|x\rangle = \sigma^*$, we yield the equation for the effective field

$$\sigma^* = \sigma^0 + \int S\left(x-x'\right)\Lambda\left\langle Z\left(x'\right)L\left(x';x\right)|x\right\rangle dx'\sigma^*. \tag{14.129}$$

The conditional mean in this equation is

$$\left\langle Z(x')L(x';x)|x\right\rangle = \frac{1}{\langle L(x)\rangle}\left\langle Z(x')L(x';x)L(x)\right\rangle =$$

$$= \frac{1}{\langle L(x)\rangle}\lim_{W\to\infty}\frac{1}{W}\int\limits_W \sum_{\substack{i,j\\(i\neq j)}} Z(x')l_i(x')l_j(x)dx' = \frac{\pi l}{2s_0}\Psi(x-x'). \tag{14.130}$$

Here, the function $\Psi(x)$ satisfies the conditions

$$\Psi(0) = 0, \lim_{W\to\infty} \frac{1}{W} \int_W \Psi(x)dx = 1. \tag{14.131}$$

This function is the correlation hole for a typical crack. Then, the solution of the Eq. (14.129) for the effective field is

$$\sigma^* = (I + \tau D^* \overline{\Lambda})^{-1} \sigma^0, \tag{14.132}$$

$$D^* = -\int S(x)\Psi(x)dx = \int S(x)\left(1 - \Psi(x)\right)dx. \tag{14.133}$$

Here, Eq. (14.124) is taken into account. If the function $\Psi(x)$ has elliptical symmetry, tensor D^* depends on the aspect ratio γ of the correlation hole only and

$$D^* = D(\gamma), \tag{14.134}$$

where tensor $D(\gamma)$ is expressed in terms of the integral over a unit circle in 2D-space of Fourier transforms.

$$D(\gamma) = \frac{1}{2\pi} \int_{|k|=1} \widetilde{S}(\alpha^{-1}k)dk, \tag{14.135}$$

$$\widetilde{S}(k) = C^0 \widetilde{K}(k) C^0 - C^0, \ \widetilde{K}(k) = k(kC^0k)^{-1}k. \tag{14.136}$$

For an isotropic host medium,

$$\widetilde{S}(k) = -4\mu_0\kappa_0 P^1(m), m = \frac{k}{|k|},$$

and the tensor $D(\gamma)$ is presented in the P-basis as follows

$$D(\gamma) = d_1 P^1(n) + d_3\left(P^3(n) + P^4(n)\right) + d_5 P^5(n) + d_6 P^6(n), \tag{14.137}$$

$$d_1 = -4\mu_0\kappa_0\left(1 - 2f_0 + f_1\right), \ d_3 = -4\mu_0\kappa_0\left(f_0 - f_1\right), d_5 = 4d_3, \tag{14.138}$$

$$d_6 = -4\mu_0\kappa_0 f_1, \quad f_0(\gamma) = \frac{\gamma}{1+\gamma}, \quad f_1(\gamma) = \frac{\gamma(2\gamma+1)}{2(1+\gamma)^2}. \tag{14.139}$$

Here, n is the normal to the crack lines, λ_0, μ_0 are Lame's constants of the host medium. If positions of cracks are not restricted, γ tends to zero, and

$$D(\gamma) \to -4\mu_0\kappa_0 P^1(n), \tag{14.140}$$

When $\gamma = 1$, the prohibited region around each crack is circular, and tensor D is

$$D(1) = -\frac{\mu_0 \kappa_0}{2} \left(E^2 + 2E^1 \right). \tag{14.141}$$

The average strain and stress tensors in the cracked medium relate by the equation

$$\langle \varepsilon(x) \rangle = \left(B^0 + \tau \overline{\Lambda}(I + \tau D(\gamma^*) \overline{\Lambda})^{-1} \right) \langle \sigma(x) \rangle , \tag{14.142}$$

and the effective compliance tensor B^* of the 2D-medium with direct cracks of the same orientation is

$$B^* = B^0 + \tau \overline{\Lambda}(I + \tau D(\gamma^*) \overline{\Lambda})^{-1}. \tag{14.143}$$

For an isotropic medium, the tensors Λ in Eq. (14.118) and $D(\gamma)\overline{\Lambda}$ are presented in the P-basis as follows

$$\overline{\Lambda}(n) = \frac{1}{2\mu_0 \kappa_0} E^5(n) = \frac{1}{2\mu_0 \kappa_0} \left(P^5(n) + P^6(n) \right), \tag{14.144}$$

$$D(\gamma)\overline{\Lambda}(n) = \frac{d_3(\gamma)}{2\mu_0 \kappa_0} P^3(n) + \frac{d_5(\gamma)}{4\mu_0 \kappa_0} P^5(n) + \frac{d_6(\gamma)}{2\mu_0 \kappa_0} P^6(n). \tag{14.145}$$

For the plane stress state, the coefficients μ_0 and κ_0 in these equations are expressed in terms of the Young modulus $E_0$and Poisson's ratio ν_0 of the host medium

$$\mu_0 = \frac{E_0}{2(1 + \nu_0)}, \ \kappa_0 = \frac{1 + \nu_0}{2}, \tag{14.146}$$

and the effective compliance tensor of the medium takes the form

$$B^* = B^0 + \frac{2\tau}{E_0 + 2\tau d_3(\gamma)} P^5(n) + \frac{2\tau}{E_0 + 2\tau d_6(\gamma)} P^6(n). \tag{14.147}$$

Effective Young modulus of the material with cracks along the crack line coincides with the the Young modulus of the host medium E_0, and in the direction orthogonal to the crack line Young modulus E_2^* is

$$\frac{E_2^*}{E_0} = 1 - \frac{2\tau}{1 + 2\tau(1 - f_1(\gamma))}. \tag{14.148}$$

The effective shear modulus μ^* has the form

$$\frac{\mu^*}{\mu_0} = 1 - \frac{\tau}{\tau + (1+\nu_0)(1+2\tau f_2(\gamma))}, \tag{14.149}$$

$$f_2(\gamma) = f_0(\gamma) - f_1(\gamma) = \frac{\gamma}{2(1+\gamma)^2}. \tag{14.150}$$

In Fig. 14.8, dependencies of the effective Young modulus along the axis x_2 orthogonal to the crack lines and the effective shear modulus μ_* for the medium with a random set of cracks of the same orientations are shown for the values of the parameter $0 < \gamma < 1$, $\nu_0 = 0.3$.

For direct cracks of various orientations, the assumption that the effective field $\sigma^* = \sigma^*(n)$ depends on the crack orientation results in the equation for the effective compliance tensor B^* similar to Eq. (14.90) for 3D-case

$$B^* = B^0 + \tau(I + \tau\langle\overline{\Lambda}(n)D(\gamma, n)\rangle)^{-1}\langle\overline{\Lambda}(n)\rangle. \tag{14.151}$$

Here, the average is preformed over the crack orientations.

For isotropic distribution of direct cracks over the orientations, the effective compliance tensor of the cracked medium takes the form

$$B^* = \left(\frac{1+\nu_0}{E_0} + \frac{2\tau}{E_0(2-\tau f_0(\gamma))}\right)E^1 +$$

$$+\left(-\frac{\nu_0}{E_0} + \frac{2\tau^2 f_0(\gamma)}{E_0(2-\tau f_0(\gamma))(4-5\tau f_0(\gamma))}\right)E^2. \tag{14.152}$$

It follows from these equations that effective Young modulus and Poisson ratio of the cracked media are

$$\frac{E_*}{E_0} = \left(1 + \frac{8\tau(1-\tau f_0(\gamma))}{(2-\tau f_0(\gamma))(4-5\tau f_0(\gamma))}\right)^{-1}, \tag{14.153}$$

$$\frac{\nu_*}{\nu_0} = \left(1 - \frac{2\tau(\tau f_0(\gamma) + 4\nu_0(1-\tau f_0(\gamma)))}{2\tau^2 f_0(\gamma) - \nu_0(2-\tau f_0(\gamma))(4-5\tau f_0(\gamma)))}\right)^{-1}, \tag{14.154}$$

where $f_0(\gamma)$ is in Eq. (14.139). For $\gamma = 0$ (no restrictions on the crack positions),

$$\frac{E_*}{E_0} = \frac{\nu_*}{\nu_0} = \frac{1}{1+\tau}. \tag{14.155}$$

For $\gamma = 1$, the correlation hole is circular, and

$$\frac{E_*}{E_0} = \left(1 + \frac{16\tau(2-\tau)}{(4-\tau)(8-5\tau)}\right)^{-1}, \tag{14.156}$$

$$\frac{\nu_*}{\nu_0} = \left(1 - \frac{4\tau(\tau + 4\nu_0(2-\tau))}{4\tau^2 - \nu_0(4-\tau)(8-5\tau)}\right)^{-1}. \tag{14.157}$$

In Fig. 14.9, dependencies of the effective Young modulus and Poisson ratio for the plane with homogeneous distribution of cracks over orientations are shown for the medium with Poisson's ratio $\nu_0 = 0.45$, the points are experimental data presented in [2]. Note that for the crack sets studied in [2], the best predictions of the EFM are obtained for $\gamma = 1$, corresponding to the model of a crack set with restrictions on the cracks intersections, and circular shape of the correlation hole.

For a plane with homogeneous distribution of direct cracks over the orientations, the effective medium method (EMM) results in the equation for the effective Young modulus of the cracked medium

$$\frac{E^*}{E_0} = 1 - \tau, \tag{14.158}$$

and the differential effective medium method (DEMM) results in the formula

$$\frac{E^*}{E_0} = \exp(-\tau). \tag{14.159}$$

In the Fig. 14.10, the dashed line corresponds to the normalized effective Young modulus obtained by the EMM, the dashed-point line–to the DEMM, and two solid

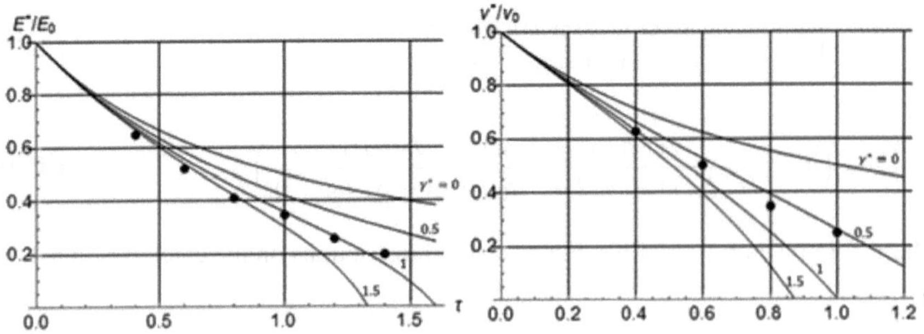

Fig. 14.9 Dependencies of the effective Young modulus E^* and Poisson ratio ν^* for the plane with homogeneous distribution of straight-line cracks over orientations for various values of the parameter γ^*; Poisson's ratio of the host medium is $\nu_0 = 0.45$; points are experimental data presented in [2]

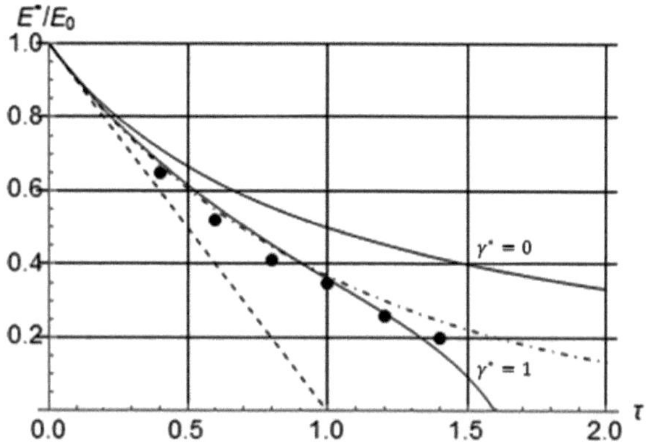

Fig. 14.10 Dependence of the effective Young modulus of the plane containing straight line cracks homogeneously distributed over orientations; the dashed line corresponds to the effective medium method, the dashed-point line–to the differential EMM, and other two lines are prediction of the effective field method s for $\gamma^* = 0$ (non-interacting cracks) and $\gamma^* = 1$. The black dots are experimental data presented in [2]

lines are the predictions of the EFM for $\gamma = 0$ and $\gamma = 1$. Black dots in this figure are experimental data presented in [2]. Thus, the differential effective medium method and the EFM with $\gamma = 1$ correspond better to the experimental data.

References

1. Prokhorov Y, Rosanov Y (1969) The probability theory: basic concepts, limit theorems, random processes. Springer, Berlin
2. Vavakin A, Salganik R (1975) Effective characteristics of heterogeneous solids with isolated inhomogeneities. Proc Acad Sci USSR Mech Solids 2:95–107

Chapter 15
Elastic Medium Reinforced with Thin Stiff Flakes and Bands

Abstract Composites reinforced with thin inclusions which elastic moduli are much larger than the moduli of the matrix phase are considered. Such inclusions are characterized by two small parameters: the "geometrical" parameter δ_1 that is the ratio of the minimal and maximal linear sizes of the inclusions, and parameter $\delta_2$that is the ratio of the characteristic elastic moduli of the matrix and inclusion materials. The principal terms of the asymptotic expansions of elastic fields in the medium with thin stiff inclusions are used for the solution of the homogenization problem by the effective field method. The method predictions are compared with experimental data.

Keywords Thin stiff inclusions · Effective medium method · Elliptical stiff flakes · Flake orientations · Long stiff bands (ribbons) · Effective elastic properties

Thin inclusions which elastic moduli are much larger than the moduli of the matrix phase are often used for reinforcement of composites with polymer matrices. Such inclusions are characterized by two small parameters: the "geometrical" parameter δ_1 that is the ratio of the minimal and maximal linear sizes of the inclusions, and parameter δ_2 that is the ratio of the characteristic elastic moduli of the matrix and inclusion materials. The principal terms of the asymptotic expansions of elastic fields in the medium with a thin stiff inclusions over the parameters δ_1, δ_2 were obtained in Chap. 6. By the solution of the homogenization problem, only these terms can be taken into account for the approximation of elastic field in the composites reinforced with thin stiff flakes.

15.1 The One-Particle Problem for a Thin Stiff Flake

The one-particle problem for a thin stiff flake in a homogeneous elastic host medium subjected to a constant external strain field ε^0 is considered in Sect. 6.3.2. With accuracy to the principal terms of solution of this problem with respect to small parameters δ_1, δ_2, the strain and stress fields in the medium are approximated by Eqs. (6.60), (6.61)

S. Kanaun, *Self-Consistent Methods for Composites*, Solid Mechanics and Its Applications 282, https://doi.org/10.1007/978-3-031-96739-9_15

$$\varepsilon(x) = \varepsilon^0 - \int_{\Omega} K(x - x')\bar{\sigma}(x')d\Omega', \tag{15.1}$$

$$\sigma(x) = \sigma^0 - \int_{\Omega} S(x - x')B^0\bar{\sigma}(x')d\Omega'. \tag{15.2}$$

Here, Ω is the middle surface of a thin inclusion, the kernels $K(x)$ and $S(x)$ are expressed in terms of second derivatives of Green's function of the host medium with the elastic compliance tensor B^0. The tensor $\bar{\sigma}(x)$ in Eqs. (15.1), (15.2) is the integral stress tensor over the inclusion cross section

$$\bar{\sigma}(x, h) = \int_{-h(x)/2}^{h(x)/2} \sigma(x + n(x)z)dz\,, \quad x \in \Omega. \tag{15.3}$$

Here, $h(x)$ is the transverse size of the inclusion at point x of its middle surface, z is the coordinate along the normal $n(x)$ to Ω. Let $\bar{\sigma}(x)$ be the principal term of the function $\bar{\sigma}(x, h)$ with respect to the parameters δ_1, δ_2

$$\overline{\sigma}(x) = \lim_{h,B\to 0} \overline{\sigma}(x, h). \tag{15.4}$$

It is shown in Sect. 6.3.2 that tensor $\overline{\sigma}(x)$ belongs to the middle surface Ω of the inclusion, and therefore,

$$n_\beta(x)\bar{\sigma}_{\alpha\beta}(x) = 0, \quad \Theta_{\alpha\beta\lambda\mu}(n)\bar{\sigma}_{\lambda\mu}(x) = \bar{\sigma}_{\alpha\beta}(x). \tag{15.5}$$

$\Theta(n)$ is the projector on the tangent plane to Ω at point x with the normal $n = n(x)$

$$\Theta(n) = E^1 - 2E^5(n) + E^6(n) = P^1(n). \tag{15.6}$$

Here, $E^i(n)$ are the elements of the E-basis (Appendix A1). Tensor $\bar{\sigma}(x)$ is the solution of the integral equation on the surface Ω

$$\mu_{\alpha\beta\lambda\mu}(x)\bar{\sigma}_{\lambda\mu}(x) + \int_{\Omega} U_{\alpha\beta\lambda\mu}(x, x')\bar{\sigma}_{\lambda\mu}(x')d\Omega' = \Theta_{\alpha\beta\lambda\mu}(x)\varepsilon^0_{\lambda\mu}\,, \quad x \in \Omega, \tag{15.7}$$

$$\mu_{\alpha\beta\lambda\mu}(x) = \frac{\Theta_{\alpha\beta\nu\rho}(x)B_{\nu\rho\tau\delta}\Theta_{\tau\delta\lambda\mu}(x)}{h(x)}, \tag{15.8}$$

$$U_{\alpha\beta\lambda\mu}(x, x') = \Theta_{\alpha\beta\nu\rho}(x)K_{\nu\rho\tau\delta}(x - x')\Theta_{\tau\delta\lambda\mu}(x'), \tag{15.9}$$

If $B \to 0$, $\mu(x) \to 0$ and Eqs. (15.1), (15.2), and (15.7) corresponds to the one-particle problem for the medium with a non-stretchable membrane Ω embedded into an elastic homogeneous medium.

For a thin ellipsoidal inclusions, the inclusion transverse size $h(x)$ is

$$h(x) = 2h_0 z(x), \ z(x) = \sqrt{1 - \left(\frac{x_1}{a_1}\right)^2 - \left(\frac{x_2}{a_2}\right)^2}, \tag{15.10}$$

where a_1, a_2, h_0 are the semiaxes of the thin ellipsoid, $h_0 \ll a_1, a_2$, and x_1, x_2 are the Cartesian coordinates along the ellipsoid semiaxes a_1, a_2. In is shown in Chap. 6 that for a thin stiff ellipsoidal inclusion subjected to a constant external strain field ε^0, the solution of Eq. (15.7) has the form

$$\overline{\sigma}(x) = \widehat{\sigma}\, z(x), \tag{15.11}$$

where $\widehat{\sigma}$ is a constant tensor. Equation for $\widehat{\sigma}$ follows from the integral equation (15.7) in the form

$$\widehat{\sigma} = \left(\frac{\Theta(n)B\Theta(n)}{2h_0} + U^0\right)^{-1} \varepsilon^0, \tag{15.12}$$

where the tensor U^0 is the integral over the inclusion middle surface Ω. It is shown in Sect. 6.5 (Eqs. (6.87), (6.88)) that tensor U^0 is presented in the form

$$U^0 = \int_{\Omega} U(x) z(x) d\Omega = \frac{1}{4} \int_0^{2\pi} \widetilde{U}\left(\frac{\cos\varphi}{a_1}, \frac{\sin\varphi}{a_2}\right) d\varphi, \tag{15.13}$$

$$\widetilde{U}\,(k_1, k_2) = \frac{1}{2\pi} \int_{-\infty}^{\infty} \Theta(n) \widetilde{K}(k_1, k_2, k_3) \Theta(n) dk_3. \tag{15.14}$$

Here, $\widetilde{K}(k_1, k_2, k_3)$ is the Fourier transform of the function $K(x_1, x_2, x_3)$.

For an isotropic host medium,

$$\widetilde{K}(k_1, k_2, k_3) = \frac{1}{\mu_0} \left(E^5(m) - \kappa_0 E^6(m)\right), \ m_\alpha = \frac{k_\alpha}{|k|}, \tag{15.15}$$

and the non-zero components of the tensor $\widetilde{U}\,(k_1, k_2)$ are

$$\widetilde{U}_{1111} = \frac{k_1^2(2k_2^2 + (2 - \kappa_0)k_1^2)}{4\mu_0(k_1^2 + k_2^2)^{3/2}}, \ \widetilde{U}_{1122} = -\frac{\kappa_0 k_1^2 k_2^2}{4\mu_0(k_1^2 + k_2^2)^{3/2}}, \tag{15.16}$$

$$\widetilde{U}_{2222}=\frac{k_2^2(2k_1^2+(2-\kappa_0)k_2^2)}{4\mu_0(k_1^2+k_2^2)^{3/2}},\ \widetilde{U}_{1212}=\frac{\left(k_1^2+k_2^2\right)^2-2\kappa_0k_1^2k_2^2}{8\mu_0(k_1^2+k_2^2)^{3/2}},\tag{15.17}$$

The integral in Eq. (15.13) is expressed in terms of the complete elliptic integrals

$$U^0_{1111}=\frac{\gamma^{3/2}}{4\sqrt{a_1a_2}\mu_0}\left((\eta(\kappa_0-2)-2\kappa_0)\,E(\eta)+2(\kappa_0\gamma^2+\eta)K\,(\eta)\right),\tag{15.18}$$

$$U^0_{2222}=\frac{\gamma^{-1/2}}{4\sqrt{a_1a_2}\gamma\eta^2\mu_0}\left((\eta(2+\kappa_0)-2\kappa_0)E(\eta)-2\gamma(\eta-\kappa_0)K(\eta)\right),\tag{15.19}$$

$$U^0_{1122}=\frac{\gamma^{3/2}\kappa_0}{4\sqrt{a_1a_2}\eta^2\mu_0}(2E(\eta)+(\eta-2)K(\eta)),\tag{15.20}$$

$$U^0_{1212}=\frac{\gamma^{-1/2}}{8\sqrt{a_1a_2}\eta^2\mu_0}\left((\eta^2+4\kappa_0\gamma^2)\mathrm{E}(\eta)+2\gamma^2(\eta-2)\mathrm{K}(\eta)\right).\tag{15.21}$$

Here, $\eta=1-\gamma^2$, $\gamma=a_2/a_1$, $\mathrm{K}(\eta)$, $\mathrm{E}(\eta)$ are the complete elliptic integrals of the first and second kind. In the T-basis (Appendix A.4), tensor U^0 is presented as follows

$$U^0=U^0_{1111}T^1+U^0_{2222}T^2+U^0_{1122}\left(T^4+T^5\right)+U^0_{1212}T^{10},\tag{15.22}$$

where

$$T^1_{\alpha\beta\gamma\delta}=n^1_\alpha n^1_\beta n^1_\gamma n^1_\delta,\ T^2_{\alpha\beta\gamma\delta}=n^2_\alpha n^2_\beta n^2_\gamma n^2_\delta,\tag{15.23}$$

$$T^4_{\alpha\beta\gamma\delta}=n^1_\alpha n^1_\beta n^2_\gamma n^2_\delta,\ T^5_{\alpha\beta\gamma\delta}=n^2_\alpha n^2_\beta n^1_\gamma n^1_\delta,\tag{15.24}$$

$$T^{10}_{\alpha\beta\gamma\delta}=4n^1_\alpha n^2_\beta n^1_\gamma n^2_\delta\big|_{(\alpha\beta)(\gamma\delta)}\,.\tag{15.25}$$

Here, n^1, n^2, n^3 are the basis vectors of the Cartesian system x_1, x_2, x_3.

For a thin spheroidal inclusion ($a_1=a_2=a$) in an isotropic host medium, the non-zero components of the tensor U^0 take the forms

$$U^0_{1111}=U^0_{2222}=\frac{\pi(8-3\kappa_0)}{64a\mu_0},\ \kappa_0=\frac{\lambda_0+\mu_0}{\lambda_0+2\mu_0},\tag{15.26}$$

$$U^0_{1122}=-\frac{\pi\kappa_0}{64a\mu_0},\ U^0_{1212}=\frac{\pi(4-\kappa_0)}{64a\mu_0},\tag{15.27}$$

and in the P -basis, the tensor U^0 is presented as follows

$$U^0(n) = \frac{\pi}{32a\mu_0}\left((4-\kappa_0)P^1(n) - \frac{\kappa_0}{2}P^2(n)\right). \tag{15.28}$$

Finally, the solution of the one-particle problem for an ellipsoidal stiff flakes is presented in Eqs. (15.1), (15.2), where

$$\overline{\sigma}(x) = M(n)\varepsilon^0 z(x), \quad z(x) = \sqrt{1-\left(\frac{x_1}{a_1}\right)^2 - \left(\frac{x_1}{a_1}\right)^2}, \tag{15.29}$$

$$M = \left(\frac{1}{2h_0}\Theta(n)B\Theta(n)+U^0\right)^{-1}. \tag{15.30}$$

Note that tensors $\Theta(n)B\Theta(n)$ and U^0 belong to the inclusion surface Ω, and inversion in Eq. (15.30) is performed in the subspace of such tensors. The unit tensor of this subspace is the element $P^1(n)$ of the P-basis in Eq. (15.6).

15.2 The Effective Field Method for the Medium with Multiple Stiff Flakes

For integral presentation of the strain and stress fields in the medium with multiple stiff flakes, we introduce a generalized functions $\Omega_i(x)$ concentrated on the middle surface Ω_i of the ith flake, $i = 1, 2, ...$ The function is defined in Eq. (3.5).

Let a homogenous elastic medium containing a random set of identical thin ellipsoidal flakes of the same orientation. The medium is subjected to a constant external strain field ε^0 at infinity. According to the principal hypothesis of the effective field method, the local effective strain field ε^* that acts on each flake is constant and the same for all the flakes. For this hypothesis, strain $\varepsilon(x)$ and stress $\sigma(x)$ tensors in the heterogeneous medium are approximated by the equations that follow from Eqs. (15.1), (15.2) and (15.29)

$$\varepsilon(x) = \varepsilon^0 - \int K(x-x')MZ(x')\Omega(x')dx'\varepsilon^*, \tag{15.31}$$

$$\sigma(x) = \sigma^0 - \int S(x-x')B^0MZ(x')\Omega(x')dx'\varepsilon^*. \tag{15.32}$$

Here, $\Omega(x) = \sum_i \Omega_i(x)$ is the generalized function concentrated on the flake middle surfaces. The tensor M is defined in Eq. (15.30), and function $Z(x)$ coincides with the function $z(x)$ in Eq. (15.10) in middle surface Ω_i of each flake. Then, we introduce the generalized function

$$\Omega(x'; x) = \sum_{i \neq k} \Omega_i(x') \text{ when } x \in \Omega_k. \tag{15.33}$$

This function allows expressing the local exciting strain field $\varepsilon^*(x)$ that acts at any point x of the surface $\Omega = \bigcup_i \Omega_i$ in the form

$$\varepsilon^*(x) = \varepsilon^0 - \int K\left(x - x'\right) M\left(x'\right) \Omega\left(x; x'\right) dx' \varepsilon^*, \quad x \in \Omega. \tag{15.34}$$

After averaging this equation by the condition that point x belongs to Ω and identifying the mean $\langle \varepsilon^*(x)|x \rangle$ with ε^*, we yield a closed equation for the effective strain field ε^* acting on each flake

$$\varepsilon^* = \varepsilon^0 - \int K\left(x - x'\right) M(n) \left\langle Z(x') \Omega\left(x; x'\right) |x \right\rangle dx' \varepsilon^*. \tag{15.35}$$

Here, symbol $\langle \cdot | x \rangle$ means averaging by the condition that $x \in \Omega$. The mean under the integral in this equation is presented in the form

$$\left\langle Z(x') \Omega\left(x; x'\right) |x \right\rangle = \langle Z(x) \Omega(x) \rangle \Psi\left(x - x'\right). \tag{15.36}$$

$$\Psi\left(x - x'\right) = \frac{\left\langle Z(x') \Omega\left(x; x'\right) |x \right\rangle}{\langle Z(x) \Omega(x) \rangle}. \tag{15.37}$$

For ergodic properties of the random flake set, we have

$$\langle Z(x) \Omega(x) \rangle = \lim_{W \to \infty} \frac{1}{W} \sum_{i=1}^{N} \int_W z_i(x) \Omega_i(x) dx = \frac{2\pi}{3} \frac{a_1 a_2}{v_0}, \tag{15.38}$$

where W is a region that in the limit $W \to \infty$ occupies the entire 3D-space, N is the number of flakes in W, and v_0 is the volume that falls on one flake. The function $z_i(x)$ has the form in Eq. (15.10) for each flake. The function $\Psi(x)$ in Eq. (15.37) is considered in Sect. 14.1 dedicated to the homogenization problem for a cracked medium. This function depends on the geometrical peculiarities of the spatial distribution of flakes in the host medium. From the definition of the function $\Omega(x'; x)$ and Eq. (15.37) follow that $\Psi(x)$ has the properties

$$\Psi(0) = 0, \quad \lim_{W \to \infty} \int_W \Psi(x) dx = 1. \tag{15.39}$$

Function $\Psi(x)$ can be interpreted as the correlation hole for a typical flake in the medium with multiple flakes. If there exists a linear transformation α of the x-space that converts the function $\Psi(x)$ into a spherically symmetric one, then the following equations hold

$$y = \alpha^{-1}x\,, \quad \Psi(\alpha y) = \Psi'(|y|)\,. \tag{15.40}$$

In this case, the symmetry of the function $\Psi(x)$ coincides with the symmetry of an ellipsoid with the semiaxes $\alpha_1, \alpha_2, \alpha_3$ and its surface is defined by the equation

$$\left|\alpha^{-1}x\right| = 1. \tag{15.41}$$

After substituting Eqs. (15.36) and (15.38) into Eq. (15.35), the equation for the effective strain field ε^* takes the form

$$\varepsilon^* = \varepsilon^0 + \tau A^* \overline{M}(n)\varepsilon^*, \; \tau = \frac{4}{3}\frac{(a_1a_2)^{3/2}}{v_0}, \tag{15.42}$$

$$\overline{M}(n) = \frac{\pi}{2\sqrt{a_1a_2}}M(n), \; A^* = -\int K(x)\Psi(x)dx, \tag{15.43}$$

Further, we assume that external strain field ε^0 is fixed in the problem, and therefore, Eqs. (10.31) for actions of the operators K and S on constants hold

$$\int K(x-x')dx' = 0, \; \int S(x-x')dx = -C^0. \tag{15.44}$$

For the function $\Psi(x)$ of ellipsoidal symmetry, the integral (15.43) takes is

$$A^* = \int K(x)\,(1 - \Psi(x))\,dx = A(\alpha), \tag{15.45}$$

where the tensor $A(\alpha)$ is the integral over the unit sphere in the k-space of the Fourier transforms

$$A(\alpha) = \frac{1}{4\pi}\int\limits_{|k|=1} \widetilde{K}(\alpha^{-1}k)d\Omega_k. \tag{15.46}$$

For an isotropic host medium, the tensor $A(\alpha)$ is defined in Eq. (4.126). After solving Eq. (15.42) for ε^*, we obtain the effective strain tensor acting on each flake in the form

$$\varepsilon^* = \left(I - \tau A(\alpha)\overline{M}(n)\right)^{-1}\varepsilon^0. \tag{15.47}$$

For calculation of the mean strain $\langle\varepsilon\rangle$ and stress $\langle\sigma\rangle$ fields in the medium with multiple flakes, Eqs. (15.31), (15.32) for $\varepsilon(x)$ and $\sigma(x)$ should be averaged over the ensemble realizations of the random flake set. Taking into account Eq. (15.38) for $\langle Z(x)\Omega(x)\rangle$, we yield for $\langle\varepsilon\rangle$ and $\langle\sigma\rangle$ the equations

$$\langle\varepsilon\rangle = \varepsilon^0 - \tau \int K\left(x - x'\right)dx'\overline{M}\varepsilon^*, \tag{15.48}$$

$$\langle\sigma\rangle = \sigma^0 - \tau \int S\left(x - x'\right)dx' B^0\overline{M}\varepsilon^*. \tag{15.49}$$

For Eqs. (15.44) that hold for a fixed external strain field, we obtain finally

$$\langle\varepsilon\rangle = \varepsilon^0 \,, \quad \langle\sigma\rangle = C^*\langle\varepsilon\rangle, \tag{15.50}$$

$$C^* = C^0 + \tau\overline{M}(n)\left(I - \tau A(\alpha)\overline{M}(n)\right)^{-1}. \tag{15.51}$$

Here, C^* is the effective elastic stiffness tensor of the composites with a set of identical thin flake of the same orientations.

For homogeneous spatial distributions of flakes, it is reasonable to assume that the ellipsoid in Eq. (15.41) that defines symmetry of the correlation hole for a typical flake is congruent to the flake, and therefore, the directions of the principal axes of the correlation hole coincide with the flake principal axes. The tensor $A(\alpha)$ in Eq. (15.51) depends on the aspect ratios $\gamma_1^* = \alpha_3/\alpha_1, \gamma_2^* = \alpha_3/\alpha_2$ of the correlation hole. If there is no restriction on the inclusion of mutual positions, the flakes can be arbitrarily close to each other, and the parameter α_3 tends to zero, and

$$\lim_{\alpha_3\to 0} A(\alpha) = \frac{1}{\mu_0}\left(E^5(n) - \kappa_0 E^6(n)\right). \tag{15.52}$$

Here, tensors $E^5(n)$ and $E^6(n)$ belong to the subspace orthogonal to the tensor $\overline{M}(n)$ in Eqs. (15.30), (15.43) and therefore,

$$\lim_{\alpha_3\to 0} A(\alpha)\overline{M} = 0. \tag{15.53}$$

In this case, the effective strain tensor ε^* in Eq. (15.47) acting on each flake coincides with the external field ε^0 applied to the medium, and the effective stiffness tensor of the composite is a linear function of the flake density parameter τ

$$\varepsilon^* = \varepsilon^0, \; C^* = C^0 + \tau\overline{M}. \tag{15.54}$$

Thus for $\alpha_3 \to 0$, flake interactions absent.

Suppose that the material of flakes is transversely isotropic with the isotropy plane coincided with the middle surface Ω of the flakes. If E_{11} and ν_{12} be the Young modulus and Poisson ratio of the inclusion in the plane x_1, x_2, the tensor $\Theta(n)C^{-1}\Theta(n)$ in Eq. (15.30) for the tensor M takes the form

$$\Theta(n)B\Theta(n) = \frac{1-\nu_{12}}{2E_{11}}P^2(n) + \frac{1+\nu_{12}}{E_{11}}\left(P^1(n) - \frac{1}{2}P^2(n)\right), \tag{15.55}$$

where $P^i(n)$ are the elements of the P-basis (Appendix A.2). For thin spheroidal flakes with aspect ratios h_0/a, the tensor $M = M(n)$ in Eq. (15.30) has the form

$$\overline{M}(n) = M_1 P^2(n) + M_2\left(P^1(n) - \frac{1}{2}P^2(n)\right), \tag{15.56}$$

$$M_1 = 4\mu_0\left(2-\kappa_0 + 4\xi\frac{1-\nu_{12}}{1+\nu_{12}}\right)^{-1}, \quad M_2 = 16\mu_0(4-\kappa_0+\xi)^{-1}, \tag{15.57}$$

$$\xi = \frac{a}{h_0}\frac{\mu_0}{\mu_{12}}, \quad \mu_{12} = \frac{E_{11}}{2(1+\nu_{12})}. \tag{15.58}$$

The tensor $A(\alpha)$ in Eq. (15.46) is a function of the aspect ratio $\gamma_* = \alpha_3/\alpha$ ($\alpha_1 = \alpha_2 = \alpha$) of the correlation hole defined by the function $\Psi(x)$. In the P-basis, the tensor $A(\alpha) = A(n, \gamma)$ takes the form

$$A(n,\gamma) = A_1 P^2(n) + A_2\left(P^1(n) - \frac{1}{2}P^2(n)\right) +$$

$$+ A_3\left(P^3(n) + P^4(n)\right) + A_5 P^5(n) + A_6 P^6(n), \tag{15.59}$$

$$A_1 = \frac{1}{2\mu_0}(\beta_1 - 4\kappa_0\beta_2), \quad A_2 = \frac{1}{\mu_0}(\beta_1 - 2\kappa_0\beta_2), \tag{15.60}$$

$$A_3 = -\frac{\kappa_0}{\mu_0}(\beta_1 - 4\beta_2), \quad A_5 = \frac{1}{\mu_0}(1-\beta_1-4\kappa_0(\beta_1-4\beta_2)), \tag{15.61}$$

$$A_6 = \frac{1}{\mu_0}(1-\kappa_0 - 2(1-2\kappa_0)\beta_1 - 8\kappa_0\beta_2), \tag{15.62}$$

$$\beta_1 = \frac{g(\gamma)-\gamma^2}{2(1-\gamma^2)}, \quad \beta_2 = \frac{1}{16(1-\gamma^2)^2}(\gamma^2(1+2\gamma^2) + (1-4\gamma^2)g(\gamma)), \tag{15.63}$$

$$g(\gamma) = \frac{\gamma}{\sqrt{1-\gamma^2}}\arccos\sqrt{1-\gamma^2}, \quad \gamma < 1. \tag{15.64}$$

The effective elastic constants of the composite that consists of a polymer matrix reinforced with thin mica flakes of the same orientations were studied experimentally in [1–3]. In Fig. 15.1, these experimental data are presented together with the theoretical predictions based on the effective field method. In the experiments, the so-called "flexural" modulus of elasticity of composites in the plane parallel to the

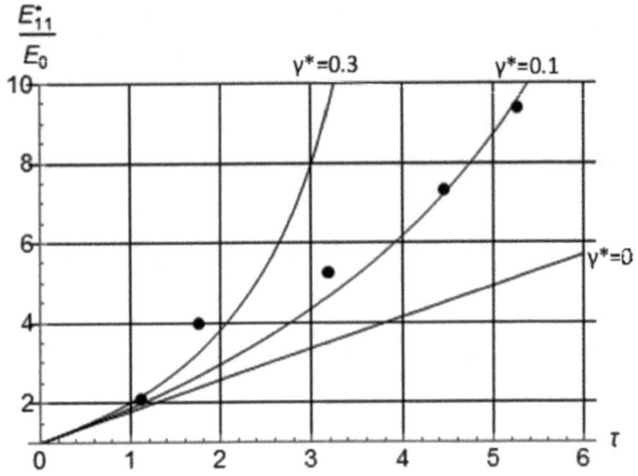

Fig. 15.1 Comparison of the EFM predictions with experimental data for the elastic modulus E_{11}^* of plastics reinforced with mica flakes that are parallel to the plane x_1, x_2, $\tau = 4a^3/(3v_0)$ is the flake density parameter, γ^* is the aspect of the correlation hole $\Psi(x)$

flake orientations was measured. The matrix phase was the polyester resin ($E_0 = 3.5$ GPa, $\nu_0 = 0.4$) filled with mica flakes with average aspect ratio $h_0/a = 0.025$, $E_{11} = 175\,\text{GPa}$, $\nu_{12} = 0.3$. The parameter τ relates to the flake volume fraction p by the equation $\tau = ap/(\pi h_0)$.

Black dots in Fig. 15.1 are the experimental values of E_{11}^*. The calculation are performed for the correlation hole aspect ratio $\gamma = \gamma_* = 0, 0.1, 0.2$. The EFM predictions correspond better to experiments for $\gamma_* = 0.1$.

15.3 Elastic Media with Thin Stiff Flakes of Random Orientations

For the medium with stiff flakes of various orientations, the principal hypothesis of the EFM is modified, and as in the case of cracked medium, the effective field ε^* acting on the flakes is assumed to be dependent on the flake orientation. The derivation is similar to the case of cracked medium, and the effective stiffness tensor of the composite with spheroidal flakes of random orientations takes the form

$$C^* = C^0 + \tau\left(I - \tau\langle\overline{M}(n)A(n)\rangle\right)^{-1}\langle\overline{M}(n)\rangle, \tag{15.65}$$

where the average is performed over the flake orientations n, and n is unit vector orthogonal to the flake middle surface.

The averaged over orientations values tensor $M(n)$ and product $M(n)A(n)$ in Eq. (15.65) have forms

$$\langle\overline{M}(n)\rangle = M_1\langle P^2(n)\rangle + M_2\left\langle P^1(n) - \frac{1}{2}P^2(n)\right\rangle, \tag{15.66}$$

$$\langle \overline{M}(n)A(n)\rangle = 2M_1A_1(\gamma^*)\langle P^2(n)\rangle +$$

$$+ M_2A_2(\gamma^*)\left\langle P^1(n) - \frac{1}{2}P^2(n)\right\rangle + 2M_1A_3(\gamma^*)\langle P^3(n)\rangle. \tag{15.67}$$

For a homogeneous flake distribution over the orientations, the mean of tensors $P^i(n)$ in this equation are

$$\langle P^2(n)\rangle = \frac{2}{15}\left(E^1 + 3E^2\right), \quad \langle P^3(n)\rangle = \frac{2}{15}\left(2E^2 - E^1\right), \tag{15.68}$$

$$\left\langle P^1(n) - \frac{1}{2}P^2(n)\right\rangle = \frac{2}{15}\left(3E^1 - E^2\right), \tag{15.69}$$

where E^1, E^2 are the elements of the E-basis. As a result, the effective stiffness tensor C^*in Eq. (15.65) is isotropic, and it is presented in the form

$$C^* = K_*E^2 + 2\mu_*\left(E^1 - \frac{1}{3}E^2\right). \tag{15.70}$$

Here, the effective bulk modulus K_* and shear modulus μ_* of the composite are

$$K_* = K_0 + \frac{4\tau M_1}{3(3 - 4\tau M_1(2A_1(\gamma_*) + A_3(\gamma_*))}, \tag{15.71}$$

$$\mu_* = \mu_* + \frac{\tau(M_1 + 3M_2)}{(15 - 2\tau(2M_1(A_1(\gamma_*) - A_3(\gamma_*) + 3M_2A_2(\gamma_*))}. \tag{15.72}$$

For the polyester resin filled with mica flakes considered in previous section, dependencies of the effective Young modulus E^* and Poisson's ratio ν_* of the composite are presented in Fig. 15.2 for the aspect ratios of the correlation hole $\gamma_* = 0, 0.3, 1$.

$$E_* = \frac{9K_*}{3K_* + \mu_*}, \quad \nu_* = \frac{3K_* - 2\mu_*}{6K_* + 2\mu_*}. \tag{15.73}$$

15.4 Elastic Medium Reinforced with Thin Stiff Bands (Ribbons)

Let Ω be the middle surface of an infinite band with a constant width $2a$. The band spreads along the x_2-axis and its thickness is $h(x_1)$, $h/a \ll 1$. The strain and stress fields in a homogeneous medium with an isolated band are approximated by Eqs. (15.1), (15.2), where tensor $\overline{\sigma}$ is the solution of the integral equation (15.7)

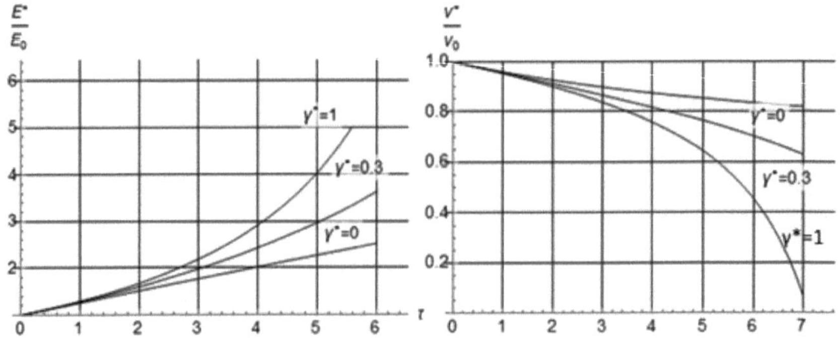

Fig. 15.2 Dependencies of the effective Young modulus E^* and Poisson ratio ν^* for the polyether resin filled with circular flakes homogeneously distributed over the orientations, elastic properties of the host medium and the flakes are the same as in Fig. 15.1

$$\mu(x)\bar{\sigma}(x) + \int_{\Omega} U(x - x')\bar{\sigma}(x')d\Omega' = \Theta(n)\varepsilon^0, \; x \in \Omega, \tag{15.74}$$

Here, $\Theta(n)$ is the projector on the surface Ω, n is the normal vector to Ω and

$$\mu(x) = \frac{1}{h(x)}\Theta(n)B\Theta(n), \tag{15.75}$$

B is the elastic compliance tensor of the band material, and $C^0 B = O(\delta_2)$, $\delta_2 \ll 1$, C^0 is the elastic stiffness tensor of the host medium.

For a band with an elliptical cross section, the function $h(x)$ is

$$h(x_1) = 2h_0 z(x_1), \; z(x_1) = \sqrt{1 - \frac{x_1^2}{a^2}}, \tag{15.76}$$

and the solution of the integral equation (15.74) has the form

$$\bar{\sigma}(x_1) = M(n)\varepsilon^0 z(x_1), \tag{15.77}$$

where the constant tensor $M(n)$ is defined by the equation

$$M(n) = \left(\frac{1}{2h_0}\Theta(n)B\Theta(n) + U^0\right)^{-1}. \tag{15.78}$$

For a thin ellipsoidal inclusion, tensor U^0 is defined in Eq. (15.13), and for a band ($a_2 \to \infty$), U^0 is the integral

$$U^0 = \frac{1}{4}\int_0^{2\pi} \widetilde{U}\left(\frac{\cos\varphi}{a}, 0\right) d\varphi. \tag{15.79}$$

Because the function $\widetilde{U}(k_1, k_2)$ is homogeneous of the order of 1, the equation

$$\widetilde{U}(\lambda k_1, 0) = |\lambda| \widetilde{U}(k_1, 0) \tag{15.80}$$

holds for any scalar parameter λ. Therefore, we have

$$\widetilde{U}\left(\frac{\cos\varphi}{a}, 0\right) = \frac{|\cos\varphi|}{a}\widetilde{U}(1, 0), \tag{15.81}$$

and after integration in Eq. (15.79), we yield

$$U^0 = \frac{1}{a}\widetilde{U}(1, 0). \tag{15.82}$$

For an isotropic medium, the non-zero components of the tensor $\widetilde{U}(k_1, k_2)$ are

$$\widetilde{U}_{1111} = \frac{k_1^2(2k_2^2 + (2-\kappa_0)k_1^2)}{4\mu_0(k_1^2+k_2^2)^{3/2}}, \quad \widetilde{U}_{1122} = -\frac{\kappa_0 k_1^2 k_2^2}{4\mu_0(k_1^2+k_2^2)^{3/2}},$$
$$\widetilde{U}_{2222} = \frac{k_2^2(2k_1^2 + (2-\kappa_0)k_2^2)}{4\mu_0(k_1^2+k_2^2)^{3/2}}, \quad \widetilde{U}_{1212} = \frac{\left(k_1^2+k_2^2\right)^2 - 2\kappa_0 k_1^2 k_2^2}{8\mu_0(k_1^2+k_2^2)^{3/2}}, \tag{15.83}$$

and therefore,

$$U^0 = \frac{(2-\kappa_0)}{4a\mu_0}T^1 + \frac{1}{8a\mu_0}T^{10}, \tag{15.84}$$

where the elements of the T-basis T^1, T^{10} are defined in Eqs. (15.23)–(15.25), n^1, n^2, n^3 are unit vectors directed along the x_1, x_2 and x_3-axes. Because tensor $\Theta(n)B\Theta(n)$ in the T-basis has the form

$$\Theta(n)B\Theta(n) = \frac{1}{E_{11}}\left(T^1 + T^2\right) - \frac{\nu_{12}}{E_{11}}(T^4 + T^5) + \frac{1}{4\mu_{12}}T^{10}, \tag{15.85}$$

the tensor $M(n)$ in Eq. (15.77) is

$$M(n) = 2a\overline{M}(n), \tag{15.86}$$

$$\overline{M}(n) = 4\mu_0\left(M_1 T^1 + M_2 T^2 + M_4(T^4 + T^5) + M_{10}T^{10}\right), \tag{15.87}$$

$$M_1 = \frac{(2-\kappa_0)(1+\nu_{12})+\zeta}{\zeta(2-\kappa_0+(1-\nu_{12})\zeta)}, \quad M_2 = \frac{1}{2-\kappa_0+(1-\nu_{12})\zeta}, \tag{15.88}$$

$$M_4 = \frac{\nu_{12}}{2-\kappa_0+(1-\nu_{12})\zeta}, \quad M_{10} = \frac{1}{2(1+\zeta)}, \quad \zeta = \frac{a\mu_0}{2h\mu_{12}}. \tag{15.89}$$

For a homogeneous isotropic host medium with the bands of the same orientations, the mean strain and stress tensor are approximated by the equations

$$\langle \varepsilon(x) \rangle = \varepsilon^0 - \int K(x-x')dx' M(n) \langle Z(x)\Omega(x) \rangle \varepsilon^*, \tag{15.90}$$

$$\langle \sigma(x) \rangle = \sigma^0 - \int S(x-x')dx' B^0 M(n) \langle Z(x)\Omega(x) \rangle \varepsilon^*. \tag{15.91}$$

For the medium with the bands, the mean $\langle Z(x)\Omega(x) \rangle$ in these equations is calculated as follows

$$\langle Z(x)\Omega(x) \rangle = \lim_{W\to\infty} \frac{1}{W} \int_W Z(x)\Omega(x)dx =$$

$$= \lim_{W\to\infty} \sum_{i=1}^{N} \int_{\Omega_i} z^i(x)d\Omega = \frac{\pi a}{2} \lim_{SL\to\infty} \frac{NL}{SL} = \frac{\pi a}{2s_0}. \tag{15.92}$$

Here, the region W is a cylinder of the length L with the generatrix in the direction of bands (x_2-axis) and the cross section S such as in the limit $W \to \infty$, the cylinder occupies the entire space, s_0 is the area that falls on one band in the plane orthogonal to the x_2-axis. As a result, the averages in Eqs. (15.90), (15.91) are

$$\langle \varepsilon(x) \rangle = \varepsilon^0, \quad \langle \sigma(x) \rangle = \sigma^0 + \tau \overline{M}(n)\varepsilon^*, \quad \tau = \frac{\pi a^2}{s_0}. \tag{15.93}$$

It is assumed that the external field ε_0 is fixed in the problem, and therefore, Eqs. (15.44) hold.

The equation for the effective field is derived similarly to the case of thin stiff flakes. If the correlation hole for a typical band has the form of an elliptical cylinder with the aspect ratio γ^* that is congruent to the band, the tensor ε^* takes the form

$$\varepsilon^* = (I - \tau A(\gamma^*)M)^{-1}\varepsilon_0, \tag{15.94}$$

were tensor $A(\gamma)$ is the following integral

$$A(\gamma) = \frac{1}{4\pi} \int_0^{2\pi} d\varphi \int_0^{\pi} \widetilde{K}(m_1, m_2, m_3) \sin\theta d\theta, \tag{15.95}$$

$$m_1 = \frac{\gamma \cos\varphi}{\sqrt{(\gamma\cos\varphi)^2 + (\sin\varphi)^2}}, m_2 = 0, m_3 = \frac{\sin\varphi}{\sqrt{(\gamma\cos\varphi)^2 + (\sin\varphi)^2}}. \quad (15.96)$$

For an isotropic host medium, the tensor $A(\gamma)$ in the T-basis takes the form

$$A(\gamma) = A_1T^1 + A_3T^3 + A_6(T^6 + T^7) + A_{10}T^{10} + A_{11}T^{11} + A_{12}T^{12}, \quad (15.97)$$

$$A_1 = \frac{\gamma(2 - \kappa_0 + 2\gamma(1 - \kappa_0))}{2(1+\gamma)^2}, \quad A_3 = \frac{1}{1+\gamma} - \frac{\kappa_0(2+\gamma)}{2(1+\gamma)^2}, \quad (15.98)$$

$$A_6 = -\frac{\kappa_0\gamma}{2(1+\gamma)^2}, \quad A_{10} = \frac{\gamma}{4(1+\gamma)}, \quad (15.99)$$

$$A_{11} = \frac{1}{4} - \frac{\kappa_0\gamma}{2(1+\gamma)^2}, \quad A_{12} = \frac{1}{4(1+\gamma)}. \quad (15.100)$$

It follows from Eqs. (15.93), (15.94) that the effective stiffness tensor C^* of the composite has the form

$$C^* = C^0 + c_1^*T^1 + c_2^*T^2 + c_4^*(T^4 + T^5) + c_{10}^*T^{10}, \quad (15.101)$$

$$c_1^* = \frac{\tau M_1}{1 - \tau A_1 M_1}, \quad c_2^* = \tau\left(M_2 + \frac{\tau A_1 M_4^2}{1 - \tau A_1 M_1}\right), \quad (15.102)$$

$$c_4^* = \frac{\tau M_4}{1 - \tau A_1 M_1}, \quad c_{10}^* = \frac{\tau M_{10}}{1 - 4\tau A_{10} M_{10}}. \quad (15.103)$$

Note that in the T-basis, the elastic stiffness tensor C^0 of an isotropic medium is

$$C^0 = (\lambda_0 + 2\mu_0)(T^1 + T^2 + T^3) +$$

$$+ \lambda_0(T^4 + T^5 + T^6 + T^7 + T^9) + \mu_0(T^{10} + T^{11} + T^{12}). \quad (15.104)$$

Thus, for composites reinforced with long stiff bands, tensor C^* is orthotropic, and it contains 6 independent constants.

Figures 15.3 and 15.4 present theoretical dependencies of the effective elastic moduli E_{11}^*, E_{22}^*, and E_{33}^* in the directions of x_1, x_2 and x_3-axis on the parameter $\tau = \pi a^2/s_0 = p(a/h_0)$ (p is the volume fraction of the bands), for an isotropic host medium with the Young modulus $E_0 = 2$ GPa and the Poisson ratio $\nu_0 = 0.25$ reinforced by glass bands with the aspect ratios $\gamma = h_0/a = 1/230$, $E_{11} = 59$ GPa, $\nu_{12} = 0.26$. Black point in these figures are experimental data presented in [4]. For $\gamma_* = 0$, the theoretical dependences of $E_{11}^*(\tau)$, $E_{22}^*(\tau)$ and describe the experimental

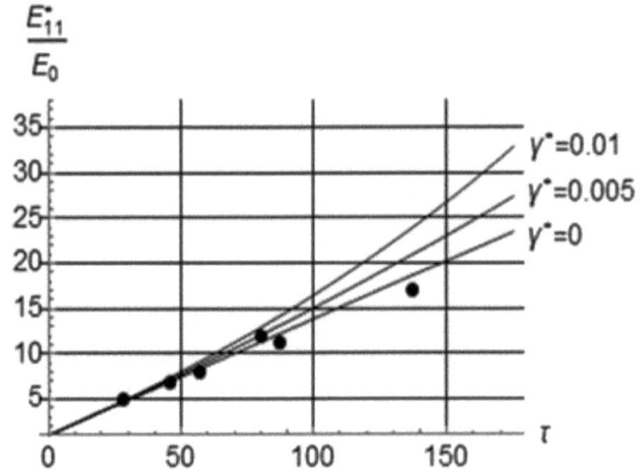

Fig. 15.3 Dependencies of the effective Young modulus E_{11} in the directions of x_1-axis on the parameter $\tau = \pi a^2/s_0 = p(a/h)$ (p is the volume fraction of the bands), for an isotropic host medium with the Young modulus $E_0 = 2GPa$ and the Poisson ratio $\nu_0 = 0.25$ reinforced with the glass bands with the aspect ratios $\gamma = h_0/a = 1/230$, $E_{11} = 59GPa$, $\nu_{12} = 0.26$; black points are experimental data presented in [65], γ^* is the aspect ratio of the correlation hole

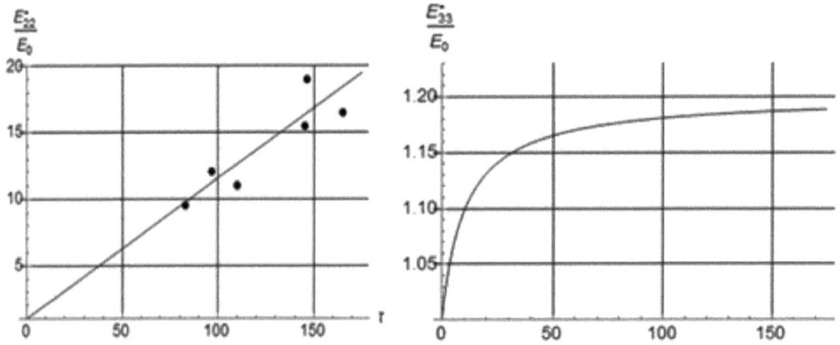

Fig. 15.4 The same as in Fig. 15.3 for Young moduli E_{22}, E_{33}

data with acceptable accuracy. Young's moduli E^*_{22} along the bands and E^*_{33} orthogonal to the bands are not sensitive to the change of the correlation hole aspect γ_* in the interval $0 < \gamma_* < 0.1$.

The proposed solution of the homogenization problem for the composites reinforced with thin stiff inclusions is based on two types of simplifications. The first one is the use of the principal terms in the asymptotic solution of the one-particle problem with respect to the small parameters of the problem $\delta_1 = h/a$ and $\delta_2 = E_0/E_{11}$. The corresponding errors have the order of $\max(\delta_1, \delta_2)$. The second simplification relates to the hypothesis of the quasicrystalline approximation for the description of interactions between inclusions. Analysis of experimental data shows that the overall effect of the interactions in the case of the stiff flakes (bands) is relatively weak. As a result, the dependence of the effective elastic moduli on the volume fraction of the filler is quasi-linear. In the framework of the EFM, the influence of the interactions correlates with the values of the parameter γ^* that describes the symmetry of the correlation hole $\Psi(x)$. For non-interacting inclusions $\gamma_* = 0$, and the relative error is minimal. The value of γ_* should be determined from the statistical analysis of the spatial distribution of the fillers in the actual composite material.

References

1. Lusis J, Woodhams R, Xanthos M (1973) The effect of flake aspect ratio on the flexural properties of mica reinforced plastics. Polym Eng Sci 13:139–145
2. Okuno K, Woodhams R (1975) Mica reinforced polypropylene. Polym Eng Sci 15:308–315
3. Okuno K (1980) On the flexural modulus of the flake-filled plastics. Kobunshi Ronbunshu 37:789–796
4. Galati S (1981) Ribbon fillers. Fillers for polymer composite materials. Nauka, Moscow, pp 429–468

Chapter 16
Composites Reinforced with Short Axisymmetric Fibers

Abstract Matrix composite materials reinforced with short axisymmetric fibers which elastic moduli are much larger than the moduli of the host medium are considered. The asymptotic solutions of the one-particle problems for a stiff fiber of various shapes are used in the framework of the effective field method for the calculation of the effective conductive tensors of the medium with fibers of large conductivity and the effective elastic stiffness tensors of the medium containing random sets of stiff axisymmetric fibers.

Keywords Stiff axisymmetric fiber · Asymptotic solution of one-particle problem · Effective field method · Fibers of various orientations · Effective elastic properties

In this chapter, we consider matrix composite materials reinforced with short axisymmetric fibers which elastic moduli are much larger than the moduli of the host medium. The asymptotic solutions of the one-particle problems for a stiff fiber of various shapes in a homogeneous medium subjected to a constant external field are considered in Chap. 7. These solutions are used in the framework of the effective field method for calculation of the effective conductive tensors of the medium with fibers of large conductivity and the effective elastic stiffness tensors of the medium containing random sets of stiff axisymmetric fibers.

16.1 Homogeneous Conductive Medium with a Random Set of Fibers of Large Conductivity

We consider a homogeneous medium with the conductivity tensor C^0 containing a random set of short axisymmetric fibers with conductivity tensors C. For a fiber of large conductivity, the solution of the one-particle problem depends on two small parameters

$$\delta_1 = O\left(\frac{a}{l}\right),\ \delta_2 = O(C^0 C^{-1}).$$

S. Kanaun, *Self-Consistent Methods for Composites*, Solid Mechanics and Its Applications 282, https://doi.org/10.1007/978-3-031-96739-9_16

Here a is the maximal fiber radius, $2l$ is the fiber length, $\delta_1, \delta_2 \ll 1$. For the medium with an isolated fiber subjected to a constant external field e^0 (j^0) at infinity, electric field intensity $e(x)$ and electric current $j(x)$ are presented in the following integral forms (Sect. 4.10)

$$e(x) = e^0 - \int_v K(x - x')j(x')dx', \tag{16.1}$$

$$j(x) = j^0 - \int_v S(x - x')B^0 j(x')dx. \tag{16.2}$$

Here, v is the region of the fiber, $B^0 = (C^0)^{-1}$. It is taken into account that for $C^0C^{-1} = O(\delta_2)$,

$$(C - C^0)C^{-1} \approx I, \; B^1 = C^{-1} - (C^0)^{-1} \approx -B^0. \tag{16.3}$$

The principal term of the current $j(x)$ inside the fiber with respect to the parameters δ_1, δ_2 is defined in Eq. (7.152)

$$j(z) = cf(\varsigma)(n \otimes n) \cdot e^0, \; \varsigma = \frac{z}{l}, \tag{16.4}$$

where z is the coordinate along the fiber with the origin at the fiber center, and n is unit vector of the z-axis. The function $f(\varsigma)$ depends on the fiber shape, and it is defined in Eqs. (7.118)–(7.121) for cylindrical, ellipsoidal, and double conical fibers. The principal terms of the asymptotic solution for the current inside the fiber is constant in each fiber cross section, and therefore, after integration over the cross-sections, the principal terms of the field intensity and the current in the medium follow from Eqs. (13.124), (13.125) in the forms of 1D-integrals

$$e(x_1, x_2, z) = e^0 - \pi a^2 \int_l K(x_1, x_2, z - z')\Lambda(z')dz'e^0, \tag{16.5}$$

$$j(x_1, x_2, z) = j^0 - \pi a^2 \int_l S(x_1, x_2, z - z')B^0\Lambda(z')dz'e^0, \tag{16.6}$$

where l is the fiber axis, $x_1, x_2, z = x_3$ are the Cartesian coordinates with the origin in the fiber center,

$$\Lambda(z) = c\alpha(\varsigma)^2 f(\varsigma)(n \otimes n), \; \alpha(\varsigma) = \frac{r(\varsigma)}{a}, \tag{16.7}$$

$r(\varsigma)$ is the fiber radius, and $r(0) = a$. The fiber material is isotropic with the conductivity coefficient c.

Let the medium contain a homogeneous in space random set of identical fibers of the same orientations. According to the hypothesis of the effective field method, each fiber in the set is subjected to a local external field e^* that is constant and the same for all the fibers. For integral presentation of the field in the conductive medium with the fibers, we introduce the generalized functions $l_i(x)$ concentrated on the axis l_i of the ith fiber, $i = 1, 2, ...$ Such a function acts on an arbitrary smooth function $\varphi(x)$ by the rule

$$\int \varphi(x)l_i(x)dx = \int_{l_i} \varphi(x)dl. \tag{16.8}$$

Then, the fields in the medium with a set of fibers can be approximated by the equations

$$e(x) = e^0 - \pi a^2 \int K(x - x')\Lambda(x')L(x')dx'e^*, \tag{16.9}$$

$$j(x) = j^0 - \pi a^2 \int S(x - x_1')B^0\Lambda(x')L(x')dx'e^*. \tag{16.10}$$

Here,

$$L(x) = \sum_i l_i(x), \tag{16.11}$$

and $\Lambda(x)$ coincides with the function $\Lambda(z)$ in Eq. (16.7) at the axis of each fiber. Then, we introduce the function

$$L(x', x) = \sum_{i \neq j} l_i(x) \text{ if } x \in l_j. \tag{16.12}$$

Using this function, the local external field $e^*(x)$ acting on point $x \in L = \bigcup_i l_i$ is presented in the form

$$e^*(x) = e^0 - \pi a^2 \int K(x - x')\Lambda(x')L(x', x)dx'e^*. \tag{16.13}$$

Here, integration is spread to the entire space. After averaging the field $e^*(x)$ over the ensemble of realizations of the fiber set by the condition that $x \in L$, we obtain

$$\langle e^*(x)|x\rangle = e^0 - \pi a^2 \int_L K(x - x')\langle \Lambda(x')L(x', x)|x\rangle dx'e^*. \tag{16.14}$$

Accepting that $\langle e^*(x)|x\rangle = e^*$, we obtain a closed equation for the effective field e^* acting on each fiber.

Using the ergodic property of the fiber set, the mean $\pi a^2 \langle \Lambda(x)L(x)\rangle$ is calculated as follows

$$\pi a^2 \langle \Lambda(x)L(x)\rangle = \lim_{W\to\infty} \frac{\pi a^2}{W} \int_W \Lambda(x)L(x)dx =$$

$$= \left(\lim_{W\to\infty} \frac{\pi a^2}{W} \sum_{i=1}^{N} \int_{-1}^{1} \alpha(\varsigma)^2 f(\varsigma)d\varsigma \right) c\,(n \otimes n) = p\overline{\Lambda}(n), \tag{16.15}$$

where W is a domain in 3D-space that occupied the entire space in the limit $W \to \infty$, N is the number of fibers in W,

$$\overline{\Lambda}(n) = \frac{c}{\beta} \left(\int_{-1}^{1} \alpha(\varsigma)^2 f(\varsigma)d\varsigma \right) (n \otimes n), \tag{16.16}$$

$$p = \frac{v}{v_0},\ v = \pi a^2 \beta,\ \beta = \int_{-1}^{1} \alpha(\varsigma)^2 d\varsigma. \tag{16.17}$$

Here, v is the volume of a fiber, v_0 is the volume fallen at one fiber, p is the fiber volume fraction. Similarly the conditional mean in Eq. (13.138) takes the form

$$\pi a^2 \langle \Lambda(x') L(x'; x)\,|x\rangle = p\overline{\Lambda}(n)\Psi(x - x'), \tag{16.18}$$

$$\Psi(x - x') = \frac{\langle L(x'; x)\,|x\rangle}{\langle L(x)\rangle}, \tag{16.19}$$

The function $\Psi(x)$ (the correlation hole for a typical fiber) has the properties

$$\Psi(0) = 0,\ \lim_{W\to\infty} \frac{1}{W} \int_W \Psi(x) = 1. \tag{16.20}$$

If there is an ellipsoidal region around each fiber where the probability to meet other fibers is small, the function $\Psi(x)$ (the correlation hole) has the symmetry of an ellipsoid of revolution with the axis n and the semiaxes $\alpha_1 = \alpha_2 = \alpha$ and α_3. In this case, the integral in Eq. (13.138) takes the form

$$\int K(x)\Psi(x)dx = -\int K(x)\,(1 - \Psi(x))\,dx = -A(\gamma_*), \tag{16.21}$$

where the tensor $A(\gamma_*)$ is defined in Eq. (4.140) and $\gamma_* = \alpha/\alpha_3$ is the aspect ratio of the correlation hole. Here, it is assumed that external current e^0 is fixed in the problem, and therefore,

$$\int K(x)dx = 0, \int S(x)dx = -C^0. \tag{16.22}$$

It follows from Eq. (16.14) that the effective field e^* acting on each fiber has the form

$$e^* = (I - pA(\gamma_*)\overline{\Lambda}(n))^{-1}e^0. \tag{16.23}$$

Averaging Eqs. (16.9), (16.10) over the ensemble of realizations of the random set of fibers we obtain

$$\langle e(x)\rangle = e^0, \tag{16.24}$$

$$\langle j(x)\rangle = j^0 - \pi a^2 \int_l S(x - x')B^0 \langle \Lambda(x')L(x')\rangle dx' e^* =$$

$$= j^0 + p\overline{\Lambda}(n)e^* = C^0e^0 + p\overline{\Lambda}(n)(I - pA(\gamma^*)\overline{\Lambda}(n))^{-1}e^0. \tag{16.25}$$

Here, Eq. (16.22) is taken into account. Thus the equation that relates the means $\langle j(x)\rangle$ and $\langle e(x)\rangle$ has the form

$$\langle j(x)\rangle = C^* \langle e(x)\rangle, \tag{16.26}$$

where C^* is the effective conductivity tensor of the medium with a set of the fibers

$$C^* = C^0 + p\overline{\Lambda}(n)(I - pA(\gamma_*)\overline{\Lambda}(n))^{-1}. \tag{16.27}$$

16.2 Elastic Medium Reinforced with Stiff Axisymmetric Fibers of the Same Orientation

For an elastic medium with stiff axisymmetric fibers, the effective field method results in the effective stiffness tensor C^* of the composite that is similar to the case of conductivity (16.27). The second rank tensors $\overline{\Lambda}(n)$ and $A(\gamma_*)$ in this equation should be changed with their fourth rank analogies in the case of elasticity. The one-particle problem for a stiff elastic fiber is considered in Chap. 7. The asymptotic solution of this problem for an isotropic host medium and transverse isotropic fibers results in the transverse isotropic tensor $\overline{\Lambda}(n)$ in Eq. (16.27). In the P-basis, this tensor takes the form

$$\overline{\Lambda}(n) = \Lambda_1 P^2(n) + \Lambda_2\left(P^1(n) - \frac{1}{2}P^2(n)\right) +$$

$$+ \Lambda_3\left(P^3(n) + P^4(n)\right) + \Lambda_5 P^5(n) + \Lambda_6 P^6(n), \tag{16.28}$$

Here, n is a unit vector of the fiber symmetry axis, the scalar coefficients in this equation are

$$\Lambda_1 = \frac{\mu_0(1-\nu_0)}{1-2\nu_0}, \quad \Lambda_2 = 8\mu_0\frac{1-\nu_0}{3-4\nu_0}, \tag{16.29}$$

$$\Lambda_3 = \frac{\mu_0}{1-2\nu_0}\left(1-\varphi\left(q\right)\left(1-2\left(1-\nu_0\right)\nu_{1n}\frac{E_n}{E_1}\right)\right), \tag{16.30}$$

$$\Lambda_5 = 8\mu_0\,, \quad \Lambda_6 = E_n\varphi\left(q\right)\,, \tag{16.31}$$

where E_0, ν_0 are Young's modulus and Poisson's ratio of the medium, E_n and E_1 are Young's moduli of the fiber material along its fiber axis n and in the transverse direction, ν_{1n} is Poisson's ratio in the transverse direction. The function $\varphi(q)$ has the form

$$\varphi\left(q\right) = \left(\int_{-1}^{1}\alpha^2\left(\varsigma\right)d\varsigma\right)^{-1}\int_{-1}^{1}\alpha^2\left(\varsigma\right)f\left(\varsigma, q\right)d\varsigma\,, \tag{16.32}$$

$$q^2 = -\frac{\mathrm{E}_0}{\mathrm{E}_n}\frac{\left(1+\nu_0\right)^{-1}}{\delta_1^2\ln\delta_1} = -\frac{2\mu_0}{\mathrm{E}_n\delta_1^2\ln\delta_1}, \quad \delta_1 = \frac{a}{l}. \tag{16.33}$$

The function $f(\varsigma, q)$ in Eq. (16.32) depends on the fiber shape, and its explicit equations for fiber particular shapes have forms in Eqs. (7.118)–(7.121). For these types of fibers, the functions $\varphi(q)$ are as follows

(a) Cylindrical fibers $\alpha(\varsigma) = 1$,

$$\varphi\left(q\right) = 1 - \frac{\mathrm{th}\, q}{q}. \tag{16.34}$$

(b) Prolate spheroidal fibers $\alpha(\varsigma) = \sqrt{1-\zeta^2}$,

$$\varphi\left(q\right) = \frac{q^2}{2+q^2}. \tag{16.35}$$

(c) Fibers in the form of a double cone $\alpha(\xi) = 1 - |\xi|$,

$$\varphi\left(q\right) = \frac{q^2\beta\left(5+\beta\right)}{\left(q^2-2\right)\left(2+\beta\right)\left(3+\beta\right)}, \quad \beta = \frac{1}{2}\left(\sqrt{1+8q}-3\right). \tag{16.36}$$

The function $\varphi(q)$ characterizes the "underloading" of a fiber of finite length in comparison with the fiber of infinite length in the composite. The graphs of these functions are presented in Fig. 16.1 by the lines 1(a), 2(b), and 3(c).

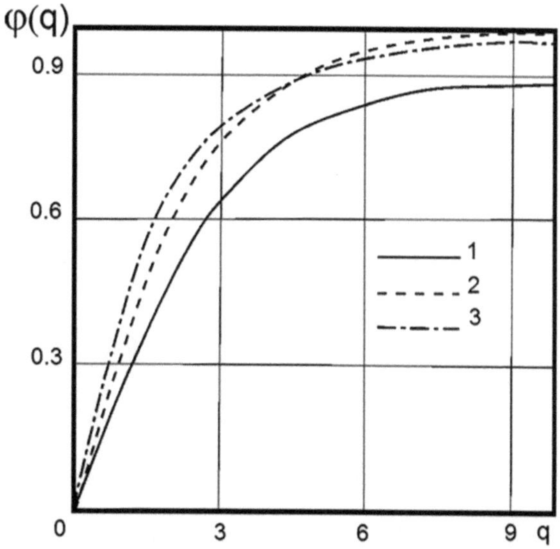

Fig. 16.1 Function $\varphi(q)$ for fibers of various forms

The tensor $A(\gamma)$ for the correlation hole $\Psi(x)$ with the aspect ratio $\gamma_* < 1$ has the form

$$A(\gamma) = A_1 P^2(n) + A_2\left(P^1(n) - \frac{1}{2}P^2(n)\right) +$$

$$+ A_3\left(P^3(n) + P^4(n)\right) + A_5 P^5(n) + A_6 P^6(n), \tag{16.37}$$

$$A_1 = \frac{1}{2\mu_0}((1-\kappa_0)\, f_0 + \kappa_0 f_1), \quad A_2 = \frac{1}{2\mu_0}((2-\kappa_0)\, f_0 + \kappa_0 f_1), \tag{16.38}$$

$$A_3 = -\frac{\kappa_0}{\mu_0} f_1, \quad A_5 = \frac{1}{\mu_0}(1 - f_0 - 4\kappa_0 f_1), \tag{16.39}$$

$$A_6 = \frac{1}{\mu_0}((1-\kappa_0)(1 - 2f_0) + 2\kappa_0 f_1), \tag{16.40}$$

$$f_0 = \frac{1-g}{2(1-\gamma^2)}, \quad f_1 = \frac{1}{4(1-\gamma^2)^2}\left((2+\gamma^2)\, g - 3\gamma^2\right), \tag{16.41}$$

$$g = \frac{\gamma^2}{2\sqrt{1-\gamma^2}} \ln \frac{1+\sqrt{1-\gamma^2}}{1-\sqrt{1-\gamma^2}}. \tag{16.42}$$

In the P-basis ($P^i = P^i(n)$), the tensor $H(n) = I - pA(\gamma_*)\overline{\Lambda}(n)$ in Eq. (16.27) takes the form

$$H(n) = h_1 P^2 + h_2 \left(P^1 - \frac{1}{2} P^2 \right) + h_3 P^3 + h_4 P + h_5 P^5 + h_6 P^6, \tag{16.43}$$

$$h_1 = \frac{1}{2} - 2p\Lambda_1 A_1 - p\Lambda_3 A_3, \quad h_2 = 1 - p\Lambda_2 A_2, \tag{16.44}$$

$$h_3 = -p\left(2\Lambda_1 A_3 + \Lambda_3 A_6\right), \quad h_4 = -p\left(2\Lambda_3 A_1 + \Lambda_6 A_3\right), \tag{16.45}$$

$$h_5 = 2 - \frac{1}{2} p\Lambda_5 A_5, \quad h_6 = 1 - 2p\Lambda_3 A_3 - p\Lambda_6 A_6, \tag{16.46}$$

where the coefficients A_i and Λ_i are defined in Eqs. (16.38)–(16.42) and (16.29)–(16.31).

The tensor C^* in Eq. (16.27) is transversely isotropic and in the P-basis ($P^i = P^i(n)$) it takes the form

$$C^*(n) = k_* P^2 + 2m_* \left(P^1 - \frac{1}{2} P^2 \right) + l_* \left(P^3 + P^4 \right) + 4\mu_* P^5 + n_* P^6, \tag{16.47}$$

$$k_* = \frac{E_0}{2\left(1 + \nu_0\right)\left(1 - 2\nu_0\right)} + \frac{p}{\Delta}\left(\Lambda_1 h_6 - \Lambda_3 h_3\right), \tag{16.48}$$

$$m_* = \frac{E_0}{2\left(1 + \nu_0\right)} + p\frac{\Lambda_2}{2h_2}, \tag{16.49}$$

$$l_* = \frac{E_0 \nu_0}{\left(1 + \nu_0\right)\left(1 - 2\nu_0\right)} + \frac{p}{\Delta}\left(\Lambda_3 h_6 - \Lambda_6 h_3\right), \tag{16.50}$$

$$\mu_* = \frac{1}{2}\left(\frac{E_0}{1 + \nu_0} + p\frac{\Lambda_5}{h_5} \right), \tag{16.51}$$

$$n_* = \frac{E_0\left(1 - \nu_0\right)}{\left(1 + \nu_0\right)\left(1 - 2\nu_0\right)} + \frac{2}{\Delta} p\left(\Lambda_6 h_1 - \Lambda_3 h_4\right), \tag{16.52}$$

$$\Delta = 2\left(h_1 h_6 - h_3 h_4\right). \tag{16.53}$$

Figure 16.2 presents the dependencies of Young's modulus E_m^* (along the axis of reinforcement) on the volume fraction p of the fibers for the polyamide thermoplastic ($E_o = 2.5GPa$, $\nu_0 = 0.35$), reinforced with unidirected cylindrical glass fibers ($E_m = 70GPa$, $\nu_{1m} = 0.22$, $\delta_1 = 0.04$). The most satisfactory agreement of Eq. (16.47) (the solid lines in Fig. 16.2) with experimental data [1] (light circles) is achieved if the parameter $\gamma^* = \gamma$ in (16.37) is taken equal to δ_1 (in Fig. 16.2, line 1 corresponds to $\gamma^* = \delta_1$, line 2 to $\gamma^* = 2\delta_1$).

Figure 16.3 presents the graphs of the Young moduli $E_1^*(p)$ and $E_3^*(p)$ for the composite reinforced with a periodic lattice of unidirected fibers in the form of

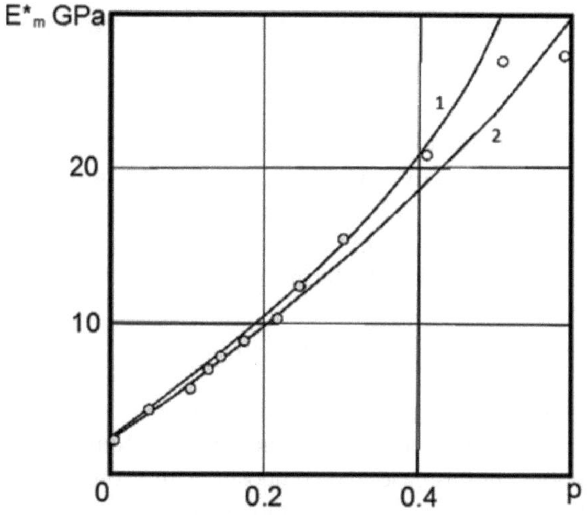

Fig. 16.2 The dependence of Young's modulus E_m^* (along the axis of reinforcement) on the volume fraction p of the fibers for the polyamide thermoplastic ($E_0 = 2.5GPa$, $\nu_0 = 0.35$), reinforced by unidirectional cylindrical glass fibers ($E_m = 70GPa$, $\nu_{1m} = 0.22$, $\delta_1 = 0.04$). The most satisfactory agreement of Eq. (16.47) with experimental data [1] (light circles) is achieved if the parameter $\gamma^* = \gamma$ in Eq. (16.37) is taken equal to δ_1 (line 1 corresponds to $\gamma^* = \delta_1$, line 2 to $\gamma^* = 2\delta_1$)

Fig. 16.3 Comparison of the approximate and exact values of the effective Young modulus of the medium reinforced with a regular lattice of cylindrical fibers

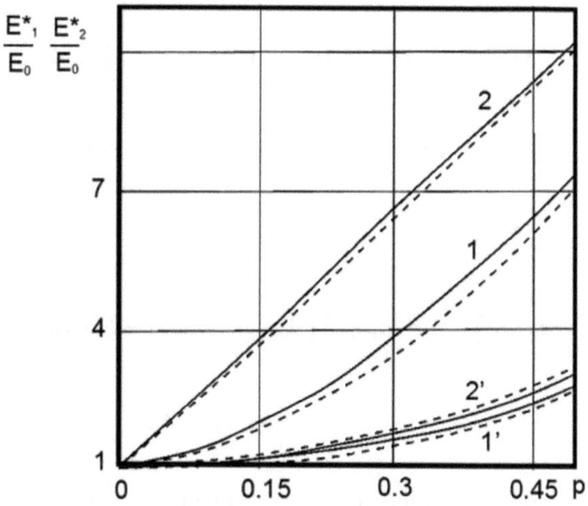

parallelepipeds. For an isotropic medium and the inclusions ($E_0/E_m = 0.05$, $\nu_m = 0.3$, $\nu_0 = 0.35$), these moduli were calculated in [2] by the finite element method. Solid lines $1(E_3^*)$ and $1'(E_1^*)$ correspond to $\delta_1 = 0.1$; $2(E_3^*)$ and $2'(E_1^*) - \delta_1 = 0.01$. When $\delta_1 = 0.1$, Eq. (16.47) is in satisfactory agreement with the exact solution [2] if $\gamma_* = 3\delta_1$ (the dashed lines in Fig. 16.3). When $\delta_1 = 0.01$ and $\gamma_* = O(\delta_1)$, Eq. (16.47) practically coincided with the results obtained in [2].

16.3 Elastic Medium Reinforced with Axisymmetric Fibers of Random Orientations

For the composites with random distribution of fiber over the orientations, the principal hypothesis of the effective method is changed. As in the case of ellipsoidal inclusions, the effective field is assumed to be dependent on the fiber orientation $\varepsilon^* = \varepsilon^*(n)$, where n is unit vector of the fiber symmetry axis. The effective stiffness tensor C^* of the composite takes the form similar to in Eq. (13.111)

$$C^* = C^0 + p(I - p\langle A(\gamma^*, n)\overline{\Lambda}(n)\rangle)^{-1}\langle\overline{\Lambda}(n)\rangle . \tag{16.54}$$

Here, symbol $\langle\cdot\rangle$ means averaging over the orientations.

Let the host medium be isotropic and the fibers be homogeneously distributed over the orientations. In this case, the composite is macroscopically isotropic, and its effective bulk K_* and shear μ_*, elastic moduli are

$$K_* = K_0 + \frac{\Lambda_1}{1 - j_1}, \quad \mu_* = \mu_0 + \frac{\Lambda_2}{1 - j_2}, \tag{16.55}$$

$$\Lambda_1 = \frac{p}{9}(4(\Lambda_1 + \Lambda_3) + \Lambda_6), \tag{16.56}$$

$$\Lambda_2 = \frac{p}{15}\left(\Lambda_1 + 3\Lambda_2 - 2\left(\Lambda_3 - \frac{3}{4}\Lambda_5\right) + \Lambda_6\right), \tag{16.57}$$

$$j_1 = \frac{p}{3}(8\Lambda_1 a_1 + 6\Lambda_3 a_3 + 4(\Lambda_1 a_3 + \Lambda_3 a_1) +$$

$$+2(\Lambda_3 a_6 + a_3\Lambda_6) + \Lambda_6 a_6), \tag{16.58}$$

$$j_2 = \frac{2p}{15}(2\Lambda_1 a_1 + 3(\Lambda_3 a_3 + \Lambda_2 a_2) - 2(\Lambda_1 a_3 + \Lambda_3 a_1) -$$

$$-(\Lambda_3 a_6 + \Lambda_6 a_3) + +\frac{3}{4}a_5\Lambda_5 + \Lambda_6 a_6\Big), \tag{16.59}$$

where the coefficients Λ_i are defined in Eqs. (16.29)–(16.31) and the coefficients a_i are in Eqs. (16.38)–(16.40).

Figure 16.4 presents theoretical (solid lines) and experimental (light circles [3]) dependencies of Young's modulus E_* and Poisson's ratio ν_* on the volume fraction p of the fibers for a composite with a thermoset plastic matrix ($E_0 = 2.25GPa$, $\nu_0 = 0.4$) reinforced by randomly oriented cylindrical glass fibers ($E_m = 77\ GPa$, $\nu_m = 0.25$, $\delta_1 = 0.002$). The solid lines correspond to Eqs. (16.55)–(16.59) when $\gamma_* = 2\delta_1$.

The results of this section lead us to the following conclusions.

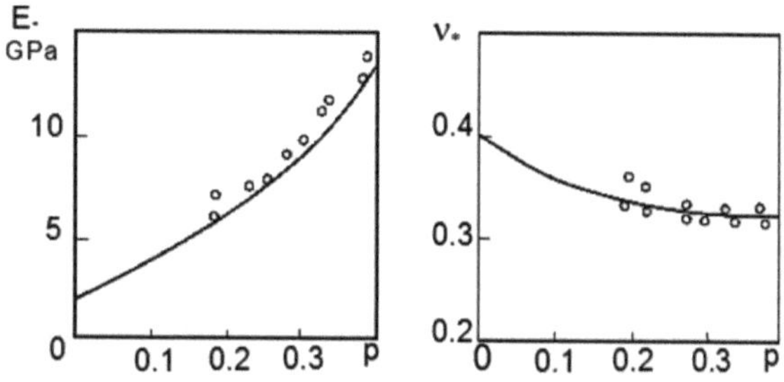

Fig. 16.4 Comparison of the theoretical predictions and experimental data for Young's modulus and Poisson's ratio of a composite reinforced with a random set of cylindrical fibers homogeneously distributed over orientations

1. For very long fibers ($q > 10$) or very stiff fibers ($q < 0.1$), the shape of the fibers practically does not affect the effective elastic stiffness tensor C^* of the composite. The tensor C^* depends on the fiber volume fraction p for $q > 10$, or on the dimensionless parameter $\chi = -l^3/(v_0 \ln \delta_1)$ for $q < 0.1$.

2. In the composites reinforced with fibers in the forms of ellipsoids or double cones, the material of the filler is used more rationally than in composites with cylindrical fibers. This fact follows from the graphs of the function $\varphi(q)$ for the cylindrical, ellipsoidal, and double cone fibers in Fig. 16.1. The lines 2, 3 for ellipsoidal and double cone fibers in Fig. 16.1 are above such a graph for the cylindrical fiber (line 1).

3. Comparisons of predictions of the EFM with experimental data show that the parameter γ_* (the aspect ratio of the correlation hole) has the order of the average aspect δ_1 of the fibers. The parameter γ^* is maximal for a spatially correlated (regular) system of fibers ($\gamma_* \approx 3\delta_1$), and it turns out to be smaller ($\gamma_* = (1 \div 2)\delta_1$) for random sets of fibers. This fact corresponds to the geometrical meaning of the parameter γ_* that relates to the probability to meet another fiber in the vicinity of a fiber. For stochastic fiber sets, when the fiber axes can approach each other on the distance of the order of their diameters ($2a$), the aspect γ_* has the order of δ_1. For regular lattices of the fiber, the value of γ_* is defined by the ratio of the characteristic sizes of the elementary cell of the fiber lattice, and γ_* exceeds δ_1. The value of γ_* depends on the technological peculiarities of the composite formation and has to be determined from the stochastic analysis of the actual composite microstructure.

16.4 Notes

Self-consistent methods can be used for the evaluation of the effective physical property of composites in the case of coupled physical fields. Piezoelectric composites are important examples of materials where coupled fields are to be considered. The constitutive equations of such materials contain, in addition to stress and strain tensors, the vectors of the electric field and electric displacement. In a number of works [4–9], the one-particle problem for an ellipsoidal inclusion in a piezoelectric material was solved explicitly for constant external fields. These solutions were used in the framework of self-consistent schemes for the calculation of overall electroelastic properties of piezoelectric heterogeneous materials containing random sets of ellipsoidal inclusions. In [10], piezoelectric materials reinforced by piezoelectric fibers were considered, and their effective electroelastic characteristics were calculated by the Mori-Tanaka method. The EMM was used in [11–13] for piezoelectric materials containing ellipsoidal inclusions. In [14, 15], the EFM was applied to the calculation of the effective properties of composites with an isotropic matrix and spheroidal piezoelectric inclusions. Piezoelectric materials are called piroactive if temperature affects their electric and elastic properties. The effective electro-thermo-elastic characteristics of such materials reinforced by piroactive continuous fibers were obtained in [16, 17] by the effective field method. The same method was used in [18, 19] for piezoelectric polycrystals.

Other coupled problems that have important applications in geomechanics are the problems of poroelasticity. In these problems, the medium contains an interconnected set of pores filled with liquid. This theory was developed in the middle of the last century in [20, 21] and was widely applied to modeling the mechanical properties of porous rocks filled with fluids. These materials can be modeled by a homogeneous matrix containing regions with different poroelastic properties. If these regions are ellipsoids, the solution of the problem for an isolated ellipsoid in a constant external field can be constructed analytically. This solution is the basis for the application of self-consistent methods to this case. The effective medium theories for the multi-component poroelastic composites were developed in [22]. The effective poroelastic characteristics of such materials were calculated by the effective field method in [23]. A series of cross-property relations (in addition to those considered in Sect. 10.3) for inhomogeneous poroelastic materials were discovered in [23–26]. Note that the mathematical analogy of poroelastic and coupled thermo-elastic theories allows one to interpret the results obtained for poroelastic inhomogeneous materials in terms of the theory of thermo-elasticity (see, for example, [25]). Poro-thermo-elastic composites were considered in [26].

References

1. Ramsteiner F (1981) Elastic behavior of unidirectional short-fiber reinforced thermoplastics. Composites 12:65–71
2. Aboudi J (1983) The effective moduli of short-fiber composites. Int J Solids Struct 19:693–707
3. Manera M (1977) Elastic properties of randomly oriented short fiber-glass composites. J Compos Mater 11:235–247
4. Wang B (1992) Three-dimensional analysis of a flat elliptical crack in a piezoelectric material. Int J Eng Sci 30:781–791
5. Benveniste Y (1992) The determination of the elastic and electric fields in a piezoelectric inhomogeneity. J Appl Phys 72:1086–1095
6. Dunn M (1994) Electroelastic Green's functions for transversely isotropic media and their application to the solution of inclusion and inhomogeneity problems. Int J Eng Sci 32:119–131
7. Kuhn R, Michelitsch T, Levin V, Vakulenko A (2002) The analogue of the Eshelby tensor for a piezoelectric spheroidal inclusion in the hexagonal electroelastic medium. ZAMP 53:584–602

8. Levin V, Michelitsch T, Sevostianov I (2000) Spheroidal inhomogeneity in a transversely isotropic piezoelectric medium. Arch Appl Mech 70:673–693
9. Michelitsch T, Levin V (2000) Inclusions and inhomogeneities in electroelastic media with hexagonal symmetry. Eur Phys J B 14:527–533
10. Chen T (1994) Micromechanical estimates of the overall thermoelectroelastic moduli of multiphase fibrous composites. Int J Solids Struct 31:3099–3111
11. Wang B (1994) Effective behavior of piezoelectric composites. Appl Mech Rev 47:112–121
12. Dunn M, Taya M (1993) Micromechanical predictions of effective electroelastic moduli of piezoelectric composites. Int J Solids Struct 30:161–175
13. Dunn M, Taya M (1993) Analysis of piezoelectric composite materials containing ellipsoidal inhomogeneities. Proc Royal Soc Lond A 443:265–287
14. Levin V (1996) On the effective properties of piezoactive matrix composite materials. Appl Math and Mech (PMM) 60:313–332
15. Levin V, Rakovskaja M, Kreer W (1999) The effective thermoelectroelastic properties of microinhomogeneous materials. Int J Solids Struct 36:2683–2705
16. Levin V, Luchaninov A (2001) On the effective properties of thermo-piezoelectric matrix composites. J Phys D Appl Phys 34:1–6
17. Levin V, Luchaninov A (2004) Effective properties of matrix piro-active composites. Proc Acad Sci USSR Mech Solids 5:44–53
18. Levin V, Rakovskaja M (1997) The effective constants of thermopiezoactive polycrystals. Proc Acad Sci USSR Mech Solids 5:100–109
19. Levin V (1999) The microstresses evaluation in piezoelectric polycrystals. Int J Fract 97:113–123
20. Biot M (1941) General theory of three-dimensional consolidation. J Appl Phys 12:155–164
21. Biot M (1962) Mechanics of deformation and acoustic propagation in porous media. J Appl Phys 33:1482–1498
22. Berryman J (2006) Effective medium theories for multicomponent poroelastic composites. J Eng Mech May:519–531
23. Levin V, Alvarez-Tostado J (2004) On the effective constants of inhomogeneous poroelastic medium. Sci Eng Compos Mater 11:35–46
24. Berryiman J, Milton G (1991) Exact results for generalized Gassmann's equations in composite porous media with two constituents. Geophys 56:1950–1960
25. Norris A (1992) On the correspondence between poroelasticity and thermoelasticity. J Appl Phys 71:1138–1141
26. Levin V, Alvarez-Tostado J (2006) Explicit effective constants for an inhomogeneous porothermoelastic medium. Arch Appl Mech 76:199–214

Chapter 17
Hybrid Composite Materials

Abstract The effective field method is applied to calculation of the effective conductive and elastic properties of matrix composites with different populations of inclusions. The shapes and conductive (elastic) properties of the inclusions are the same inside each population but they are different for inclusions of different populations. Applications of the simplest version of the effective field method (or Mori-Tanaka method) results in the non-symmetric effective conductivity tensor and the elastic stiffness tensor of the composite. The version of the effective field method that allows correcting this drawback is presented in this chapter. In this version, the effective fields that act on the inclusions of different populations are assumed to be different. The effective field method requires construction of two-point correlation functions for random sets of inclusions in hybrid composites. Boolean models of the inclusion sets are used for construction of the correlation functions. The effective elastic properties of a homogeneous host medium with infinite cylindrical fibers and spherical inclusions, cracks, and spherical pores are considered in detail.

Keywords Matrix composites · Effective exiting fields · Boolean models of inclusion populations · Correlation functions of random sets of inclusions · Effective elastic properties

In this chapter, the effective field method is applied to calculation of the effective conductive and elastic properties of matrix composite materials with different populations of inclusions. The shapes and conductive (elastic) properties of the inclusions are the same inside each population but they are different for the inclusions of different populations. Applications of the simplest version of the effective field method (or Mori-Tanaka method) results in the non-symmetric effective conductivity tensor and the elastic stiffness tensor that is not symmetric with respect to permutation of the index pairs (Sect. 7.6). The symmetry of the effective property tensors is a consequence of existence of the energy density of the composite that is a one-valued function of the mean field intensity (strain tensor). Existence of such a function follows from the basic principles of thermodynamics. Lack of the symmetry is a drawback of the effective field method.

The version of the EFM that allows correcting this drawback is presented in this chapter. The effective property tensors obtained in the framework of this version have

S. Kanaun, *Self-Consistent Methods for Composites*, Solid Mechanics and Its Applications 282, https://doi.org/10.1007/978-3-031-96739-9_17

necessary symmetry with respect to the indices. The equations for these tensors contain specific correlation functions that depend on statistical properties of the random sets of inclusions in the composite. Construction of these functions is performed for Boolean-type models of random sets of inclusions. The cases of a homogeneous host medium with infinite cylindrical fibers and spherical inclusions, cracks, and spherical pores are considered.

17.1 Two Populations of Inclusions in a Homogeneous Host Medium

Let a homogeneous host medium contain two populations of inclusions. In each population, the properties and shapes of inclusions are the same but they are different for inclusions of different populations. First, we consider the composite with regular lattices of two populations of inclusions in the host medium. In Fig. 17.1, gray ellipses and black circles indicate two different populations of the inclusions. It is seen that the nearest neighbors of each inclusion of population 1 (ellipses) are the inclusions of population 2 (circles), and *vice versa*. Therefore, by application of a constant external filed, the local exciting fields that act on the inclusions of the families 1 and 2 are different. Note that the local exciting field is the field that would have existed in the host medium if the states of all the inclusions are frozen and the considered inclusion is changed with the host medium. By randomization of a regular lattice of inclusions, the effective exciting fields acting on the inclusions of different populations remain different if the inclusion mixture is well-stirred and agglomerations of inclusions of the same population do not occur.

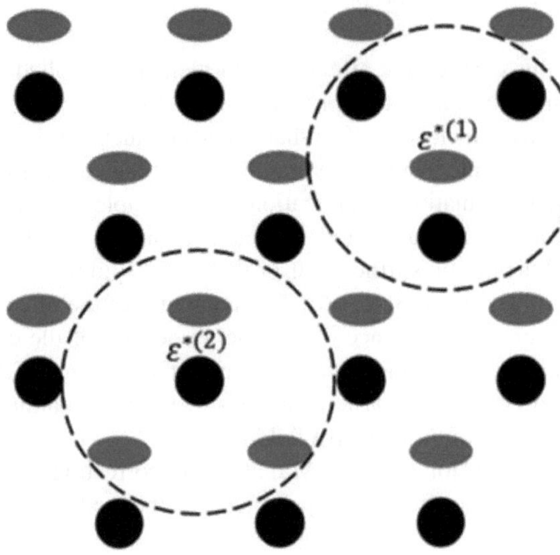

Fig. 17.1 Two regular populations of different inclusions in a homogeneous medium subjected to a constant external strain field ε^0, the nearest neighbors of an inclusion of population 1 are the inclusions of population 2 and vice versa, therefore, the effective fields $\varepsilon^{*(1)}$ and $\varepsilon^{*(2)}$ acting on the inclusion of different populations are different

The version of the effective field method for the medium with two different populations of inclusions is based on the following hypotheses.

H_1. *The local exciting field (effective field) $\varepsilon_*(x)$ acting on each inclusion in the composite is constant in the region occupied by the inclusion.*

H_2. *The effective field is the same for the inclusions of the same population but it is different for inclusions of different populations.*

Let $V_1(x)$, $V_2(x)$ be the characteristic functions of the regions V_1, V_2 occupied by the inclusions of two different populations

$$V_1(x) = \sum_j v_j^{(1)}(x), \quad V_2(x) = \sum_j v_j^{(2)}(x), \tag{17.1}$$

$$V(x) = V_1(x) + V_2(x). \tag{17.2}$$

Here, $v_j^{(i)}(x)$ is the characteristic function of the region $v_j^{(i)}$ occupied by the jth inclusion of the ith population, $V(x)$ is the characteristic function of the region V occupied by the inclusions of both populations. We also introduce the functions

$$V_1(x'; x) = \sum_{i \neq k} v_i^{(1)}(x'), \quad \text{if } x \in v_k^{(1)}, \tag{17.3}$$

$$V_2(x'; x) = \sum_{j \neq m} v_j^{(2)}(x'), \quad \text{if } x \in v_m^{(2)}. \tag{17.4}$$

Using these functions, the integral equation for the local external field $\varepsilon^*(x)$ acting on point x in the region occupied by an inclusion is presented in the form

$$\varepsilon^*(x) + \int K(x - x')C^{1(1)}\Lambda^{(1)}(x')\varepsilon^*(x')V_1(x'; x)dx' +$$

$$+ \int K(x - x')C^{1(2)}\Lambda^{(2)}(x')\varepsilon^*(x')V_2(x'; x)dx' = \varepsilon_0(x), x \in V. \tag{17.5}$$

Here, $C^{1(1)} = C^{(1)} - C^0$, $C^{1(2)} = C^{(2)} - C^0$, $C^{(i)}$ is the elastic stiffness tensor of the inclusions of the ith population, C^0 is the elastic stiffness tensor of the host medium, $K(x)$ is the second derivatives of Green's function of the host medium. The functions $\Lambda^{(1)}(x)$ and $\Lambda^{(2)}(x)$ relate the strain fields inside the inclusions of each population to the local external field $\varepsilon^*(x)$ acting on these inclusions

$$\varepsilon(x) = \Lambda^{(1)}(x)\varepsilon^*(x), \quad x \in V_1, \quad \varepsilon(x) = \Lambda^{(2)}(x)\varepsilon^*(x), \quad x \in V_2, \tag{17.6}$$

$$\Lambda^{(i)}(x) = \sum_j \Lambda_j^{(i)}(x)v_j^{(i)}(x), \quad i = 1, 2. \tag{17.7}$$

Here, the functions $\Lambda_j^{(i)}(x)$ are to be constructed from the solution of the one-particle problem for the inclusion $v_j^{(i)}$ subjected to the external strain field $\varepsilon^*(x)$.

The effective fields $\varepsilon^{*(1)}$ and $\varepsilon^{*(2)}$ that act on the inclusions of different populations are the following conditional means

$$\varepsilon^{*(1)}(x) = \left\langle \varepsilon^*(x)|x \in V_1 \right\rangle, \ \varepsilon^{*(2)}(x) = \left\langle \varepsilon^*(x)|x \in V_2 \right\rangle, \tag{17.8}$$

where $\varepsilon^*(x)$ is the local external field at point $x \in V$. For the hypothesis H_1, the effective fields $\varepsilon^{*(i)}$ are constant, and therefore, the functions $\Lambda^{(1)}(x)$, $\Lambda^{(2)}(x)$ are defined from the solution of the one-particle problems for the inclusions in each population subjected to constant external fields. Integral equations of these problems have the forms

$$\varepsilon(x) + \int\limits_{v_j^{(i)}} K(x - x')C^{1(i)}\varepsilon(x')dx' = \varepsilon^{*(i)}, \qquad i = 1, 2; \ j = 1, 2, 3, ... \tag{17.9}$$

The solutions of these equations can be presented in the form

$$\varepsilon(x) = \Lambda_j^{(i)}(x)\varepsilon^{*(i)}, \ x \in v_j^i, i = 1, 2. \tag{17.10}$$

For ellipsoidal inclusions, $\Lambda_j^{(i)}$ are constant tensors that are the same for the inclusions of the same population but they are different for inclusions of different populations.

In order to obtain a system of equations for the effective fields $\varepsilon^{*(1)}$ and $\varepsilon^{*(2)}$, we average both parts of Eq. (17.5) first by the conditions $x \in V_1$ and then, by the condition $x \in V_2$. As a result, we obtain the system

$$\begin{aligned} &\varepsilon^{*(1)} + p_1 \int K(x - x')P^1\varepsilon^{*(1)}\Psi_{11}(x - x')dx' + \\ &+ p_2 \int K(x - x')P^2\varepsilon^{*(2)}\Psi_{12}(x - x')dx' = \varepsilon_0, \end{aligned} \tag{17.11}$$

$$\begin{aligned} &\varepsilon^{*(2)} + p_1 \int K(x - x')P^1\varepsilon^{*(1)}\Psi_{21}(x - x')dx' + \\ &+ p_2 \int K(x - x')P^2\varepsilon^{*(2)}\Psi_{22}(x - x')dx' = \varepsilon_0. \end{aligned} \tag{17.12}$$

Here, $p_1 = \langle V_1(x) \rangle$, $p_2 = \langle V_2(x) \rangle$ are the volume fractions of the inclusions of each population;

$$P^i = C^{1(i)}\Lambda^{(i)}, \qquad i = 1, 2. \tag{17.13}$$

$$\Lambda^{(i)} = \left(I + A^i C^{1(i)}\right)^{-1}, \quad A^i = A(a_i), \tag{17.14}$$

where a_i is the linear transform of spatial coordinate that converts ellipsoids of the ith population into unit spheres. It is taken into account that the effective field $\varepsilon^{*(1)}$ acts on the inclusions of the first population and the effective field $\varepsilon^{*(2)}$ acts on the inclusions of the second population. The correlation functions $\Psi_{ij}(x-x')$ in Eqs. (17.11) and (17.12) are defined by the equations

$$\Psi_{ij}(x-x') = \frac{1}{p_j}\left\langle V_j(x';x)|x \in V_i\right\rangle, \qquad i,j=1,2. \tag{17.15}$$

These functions have the properties of the correlation function $\Psi(x)$ in Eq. (13.29) for composites with one family of inclusions

$$\Psi_{ij}(0)=0, \lim_{W\to\infty}\frac{1}{W}\int_W \Psi_{ij}(x)dx = 1. \tag{17.16}$$

After calculating the integrals in Eqs. (17.11) and (17.12), we obtain the following linear algebraic system for the constant tensors $\varepsilon_*^{(1)}, \varepsilon_*^{(2)}$

$$(I - p_1A^{11}P^1)\varepsilon^{*(1)} - p_2A^{12}P^2\varepsilon^{*(2)} = \varepsilon^0, \tag{17.17}$$

$$-p_1A^{21}P^1\varepsilon^{*(1)} + (I - p_2A^{22}P^2)\varepsilon^{*(2)} = \varepsilon^0, \tag{17.18}$$

$$A^{ij} = \int K(x)\Phi_{ij}(x)\mathrm{d}x, \qquad \Phi_{ij}(x) = 1 - \Psi_{ij}(x), \qquad i,j=1,2. \tag{17.19}$$

Here, it is assumed that the external strain field ε^0 is fixed in the problem and therefore, the regularization (13.31) holds

$$\int K(x-x')dx = 0. \tag{17.20}$$

The solution of the system (17.17), (17.18) is presented in the form

$$\varepsilon^{*(1)} = \Lambda^{*(1)}\varepsilon_0, \qquad \varepsilon^{*(2)} = \Lambda^{*(2)}\varepsilon_0, \tag{17.21}$$

$$\Lambda^{*(1)} = \left(I - p_1A^{11}P^1 - p_2A^{12}P^2D_2^{-1}D_1\right)^{-1}, \tag{17.22}$$

$$\Lambda^{*(2)} = \left(I - p_2A^{22}P^2 - p_1A^{21}P^2D_1^{-1}D_2\right)^{-1}, \tag{17.23}$$

$$D_1 = I - p_1(A^{11} - A^{21})P^1, \qquad D_2 = I - p_2(A^{22} - A^{12})P^2. \tag{17.24}$$

In the framework of the effective field method, the strain and stress field in the medium with two populations of inclusion is approximated by the equations

$$\varepsilon(x) = \varepsilon^0 - \int K(x - x')q(x')dx', \tag{17.25}$$

$$\sigma(x) = \sigma^0 - \int S(x - x')B^0 q(x')dx', \tag{17.26}$$

where

$$q(x) = C^{1(1)}\Lambda^{(1)}\varepsilon^{*(1)}V_1(x) + C^{1(2)}\Lambda^{(2)}\varepsilon^{*(2)}V_2(x). \tag{17.27}$$

After averaging of $\varepsilon(x)$ and $\sigma(x)$ over ensemble realizations of the random set of inclusions, we yield

$$\langle\varepsilon(x)\rangle = \varepsilon^0, \tag{17.28}$$

$$\langle\sigma(x)\rangle = \sigma^0 - \int S(x - x')dx' B^0 \langle q(x)\rangle = \sigma^0 + \langle q(x)\rangle. \tag{17.29}$$

Here, it is taken into account that for a homogeneous in space random set of inclusion, the mean $\langle q(x)\rangle$ is constant, and for a fixed external strain field ε^0, we have

$$\int K(x - x')dx' = 0, \int S(x - x')dx' = -C^0. \tag{17.30}$$

Because

$$\langle q(x)\rangle = \left(p_1 P^1 \Lambda^{*(1)} + p_2 P^2 \Lambda^{*(2)}\right)\varepsilon_0, \tag{17.31}$$

and $\sigma^0 = C^0\varepsilon^0$ we obtain from Eqs. (17.28), (17.29), and (17.31) the relation between the means $\langle\sigma(x)\rangle$ and $\langle\varepsilon(x)\rangle$

$$\langle\sigma(x)\rangle = C^* \langle\varepsilon(x)\rangle. \tag{17.32}$$

Here, C^* is the effective elastic stiffness tensor C_* of the composite and

$$C^* = C^0 + p_1 P^1 \Lambda^{*(1)} + p_2 P^2 \Lambda^{*(2)}. \tag{17.33}$$

The tensors $\Lambda^{*(1)}$, $\Lambda^{*(2)}$ are defined in Eqs. (17.22)–(17.24):

$$\Lambda^{*(1)} = \left(I - p_1 A^{11} P^1 - p_2 A^{12} P^2 D_2^{-1} D_1\right)^{-1}, \tag{17.34}$$

$$\Lambda^{*(2)} = \left(I - p_2 A^{22} P^2 - p_1 A^{21} P^2 D_1^{-1} D_2\right)^{-1}, \tag{17.35}$$

$$D_1 = I - p_1(A^{11} - A^{21})P^1, \qquad D_2 = I - p_2(A^{22} - A^{12})P^2. \tag{17.36}$$

We consider symmetry of the of the fourth rank tensors in Eq. (17.33) with respect to index permutations. For ellipsoidal inclusions, the tensors P^1, P^2 are defined by the equations

$$P^i = C^{1(i)} \Lambda^{(i)}, \quad \Lambda^{(i)} = \left(I + A^i C^{1(i)}\right)^{-1}, \quad A^i = A(a_i), \quad i = 1, 2, \tag{17.37}$$

where the tensors $C^{1(i)}$ and A^i are symmetric with respect to permutations of the index pairs as well as the index permutations inside the pairs. Formal expanding of the tensor P^i in the series

$$P^i = C^{1(i)} - C^{1(i)} A^i C^{1(i)} + C^{1(i)} A^i C^{1(i)} A^i C^{1(i)} - .. \tag{17.38}$$

show that P^i has symmetry of the elastic stiffness tensors with respect to permutation of index pairs.

After expanding the tensor C^* in Eq. (17.33) into the series with respect to the inclusion volume fractions p_1, p_2, and keeping the terms of the order of p_i^2, we obtain

$$C^* = C^0 + p_1 P^1 + p_2 P^2 + p_1^2 P^1 A^{11} P^1 + p_2^2 P^2 A^{22} P^2 +$$

$$+ p_1 p_2 (P^1 A^{12} P^2 + P^2 A^{21} P^1) + ... \tag{17.39}$$

Here, the terms proportional to p_1, p_2, p_1^2, and p_2^2 have the symmetry of elastic stiffness tensors with respect to index permutations because tensors P^i and A^{ii} have such symmetry. The last term in (17.39) has this symmetry if

$$A^{12} = A^{21}. \tag{17.40}$$

To prove this equation, we consider Eq. (17.19) for tensors A^{12} and A^{21} in detail. Functions $\Psi_{12}(x)$ and $\Psi_{21}(x)$ in this equation are defined in Eq. (17.13) and they can be presented in the forms

$$\Psi_{12}(x - x') = \frac{1}{p_2} \left\langle V_2(x'; x) | x \in V_1 \right\rangle = \frac{1}{p_1 p_2} \left\langle V_2(x') V_1(x) \right\rangle, \tag{17.41}$$

$$\Psi_{21}(x - x') = \frac{1}{p_1} \left\langle V_1(x'; x) | x \in V_2 \right\rangle =$$

$$= \frac{1}{p_1 p_2} \left\langle V_1(x') V_2(x) \right\rangle = \Psi_{12}(x' - x). \tag{17.42}$$

Here, it is take into account that if point x is inside the inclusion of the population V_1, it cannot belong to the region V_2, and *vice versa*. This equation yields the general property of the cross-correlation functions Ψ_{12} and Ψ_{21}

$$\Psi_{12}(x) = \Psi_{21}(-x) \tag{17.43}$$

for homogeneous in space random populations of inclusions. Because $K(x)$ is an even function of the argument $K(-x) = K(x)$, from Eq. (17.19), (17.43) we have

$$A^{12} = \int K(x)\,(1 - \Psi_{12}(x))\,dx =$$

$$= \int K(x)\,(1 - \Psi_{21}(-x))\,dx = \int K(-x)\,(1 - \Psi_{21}(x))\,dx = A^{21}. \tag{17.44}$$

Thus, Eq. (17.40) is a consequence of the general property (17.43) of the cross-correlation functions $\Psi_{ij}(x)$. As a result, all the terms on the right-hand side of Eq. (17.39) have symmetry of elastic stiffness tensors. Other terms in expansion (17.39) have symmetry of the elastic stiffness tensor with respect to index permutations if Eq. (17.40) holds. For instance, the terms of the order p_i^3 of this expansion are

$$C^* = ... + p^3 P^1 A^{11} P^1 A^{11} P^1 + (p_1)^2\, p_2 (P^1 A^{11} P^1 A^3 P^2 +$$

$$+ P^2 A^3 P^1 A^{11} P^1 + P^1 A^3 P^2 A^3 P^1) + p_1\,(p_2)^2\,(P^2 A^{22} P^2 A^3 P^1 +$$

$$+ P^1 A^3 P^2 A^{22} P^2 + P^2 A^3 P^1 A^3 P^2) + (p_2)^3\, P^2 A^{22} P^2 A^{22} P^2 + ..., \tag{17.45}$$

$$A^3 = A^{12} = A^{12}. \tag{17.46}$$

Direct checking shows that all the terms in this equation are symmetric with respect to permutation of index pairs as well as the indices inside pares.

Let us assume for a moment that the following equations hold

$$A^{11} = A^{21} = A^1, \qquad A^{22} = A^{12} = A^2. \tag{17.47}$$

In that case, Eqs. (17.17) and (17.18) are dramatically simplified, and we obtain the following equation for the overall elastic stiffness: tensor of the composite ($p_0 = 1 - p_1 - p_2$)

$$C^* = C^0 + \left(p_1 C^{1(1)} \Lambda^{(1)} + p_2 C^{1(2)} \Lambda^{(2)}\right) \left(p_0 I + p_1 \Lambda^{(1)} + p_2 \Lambda^{(2)}\right)^{-1}. \tag{17.48}$$

This equation coincides with the result of the application of Mori-Tanaka method to three-phase composite materials (see Sect. 7.6), and this tensor is not symmetric with respect to permutation of index pares. The reason for such a nonsense is quite clear: the assumptions (17.47) violate the general property (17.40) of the tensors A^{12}, A^{21}, and the penalty for the simplicity of Eq. (17.48) is the absence of the correct symmetry of the effective elastic stiffness tensor C^*.

17.2 Homogeneous Elastic Media Containing Ellipsoidal Inclusions and Elliptical Cracks

We consider the homogenization problem for an anisotropic host medium containing random sets of identical ellipsoidal inclusions and elliptical cracks of comparable sizes. The orientations of cracks and inclusions can be different. Positions of cracks and inclusions are statistically independent, but their intersections (overlapping) are prohibited. The medium is subjected to a constant external stress field σ^0 at infinity.

For solution of the homogenization problem, we assume that a constant effective stress field σ^{*p} acts on each ellipsoidal inclusions, a constant effective stress field $\sigma^{*\tau}$ acts on each cracks, $\sigma^{*p} \neq \sigma^{*\tau}$. In this approximation, the equations for strain and stress fields in the medium follow from Eqs. (13.49), (13.50) and (14.28), (14.29)

$$\varepsilon(x) = \varepsilon^0 + \int K(x - x')C^0 P^\sigma V(x')dx'\sigma^{*p} +$$

$$+ \int K(x - x')C^0 \Lambda(n) Z(x') \Omega(x') dx' \sigma^{*\tau}, \tag{17.49}$$

$$\sigma(x) = \sigma^0 + \int S(x - x')P^\sigma V(x')dx'\sigma^{*p} +$$

$$+ \int S(x - x')\Lambda(n) Z(x') \Omega(x') dx' \sigma^{*\tau}. \tag{17.50}$$

Here, $V(x)$ is the characteristic function of the region V occupied by the inclusions, $\Omega(x)$ is the delta-function concentrated on crack surfaces Ω. The tensors P^σ is

$$P^\sigma = B^1(I - D(a)B^1)^{-1},\ B^1 = B - B^0, \tag{17.51}$$

where B and B^0 are elastic compliance tensors of the inclusions and the host medium. Tensor $D(a)$ is defined in Eq. (4.92), and it depends on the second rank tensor a that relates to the shape and orientation of the ellipsoidal inclusion. For a spherical inclusion of radius a, $a_{\alpha\beta} = a\delta_{\alpha\beta}$. Tensor $\Lambda(n)$ in Eqs. (17.49), (17.50) is

$$\Lambda(n)_{\alpha\beta\gamma\delta} = n_\alpha (T^0)^{-1}_{\alpha\beta} n_\beta |_{(\alpha\beta)(\gamma\delta)}, \tag{17.52}$$

were T^0 is in Eq. (5.52) and function $Z(x)$ coincides with $z(x_1, x_2)$ in Eq. (5.24) on the surface of each crack

$$z(x_1, x_2) = \sqrt{1 - \frac{x_1^2}{a_1^2} - \frac{x_2^2}{a_2^2}}. \tag{17.53}$$

Here, a_1, a_2 are crack semiaxes, x_1, x_2 are local Cartesian coordinates along the crack semiaxes.

For integral presentation of the local exciting fields acting on the crakes and inclusions, we introduce the functions

$$V(x'; x) = \sum_{i \neq k} v_i(x'), \quad \text{if } x \in v_k, \tag{17.54}$$

$$\Omega(x'; x) = \sum_{j \neq k} \Omega_j(x'), \ \text{if } x \in \Omega_k. \tag{17.55}$$

Here, $v_i(x)$ is the characteristic function of the region occupied by the ith inclusion, and $\Omega_j(x)$ is the delta-function concentrated an the surface Ω_j of the jth crack. Then, the local external stress field that acts on the points x inside the inclusions (on the region V) is presented in the form

$$\sigma^{*p}(x) = \sigma^0 + \int S(x - x')V(x'; x)dx' \, P^{\sigma}\sigma^{*p} +$$
$$+ \int S(x - x')Z(x')\Omega(x')dx' \Lambda\sigma^{*\tau}, x \in V. \tag{17.56}$$

A similar field acting on the points at the crack surfaces (on the region Ω) is

$$\sigma^{*\tau}(x) = \sigma^0 + \int S(x - x')V(x')dx' \, P^{\sigma}\sigma^{*p} +$$
$$+ \int S(x - x')Z(x')\Omega(x'; x)dx' \Lambda\sigma^{*\tau}, x \in \Omega. \tag{17.57}$$

The effective fields σ^{*p} and $\sigma^{*\tau}$ are the following conditional means

$$\sigma^{*p} = \langle \sigma^{*p}(x) | x \in V \rangle, \sigma^{*\tau} = \langle \sigma^{*\tau}(x) | x \in \Omega \rangle, \tag{17.58}$$

where $\sigma^{*p}(x)$ and $\sigma^{*\tau}(x)$ are actual exciting stress fields that act on the inclusions and cracks in the composite. After averaging Eqs. (17.56), (17.57) over the ensemble realizations of the inclusion and crack sets by the conditions that $x \in V$ and $x \in \Omega$, we obtain a closed system of equations for the effective fields σ^{*p} and $\sigma^{*\tau}$

$$\sigma^{*p} = \sigma^0 + p \int S(x - x')\Psi^{pp}(x - x')dx' \, P^{\sigma}\sigma^{*p} +$$
$$+ \tau \int S(x - x')\Psi^{p\tau}(x - x')dx' \overline{\Lambda}\sigma^{*\tau}, \tag{17.59}$$

$$\sigma^{*\tau} = \sigma^0 + p \int S(x - x')\Psi^{\tau p}(x - x')dx' \, P^{\sigma}\sigma^{*p} +$$

$$+\tau \int S(x-x')\Psi^{\tau\tau}(x-x')dx'\overline{\Lambda}\sigma^{*\tau}. \tag{17.60}$$

Here, p is the inclusion volume fraction, and τ is the crack density parameter

$$\tau = \frac{2}{3}\frac{(a_1 a_2)^{3/2}}{v_0}, \quad \overline{\Lambda} = \frac{\pi}{\sqrt{a_1 a_2}}\Lambda, \tag{17.61}$$

where v_0 is the volume that falls on one crack.

The correlation functions $\Psi^{pp}(x)$, $\Psi^{p\tau}(x)$, $\Psi^{\tau p}(x)$, and $\Psi^{\tau\tau}(x)$ are defined by the equations

$$\Psi^{pp}(x-x') = \frac{\langle V(x';x)|x \in V\rangle}{\langle V(x)\rangle}, \tag{17.62}$$

$$\Psi^{p\tau}(x-x') = \frac{\langle Z(x')\Omega(x')|x \in V\rangle}{\langle Z(x)\Omega(x)\rangle}, \tag{17.63}$$

$$\Psi^{\tau p}(x-x') = \frac{\langle V(x')|x \in \Omega\rangle}{\langle V(x)\rangle}, \tag{17.64}$$

$$\Psi^{\tau\tau}(x-x') = \frac{\langle Z(x')\Omega(x';x)|x \in \Omega\rangle}{\langle Z(x)\Omega(x)\rangle}. \tag{17.65}$$

For statistically independent positions of the inclusions, the function $\Psi^{pp}(x)$ (the correlation hole for inclusions) has the symmetry that coincides with the inclusion symmetry, and therefore, $\alpha = a$. As a result, the first integral in Eq. (17.59) take the forms

$$\int S(x)\Psi^{pp}(x)dx = -D(a). \tag{17.66}$$

For statistically independent crack positions, the semiaxis α_3 of the correlation hole $\Psi^{\tau\tau}(x)$ tends to zero and we have

$$\int S(x)\Psi^{\tau\tau}(x)dx = -\lim_{\alpha_3\to 0} D(\alpha) = -\widetilde{S}(n), \tag{17.67}$$

Here, n is the normal to the crack surfaces. Because $\widetilde{S}(n)\Lambda(n) = 0$, the last integral in Eq. (17.60) vanishes

$$\tau \int S(x-x')\Psi^{\tau\tau}(x)dx\overline{\Lambda}(n)\sigma^{*\tau} = -\tau\widetilde{S}(n)\overline{\Lambda}(n)\overline{\sigma}^{*\tau} = 0. \tag{17.68}$$

For statistically independent sets of inclusions and cracks, the function $\Psi^{p\tau}(x)$ and $\Psi^{\tau p}(x)$ in Eqs. (17.63), (17.64) relate by the equation

$$\Psi^{\tau p}(x) = \Psi^{p\tau}(-x), \tag{17.69}$$

and the correlation hole $\Psi^{\tau p}(x)$ have shape of an ellipsoidal inclusion. For $S(x)$ is an even function of x, we have

$$\int S(x)\Psi^{\tau p}(x)dx = \int S(x)\Psi^{p\tau}(x)dx = -D(a) \tag{17.70}$$

Finally, the system for the effective fields $\overline{\sigma}^{*p}$ and $\overline{\sigma}^{*\tau}$ follows from Eqs. (17.59), (17.60) in the form

$$\sigma^{*p} = \sigma^0 - pD(a)\, P^{\sigma}\sigma^{*p} - \tau D(a)\overline{\Lambda}\sigma^{*\tau}, \tag{17.71}$$

$$\sigma^{*\tau} = \sigma^0 - pD(a)\, P^{\sigma}\sigma^{*p}. \tag{17.72}$$

The solution of this system is

$$\sigma^{*p} = (I + pD(a)QD(a)P^{\sigma})^{-1}\, D(a)Q\sigma^0, \tag{17.73}$$

$$\sigma^{*\tau} = \left(I - pD(a)P^{\sigma}\,(I + pD(a)QD(a)P^{\sigma})^{-1}\, D(a)Q\right)\sigma^0, \tag{17.74}$$

$$Q = D(a)^{-1} - \tau\overline{\Lambda}, \tag{17.75}$$

For a fixed external stress field σ^0, we have

$$\int K(x - x')dx' = B^0, \quad \int S(x - x')dx' = 0, \tag{17.76}$$

and because

$$\langle V(x)\rangle = p, \quad \langle Z(x)\Omega(x)\rangle = \frac{2}{3}\frac{\pi a_2 a_2}{v_0}, \tag{17.77}$$

the mean strain and stress tensors in Eqs. (17.49), (17.50) are

$$\langle\varepsilon\rangle = B^0\sigma^0 + pP\sigma^{*p} + \tau\overline{\Lambda}\sigma^{*\tau} = B^*\langle\sigma\rangle\,, \quad \langle\sigma\rangle = \sigma^0. \tag{17.78}$$

From these equations and Eqs. (17.71), (17.72), we obtain the tensor of the effective elastic compliance of the composite with ellipsoidal inclusions and elliptical cracks

$$B^* = B^0 + \tau\overline{\Lambda} + pQTQ\,(I + pTQ)^{-1}, \quad T = D(a)P^{\sigma}D(a). \tag{17.79}$$

The tensors Λ, Q, T in this equation have symmetry of elastic compliance tensors with respect to permutations of the tensor indices. As a result, B^* is also the tensor of such a symmetry. In order to prove this, we can expand the last term in Eq. (17.79) for B^* in series with respect to the volume fraction of inclusions p

$$QTQ\,(I + pTQ)^{-1} = QTQ - pQTQTQ + p^2QTQTQTQ - \dots \qquad (17.80)$$

The structure of all the terms in this series show that B^* has symmetry as elastic compliance tensors with respect to permutation of indices.

Remark. If we assume that correlation functions $\Psi^{p\tau}(x)$ and $\Psi^{\tau\tau}(x)$ coincide (that is physically incorrect!), the last integral in Eq. (17.59) vanishes, and the effective fields $\sigma^{*\tau}$ and σ^{*p} coincide

$$\sigma^{*\tau} = \sigma^{*p} = \sigma^0 - pD(a)\,P\sigma^{*p}. \qquad (17.81)$$

This equation results in the following tensor of the effective elastic compliance B^* of the hybrid composite

$$B^* = B^0 + (pP + \tau\overline{\Lambda})\,(I + pD(a)P)^{-1}. \qquad (17.82)$$

This equation for B^* is simpler than Eq. (17.79), but the tensor B^* in Eq. (17.82) it is not symmetric with respect to permutation of index pears that can be easy seen by series expansion of the right-hand side of (17.82) with respect to inclusion volume fraction p and crack density τ.

17.3 Isotropic Elastic Media with Spherical Pores and Penny-Shape Cracks of the Same Orientations

Elastic medium containing spherical pores and cracks is often used as a model for simulation of effective elastic properties of rock materials. By solution of the homogenization problem for such a hybrid composite, at least three homogenization schemes can be proposed (Fig. 17.2). If the pore sizes are much smaller than the crack sizes (Fig. 17.2a), the effective stiffness tensor can be calculated by a two-step homogenization procedure. First the effective stiffness tensor of the host medium containing pores only is calculated. Then, a random set of cracks is imposed into the effective medium with the elastic properties found at the first step. After solution of the homogenization problem for the medium with cracks, we obtain the effective stiffness tensor of the hybrid composite. At each step the homogenization process, the medium with one type of heterogeneities is considered. Thus application of the effective field method results in a physically correct effective elastic stiffness tensor. If the crack sizes are much smaller than inclusion sizes (Fig. 17.2b), the order of application of the homogenization procedures should be changed. At the first step,

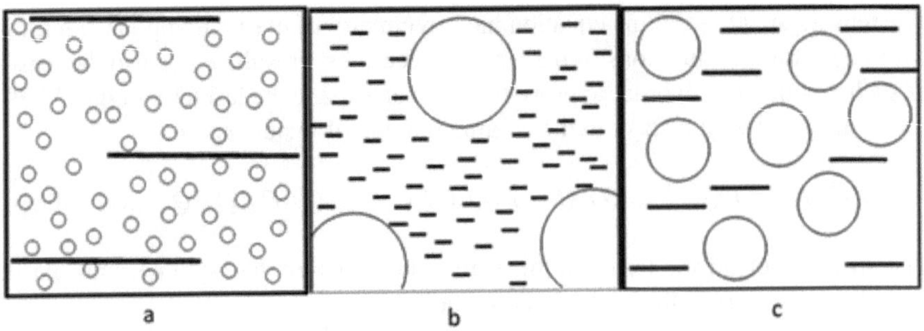

Fig. 17.2 Three types of hybrid composites containing spherical inclusions and elliptical cracks: **a** inclusion sizes are much smaller than crack sizes; **b** crack sizes are much smaller than the inclusion sizes; **c** inclusions and cracks have comparable sizes

the homogeneous host medium containing cracks only is considered, and then, the homogenization problem for a random set of pores in the effective medium obtained after the first step has to be solved. If inclusions and cracks have comparable sizes Fig. 17.2c, the version of the effective field method developed in the previous sections for homogenization of hybrid composites can be used. This method provides the effective stiffness tensor that has correct symmetry with respect to indices. Note that these three homogenization schemes result in different effective stiffness tensors of the composite, and in this section, we consider these differences in detail.

1. For Scheme 1 (Fig. 17.2a), first we consider the original isotropic host medium with random set of spherical inclusions. Application of the effective field method results in the effective elastic compliance tensor B^* of the composite in Eq. (13.149)

$$B^* = \frac{1}{3K_*}E^2 + \frac{1}{2\mu^*}\left(E^1 - \frac{1}{3}E^2\right), \tag{17.83}$$

where the bulk K_* and shear μ_* moduli have forms

$$K_* = K_* + \frac{pK_0(K - K_0)}{K_0 + (1-p)s_1(K - K_0)}, \quad s_1 = \frac{3K_0}{3K_0 + 4\mu_0}, \tag{17.84}$$

$$\mu_* = \mu_* + \frac{p\mu_0(\mu - \mu_0)}{\mu_0 + (1-p)s_2(\mu - \mu_0)}, \quad s_2 = \frac{1}{5}(3 - s_1). \tag{17.85}$$

Here, K_0 and μ_0 are the bulk and shear moduli of the host medium, K, μ are the same for the inclusions, and for porous medium, $K = \mu = 0$.

At the second step, the system of parallel penny-shape cracks is imposed into the medium with the effective elastic compliance tensor B^* in Eq. (17.83). This problem is considered in Sect. 14.3, and in the P -basis, the effective compliance tensor of the hybrid composite takes the form

$$B^{*(1)} = B^* + \frac{8\tau}{\mu_*(1+2\kappa_*)} P^5(n) + \frac{2\tau}{\mu_*\kappa_*} P^6(n), \tag{17.86}$$

$$\kappa_* = \frac{\lambda_* + \mu_*}{\lambda_* + 2\mu_*}, \quad \lambda_* = K_* - \frac{2}{3}\mu_*. \tag{17.87}$$

Here, λ_*, μ_* are Lame's parameters of the isotropic medium obtained after the first step of the homogenization procedure, n is the normal vector to the crack planes. Finally for Scheme 1, the effective stiffness tensor $C^{*(1)} = \left(B^{*(1)}\right)^{-1}$ of the hybrid composite is

$$C^{*(1)} = c_1^* P^2(n) + c_2^* \left(P^1(n) - \frac{1}{2} P^2(n) \right) +$$

$$+ c_3^*(P^3(n) + P^4(n)) + c_5^* P^5(n) + c_6^* P^6(n), \tag{17.88}$$

where the coefficients c_i^* are defined by the equations

$$c_1^* = \lambda_* + \mu_* - \frac{2\tau\lambda_*^2}{\kappa_*\mu_* + 2\tau(\lambda_* + 2\mu_*)}, \quad c_1^* = 2\mu_*, \tag{17.89}$$

$$c_3^* = \lambda_* - \frac{2\tau\lambda_*(\lambda_* + 2\mu_*)}{\kappa_*\mu_* + 2\tau(\lambda_* + 2\mu_*)}, \quad c_5^* = 4\mu_* - \frac{32\tau\mu_*}{1 + 2\kappa_* + 8\tau}, \tag{17.90}$$

$$c_6^* = \lambda_* + 2\mu_* - \frac{2\tau(\lambda_* + 2\mu_*)^2}{\kappa_*\mu_* + 2\tau(\lambda_* + 2\mu_*)}. \tag{17.91}$$

2. At the first step of Scheme 2, the homogenization problem should be solved for the host medium containing penny-shape cracks of the same orientations. The effective stiffness tensor C^* of such a medium has form (17.88)–(17.91), where Lame's parameters λ_*, μ_* should be changed with the parameters λ_0, μ_0 of the host medium. Then, a set of spherical inclusions is imposed into the effective medium with the elastic stiffness tensor in Eq. (17.88) that has transverse isotropic symmetry. Scheme 2 results in the effective elastic stiffness tensor of the hybrid composite that follows from Eq. (13.71)

$$B^{*(2)} = B^* + pP^{\sigma*}(I + pD^*P^{\sigma*})^{-1}, \tag{17.92}$$

$$P^{\sigma*} = B^{*1}(I - D^*B^{*1})^{-1}, \; B^{*1} = B - B^*, \tag{17.93}$$

For pores $B^{-1} \to 0$,

$$\lim_{B\to\infty} P^{\sigma*} = -(D^*)^{-1}, \tag{17.94}$$

and we obtain for the effective compliance tensor $B^{*(2)}$ of the hybrid composite the equation

$$B^{*(2)} = B^* - \frac{p}{1-p}(D^*)^{-1}. \tag{17.95}$$

Thus, the problem is reduced for calculation of the tensor D^* for a spherical inclusion in the transverse isotropic medium. Taking into account Eqs. (4.89) and (4.98), this tensor is defined by the equation

$$D^* = C^* A^* C^* - C^*, \tag{17.96}$$

where A^* is the integral over the unit sphere

$$A^*_{\alpha\beta\gamma\delta} = \frac{1}{4\pi} \int_0^{2\pi} d\varphi \int_0^{\pi} \widetilde{K}_{\alpha\beta\gamma\delta}(k_1, k_2, k_3) \sin\theta d\theta, \tag{17.97}$$

$$k_1 = \cos\varphi \sin\theta,\ k_2 = \sin\varphi \sin\theta,\ k_3 = \cos\theta.$$

Here, function $\widetilde{K}(k)$ is defined by the equations

$$\widetilde{K}_{\alpha\beta\gamma\delta}(k) = k_\alpha \widetilde{G}_{\beta\gamma} k_\delta|_{(\alpha\beta)(\gamma\delta)}, \tag{17.98}$$

where $\widetilde{G}_{\alpha\beta}(k)$ is Fourier transform of Green's function of the transverse isotropic media with elastic stiffness tensor C^* obtained after the first step of the homogenization procedure. The explicit form of $\widetilde{G}_{\alpha\beta}(k)$ for a transverse isotropic medium is

$$\widetilde{G}_{\alpha\beta}(k) = g_1\delta_{\alpha\beta} + g_2 n_\alpha n_\beta +$$

$$+ g_3(n_\alpha m_\beta + n_\beta m_\alpha) + g_4 m_\alpha m_\beta,\ \ m_\alpha = \frac{k_\alpha}{|k|}. \tag{17.99}$$

Here, n is unit vector orthogonal to the isotropy plane of the tensor C^*, and the coefficients g_i are

$$g_1 = \frac{1}{l_1},\ g_2 = \frac{l_3^2 - l_2(l_1 + l_4)}{\Delta}, \tag{17.100}$$

$$g_3 = \frac{-l_1 l_3 + (l_2 l_4 - l_3^2)(m \cdot n)}{\Delta},\ \ g_4 = \frac{l_3^2 - (l_1 + l_2) l_4}{\Delta}, \tag{17.101}$$

$$\Delta = l_1(l_1(l_1 + l_2 + l_4 + 2l_3(m \cdot n)) - (l_3^2 - l_2 l_4)(1 - (m \cdot n)^2)). \tag{17.102}$$

The coefficients l_i relate to elastic constants of the effective stiffness tensor C^* obtained at the first step of the homogenization process

$$l_1 = \frac{1}{4}(2c_2^* + (c_5^* - 2c_2^*)(m \cdot n)^2), \tag{17.103}$$

$$l_2 = \frac{1}{4}(c_5^* - 2c_2^* + 2(2c_1^* + c_2^* - 4c_3^* - 2c_5^* + 2c_6^*)(m \cdot n)^2), \tag{17.104}$$

$$l_3 = \frac{1}{4}(-4c_1^* + 4c_3^* + c_5^*)(m \cdot n), l_4 = c_1^*. \tag{17.105}$$

In this equations, c_i^* are the components of the elastic stiffness C^* in Eq. (17.89)–(17.91), where Lame's parameters λ_*, μ_* should be changed with the parameter λ_0, μ_0 of the host medium. For given elastic constants c_i^*, regular integral (17.97) can be calculated numerically. Tensor A^* has transverse isotropic symmetry, and in the P-basis, it is presented in the form

$$A^* = A_1^* P^2(n) + A_2^* \left(P^1(n) - \frac{1}{2} P^2(n) \right) +$$

$$+ A_3^*(P^3(n) + P^4(n)) + A_5^* P^5(n) + A_6^* P^6(n), \tag{17.106}$$

where the coefficients A_i^* relate to the components of the tensor A^*

$$A_1^* = \frac{1}{2}\left(A_{1111}^* + A_{1122}^*\right), \; A_2^* = A_{1111}^* - A_{1122}^*, \tag{17.107}$$

$$A_3^* = A_{1133}^*, \; A_5^* = 4A_{1313}^*, \; A_6^* = A_{3333}^*. \tag{17.108}$$

3. For spherical inclusions and cracks of comparable sizes (Scheme 3, Fig. 17.2c), the effective compliance tensor of the hybrid composite $B^{*(3)}$ takes the form in Eq. (17.79).

$$B^{*(3)} = B^0 + \tau\overline{\Lambda} + pQTQ\,(I + pTQ)^{-1}, \tag{17.109}$$

$$T = D^0 P^\sigma D^0, \; Q = \left(D^0\right)^{-1} - \tau\overline{\Lambda}, \tag{17.110}$$

For pores, $T = -D^0$ and

$$B^{*(3)} = B^0 + \tau\overline{\Lambda} - pQD^0Q\left(I - pD^0Q\right)^{-1}. \tag{17.111}$$

In this equation, tensor D^0 corresponds to spherical inclusion in an isotropic medium in Eq. (4.152). In the P-basis, this tensor and tensor $\overline{\Lambda}(n)$ are presented as follows

$$D^0 = \left(d_1 + \frac{d_2}{6}\right) P^2(n) + d_2 \left(P^1(n) - \frac{1}{2} P^2(n)\right) +$$

$$+ \left(d_1 - \frac{1}{3} d_2\right) \left(P^3(n) + P^4()\right) + 2d_2 P^5(n) + \left(d_1 + \frac{2}{3} d_2\right) P^6(n), \quad (17.112)$$

$$d_1 = \frac{4}{9} \mu_0 \left(1 - 4\kappa_0\right), \ d_2 = -\frac{2}{15} \mu_0 \left(5 + 4\kappa_0\right), \quad (17.113)$$

$$\Lambda(n) = \frac{8}{\mu_0(1 + 2\kappa_0)} P^5(n) + \frac{4}{\mu_0 \kappa_0} P^6(n). \quad (17.114)$$

For calculation of the effective compliance tensor $B^{*(3)}$ from Eq. (17.111), algebra of the P-tensors in Appendix A.2 can be used.

We consider an example of an isotropic host medium with the bulk and shear moduli $K_0 = 50GPa$, $\mu_0 = 30GPa$. The medium contains spherical voids and penny-shape cracks of the same orientations. Predictions of the three homogenization schemes for the effective stiffness tensor of the composite are shown in Figs. 17.3, 17.4, 17.5, 17.6, 17.7, and 17.8 for two values of pore volume fractions $p = 0.2$ and $p = 0.4$, and crack density τ in the interval $0 < \tau < 1$. In these figures, dependencies of the matrix components $c^*_{11}, c^*_{12}, c^*_{13}, c^*_{44}$ of the effective stiffness tensors C^* of the composites on the parameters p and τ are presented. This components relate with the coefficients of the effective stiffness tensor in the P-basis as follows

$$c^*_{11} = c^*_1 + \frac{c^*_2}{2}, c^*_{12} = c^*_1 - \frac{c^*_2}{2}, c^*_{13} = c^*_3, c^*_{33} = c^*_6, c^*_{44} = \frac{c^*_5}{4}. \quad (17.115)$$

It is seen from Figs. 17.3, 17.4, 17.5, 17.6, 17.7, and 17.8 that predictions of three homogenization schemes for the effective elastic parameters of the hybrid composites deviate substantially. The most notable differences are for the coefficients c^*_{33} and

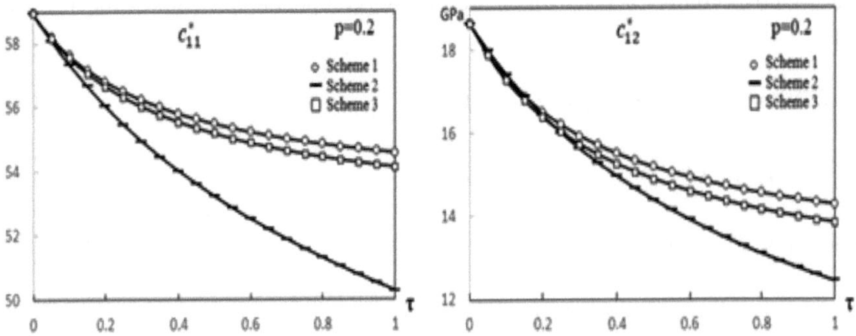

Fig. 17.3 Predictions of three homogenization schemes for an isotropic medium containing spherical pores of volume fraction $p = 0.2$ and penny-shape cracks of density τ, C^*_{11}, C^*_{12} are the coefficients in the matrix presentation of the affective stiffness tensor C^* of the hybrid composite

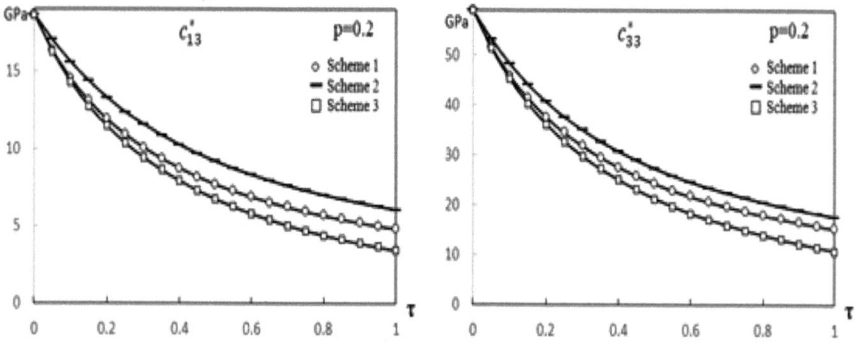

Fig. 17.4 The same as in Fig. 17.3 for the coefficients C^*_{13} and C^*_{33}

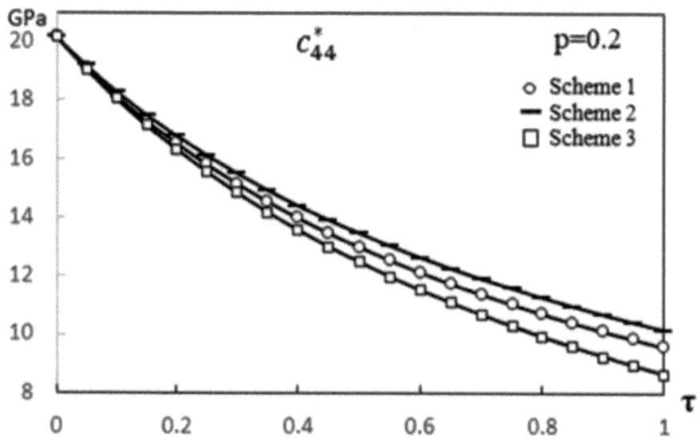

Fig. 17.5 The same as in Fig. 17.3 for the coefficient C^*_{44}

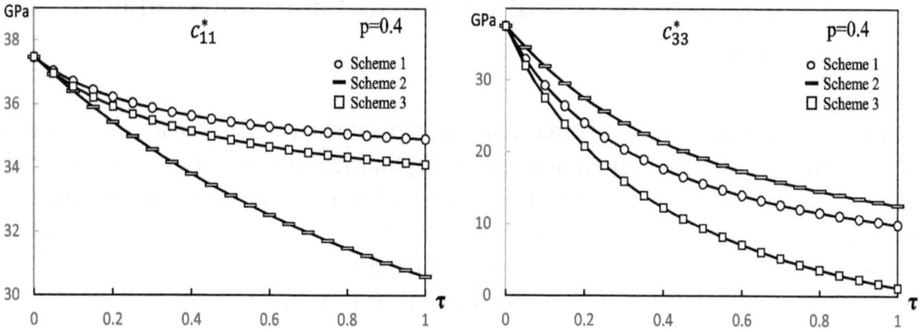

Fig. 17.6 The same as in Fig. 17.3 for the pore volume fraction $p = 0.4$

c^*_{44} that are proved to be much smaller for Scheme 3 than for Schemes 1 and 2. This is the result of stronger interactions between inclusions and crakes predicted by the EFM for Scheme 3.

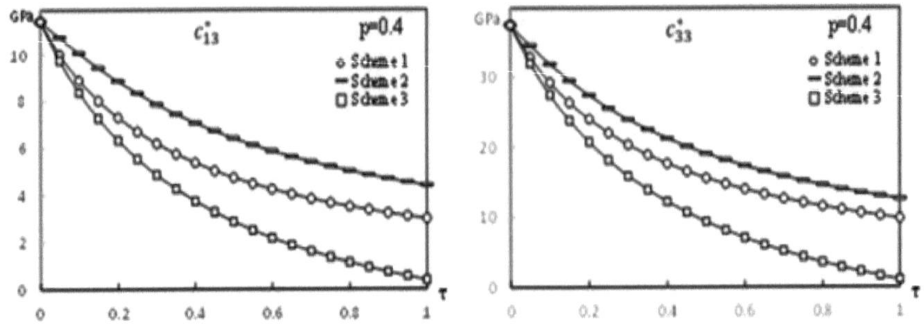

Fig. 17.7 The same as in Fig. 17.4 for the pore volume fraction $p = 0.4$

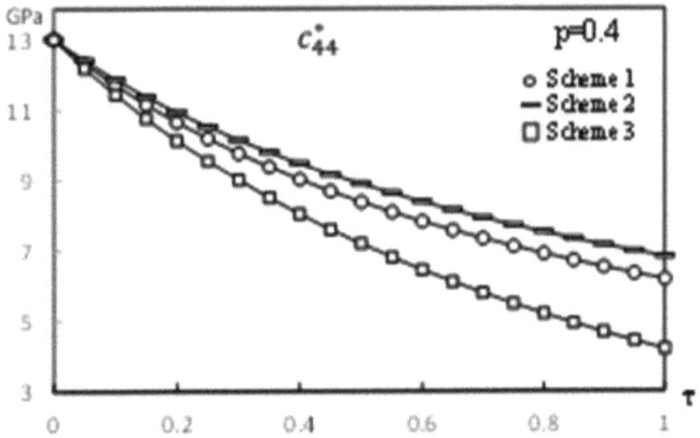

Fig. 17.8 The same as in Fig. 17.5 for the pore volume fraction $p = 0.4$

17.4 Two-Point Correlation Functions for a Hybrid Composite with Sets of Cylindrical and Spheroidal Inclusions[1]

In this section, we consider the three-phase composite that consists of a homogeneous host medium, a set of infinite unidirectional cylindrical fibers and a set of spheroids of the same orientations. The directions of the fiber axes and spheroid rotation axes coincide. For the application of Eqs. (17.33)–(17.36) of the effective field method to this composite material, we have to calculate tensors A^{11}, A^{22}, $A^{12} = A^{21}$ in Eqs. (17.34)–(17.36). These tensors are the integrals in Eqs. (17.19)

$$A^{ij} = \int K(x)\Phi_{ij}(x)dx, i, j = 1, 2. \tag{17.116}$$

where $\Phi_{ij}(x) = 1 - \Psi_{ij}(x)$, and $\Psi_{ij}(x)$ are the correlation functions of the random set of inclusions in Eq. (17.15). These functions depend on the statistical properties of the random sets of cylinders and spheroids. Thus for calculation of $\Psi_{ij}(x)$, we have to introduce probabilistic models of the inclusion distributions. Further, Boolean-type random models of cylindrical fibers and spheroids in 3D-space are considered.

[1] This section is written by D. Jeulin.

17.4.1 Correlation Functions of Boolean Random Sets of Spheres and Cylinders

Let us implant in a finite region W of 3D(2D)-space a number of spheres of the same volume v so that the positions of their centers are independent and homogeneously distributed in W (Poisson process). If $W \to \infty$, and the region W occupies the entire space in this limit, we obtain Boolean random set of the spheres in this space.

We consider statistical characteristics of Boolean random set of inclusions and start with the probability of point x to be inside an arbitrary inclusion. This probability coincides with the inclusion volume fraction p. For one sphere, the probability not to cover a point x inside W is

$$q = 1 - \frac{v}{W} \tag{17.117}$$

because of a homogeneous distribution of the sphere center inside W. If we implants n spheres inside W, the point x will be not covered by the spheres with the probability

$$q = \left(1 - \frac{v}{W}\right)^n \tag{17.118}$$

because each implantation is independent. Let us introduce a fixed parameter

$$\lambda = \frac{n}{W} \tag{17.119}$$

that is called the intensity of Poisson point process. Thus $n = \lambda W$, and in the limit $W \to \infty$, we obtain

$$q = \lim_{W \to \infty} \left(1 - \frac{v}{W}\right)^{\lambda W} = e^{-\lambda v}, \tag{17.120}$$

where q is the probability that a point x is not covered by the inclusions. As a result, the probability for x to be inside inclusion is

$$p = 1 - q = 1 - e^{-\lambda v}. \tag{17.121}$$

In the next step, we consider the correlation functions of Boolean's random sets of inclusions . Because of independency of the implantations, the probability for two points x and y to be inside inclusions is p^2, where p has form in Eq. (17.121). In the effective field method, the function $\Psi(x)$ relates to the probability of two points x and y to be covered by different spheres. First we consider the probability for two points to be covered by one sphere. It happens if the center of the sphere is inside the region of intersection of two spheres which centers are at the points x and y. Let this region be w. Thus, the probability that point x is covered by an inclusion but at the same time point y is not covered by this inclusions is $1 - e^{-\lambda(v-w)}$. The same probability but for the point y is also $1 - e^{-\lambda(v-w)}$. As the result, the probability C of the point x and y to be covered by different inclusions is

$$C(x, y) = \left(1 - e^{-\lambda(v-w)}\right) e^{-\lambda w}. \tag{17.122}$$

The multiplier $e^{-\lambda w}$ is the probability that a sphere does not cover both points x and y. The region w is described by the function

$$w(x, y) = vh(|x - y|), \tag{17.123}$$

where $h(r)$ is the volume of the intersection of two spheres of radii 1 with the centers located at points y and x. In 2D-case, this function is

$$h(\rho) = \begin{cases} \dfrac{2}{\pi}\left(\arccos(\dfrac{\rho}{2}) - \dfrac{\rho}{2}\sqrt{1 - \dfrac{\rho}{2}}\right) & \text{if} \quad \rho < 2; \\ h(\rho) = 0 \quad \text{if} \quad \rho \geq 2, \end{cases} \tag{17.124}$$

In 3D-case, we have

$$h(R) = H(R) = 1 - \frac{3}{4}R + \frac{1}{16}R^3 \quad \text{if } R < 2; \quad H(R) = 0 \quad \text{if} \quad R \geq 2. \tag{17.125}$$

In these equations,

$$\rho = \sqrt{x_1^2 + x_2^2},\ R = \sqrt{x_1^2 + x_2^2 + x_3^2}. \tag{17.126}$$

Using the expression for $h(r)$ and $H(R)$, the function $C(x, y)$ takes the form

$$C(x, y) = C(R) = \left(1 - q^{(1-h(R))}\right)^2 q^{h(R)}, \quad q = e^{-\lambda v}. \tag{17.127}$$

The specific correlation function $\Phi(R)$ that presents in the equations of the effective field method relates to the function $C(R)$ by the equation

$$\Phi(R) = 1 - \frac{1}{p^2}C(R). \tag{17.128}$$

17.4.2 Random Models of Two Populations of Inclusions

We consider spherical inclusions and infinite cylindrical inclusions of the same orientation in 3D-space. The change of spheres to spheroids is obtained by affine coordinate transformation applied to the two-point probability functions. The functions $\Phi_{ij}(x) = 1 - \Psi_{ij}(x)$ in the equations for the effective elastic properties of three-phase composites are deduced from the following probabilities:

$$C_{cc}(x) = P(y \in C_1, y + x \in C_2), \tag{17.129}$$

$$C_{ss}(x) = P(y \in S_1, y + x \in S_2), \tag{17.130}$$

$$C_{cs}(x) = P(y \in C, y + x \in S). \tag{17.131}$$

The right-hand sides of these equations are: probability for the points x and $x + y$ to belong to two separate cylinders C_1 and C_2, two separate spheres S_1 and S_2, or to sphere S and to cylinder C. For instance, for the function $\Phi_{cc}(x)$, $\Phi_{ss}(x)$, $\Phi_{cs}(x)$, we have

$$\Phi_{cc}(x) = 1 - \Psi_{cc}(x) = 1 - \frac{1}{p_c^2} C_{cc}(x), \tag{17.132}$$

$$\Phi_{ss}(x) = 1 - \Psi_{ss}(x) = 1 - \frac{1}{p_s^2} C_{ss}(x), \tag{17.133}$$

$$\Phi_{cs}(x) = 1 - \Psi_{cs}(x) = 1 - \frac{1}{p_c p_s} C_{cs}(x). \tag{17.134}$$

For the construction of various models combining spheres and parallel cylinders, we make use of two Poisson point processes: a process in the three-dimensional space R^3 with intensity (average number of points per unit volume) λ_s, for the implantation of spheres S (with volume V_s), and a process in R^2 (orthogonal to the axis of cylinders) with intensity λ_c for the implantation of disks D (with area A_c) that are the sections of cylinders by an orthogonal plane. We denote $S(y)$, $S(y + x)$, $D(y)$, $D(y + x)$ the spheres located at points x and $x + y$, and the disks located at the projections of x and $x + y$ on the horizontal plane. The complementary set of A is denoted as A^c.

Boolean model is obtained by implantation of spheres and cylinders using Poisson point processes . We first give the priority to cylinders when there is an overlap between a cylinder and spheres. For this model, the probabilities $q_c = 1 - p_c = 1 - P(y \text{ belongs to cylinders})$ and $q_s = 1 - p_s = 1 - P(y \text{ belongs to spheres})$ are given by

$$q_c = \exp(-\lambda_c A_c), \quad q_s = 1 - (1 - q_{cs})q_c, \quad q_{cs} = \exp(-\lambda_s V_s), \tag{17.135}$$

since

$q_c = P(D(y)$ contains no Poisson point of the second process,
$q_{cs} = P(S(y)$ contains no Poisson point of the first process.
To calculate the probability $C_{cc}(x)$, we have to consider the independent events:
$D(y) \cap D(y + x)^c$ contains at least one Poisson point,
$D(y)^c \cap D(y + x)$ contains at least one Poisson point,
$D(y) \cap D(y + x)$ contains no Poisson point.

Therefore, using the function $h(\rho)$ in Eq. (17.124), we have

$$C_{cc}(x) = C_{cc}(\rho) = \left(1 - {q_c}^{1-h(\rho)}\right)^2 {q_c}^{h(\rho)}. \tag{17.136}$$

The probability $C_{ss}(x)$ is derived from the independent events:
$S(y) \cap S(y+x)^c$ contains at least one Poisson point,
$S(y)^c \cap S(y+x)$ contains at least one Poisson point,
$S(y) \cap S(y+x)$ contains no Poisson point,
$D(y) \cup D(y+x)$ contains no Poisson point of the second process.
Using the expression for $H(R)$ in Eq. (17.125), we obtain

$$C_{ss}(x) = \left(1 - {q_{cs}}^{1-H(R)}\right)^2 q_{cs}^{H(R)} {q_c}^{2-h(\rho)}. \tag{17.137}$$

The probability $C_{cs}(x)$ is calculated from the independent events:
$D(y) \cap D(y+x)^c$ contains at least one Poisson point of the second process,
$D(y+x)$ contains no Poisson point of the second process,
$S(y+x)$ contains at least one Poisson point.
Then, we obtain

$$C_{cs}(x) = \left(1 - {q_c}^{1-h(\rho)}\right) q_c(1 - q_s). \tag{17.138}$$

The second version of the model gives the priority to spheres over cylinders. For this case, the same derivation can be used. After permutation of expressions concerning spheres and cylinders in previous equations, we yield

$$q_s = \exp(-\lambda_s V_s), \quad q_c = 1 - (1 - q_{sc})q_s, \tag{17.139}$$

$$q_{sc} = \exp(-\lambda_c A_c), \tag{17.140}$$

$$C_{cc}(x) = C_{cc}(\rho) = \left(1 - {q_{sc}}^{1-h(\rho)}\right)^2 q_{sc}^{h(\rho)} {q_s}^{2-H(R)}, \tag{17.141}$$

$$C_{ss}(x) = \left(1 - {q_s}^{1-H(R)}\right)^2 {q_s}^{H(R)}, \tag{17.142}$$

$$C_{cs}(x) = \left(1 - {q_s}^{1-H(R)}\right) q_s(1 - q_c). \tag{17.143}$$

The dilution Boolean model is obtained by counting the number of grains (inclusions) $N(y) = N_c(y) + N_s(y)$ covering each point y in space, $N_c(y)$ and $N_s(y)$ being the numbers of cylinders and of spheres covering y. For the purpose of the present application, we keep all points y where $N(y) = 1$. Since $N(y)$ is a Poisson variable with average

$$\theta = \theta_s + \theta_c, \theta_s = \lambda_s V_s, \ \theta_c = \lambda_c A_c, \tag{17.144}$$

we have

$$p = P(N(y) = 1) = p_c + p_s = \theta \exp(-\theta). \tag{17.145}$$

Similarly, we derive the volume fractions p_c and p_s

$$p_c = P(D(y) \text{ contains one point, and } S(y) \text{ contains no point}) =$$

$$= \theta_c \exp(-\theta), \tag{17.146}$$

$$p_s = P(S(y) \text{ contains one point, and } D(y) \text{ contains no point}) =$$

$$= \theta_s \exp(-\theta). \tag{17.147}$$

To calculate $C_{ss}(x)$, we have to consider the probability of the following independent events:
$S(y) \cap S(y+x)^c$ contains one Poisson point,
$S(y)^c \cap S(y+x)$ contains one Poisson point,
$S(y) \cap S(y+x)$ contains no Poisson point,
$D(y) \cup D(y+x)$ contains no Poisson point of the second process.
Using the independence between the number of Poisson points in separate domains of space, we yield

$$C_{ss}(x) = p_s^2 \left(1 - H(R)\right)^2 \exp\left(\theta H_{cs}(x)\right). \tag{17.148}$$

For calculation of $C_{cc}(x)$, we consider the following independent events:
$D(y) \cap D(y+x)^c$ contains one Poisson point,
$D(y)^c \cap D(y+x)$ contains one Poisson point,
$D(y) \cap D(y+x)$ contains no Poisson point,
$S(y) \cup S(y+x)$ contains no Poisson point of the first process.
Therefore,

$$C_{cc}(x) = p_c^2 \left(1 - h(\rho)\right)^2 \exp\left(\theta H_{cs}(x)\right). \tag{17.149}$$

$C_{cs}(x)$ is given from the following independent events:
$D(y) \cap D(y+x)^c$ contains one Poisson point of the second process,
$D(y+x)$ contains no Poisson point of the second process,
$S(y)$ contains no Poisson point of the first process,
$S(y)^c \cap S(y+x)$ contains one Poisson point of the first process.
Then,

$$C_{cs}(x) = p_c p_s \left(1 - h(\rho)\right)\left(1 - H(R)\right) \exp\left(\theta H_{cs}(x)\right), \tag{17.150}$$

$$\theta H_{cs}(x) = \lambda_s V_s H(R) + \lambda_c A_c h(\rho) = \frac{p_s}{p} H(R) + \frac{p_c}{p} h(\rho). \tag{17.151}$$

The dead leaves model uses the order of appearance of a sequence of inclusions implanted according to a space-time Poisson process. From results given in [1], we have

$$p_s = \frac{\lambda_s V_s}{\lambda_s V_s + \lambda_c A_c} \left(1 - \exp\left(- \left(\lambda_s V_s + \lambda_c A_c\right)\right)\right), \tag{17.152}$$

$$p_c = \frac{\lambda_c A_c}{\lambda_s V_s + \lambda_c A_c} \left(1 - \exp\left(- \left(\lambda_s V_s + \lambda_c A_c\right)\right)\right). \tag{17.153}$$

These equations are obtained using the condition that during the sequence, point y is covered by some grain (sphere or cylinder, according to the considered phase), and then remains uncovered by new grains, using Markov's property of the sequence. Similarly, after a lengthy proof, one can derive the remaining probabilities,

$$C_{cc}(x) = \frac{2p_c^2}{p^2} \left(1 - h(\rho)\right) \left(\frac{1 - Q(x)}{2 - H_{cs}(x)} - \frac{Q(0) - Q(x)}{1 - H_{cs}(x)}\right), \tag{17.154}$$

$$C_{ss}(x) = \frac{2p_s^2}{p^2} \left(1 - H(R)\right) \left(\frac{1 - Q(x)}{2 - H_{cs}(x)} - \frac{Q(0) - Q(x)}{1 - H_{cs}(x)}\right), \tag{17.155}$$

$$C_{cs}(x) = \frac{p_c p_s}{p^2} \left(2 - h(\rho) - H(R)\right) \left(\frac{1 - Q(x)}{2 - H_{cs}(x)} - \frac{Q(0) - Q(x)}{1 - H_{cs}(x)}\right) \tag{17.156}$$

with $Q(x)$ and $H_{cs}(x)$ given by the equations

$$Q(x) = (1 - p)^{2 - H_{cs}(x)}, \quad H_{cs}(x) = \frac{p_c}{p} h\left(\frac{\rho}{a_c}\right) + \frac{p_s}{p} H(R). \tag{17.157}$$

17.4.3 Functions $\Phi(x)$ for Boolean's Models

We start with one population of a Boolean random set of inclusions of arbitrary shapes. The Boolean model is obtained by implantation of random "grains" v_k on the points $x^{(k)}$ of a Poisson point process of intensity λ (λ is the number of points in a unit volume of the host medium) [1, 2]:

$$V = \bigcup_k v_{k, x^{(k)}}. \tag{17.158}$$

For instance, for a population of spheroidal inclusions of the same orientation, the Boolean model is obtained by selecting for v_k a spheroid with semiaxis a_s and l_s

(l_s is the semiaxis in the direction of the rotation axis x_3) with the center at point $x^{(k)}$. Random points $x^{(k)}$ are independently and homogeneously distributed in the 3D-space. For this model, the function $\Phi_{ss}(x)$ in (17.133), (17.137) takes the form

$$\Phi_{ss}(x_1, x_2, x_3) = \Phi_s(\rho, x_3) =$$

$$= 1 - \frac{1}{p_s^2}\left(1 - {q_s}^{1-H(R)}\right)^2 {q_s}^{H(R)}, \tag{17.159}$$

$$R^2 = \left(\frac{\rho}{a_s}\right)^2 + \left(\frac{x_3}{l_s}\right)^2, \quad \rho^2 = x_1^2 + x_2^2 \tag{17.160}$$

Here, p_s is the volume fraction of spheroids, $q_s = 1 - p_s$, the function $H(R)$ is defined in Eq. (17.125). It is seen that this correlation function has the symmetry of the spheroidal inclusion.

For a Boolean population of infinite cylinders of the same radius a_c orientated along the axis x_3, the corresponding correlation function takes the form (17.132), (17.136) [1]

$$\Phi_{cc}(x) = \Phi_c(\rho) = 1 - \frac{1}{p_c^2}\left(1 - q_c^{\,1-h\left(\frac{\rho}{a_c}\right)}\right)^2 q_c^{\,h\left(\frac{\rho}{a_c}\right)}. \tag{17.161}$$

Here, p_c is the volume concentration of cylinders, $q_c = 1 - p_c$, and the function h is defined in Eq. (17.124). This function clearly has cylindrical symmetry.

Now let us obtain a model of two Boolean populations of inclusions (cylinder and spheroids). In this model the centers of the spheroids are at Poisson's set of points $x^{(k)}$ with intensity λ_s in 3D-space and the axes of the cylinders intersect the plane (x_1, x_2) in a set of Poisson points with intensity λ_c. Volume concentrations of cylinders and spheroids are p_c and p_s, respectively. The directions of the rotation axes of spheroids coincide with the directions of the cylinder axes. Because of the independence of the positions of the centers of spheroids and the axes of cylinders there are possibilities of overlaps between inclusions in the same population, as well as overlaps between inclusions in different populations. To specify the model we have to define a rule of priority, namely to decide which parts of the inclusions will be kept in the intersections. Thus, parts of the inclusion of one population should be deleted from the zones of intersection. There are two possibilities here. Parts of spheroids in the intersections with cylinders are deleted (cylinder priority) or parts of cylinders are deleted from such intersections (spheroid priority). Let us consider each of these models.

Cylinder Priority The function $\Phi_{cc}(x)$ (point y and $y + x$ are inside different cylinders) takes the form (17.132), (17.136)

$$\Phi_{cc}(x) = \Phi_{cc}(\rho) = 1 - \frac{1}{p_c^2}\left(1 - q_c^{1-h\left(\frac{\rho}{a_c}\right)}\right)^2 q_c^{h\left(\frac{\rho}{a_c}\right)} \tag{17.162}$$

and $\Phi_{cc}(x)$ coincides with function Φ_c in (17.161) for one population of cylinders.

Function $\Phi_{ss}(x)$ (point y and $y + x$ inside different spheroids) takes the form that follows from (17.133), (17.150)

$$\Phi_{ss}(x) = \Phi_{ss}(\rho, x_3) = 1 - \frac{1}{p_s^2}\left(1 - q_{cs}{}^{1-H(R)}\right)^2 q_{cs}^{H(R)} q_c^{2-h\left(\frac{\rho}{a_c}\right)}, \tag{17.163}$$

$$q_{cs} = 1 - \frac{p_s}{1 - p_c}. \tag{17.164}$$

Function $\Phi_{cs}(x) = \Phi_{sc}(x)$ (point y is inside a cylinder and point $y + x$ is inside a sphere) is

$$\Phi_{cs}(x) = \Phi_{sc}(x) = 1 - \frac{1}{p_c}\left(1 - q_c^{1-h\left(\frac{\rho}{a_c}\right)}\right). \tag{17.165}$$

Because of intersections with cylinders, the actual volume fraction of spheroids p_{*s} will be less than its original value p_s

$$p_{*s} = p_s(1 - p_c). \tag{17.166}$$

The volume fraction of cylinders p_c does not change in this model.

In Fig. 17.9, a realization of this model for $p_c = p_s = 0.2$, $a_c = a_s = l_s$ is presented. In this figure, a section of such a realization in the plane orthogonal to the axes of the cylinders is shown.

Spheroid Priority In this case, the correlation functions $\Phi_{ij}(x)$ have the forms:

$$\Phi_{cc}(x) = 1 - \frac{1}{p_c^2}\left(1 - q_{sc}^{1-h\left(\frac{\rho}{a_c}\right)}\right)^2 q_{sc}^{h\left(\frac{\rho}{a_c}\right)} q_s{}^{2-H(R)}, \tag{17.167}$$

$$q_{sc} = 1 - \frac{p_c}{1 - p_s}, \tag{17.168}$$

$$\Phi_{ss}(x) = 1 - \frac{1}{p_s^2}\left(1 - q_s{}^{1-H(R)}\right)^2 q_s{}^{H(R)}, \tag{17.169}$$

$$\Phi_{cs}(x) = \Phi_{sc}(x) = 1 - \frac{1}{p_s}\left(1 - q_s{}^{1-H(R)}\right). \tag{17.170}$$

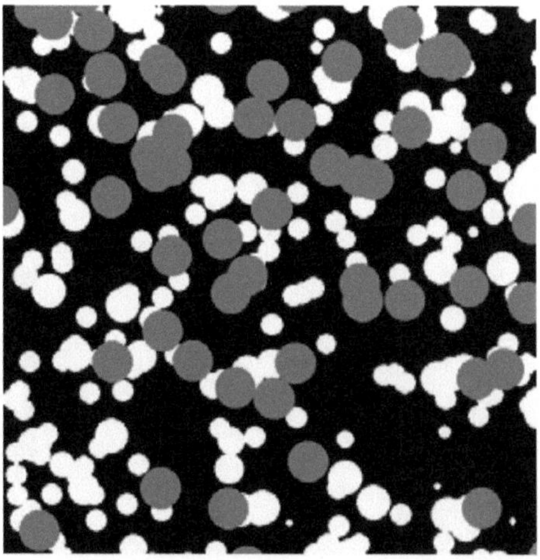

Fig. 17.9 Simulation of a section of a Boolean model of cylinders (gray) and spheres (white) with cylinder priority ,$p_c = 0.2$, $p_s = 0.1$

Fig. 17.10 Simulation of a section of a Boolean model of cylinders (gray) and spheres (white) with spheres priority, $p_c = 0.2$, $p_s = 0.1$

The actual volume fraction of cylinders p_{*c} differs from its initial fraction p_c because of their intersections with spheroids

$$p_{*c} = p_c(1 - p_s), \tag{17.171}$$

and the volume fraction of spheroids p_s is not changed.

A realization of this model for $p_c = p_s = 0.2$ and $a_c = a_s = l_s$ is presented in Fig. 17.10.

The next probabilistic model of two populations of inclusions is called the "dead leaves" model ([1, 3, 4]). As before, we consider populations of infinite cylinders and spheroids with the same orientations. Two sequences of Poisson's point processes for each population are implanted homogeneously in time. In the intersections, we

Fig. 17.11 Simulation of a section of a dead leaves model of cylinders (gray) and spheres (white) ,$p_c = 0.2$, $p_s = 0.1$

keep parts of inclusions that have appeared later in time and delete parts of all the inclusions that appeared earlier. For this model the correlation functions Φ_{ij} take the forms

$$\Phi_{cc}(x) = 1 -$$

$$- \frac{2}{p^2}\left(1 - h\left(\frac{\rho}{a_c}\right)\right)\left(\frac{1 - Q(x)}{2 - H_{cs}(x)} - \frac{Q(0) - Q(x)}{1 - H_{cs}(x)}\right), \tag{17.172}$$

$$\Phi_{ss}(x) = 1 - \frac{2}{p^2}\left(1 - H(R)\right)\left(\frac{1 - Q(x)}{2 - H_{cs}(x)} - \frac{Q(0) - Q(x)}{1 - H_{cs}(x)}\right), \tag{17.173}$$

$$\Phi_{cs}(x) = 1 -$$

$$- \frac{1}{p^2}\left(2 - h\left(\frac{\rho}{a_c}\right) - H(R)\right)\left(\frac{1 - Q(x)}{2 - H_{cs}(x)} - \frac{Q(0) - Q(x)}{1 - H_{cs}(x)}\right). \tag{17.174}$$

Here, $p = p_c + p_s$,

$$Q(x) = (1 - p)^{2 - H_{cs}(x)}, \quad H_{cs}(x) = \frac{p_c}{p} h\left(\frac{\rho}{a_c}\right) + \frac{p_s}{p} H(R). \tag{17.175}$$

A realization of this model for $p_c = p_s = 0.2$, $a_c = a_s = l_s$ is presented in Fig. 17.11.

The next probabilistic model of cylinders and spheroids has no priorities for any type of inclusions. Called the dilution model, it is obtained by implantation of independent inclusions of both populations inside a domain in 3D-space. If the inclusions intersect, the region of intersection is replaced by the matrix phase. As

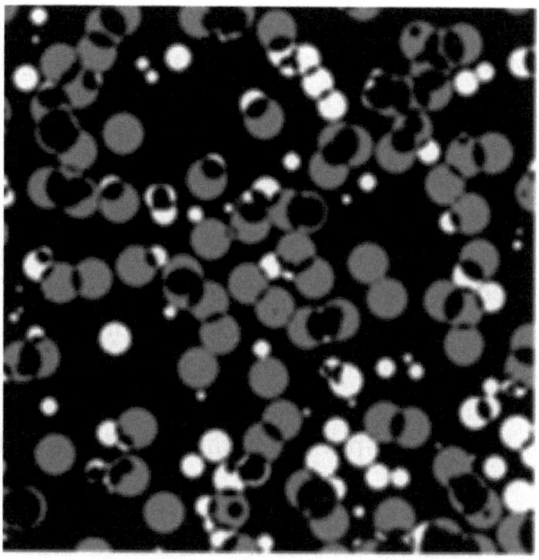

Fig. 17.12 Simulation of a section of a dilution model of cylinders (gray) and spheres (white) ,$p_c = 0.2$, $p_s = 0.1$

a result, the inclusions do not overlap in this model. The corresponding correlation functions have the following forms:

$$\Phi_{cc}(x) = 1 - \frac{2}{p^2}\left(1 - h\left(\frac{\rho}{a_c}\right)\right)^2 \exp\left(\theta H_{cs}(x)\right), \tag{17.176}$$

$$\Phi_{ss}(x) = 1 - \frac{2}{p^2}\left(1 - H(R)\right)^2 \exp\left(\theta H_{cs}(x)\right), \tag{17.177}$$

$$\Phi_{cs}(x) = 1 - \frac{2}{p^2}\left(1 - h\left(\frac{\rho}{a_c}\right)\right)\left(1 - H(R)\right) \exp\left(\theta H_{cs}(x)\right). \tag{17.178}$$

Here, parameter θ relates to the total volume fraction of inclusions $p = p_c + p_s$ by the equation

$$p = \theta \exp(-\theta). \tag{17.179}$$

The maximal value of p in this model is $p = 0.368$. This value of p corresponds to the maximum of the function on the right-hand side of Eq. (17.179). This maximum is achieved at $\theta = 1$. In Fig. 17.12, a realization of the dilution model for $p_c = 0.2$, $p_d = 0.15$, $a_c = a_s = l_s$ is shown.

It is important to notice that the various probabilistic models present very different morphologies: the two versions of Boolean model are not symmetric with respect to spheroids and cylinders that do not play the same roles as a result of the order involved by the priority rule. On the other hand, the dead leaves and the dilution model give the same roles to the two type of inclusions, which can alternatively dominate (according to their date of arrival) for the dead leaves, or which are always erased if overlaps occur for the dilution model.

The presented probabilistic models are not completely compatible with the main assumptions of the EFM because of the possibility of inclusion overlapping. Only for small volume fractions of inclusions the realizations of these models are sets of isolated cylinders and spheroids as implied in the EFM. For high volume fractions, many inclusions overlap and a typical inhomogeneity in the composite does not have a cylindrical or spheroidal form, but some agglomeration of these forms. Thus, the solutions of the one-particle problems that are used in the method and obtained for non-overlapping inclusions do not correspond completely to these models. The volume fractions of inclusions when the assumptions of the EFM are valid, in spite of the overlapping, depend on the relative sizes of cylinders and spheroids. For instance, in the Boolean model with cylinder priority the relative number of "defected" spheroids tends to zero if the radii of spheroids become much smaller than the radii of cylinders. Thus, the lower is the ratio a_s/a_c, the more reasons we have to use Boolean's model with cylinder priority in the framework of EFM.

17.5 Composites Reinforced with Unidirected Fibers and Spherical Inclusions

For the calculation of the overall elastic stiffness tensors of three-phase composites using Eqs. (17.33)–(17.36) of the effective field method, the tensors A^{ij} in these equation should be constructed. In this section, these tensors are obtained for the probabilistic models of hybrid composites considered in the previous section. For the Boolean model of cylinders and spheroids with the cylinder priority, the function $\Phi_{cc}(x)$ has the form (17.162) and it depends on the distance ρ from the axis x_3 only. Therefore this function has a cylinder symmetry and the corresponding integral in Eq. (17.116) takes the form

$$A^{cc} = \int K(x)\Phi_{cc}(x)\mathrm{d}x. \tag{17.180}$$

$$A^{cc} = A^{c} = -\frac{\lambda_0+\mu_0}{8\mu_0(\lambda_0+2\mu_0)}P_1(n) + \frac{\lambda_0+3\mu_0}{4\mu_0(\lambda_0+2\mu_0)}P_2(n) + \frac{1}{2\mu_0}P_5(n), \tag{17.181}$$

where λ_0, μ_0 are Lame's constants of the matrix , n is a unit vector of the cylinder axes, $P_i(n)$ are the elements of the P-basis.

For the calculation of the integral A^{ss} with the function $\Phi_{ss}(x)$ in Eq. (17.163), this function is presented in the form

$$\Phi_{ss}(x) = \Phi_{ss}^{(1)}(\rho) + \Phi_{ss}^{(2)}(x), \tag{17.182}$$

$$\Phi_{ss}^{(1)}(\rho) = 1 - \frac{1}{p_s^2}(1-q_{cs})^2 q_c^{\,2-h\left(\frac{\rho}{a_c}\right)}, \tag{17.183}$$

$$\Phi_{ss}^{(2)}(x) = \frac{1}{p_s^2}\left((1-q_{cs})^2 - \left(1-q_{cs}^{1-H(R)}\right)^2 q_{cs}^{H(R)}\right) q_c^{2-h\left(\frac{\rho}{a_c}\right)}. \tag{17.184}$$

Here, the function $\Phi_{ss}^{(1)}(\rho)$ depends on the distance from the $x_3$axis only, and hence it has cylindrical symmetry. The function $\Phi_{ss}^{(2)}(x)$ is equal to zero if $|x| > r_0 = \max(2a_c, 2a_s, 2l_s)$, and

$$\Phi_{ss}^{(1)}(0) = -\frac{p_c}{1-p_c}, \quad \Phi_{ss}^{(2)}(0) = \frac{1}{1-p_c}. \tag{17.185}$$

As a result, the tensor A^{ss} takes the form

$$A^{ss} = \int K(x)\left(\Phi_{ss}^{(1)}(\rho) + \Phi_{ss}^{(2)}(x)\right)dx = -\frac{p_c}{1-p_c}A^c +$$

$$+\frac{1}{1-p_c}A(a_s) + \det(a_s) vp \int K(a_s y)\Phi_{ss}^{(2)}(a_s y)dy. \tag{17.186}$$

Here, the tensor A^c coincides with (17.181), the tensor $A(a_s)$ has the form in Eq. (4.89) when a_s is the linear transformation that converts the spheroidal inclusion into a unit sphere. The last integral in Eq. (17.186) should be calculated numerically. The tensor $A^{cs} = A^{sc}$ for the function $\Phi_{cs}(\rho)$ in Eq. (17.165) with a cylinder symmetry takes the form

$$A^{cs} = \int K(x)\Phi_{cs}(\rho)dx = A^{cc}. \tag{17.187}$$

For the model with spheroid priority, the function $\Phi_{cc}(x)$ in Eq. (17.167) can be presented in the form

$$\Phi_{cc}(x) = \Phi_{cc}^{(1)}(\rho) + \Phi_{cc}^{(2)}(x), \tag{17.188}$$

$$\Phi_{cc}^{(1)}(\rho) = 1 - \frac{1}{p_c^2}\left(1 - q_{sc}^{1-h\left(\frac{\rho}{a_c}\right)}\right)^2 q_{sc}^{h\left(\frac{\rho}{a_c}\right)} q_s^{\,2}, \tag{17.189}$$

$$\Phi_{cc}^{(2)}(x) = \frac{1}{p_c^2}\left(1 - q_{sc}^{1-h\left(\frac{\rho}{a_c}\right)}\right)^2 q_{sc}^{h\left(\frac{\rho}{a_c}\right)} \left(q_s^{\,2} - q_s^{\,2-H(R)}\right), \tag{17.190}$$

where the function $\Phi_{cc}^{(1)}(\rho)$ tends to zero as $\rho \to \infty$, and $\Phi_{cc}^{(2)}(x) \to 0$ when $|x| \to \infty$,

$$\Phi_{cc}^{(1)}(0) = 1, \quad \Phi_{cc}^{(2)}(0) = 0. \tag{17.191}$$

As a result, the tensor A^{cc} for the model with the spheroid priority takes the form

$$A^{cc} = \int K(x)\Phi_{cc}(x)\mathrm{d}x = A^c + \det(a_s)\int K(a_s y)\Phi_{cc}^{(2)}(a_s y)\mathrm{d}y, \tag{17.192}$$

where the last integral converges absolutely because the function $\Phi_{cc}^{(2)}(x)$ is equal to zero if $|x| > \max(2a_c, 2a_s, 2l_s)$ and $\Phi_{cc}^{(2)}(0) = 0$.

It follows from Eqs. (17.169), (17.170) that the tensors A^{ss} and A^{cs} for the spheroid priority take the forms

$$A^{ss} = A^{cs} = A^{sc} = A(a_s). \tag{17.193}$$

Let us consider the case of very thin ellipsoidal disks ($l_s << a_s$). In this case the integrals on the right-hand sides of Eqs. (17.186), (17.192) have the order of $\frac{l_s}{a_s}$, and can be neglected. Thus, the equations for the tensors A^{ij} in the case of thin disks are simplified, and for the cylinder priority we yield

$$A^{cc} = A^{cd} = A^{dc} = A^c, \tag{17.194}$$

$$A^{dd} = -\frac{p_c}{1-p_c}A^c + \frac{1}{1-p_c}A(a_d). \tag{17.195}$$

Here, the tensor $A(a_d)$ for a disk with the aspect ratio $\gamma = 0$ takes the form

$$A(a_d) = \frac{1}{2\mu_0}P^5(n) + \frac{1}{\lambda_0 + 2\mu_0}P^6(n). \tag{17.196}$$

For disk priority, the tensor A^{cc} is in Eq. (17.192), and

$$A^{cc} = A^c,\, A^{dd} = A^{dc} = A^{cd} = A^d, \tag{17.197}$$

tensor A^d is in Eq. (17.195).

We consider the overall elastic stiffness tensor C^* of the hybrid composite when inclusions are cylinders and thin spheroids (disks). For cylinder priority, Eqs. (17.33)–(17.36) for C^* take he forms

$$C^* = C^0 + p_{*c}T^c\Lambda^{*c} + p_{*d}T^d\Lambda^{*d}, \tag{17.198}$$

$$\Lambda^{*c} = \left(I - p_{*c}A^cT^c - p_{*d}A^cT^dD_d^{-1}\right)^{-1}, \tag{17.199}$$

$$\Lambda^{*d} = \left(I - p_{*d}A^{dd}T^d - p_{*c}A^cT^cD_d\right)^{-1}, \tag{17.200}$$

$$D_d = I - p_{*d}(A^d - A^c)T^d, \quad T^c = C^{1c}\Lambda^c, \tag{17.201}$$

$$T^d = C^{1d}\Lambda^d, \quad p_{*c} = p_c, \quad p_{*d} = p_d(1 - p_c). \tag{17.202}$$

For the disk priority, Eqs. (17.33)–(17.36) and (17.197) result in the following equations for the tensor C^*

$$C^* = C^0 + p_{*c}T^c\Lambda^{*c} + p_{*d}T^d\Lambda^{*d}, \tag{17.203}$$

$$\Lambda^{*c} = \left(I - p_{*c}A^cT^c - p_{*d}A^dT^dD_c\right)^{-1}, \tag{17.204}$$

$$\Lambda^{*d} = \left(I - p_{*d}A^dT^d - p_{*c}A^dT^cD_c^{-1}\right)^{-1}, \tag{17.205}$$

$$D_c = I - p_{*c}(A^c - A^d)T^c, \quad p_{*c} = p_c(1 - p_d), \quad p_{*d} = p_d. \tag{17.206}$$

In Fig. 17.13, the dependences of Young's moduli E_{*1} in the direction orthogonal to the axes of the cylinders and E_{*3} in the direction of the cylinder axes on the actual volume concentrations of disks p_{*d} are presented for a fixed actual volume fraction of cylinders $p_{*c} = 0.2$. The bold lines with dots correspond to the results of the EFM for the Boolean model with cylinder priority, the bold lines with short horizontal bars present the results of the same method for disk priority.

Note that the effective elastic properties of three-phase composites can be calculated using the equations of the two-phase theory. We can consider first the matrix with one population of inclusions, and using equation of the EFM for the effective elastic stiffness tensor of two-phase composites. After that, we can consider the two-phase system of a matrix with the calculated elastic properties and the second population of inclusions and again apply the formulas of the two-phase theory (sequential approach, Fig. 17.14).

As a result, we obtain the effective elastic stiffness tensor of the hybrid composite that has the necessary symmetry with respect to indices. The thin line with dots in Fig. 17.13 corresponds to the sequential approach when cylinders are considered in the effective medium of the host medium with the disks, and the thin line with short bars corresponds to disks in the effective medium of the host medium with cylinders. It is seen from this figure that the result of the sequential approach strongly depends on the order of phase consideration. The sequential approaches and the EFM applied to the Boolean probabilistic models with the cylinders or disks priorities practically coincide.

The application of the sequential approach to the calculation of the effective elastic properties of hybrid composites is reasonable if the sizes of the inclusions of different populations are essentially different. Let us consider the sequence (mdc) ((matrix + disks) + cylinders). In this case the medium with disks may be considered as a homogeneous medium with respect to cylindrical inclusions if the radii a_d of disks are much smaller than the radii a_c of the cylinders ($a_d << a_c$). Note

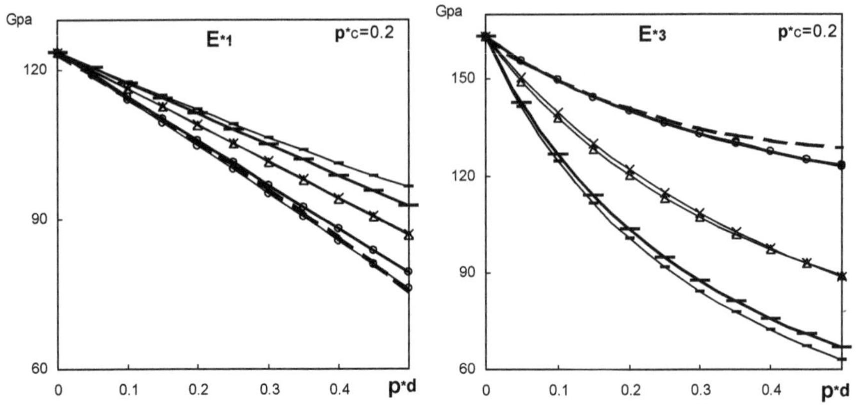

Fig. 17.13 Elastic moduli of the cylinders and disks hybrid composite, E_{*1} is Young's modulus orthogonal to cylinders, E_{*3} is Young's modulus along cylinders. Thin lines with dots correspond to cylinders in the effective medium of matrix and disks, thin lines with short bars correspond to disks in the effective medium matrix with cylinders. Bold lines with dots correspond to the EFM and the Boolean model of cylinders and disks with cylinder priority, bold lines with short bars correspond to the EFM and the same model with disk priority, lines with triangles correspond to the EFM and the dead leaves model, lines with crosses are given by the Mori-Tanaka method. Bold lines correspond to the EFM and the dilution model, dashed line is the extrapolation of the dilution model in the region $p > 0.368$

that because cylinders occupy a part of the volume of the composite, the effective volume concentration p_{*d} of disks in the final three-phase medium will be less then the original concentration p_d that was considered in the first step of the sequential approach, and $p_{*d} = p_d(1 - p_c)$. On the contrary, if we consider the sequence (mcd) ((matrix + cylinders) + disks) the results of this approach is valid, strictly speaking, in the limit $\frac{a_c}{a_d} \to 0$ and the effective volume concentration p_{*c} of cylinders in the hybrid composite is $p_{*c} = p_c(1 - p_d)$.

Consider the three-phase composite studied in [5]: a composite with isotropic matrix (Young's modulus $E_0 = 96.5$ GPa, shear modulus $\mu_0 = 37.1$ GPa), a set of infinite unidirectional isotropic cylindrical fibers ($E^{(1)} = 431$ GPa, $\mu^{(1)} = 172$ GPa) and a set of isotropic spheroids (thin spheroidal disks) ($E^{(2)} = 34.4$ GPa, $\mu^{(2)} = 14.3$ GPa). The directions of the disk rotation axes and the fiber axes coincide. For application of Eqs. (17.33)–(17.36) of the EFM to this composite material, we have to calculate tensors A^{11}, A^{22}, $A^{12} = A^{21}$ defined in Eq. (17.116). The correlation functions $\Phi_{11}(x)$, $\Phi_{22}(x)$ and $\Phi_{12}(x)$ in this equation depend on the spatial distribution of cylinders and disks in the matrix. Thus we have to introduce probabilistic models of such distributions. For the dead leaves model of cylinders and disks, the functions are Φ_{cc}, Φ_{dd}, and Φ_{cd} are defined in Eqs. (17.172)–(17.175), for this case and $p_d = p_{*d}$, $p_c = p_{*c}$, and the expressions for the tensors A^{cc}, A^{dd}, and A^{cd} follow from Eq. (17.116) in the forms

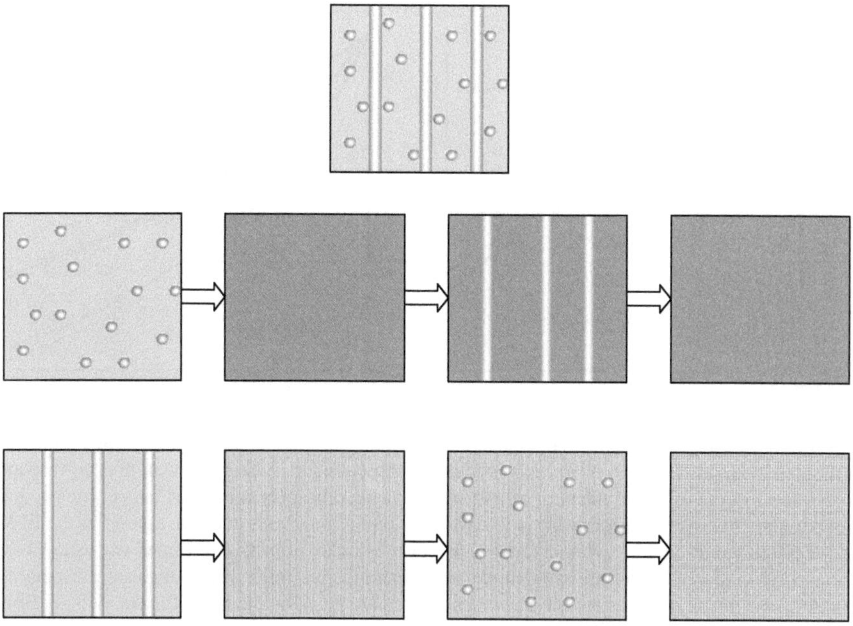

Fig. 17.14 Sequential approach for the calculation of the overall properties of hybrid composites. The final results depends on the order of the application the homogenization equations

$$A^{cc} = A^c,\, A^{dd} = (1 - g_{dd})A(a_c) + g_{dd}A(a_d), \tag{17.207}$$

$$A^{cd} = (1 - g_{cd})A(a_c) + g_{cd}A(a_d), \tag{17.208}$$

$$g_{cd} = \frac{1}{p}\left(\frac{1-(1-p)^{2-\frac{p_c}{p}}}{2p-p_c} - \frac{1-p}{p_d}\left(1-(1-p)^{\frac{p_d}{p}}\right)\right), \tag{17.209}$$

$$g_{dd} = 2g_{cd}. \tag{17.210}$$

The dependencies of Young's moduli E_{*1} and E_{*3} for this model are lines with triangles in Fig. 17.15. These lines are very close to the results of Mori-Tanaka method that are presented in Fig. 17.15 by the lines with crosses.

The dependencies of Young's moduli E_{*1} and E_{*3} for this model are lines with triangles in Fig. 17.15. These lines are very close to the results of Mori-Tanaka method that are presented in Fig. 17.15 by the lines with crosses.

For the dilution model of cylinder and disks, the tensors A^{ij} take the forms

$$A^{cc} = A^{cd} = A(a_c), \quad A^{dd} = (1 - g_{dd})A(a_c) + g_{dd}A(a_d), \tag{17.211}$$

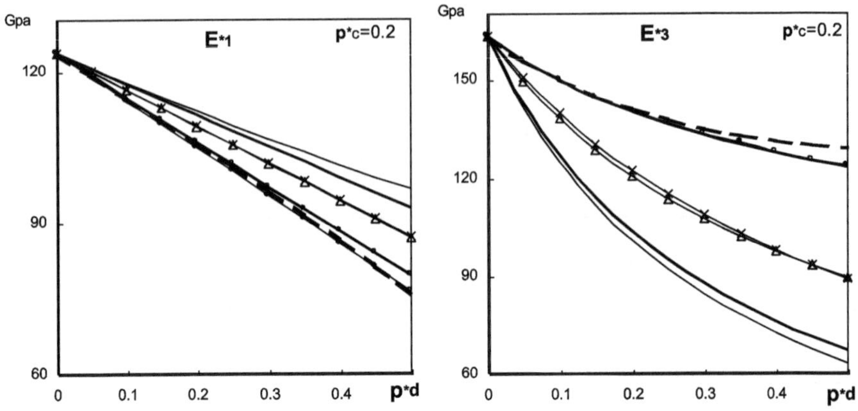

Fig. 17.15 Elastic moduli of the cylinders and disks hybrid composite, E_{*1} is Young's modulus orthogonal to cylinders, E_{*3} is Young's modulus along cylinders. Thin lines with dots correspond to cylinders in the effective medium of matrix and disks, thin lines with short bars correspond to disks in the effective medium matrix with cylinders. Bold lines with dots correspond to the EFM and the Boolean model of cylinders and disks with cylinder priority, bold lines with short bars correspond to the EFM and the same model with disk priority, lines with triangles correspond to the EFM and the dead leaves model, lines with crosses are given by the Mori-Tanaka method. Bold lines correspond to the EFM and the dilution model, dashed line is the extrapolation of the dilution model in the region $p > 0.368$

$$g_{dd} = \exp\left(p_c \frac{\theta(p)}{p}\right), \qquad p = p_c + p_d, \tag{17.212}$$

where the dependence $\theta(p)$ is defined in Eq. (17.179), and actual concentrations p_{*c}, p_{*d} of inclusions coincide with p_c, p_d. As it is mentioned above, this model serves only until the volume concentration of disks reaches $p_d = 0.168$ ($p_c = 0.2$), the bold lines in Fig. 17.15 correspond to this solution. In the region $p_c + p_d < 0.368$ the result of the dilution model may be approximated with a sufficient accuracy if the function $\theta(p)$ in Eq. (17.179) is approximated by the first terms of its Taylor series

$$\theta(p) = p + p^2 + \frac{3}{2}p^3. \tag{17.213}$$

The dashed lines in Fig. 17.15 correspond to the extrapolation of Young's moduli of the composite into the region $p_d > 0.168$ if the function $\theta(p)$ has the form (17.213).

In Fig. 17.16 the results of all approaches are compared for the composite with infinite cylinders and spherical pores. Young's modulus and Poisson's ratio of the isotropic matrix are $E_0 = 100$GPa, $\nu_0 = 0.4$ and the elastic parameters of cylinders are $E_c = 500$Gpa, $\nu_c = 0.3$. The notations for the different curves are the same as in Fig. 17.16 (the lines with short bars correspond to the Boolean model with

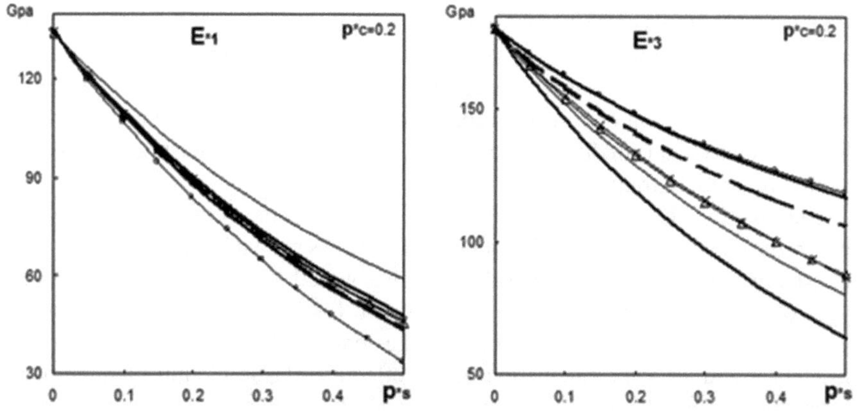

Fig. 17.16 The same dependences as in Fig. 17.15 for the hybrid composite containing cylindrical inclusions and spherical pores

sphere priority (bold lines) and to the spheres in the effective medium of matrix and cylinders (thin lines)). Note that the integrals in Eqs. (17.186), (17.192) give a small contribution to the values of the elastic constants, and neglecting these integrals changes the values of the overall elastic moduli by less than about 5%.

As for cylinder-disk composites, the results of the sequential approaches for cylinders-spheres are very close to the EFM method applied to the Boolean probabilistic models of cylinders and spheres with cylinder or sphere priorities. The dead leaves model of cylinders and spheres again turns out to be close to the results of Mori-Tanaka method.

17.6 Conclusions

Among different approximate methods for the solution of the homogenization problem, the effective field method has several important advantages. In many cases it results in simple equations for the effective conductive and elastic properties of composite materials, that are in good agreement with experimental data and numerical simulations in a wide range of its micro-parameters. The method is sufficiently flexible to be applied to composites with non-ellipsoidal inclusions.

In application to two-phase composite materials, the simplest version of the method results in Eq. (13.47) for the effective elastic stiffness tensor C^* that is not very sensitive to the statistical properties of random sets of inclusions. For instance, if we consider quasispherical inclusions in a homogeneous matrix and restrict ourselves to statistically isotropic sets of inclusions, the final equation for C^* (13.70) will be the same for all Boolean-type models similar to those considered in Sect. 17.4. This is a consequence of the property of the integral A^* in Eq. (13.23) that does not depend on the details of the behavior of the correlation function $\Psi(x)$ if this function has spherically symmetry $\Psi(x) = \Psi(|x|)$. Of course, there are some phenomena that strongly depend on microstructures and that the EFM cannot predict. For instance, existence of a percolation threshold for the overall conductivity of composites with dielectric matrix and conducting inclusions. This phenomenon must be described by another theoretical framework.

The situation changes dramatically when we consider the influence of microstructure on the elastic properties of hybrid composites. As shown in Sects. 17.3 and 17.5, the effective elastic moduli of hybrid composites with the same volume fractions and types of inclusions can differ by a factor of two for different probabilistic models of inclusions distributions. The results of the sequential approaches considered in Sects. 17.3 and 17.5 for calculation of the effective elastic moduli of three-phase composites, provide good approximations when the inclusions of two populations have essentially different sizes. Note that the sequential approaches result in substantial differences in the elastic properties of hybrid composites by exchanging the order of phase implantation. On the other hand, the EFM for Boolean models with cylinder priority and spheroid priority gives results that are very close to the sequential approaches. This fact can be considered as a verification of substantial differences in the elastic properties of hybrid composites with the same volume fractions but different types of spatial distributions of the inclusions in the host medium.

The main conclusion to this chapter can be formulated as follows. For the reliable predictions of the effective elastic stiffness tensor of hybrid composites, more precise microstructural information than for two-phase composite materials is required. The version of the EFM developed in this chapter allows accounting for this information.

This chapter is based on the publication [6].

References

1. Jeulin D (1987) Random structure analysis and modelling by mathematical morphology. In: Spencer A, Balkema A (eds) Proceedings of continuum media and discrete systems, vol 5, Nottingham, pp 217–226
2. Matheron G (1975) Random sets and integral geometry. Wiley & Sons, New York, p 425
3. Jeulin D (1993) Random models for the morphological analysis of powder. J Microscopy 172:13–21
4. Jeulin D (1997) Dead leaves models: from space tessellation to random functions. In: Jeulin D (ed) Proceedings of international conference advances theory applications random sets, Fontainebleau, 9–11 Oct 1996. World Scientific, pp 137–156
5. Benveniste Y (1991) On diagonal and elastic symmetry of the approximate effective stiffness tensor of heterogeneous media. J Mech Phys Solids 39:927–946
6. Kanaun S, Jeulin D (2001) Elastic properties of hybrid composites by the effective field approach. J Mech Phys Solids 49:2339–2367

Chapter 18
The Effective Field Method Beyond the Quasicrystalline Approximation

Abstract A modification of the effective field method that allows improving its predictions in the region of high volume fractions of inclusions is considered. In this version, the local exciting field acting on the inclusions is assumed to be a random function. The method allows deriving an infinite chain of integral equations for n-point statistical moments of the effective field. Additional assumptions allow cutting this chain at the n-th equation for the calculation of the multipoint statistical moments of the order 1, 2, ..., n. Then, the method allows the construction of the effective properties of the composites, taking into account the n-point statistical moments of the effective field. An example of the conductive medium with spherical homogeneous inclusions is considered. The two-point Percus-Yevick correlation functions of the centers of non-penetrating spheres are used in the calculations. The predictions of the method are compared with numerical simulations.

Keywords Local exiting fields · n-point correlation functions · Spherical inclusions · Perkus-Yevick correlation function · Effective conductive properties

The version of the effective field method considered in the previous chapters allows estimating the effective elastic and conductive properties of matrix composites containing inclusions of various shapes, properties, and volume fractions with the accuracy sufficient for engineering applications. But, e.g., for composites with spherical inclusions that are much stiffer than the matrix material, the method underestimates interactions between inclusions. In this case, predictions of the EFM for effective elastic constants are smaller than the values of these constants obtained in experiments or numerical simulations. The difference is notable when the volume fractions of inclusions exceed 0.3, and the error grows with the inclusion volume fraction. In this chapter, we consider a modification of the effective field method that allows improving predictions of the EFM in the region of high volume fractions of inclusions.

S. Kanaun, *Self-Consistent Methods for Composites*, Solid Mechanics and Its Applications 282, https://doi.org/10.1007/978-3-031-96739-9_18

18.1 Local Exiting Fields Acting on Inclusions in Matrix Composite Materials

We consider a homogeneous host medium with elastic stiffness tensor C^0 containing a homogeneous in space random set of isolated inclusions. The medium is subjected to a constant external strain ε^0 (stress field σ^0) at infinity. The inclusion that occupies a region v_i can be considered as an isolated one in the original host medium subjected to a local exciting strain field $\varepsilon^{*i}(x)$. If the solution of the one-particle problem for this inclusion is known, the field $\varepsilon(x)$ inside the inclusion is defined by the equation

$$\varepsilon(x) = \left(\Lambda^i \varepsilon^{*i}\right)(x) \ , \ x \in v_i \ , \ i = 1, 2, ..., \tag{18.1}$$

where Λ^i is a linear operator. The equation for the fields $\varepsilon^{*i}(x)$ follows from Eq. (13.4)

$$\varepsilon^{*i}(x) = \varepsilon^0(x) -$$

$$-\sum_{k\neq i} \int K(x - x') C^{1k}(x') \left(\Lambda^k \varepsilon^{*k}\right)(x') v_k(x') dx', \ x \in v_i, \ i = 1, 2, ... \tag{18.2}$$

Here, $C^{1i}(x) = C^i(x) - C^0$ is the perturbation of the elastic stiffness tensor inside the ith inclusion, $v_i(x)$ is the characteristic function of the region v_i. If the fields $\varepsilon^{*i}(x)$ can be found from the solution of this system, the strain and stress fields in the composite are expressed in terms of the local exiting fields $\varepsilon^{*i}(x)$ by the equations that follows from Eqs. (13.39), (13.40)

$$\varepsilon(x) = \varepsilon^0(x) - \int K(x - x') q(x') dx', \tag{18.3}$$

$$\sigma(x) = \sigma^0(x) - \int S(x - x') B^0 q(x') dx', \tag{18.4}$$

$$q(x) = \sum_i C^{1i}(x) \left(\Lambda^i \varepsilon^{*i}\right)(x) v_i(x). \tag{18.5}$$

We introduce the field $\varepsilon^*(x)$ that coincides with $\varepsilon^{*i}(x)$ when $x \in v_i$ and present the function $q(x)$ in the form

$$q(x) = \left(P\varepsilon^*\right)(x) V(x), \tag{18.6}$$

where $V(x) = \sum_i v_i(x)$, and P is the linear operator defined by the equation

$$\left(P\varepsilon^*\right)(x) V(x) = \sum_i C^{1i}(x) \left(\Lambda^i \varepsilon^{*i}\right)(x) v_i(x). \tag{18.7}$$

After introducing the functions

$$V(x'; x) = \sum_{i \neq k} v_i(x') \text{ if } x \in v_k, \tag{18.8}$$

the system (18.2) can be written as an equation for the field $\varepsilon^*(x)$ in the region $V = \bigcup_i v_i$ occupied by the inclusions

$$\varepsilon^*(x) = \varepsilon^0 - \int K\left(x - x'\right)\left(P\varepsilon^*\right)\left(x'\right) V\left(x'; x\right) dx', \; x \in V. \tag{18.9}$$

Equations (18.3), (18.4) for the strain and stress fields in the medium are expressed in terms of the field $\varepsilon^*(x)$ as follows

$$\varepsilon(x) = \varepsilon^0 - \int K\left(x - x'\right)\left(P\varepsilon^*\right)\left(x'\right) V(x') dx', \tag{18.10}$$

$$\sigma(x) = \sigma^0 - \int S\left(x - x'\right) B^0 \left(P\varepsilon^*\right)\left(x'\right) V(x') dx'. \tag{18.11}$$

For a random set of inclusions, $\varepsilon^*(x)$ is a random function. For construction of statistical moments of the function $\varepsilon^*(x)$, we introduce the following hypotheses.

H_1. *The field* $\varepsilon^*(x)$ *is constant in each region* v_i $(i = 1, 2, ...)$ *occupied by the inclusions.*

H_2. *The value of the function* $\varepsilon^{*i}(x)$ *in the region* v_i *is statistically independent of the physical and geometrical properties of the inclusion that occupies this region.*

Hypothesis H_2 implies that the local exciting field acting on an arbitrary inclusion depends weakly on its shape and properties, but it is determined by the cooperative effect of interacting inhomogeneities.

Let the inclusions be identical homogeneous ellipsoids. For hypothesis H_1 and the polynomial conservation theorem, the operator P in Eq. (18.7) is multiplication by the constant tensor P

$$\left(P\varepsilon^*\right)(x) = P\varepsilon^*(x), \quad x \in V, \tag{18.12}$$

$$P = C^1 \left(I + A(a) C^1\right)^{-1}, \quad \varepsilon^*(x) = \varepsilon^{*i}, \quad x \in v_i. \tag{18.13}$$

Here, the tensor $A(a)$ is defined in Eq. (4.89). In this case, the strain and stress tensors in Eqs. (18.10), (18.11) averaged over realizations of the random set of inclusions are

$$\langle \varepsilon(x) \rangle = \varepsilon^0 - \int K\left(x - x'\right) dx' P \left\langle \varepsilon^*(x) V(x) \right\rangle, \tag{18.14}$$

$$\langle \sigma(x) \rangle = \sigma^0 - \int S\left(x - x'\right) dx' B^0 P \left\langle \varepsilon^*(x) V(x) \right\rangle. \tag{18.15}$$

Here, we take into account that for a homogeneous in space random set of inclusions and a constant external field, the mean $\langle\varepsilon^*(x)\,V(x)\rangle$ is constant. This mean is presented in the form

$$\left\langle\varepsilon^*(x)\,V(x)\right\rangle = \left\langle\varepsilon^*(x)\,|x\right\rangle\langle V(x)\rangle = p\varepsilon^{*(1)}, \tag{18.16}$$

where

$$\varepsilon^{*(1)} = \left\langle\varepsilon^*(x)\,|x\right\rangle,\ p = \langle V(x)\rangle\,. \tag{18.17}$$

Here, $\langle\varepsilon^*(x)\,|x\rangle$ is averaging $\varepsilon^*(x)$ by the condition that $x \in V$. Then, for a fixed external strain field ε^0, Eqs. (10.31) ($\widehat{C} = C^0$) hold, and we obtain

$$\langle\varepsilon(x)\rangle = \varepsilon^0,\ \langle\sigma(x)\rangle = \sigma^0 + pP\varepsilon^{*(1)}, \tag{18.18}$$

and the homogenization problem is reduced to the calculation of the tensor $\varepsilon^{*(1)}$.

After averaging Eq. (18.9) by the condition that $x \in V$, we yield the equation for $\varepsilon^{*(1)}$

$$\varepsilon^{*(1)} = \varepsilon^0 - \int K\left(x - x'\right)P\left\langle\varepsilon^*\left(x'\right)V\left(x';x\right)|x\right\rangle dx'\,. \tag{18.19}$$

Using hypothesis H_2 on statistical independence of the field $\varepsilon^*(x)$ in the region v_k on the geometrical characteristics and properties of the kth inclusion, the conditional mean under the integral in this equation is presented in the form

$$\left\langle\varepsilon^*\left(x'\right)V\left(x';x\right)|x\right\rangle = p\varepsilon^{*(2)}\left(x',x\right)\Psi^{(1)}\left(x',x\right), \tag{18.20}$$

where

$$\varepsilon^{*(2)}\left(x',x\right) = \left\langle\varepsilon^*\left(x'\right)|x',x\right\rangle,\ \Psi^{(1)}\left(x',x\right) = \frac{\left\langle V\left(x';x\right)|x\right\rangle}{\langle V(x)\rangle}. \tag{18.21}$$

For homogeneous in space random sets of inclusion, the functions $\varepsilon^{*(2)}\left(x',x\right)$ and $\Psi^{(1)}\left(x',x\right)$ depend on the difference $x - x'$ only, and we have from Eq. (18.19)

$$\varepsilon^{*(1)} = \varepsilon^0 - p\int K\left(x - x'\right)P\varepsilon^{*(2)}\left(x - x'\right)\Psi^{(1)}\left(x - x'\right)dx'\,. \tag{18.22}$$

The mean $\varepsilon^{*(2)}\left(x - x'\right)$ is the averaged effective field $\varepsilon^*\left(x'\right)$ at point x' inside an inclusion by the condition that point x is inside another inclusion. This function characterizes pair interactions between the inclusions in the composite. For the construction of this function, we can average Eq. (18.9) by the condition that point x and x_1 are inside different inclusions

$$\varepsilon^{*(2)}(x, x_1) = \varepsilon^0 - \int K\left(x - x'\right) P \left\langle \varepsilon^* \left(x'\right) V \left(x'; x\right) | x, x_1 \right\rangle dx'. \tag{18.23}$$

The mean under the integral in this equation is presented in the form

$$\left\langle \varepsilon^* \left(x'\right) V \left(x'; x\right) | x, x_1 \right\rangle = p\varepsilon^{*(3)} \left(x', x, x_1\right) \Psi^{(2)}(x', x, x_1), \tag{18.24}$$

where

$$\varepsilon^{*(3)} \left(x', x, x_1\right) = \left\langle \varepsilon^* \left(x'\right) | x', x, x_1 \right\rangle, \tag{18.25}$$

$$\Psi^{(2)}(x', x, x_1) = \frac{\left\langle V \left(x'; x\right) | x, x_1 \right\rangle}{\langle V(x) \rangle}. \tag{18.26}$$

The object $\varepsilon^{*(3)} \left(x', x, x_1\right)$ is the effective field $\varepsilon^*(x')$ at point $x' \in V$ averaged by the condition that points x and x_1 are inside other inclusions. Then, for $\varepsilon^{*(2)}(x, x_1)$ we have the equation

$$\varepsilon^{*(2)}(x, x_1) = \varepsilon^0 - p \int K\left(x - x'\right) P\varepsilon^{*(3)} \left(x', x, x_1\right) \Psi^{(2)}(x', x, x_1) dx'. \tag{18.27}$$

For calculation of the function $\varepsilon^{*(3)} \left(x', x, x_1\right)$ we have to average Eq. (18.9) by the conditions that 3 points x, x_1, x_2 are in different inclusions. But in the right-hand side of the averaged equations, there appears the mean of $\varepsilon^*(x)$ by more complex conditions - $\varepsilon^{*(4)}(x, x_1, x_2, x_3)$ that is the average of the field $\varepsilon^*(x)$ by the condition that the point (x, x_1, x_2, x_3) are in different inclusions. Thus we obtain an infinite chain of equations that relates all the multipoint conditional means of the function $\varepsilon^*(x)$. In order to close this chain, we have to introduce additional assumptions on statistical properties of the effective field $\varepsilon^*(x)$. The simplest assumption is the "quasicrystalline approximation" [1, 2]

$$\varepsilon^{*(2)}(x, x_1) = \left\langle \varepsilon^*(x) | x; x_1 \right\rangle = \left\langle \varepsilon^*(x) | x \right\rangle = \varepsilon^{*(1)}(x). \tag{18.28}$$

Therefore, we accept that the mean effective field at a point $x \in V$ coincides with the mean of $\varepsilon^*(x)$ by the condition that point x_1 is inside another inclusion. Then, from Eq. (18.22) follows a closed equation for the mean value of the effective field $\varepsilon^{*(1)}(x)$

$$\varepsilon^{*(1)} = \varepsilon^0(x) + pA^* P\varepsilon^{*(1)}, \tag{18.29}$$

$$A^* = - \int K(x)\Psi^{(1)}(x) dx. \tag{18.30}$$

This equation coincides with Eq. (13.22) of the version of the effective field method considered in Sect. 13.1.

The next approximation for $\varepsilon^{*(1)}(x)$ can be found by closing the chain of equations for the condition means of the function $\varepsilon^*(x)$ in Eq. (18.27) at the second step by the equation

$$\varepsilon^{*(3)}\left(x', x, x_1\right) = \left\langle\varepsilon^*\left(x'\right)|x', x, x_1\right\rangle = \left\langle\varepsilon^*\left(x'\right)|x', x\right\rangle = \varepsilon^{*(2)}\left(x', x\right). \qquad (18.31)$$

Here, the function $\varepsilon^{*(2)}(x, x_1)$ is the mean effective field $\varepsilon^*(x)$ under the condition that points x and x_1 are inside different inclusions. The function $\varepsilon^{*(2)}(x, x_1) = \varepsilon^{*(2)}(x - x_1)$ characterizes pair interactions between two inclusions in the composite. When $|x_1|$ tends to infinity, influence of the inclusions at points x_1 on the inclusion at point x decays, and we have

$$\varepsilon^{*(2)}\left(x - x_1\right) \to \varepsilon^{*(1)}, \quad |x_1| \to \infty, \qquad (18.32)$$

$$\Psi^{(2)}\left(x', x, x_1\right) \to \Psi^{(1)}\left(x - x'\right). \qquad (18.33)$$

For Eqs. (18.31), (18.27) becomes a closed equation for the function $\varepsilon^{*(2)}(x, x_1)$

$$\varepsilon^{*(2)}(x, x_1) = \varepsilon^0 - p\int K\left(x - x'\right) P\varepsilon^{*(2)}\left(x', x\right)\Psi^{(2)}(x', x, x_1)dx', \qquad (18.34)$$

and Eq. (18.22) results in the following equation for $\varepsilon^{*(1)}$

$$\varepsilon^{*(1)} = \varepsilon^0 + pA^*P\varepsilon^{*(1)} - p\int K(x)\,P\left(\varepsilon^{*(2)}(x) - \varepsilon^{*(1)}\right)\Psi^{(1)}(x)dx, \qquad (18.35)$$

where the tensor A^* is in Eq. (18.30). This equation differs from Eq. (18.29) for $\varepsilon^{*(1)}$ by the integral term in the right-hand side that contains the function $\varepsilon^{*(2)}(x)$.

In order to find the function $\varepsilon^{*(2)}(x)$ from Eq. (18.34), we have to construct the correlation function $\Psi^{(2)}(x', x, x_1)$ in these equations. For simplicity, let us change the inclusions of finite sizes by equivalent point defects using the procedure described in Sect. 13.5. In this case, the functions $V(x)$ and $V(x; x')$ are replaced by the principal terms of their expansion in the series on multipoles concentrated at the inclusion centers ξ_i

$$V(x) \cong \sum_i v_i\delta\left(x - \xi_i\right); \quad V\left(x'; x\right) \cong \sum_{i \neq k} v_i\delta\left(x' - \xi_i\right) \text{ when } x = \xi_k. \qquad (18.36)$$

Here, $v_i = v$ is the volume of the i-th inclusion. In this approximation, the mean $\left\langle V\left(x'; x\right)|x, x_1\right\rangle$ takes the form

$$\left\langle V\left(x'; x\right)|x, x_1\right\rangle \cong v\left\langle X\left(x'; x\right)|x, x_1\right\rangle, \qquad (18.37)$$

where the mean in the right-hand side is

$$v\langle X\left(x;x'\right)|x,x_1\rangle = v\delta\left(x'-x_1\right) + p\Psi^{(2)}\left(x',x,x_1\right), \tag{18.38}$$

$$\Psi^{(2)}\left(x',x,x_1\right) = \frac{1}{n_0}\langle X\left(x';x,x_1\right)\rangle, \tag{18.39}$$

where n_0 is the number of inclusions in the unit volume of the medium, $\Psi^{(2)}\left(x',x,x_1\right)$ is a continuous function and

$$\Psi^{(2)}\left(x',x,x_1\right) \to \Psi^{(1)}\left(x'-x\right) \quad \text{when } |x_1| \to \infty\,. \tag{18.40}$$

Substituting Eq. (18.38) into Eq. (18.34), we yield to the equation

$$\varepsilon^{*(2)}\left(x-x_1\right) = \varepsilon^{*(1)} - vK\left(x-x_1\right)P\varepsilon^{*(2)}\left(x-x_1\right) - pJ\left(x,x_1\right), \tag{18.41}$$

where

$$J\left(x,x_1\right) = \int K\left(x-x'\right)P(\varepsilon^{*(3)}(x',x,x_1) - \varepsilon^{*(2)}\left(x',x\right)\times$$

$$\times\,\Psi^{(2)}\left(x',x,x_1\right) + \varepsilon^{*(2)}\left(x-x'\right)\left(\Psi^{(2)}\left(x',x,x_1\right) - \Psi^{(1)}\left(x-x'\right)\right)dx'. \tag{18.42}$$

Neglecting in Eq. (18.41) the term $pJ(x,x_1)$ that has the order p and vanishes when $|x_1| \to \infty$, due to Eqs. (18.32) and (18.40), we obtain the function $\varepsilon^{*(2)}(x)$ in the form

$$\varepsilon^{*(2)}(x) = \left(I + vK(x)P\right)^{-1}\varepsilon^{*(1)}. \tag{18.43}$$

This equation can also be derived in the following way. Consider two point defects with the same characteristics vP at points x_1 and x_2 in a homogeneous medium with the elastic stiffness tensor C^0, and $x = x_1 - x_2$. Let the medium be subjected to a constant external field $\varepsilon^{*(1)}$. The exiting field $\varepsilon^{*(2)}(x)$ that acts on each defect is the solution of the discrete analogy of Eq. (18.9) for the inclusions of finite sizes

$$\varepsilon^{*(2)}(x) = \varepsilon^{*(1)} - vK(x)P\varepsilon^{*(2)}(x) \tag{18.44}$$

The solution of this equation coincides with the field $\varepsilon^{*(2)}(x)$ in Eq. (18.43). The function $\varepsilon^{*(2)}(x)$ characterizes the interaction between two isolated point defects in a homogeneous elastic medium subjected to a constant external field. Existence of surrounding defects is taken into account by the external field $\varepsilon^{*(1)}$ acting on both defects.

Substituting Eq. (18.43) into Eq. (18.35) and solving the resulting equation with respect to $\varepsilon^{*(1)}$, we yield

$$\varepsilon^{*(1)} = \left(I - pA^0P + pK^0\right)^{-1} \varepsilon^0, \tag{18.45}$$

$$K^0 = \int K(x)\, P\left((I + vK(x)\, P)^{-1} - I\right) \Psi^{(1)}(x) dx\ . \tag{18.46}$$

The function under the integral in this equation can be approximated as follows

$$K(x)\, P\left((I + vK(x)\, P)^{-1} - I\right) = -vK(x)\, PK(x)\, P + O(|x|^{-9}), \tag{18.47}$$

and the equation for the tensor K^0 takes the form

$$K^0 = -v \int K(x)\, PK(x)\, P\Psi^{(1)}(x) dx. \tag{18.48}$$

Because the tensor $\varepsilon^{*(1)}$ in Eq. (18.45) is constant, from Eqs. (18.14)–(18.16) we obtain

$$\langle\sigma\rangle = C^*\langle\varepsilon\rangle\ , \quad C^* = C^0 + pP\left(I - p(A^0P - K^0)\right)^{-1}. \tag{18.49}$$

When $K^0 = 0$, this equation for the effective stiffness tensor C^* coincides with Eq. (13.47) of the EFM based on quasicrystalline approximation.

For an isotropic host medium, the tensor $K(x)$ in Eq. (18.46) is (Appendix B.2)

$$K(x) = \frac{1}{4\pi\mu_0 r^3}\Big(E^1 - 3E^5(n) - \frac{\kappa_0}{2}\big(E^2 + 2E^1 -$$
$$-3\left(E^3(n) + E^4(n) + 4E^5(n)\right) + 15E^6(n)\big)\Big)\ ,\ r = |x|,\ n = \frac{x}{r}, \tag{18.50}$$

and tensor P for isotropic homogeneous inclusions is

$$P = P_1^0 E^2 + P_2^0\left(E^1 - \frac{1}{3}E^2\right), \tag{18.51}$$

where the coefficients $P_1^0,\ P_2^0$ are

$$P_1^0 = \frac{K_0(K - K_0)}{K_0 + s_1(K - K_0)},\quad P_2^0 = \frac{2\mu_0(\mu - \mu_0)}{\mu_0 + s_2(\mu - \mu_0)}, \tag{18.52}$$

$$s_1 = \frac{3K_0}{3K_0 + 4\mu_0},\quad s_2 = \frac{6(K_0 + 2\mu_0)}{5(3K_0 + 4\mu_0)}. \tag{18.53}$$

Here, K_0, μ_0 are the bulk and shear moduli of the host medium, and K, μ are the same for the inclusions. After introducing the spherical coordinates and integrating over the unit sphere in Eq. (18.46), we obtain

$$K^0 = k_1^0 E^2 + k_2^0 \left(E^1 - \frac{1}{3} E^2 \right), \tag{18.54}$$

$$k_1^0 = \int_0^\infty K_1(\xi)\, \Psi(\xi)\, \xi^2 d\xi, \quad k_2^0 = \int_0^\infty K_2(\xi)\, \Psi(\xi)\, \xi^2 d\xi, \tag{18.55}$$

$$K_1(\xi) = -\frac{(1-\kappa_0) P_2^0}{9\mu_0^2 \xi^6} \left(3(1-4\kappa_0) P_1^0 + (1+2\kappa_0)\, P_2^0 \right), \tag{18.56}$$

$$K_2(\xi) = -\frac{P_2^0}{30\mu_0^2 \xi^6} \left(12(1-\kappa_0)^2 P_1^0 + (7 - 4\kappa_0 (2 - 7\kappa_0))\, P_2^0 \right). \tag{18.57}$$

The coefficients k_1^0 and k_2^0 depend on the form of the correlation function $\Psi(\xi) = \Psi^{(1)}(\xi)$. For point defects, the function $\Psi(\xi)$ is a normalized probability density of the defect distribution in space under the condition that the point $x = 0$ is the center of another defect. If the point defects simulate the inclusions of the same radius a, their centers cannot be closer than $2a$. Therefore, the simplest approximation of the function $\Psi(x)$ has the form

$$\Psi(r) = H\left(\frac{r}{a} - 2\right) = \begin{cases} 0, & r/a < 2 \\ 1, & r/a > 2 \end{cases}. \tag{18.58}$$

Here, $H(\xi)$ is Heaviside's function. Strictly speaking, the function $\Psi(r)$ can have this form only in the limit $p \to 0$. For finite volume fractions of inclusions, $\Psi(r/a) = 0$ if $r < 2a$, $\Psi(r/a)$ has a maximum at $r = 2a$, and $\Psi(r/a)$ tends to 1 when $r \to \infty$. A reliable approximation of the function $\Psi(r)$ was proposed in [3] and it is called the Percus-Yevick correlation function [4] (Appendix D).

In Fig. 18.1, the two-point Percus-Yevick correlation functions of the distribution of centers of spherical inclusions of unit radii for the volume fraction of inclusions $p = 0.2, 0.4, 0.6$ are shown. The maximum value of this function is reached at $r = 2a$ and it has the form [5]

$$\Psi(\xi)|_{\xi=2} = \frac{2+p}{2(1-p)^2}, \quad \xi = \frac{r}{a}. \tag{18.59}$$

The lines in Fig. 18.1 are obtained by numerical solution of the Percus-Yevick equation in [5]. For calculation of the coefficients k_1^0, k_2^0 in Eq. (18.55), we use the approximation of the function $\Psi(\xi)$ suggested in [6]

$$\Psi(\xi) = \begin{cases} 0, & \xi < 2 \\ 1 + \left(\frac{2+p}{2(1-p)^2} - 1 \right) \cos(\pi\xi) \exp(2(2-\xi)), & \xi > 2 \end{cases}. \tag{18.60}$$

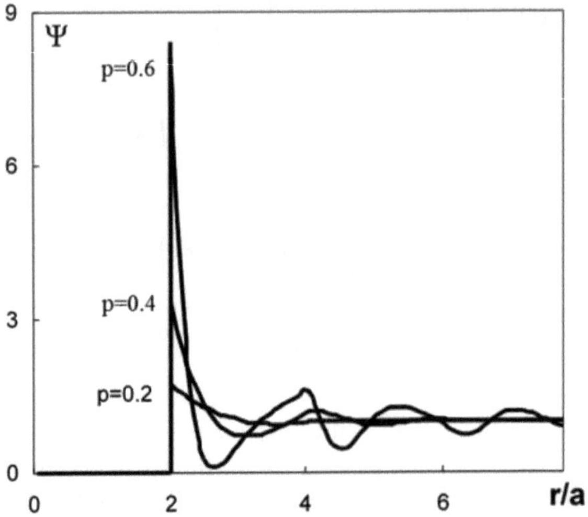

Fig. 18.1 The two-point Percus-Yevick correlation functions for the distribution of centers of spherical inclusions of unit radii in 3D-space

From Eqs. (18.49)–(18.62) follow the equations for the effective bulk K_* and shear μ_* moduli of a composite with spherical inclusions in the form

$$K_* = K_0 + \frac{pP_1}{1 - 3p\left(3A_1^0 P_1 - k_1^0\right)}, \tag{18.61}$$

$$\mu_* = \mu_0 + \frac{pP_2}{2(1 - p\left(A_2^0 P_2 - k_2^0\right))}, \tag{18.62}$$

where A_1^0, A_2^0 are the components of the tensor A^0

$$A^0 = A_1^0 E^2 + A_2^0 \left(E^1 - \frac{1}{3}E^2\right), \tag{18.63}$$

$$A_1^0 = \frac{1-\kappa_0}{9\mu_0}, \quad A_2^0 = \frac{5 - 2\kappa_0}{15\mu_0}, \quad \kappa_0 = \frac{3K_0 + \mu_0}{3K_0 + 4\mu_0}. \tag{18.64}$$

In Fig. 18.2, Eqs. (18.61), (18.62) are compared with the results of experiments for the composites with a polymeric matrix and stiff spherical inclusions presented in [7] (black dots). In these experiments, Young's moduli and Poisson's ratios of the phases are: $E/E_0 = 28.7$, $\nu = 0.33$, $\nu_0 = 0.394$. Line 1 in the figure corresponds to Eqs. (18.61), (18.62), line 2 is quasicrystalline approximation when $k_1^0 = k_2^0 = 0$.

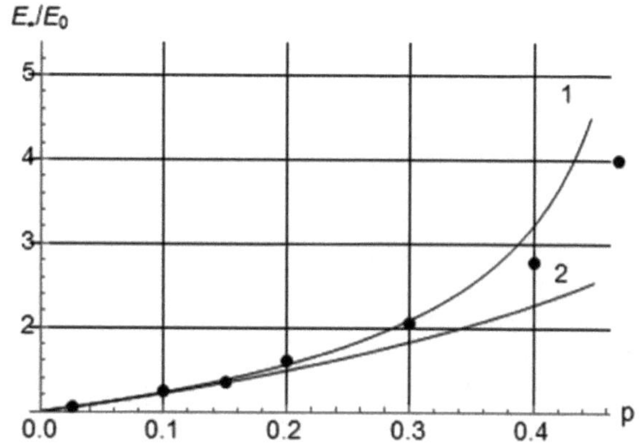

Fig. 18.2 Young modulus of the composite with polymer matrix end rigid spherical inclusions ($E/E_0 = 28.7$, $\nu = 0.33$, $\nu_0 = 0.394$); line 1 corresponds to Eqs. (18.60), (18.61), line 2 is quasicrystalline approximation when $A_1 = A_2$; black dots are experimental data presented in [7]

18.2 Conductive Properties of Composites with Large Volume Fraction of Inclusions

In the case of conductivity, the effective conductivity tensor of the matrix composite takes the form similar to in Eq. (18.49)

$$C^* = C^0 + pP\big(I - p(A^0P - K^0)\big)^{-1}. \tag{18.65}$$

with K^0 in Eq. (18.52)

$$K^0 = \int K(x)\, P\,\big((I + vK(x)\,P)^{-1} - I\big)\,\Psi^{(1)}(x)dx\;. \tag{18.66}$$

Consider a 3D-composite medium that consists of an isotropic matrix and a set of isotropic spherical inclusions. In this case.

$$C^0 = c_0 I,\;\; C = cI, \tag{18.67}$$

where c_0 and c are scalar conductivity coefficients of the matrix and inclusion materials, $I_{\alpha\beta} = \delta_{\alpha\beta}$ is unit second rank tensor. The function $K(x)$ in Eq. (18.66) has the form

$$K_{\alpha\beta}(x) = \frac{1}{4\pi c_0 r^3}\big(\delta_{\alpha\beta} - 3n_\alpha n_\beta\big), \quad r = |x|, \quad n_\alpha = \frac{x_\alpha}{r}, \tag{18.68}$$

and the tensors A^0 and P are

$$A^0 = \frac{1}{3c_0}I,\; P = P_0\delta_{ij}, \quad P_0 = \frac{3c_0c_1}{3c_0 + c_1}. \tag{18.69}$$

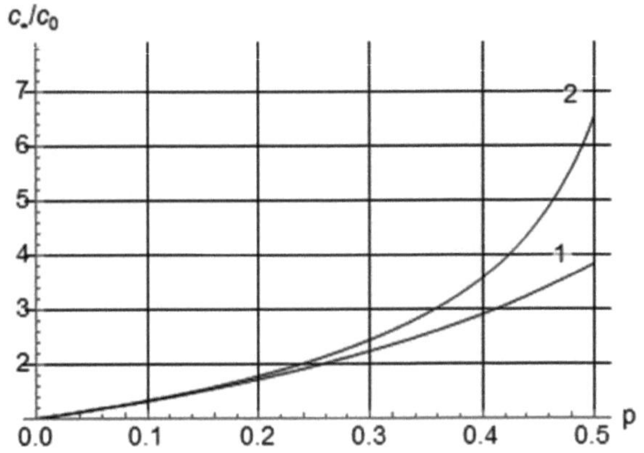

Fig. 18.3 The conductivity coefficient c^* of the matrix-inclusions composites with the contrast in the conductivity coefficients of the phases $c/c_0 = 100$; line 1 corresponds to the quasicrystalline approximation in Eq. (18.67), and line 2 is the improvement based on the two-point Perkus-Yevick correlation function in Eq. (18.66)

Integration over the unit sphere in Eq. (18.66) reduces the tensor K^0 to the 1D-integral

$$K^0 = \boldsymbol{k}_0 I, \ \boldsymbol{k}_0 = 6P_0^2 \int_2^\infty \frac{\Psi(\xi)\xi^2 d\xi}{(2P_0 - 3c_0\xi^3)(P_0 + 3c_0\xi^3)}. \tag{18.70}$$

For the composite medium is isotropic, $C_* = c_* I$ and the effective conductivity coefficient c_* takes the form

$$c_* = c_0 + \frac{pP_0}{(1 - p(A^0 P^0 - \boldsymbol{k}_0))}. \tag{18.71}$$

For $\boldsymbol{k}_0 = 0$, we yield the effective conductivity in the quasicrystalline approximation

$$c_* = c_0 + \frac{pc_0(c - c_0)}{3c_0 + (1 - p)(c - c_0)}. \tag{18.72}$$

In Fig. 18.3, the results of calculations of c_* from Eq. (18.72) (line 1) and from Eq. (18.72) (line 2) for the correlation function $\Psi(\xi)$ in Eq. (18.60) are presented; the conductivity coefficient of inclusions is $c = 100c_0$.

18.2.1 Composites with Circular Inclusions of High Conductivity in the 2D-Case

In the 2D-case, the effective conductivity of the medium with a random set of isolated inclusions keeps form in Eqs. (18.65), (18.66). For an isotropic host medium and circular isotropic inclusions, the function $K(x)$ and the tensors A^0 and P in these equations are

$$K_{\alpha\beta}(x) = \frac{1}{2\pi c_0 r^2}\left(\delta_{ij} - 2n_\alpha n_\beta\right), \; r = |x|, \; n_\alpha = \frac{x_\alpha}{r}, \tag{18.73}$$

$$A^0 = A_0 I, \; A_0 = \frac{1}{2c_0}; \quad P = P_0 I, \; P_0 = \frac{2c_0 c_1}{2c_0 + c_1}. \tag{18.74}$$

The effective medium is isotropic and its effective conductivity coefficient c_* is

$$c_* = c_0 + \frac{pP_0}{1 - p(A_0 P_0 - \mathrm{k}_0)}, \tag{18.75}$$

$$\mathrm{k}_0 = 2P_0 \int_2^\infty \frac{\Psi(\xi)\xi}{P_0^2 - 4c_0^2\xi^4} d\xi. \tag{18.76}$$

In the quasicrystalline approximation, the effective conductivity is

$$c_* = c_0 + \frac{pP_0}{1 - pA^0 P_0}, \tag{18.77}$$

and it is the Maxwell-Garnett formula in the 2D-case.

For the point defect model, the function $\Psi(\xi)$ in Eq. (18.76) is the Percus-Yevick correlation function of the centers of disks of radius a in the 2D-plane. Detailed tables of this function are presented in [8], and in Fig. 18.4, the graphs of this function for the inclusion fraction $p = 0.2, 0.4, 0.6$ are presented.

In Fig. 18.5, dependencies the effective conductivity $c_*(p)$ calculated from Eq. (18.77) (line 1) and Eq. (18.75) (line 2) for the inclusions with the conductivity $c = 100c_0$ are presented. The black dots are results of numerical simulations of the effective conductivity of composites with random sets of circular disks.

The numerical method used for the solution of the homogenization problem is based on the Fast Fourier Transform algorithm [9–11]. For the construction of the

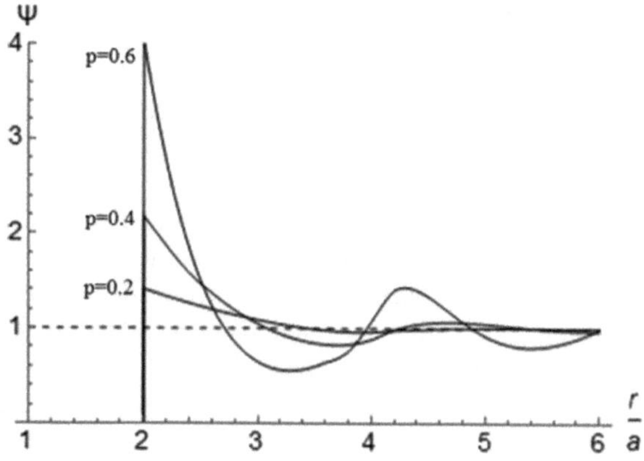

Fig. 18.4 The two-point Percus-Yevick correlation functions for the distribution of centers of circular disks 2D-space

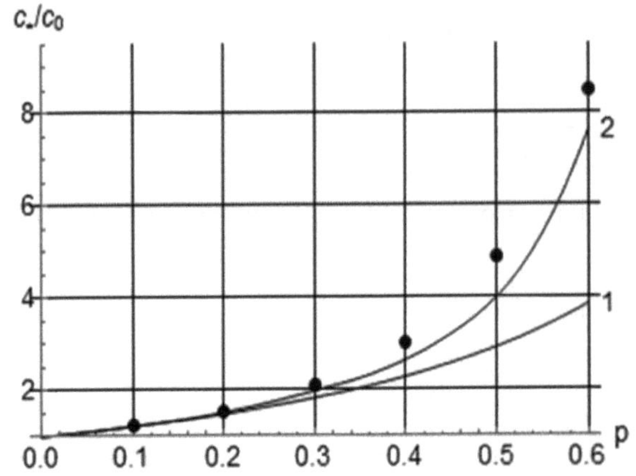

Fig. 18.5 Dependencies of the effective conductivity c^* of a two-phase composites with circular inclusions of the conductivity $c = 100c_0$ on the inclusion surface faction p, c_0 is the conductivity of the host medium, line 1 corresponds to the quasicrystalline approximation in Eq. (18.78) and line (2) to Eq. (18.76) that allows for Perkus-Yievic correlation function for the inclusion centers in the 2D-case

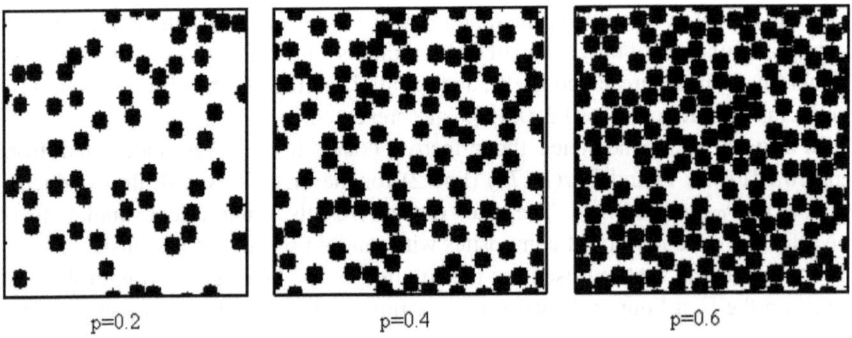

Fig. 18.6 Examples of the realizations of random fields of inclusions in the 2D-case used in the numerical calculations

numerical solutions, realizations of random sets of non-overlapping disks were generated on a square area. Then, the integral equation (11.1) for the electric field was solved for each realization using the mentioned algorithm. The method implies periodic boundary conditions on the square sizes. The effective conductivity was obtained by averaging the detailed electric fields and electric current over the considered area. For each inclusion fraction, ten realizations of the non-overlapping inclusion sets were considered. Typical realizations of random sets of inclusions for $p = 0.2, 0.4, 0.6$ are shown in Fig. 18.6. The mean values of the conductivity coefficients over 10 realization are black points in Fig. 18.5.

18.3 Discussion and Conclusion

Comparisons with experimental data and numerical simulations show that replacing the hypothesis of the quasicrystalline approximation (18.28) with the hypothesis (18.31) improves predictions of the EFM for effective conductive and elastic properties of matrix composites.

The hypotheses of the EFM simplify the solution of the homogenization problems for multi-particle systems, but these hypotheses are also the sources of possible errors. According to hypothesis H_1, the local exciting fields acting on each inclusion in the composite are constant. The actual exciting fields are not constant, but the presence of surrounding inclusions levels the averaged local effective fields and makes them closer to homogeneous in the regions of each inclusion. This is the reason for accepting hypothesis H_1 of the EFM.

The hypothesis of the quasicrystalline approximation (18.28) or more complex hypotheses similar to (18.31)

$$\varepsilon^{*(n+1)}(x_1, x_2, ..., x_n, x_{n+1}) = \varepsilon^{*(n)}(x_1, x_2, ..., x_n), \tag{18.78}$$

where

$$\varepsilon^{*(n)}(x_1, x_2, ..., x_n) = \left\langle \varepsilon^*(x_1) | x_1, x_2, ..., x_n \right\rangle, \tag{18.79}$$

close the chain of integral equations for multipoint conditional means of the effective field at the n-th equation. Each equation of this chain relates the n-point conditional mean of the effective field with the $(n+1)$-point mean

$$\varepsilon^{*(n)}(x_1, x_2, ..., x_n) = \varepsilon_0 - \tag{18.80}$$

$$- \int K(x_1 - x') P \varepsilon^{*(n+1)}(x', x_1, x_2, ..., x_n) \left\langle V(x'; x_1) | x_1, x_2, ..., x_n \right\rangle dx'.$$

Hypothesis (18.78) results in a closed equation for the n-point conditional mean $\varepsilon^{*(n)}$. By closing the chain of equations for $\varepsilon_*^{(n)}$ at the first, second, etc., steps, one can successively improve predictions of the effective properties of the composite materials. The principal difficulties of carrying out this procedure are in the construction of the $n+1$-point correlation function $\langle V(x; x_1) | x_1, x_2, ..., x_n \rangle$ of random set of inclusions in Eq. (18.80) and subsequent solution of this equation.

In a number of works, the Monte-Carlo procedure was used for the construction of two and three-point correlation functions for sets of spherical particles [12–14]. This problem has high computational cost that increases with the order of the correlation functions. The influence of the details of the correlation functions of random sets of inclusions on the effective properties of composites is substantial, as it is shown in this Section. Therefore, simple approximations of multipoint correlation functions can result in errors in the calculation of the effective property tensors. The accuracy of the method depends on the precision of the geometrical information about the microstructure of composites in terms of the multipoint correlation functions.

The contents of this chapter are based on the works [15–17]. Accounting for pair interactions between inclusions by calculation of the effective elastic stiffness tensor of matrix composites is considered in [17, 18].

References

1. Lax M (1952) Multiple scattering of waves II. The effective field in dense systems. Phys Rev 85:621–629
2. Willis J (1984) Some remarks on the application of the QCA to the determination of the overall elastic response of a matrix/ inclusion composite. J Math Phys 25:2116–2120
3. Percus J, Yevick G (1958) Analysis of classical statistical mechanics by means of collective coordinates. Phys Rev 111:1–13
4. Zaiman J (1982) The models of disorder. Mir, Moskow, p 591
5. Throop G, Bearman R (1965) Numerical solutions of the Percus-Yevick equation for the hard-sphere potential. J Chem Phys 42:2408–2411
6. Willis J (1978) Variational principles and bounds for overall properties of composites. In: Porovan E (ed) Continuum models of discrete system. Proceedings of the second international symposium on continuum models discrete systems. University Waterloo Press, Waterloo, pp 185–215
7. Smith J (1976) The elastic constants of a particulate filled glassy polymer. Poly Eng Sci 16:394–399
8. Ivanov A, Loyko V, Dik V (1998) Light propagation through dense dispersed media. Nauka y Tehnica, Minsk, p 192
9. Moulinec H, Suquet P (1994) A fast numerical method for computing the linear and nonlinear mechanical properties of composites. Compt Rend A Paris 318:1417–1423
10. Eyre D, Milton G (1999) A fast numerical scheme for computing the response of composites using grid refinement. Eur Phys J Appl Phys 6:41–47
11. Delarue A (2001) Prévision du comportement é
12. Alder B (1964) Triplet correlation in hard spheres. Phys Rev Lett 12:317–319
13. Curtin W (1990) Triplet direct correlation function in the hard-sphere fluid. J Chem Phys 93:1919–1925
14. Bildstein B, Kahi G (1994) Triplet correlation function for hard spheres: computer simulation results. J Chem Phys 100:5882–5893
15. Kanaun S (1982) Effective field method in the linear static problems of composite media. Appl Math Mech (PMM) 46:655–665
16. Kanaun S (1983) Correlation function of the stress field in an elastic medium with point defects. Appl Math Mech (PMM) 47:652–661
17. Kanaun S (1983) Elastic medium with random field of inhomogeneities. In: Kunin I (ed) The theory of elastic media with microstructure, vol. 2, Chap. 7. Springer, Berlin Heidelberg New York, pp 165–228
18. Kanaun S (1990) Self-consistent schemes of averaging in the mechanics of matrix composite materials. Mech Compos Mater 6:984–994

Chapter 19
Nonlocal Properties of Matrix Composites

Abstract The self-consistent methods are used for the construction of nonlocal effective property operators of matrix composite materials with respect to the space and time variables. Spatial nonlocality is pronounced for the external fields which characteristic scale change has the order of the correlation radius of the random set of inhomogeneities. The effective field method is used for the construction of the operator of the effective elastic stiffness that is nonlocal with respect to space variables. The effective medium method is used for the solution of the non-stationary homogenization problem in the case of composites with spherical inclusions by action of time-varying external fields. The time nonlocality can be observed by the action of space-localized impulses on composite materials. The corresponding fields are substantially different from Green's functions of the homogeneous medium with stationary parameters. The difference is notable at short time intervals from the moment of excitation, and it tends to diminish with time. The diminishing rate depends on the contrast in the thermo-conductivity properties of the matrix and inclusion materials. Examples of impulse excitations of matrix composites are considered.

Keywords Nonlocal operators of effective properties · Spatial nonlocality · Time nonlocality · Parabolic equations · Space-time Green's functions

In the previous chapters, the external field applied to heterogeneous media was assumed to be constant. As a result, the mean field intensity $\langle \varepsilon \rangle$ and flax density $\langle \sigma \rangle$ for statistically homogeneous in space random sets of heterogeneities are also constant. For the media with a linear constitutive law, the relation between these means is described by a similar linear law

$$\langle \sigma \rangle = C^* \langle \varepsilon \rangle \tag{19.1}$$

with a constant effective property tensor C^*. This linear law serves for external fields which characteristic scales of spatial variations are much larger than the variation scale of the property tensor $C(x)$ of the heterogeneous medium. The situation changes if the variation scale of the external fields is comparable or smaller than the characteristic variation scale of the tensor $C(x)$. In this case, the means $\langle \sigma \rangle$ and $\langle \varepsilon \rangle$ over

S. Kanaun, *Self-Consistent Methods for Composites*, Solid Mechanics and Its Applications 282, https://doi.org/10.1007/978-3-031-96739-9_19

ensemble realizations of random function $C(x)$ become functions of coordinates, and the relation between them is presented in the integral form [1, 2]

$$\langle\sigma\rangle(x) = \int C^*(x - x')\langle\varepsilon\rangle(x')dx'. \tag{19.2}$$

The kernel $C^*(x)$ of the integral operator in this equation is a generalized function with a finite support. Therefore, the mean flux density at point x is defined by the field intensity distribution in a finite vicinity of a point x. Nonlocal behavior of a heterogeneous medium is the reason for dispersion phenomena: dependence of the phase velocity of propagating waves on the wave frequency. It is known that the media with microstructure are always associated with spatial dispersion [1–3].

For time-varying external field, the mean field intensity and flux density turn out to be the functions of time, also, and nonlocality with respect to time appears in the constitutive equation for the effective medium

$$\langle\sigma\rangle(x,t) = \int\int C^*(x - x', t - t')\langle\varepsilon\rangle(x', t')dx'dt'. \tag{19.3}$$

In this chapter, self-consistent methods are applied for revealing the nonlocal character of the constitutive laws for composites in the case of space- and time-varying external fields.

19.1 The Effective Field Method for Composites Subjected to Varying in Space External Fields

We consider a homogeneous host medium with the property tensor C^0 containing a random set of isolated inclusions with the property tensor C. Let $\varepsilon^{*(i)}(x)$ be the local external field acting on the ith inclusion that occupies the region v_i. The actual field $\varepsilon^{(i)}(x)$ inside the inclusion is expressed in terms of the field $\varepsilon^{*(i)}(x)$ as follows

$$\varepsilon^{(i)}(x) = \left(\Lambda^i\varepsilon_i^{*(i)}\right)(x)\ ,\ x \in v_i\ ,\ i = 1, 2, ..., \tag{19.4}$$

Here, Λ^i is a linear operator that is to be constructed from the solution of the one-particle problems. Using the local external fields $\varepsilon^{*(i)}(x)$, the actual field $\varepsilon(x)$ in the composite is presented in the integral form

$$\varepsilon(x) = \varepsilon^0(x) - \sum_i \int K(x - x')\, C^1 \left(\Lambda^i\varepsilon^{*(i)}\right)(x')\, v_i\,(x')dx'. \tag{19.5}$$

The equation for the fields $\varepsilon^{*(i)}(x)$ follows from this equation in the form

$$\varepsilon^{*(i)}(x) = \varepsilon^0(x) - \sum_{j\neq i} \int K(x - x')\, C^1 \left(\Lambda^j\varepsilon^{*(j)}\right)(x')\, v_j\,(x')dx', \tag{19.6}$$

$x \in v_i \ , \quad i = 1, 2, \ldots \ .$

In these equations, $C^1 = C - C^0$ is the perturbations of the property tensor inside the inclusions.

We introduce the characteristic function $V(x)$ of the region $V = \bigcup_i v_i$ occupied by the inclusions

$$V(x) = \sum_i v_i(x), \tag{19.7}$$

and present the field intensity $\varepsilon(x)$ and flux density $\sigma(x)$ vectors in the medium in the forms

$$\varepsilon(x) = \varepsilon^0(x) - \int K(x - x') \left(P\varepsilon^*\right)(x')V(x')dx', \tag{19.8}$$

$$\sigma(x) = \sigma^0(x) - \int \mathrm{S}(x - x') \, B^0 \left(P\varepsilon^*\right)(x)V(x')\mathrm{d}x', \tag{19.9}$$

The function $(P\varepsilon^*)(x)$ in this equation coincides with $C^1 \left(\Lambda^i \varepsilon^{*(i)}\right)(x)$ in each region v_i. Then, we introduce the function

$$V\left(x'; x\right) = \sum_{i \neq j} v_i(x') \text{ if } x \in v_j \tag{19.10}$$

and present the local effective filed $\varepsilon^*(x)$ in the region V in the integral form

$$\varepsilon^*(x) = \varepsilon^0(x) - \int K(x - x') \left(P\varepsilon^*\right)(x') \, V\left(x'; x\right)dx', x \in V. \tag{19.11}$$

This equation is equivalent to Eq. (19.6).

If the inclusion set is random, $\varepsilon^*(x)$ is a random function. In order to construct the mean value of this function, we introduce the hypotheses.

H_1. *The field $\varepsilon^*(x)$ is constant in each region v_i but it can be different for different inclusions.*

H_2. *The value of the function $\varepsilon^*(x)$ in the region v_i is statistically independent of the physical and geometrical properties of the inclusion that occupies this region.*

Let the inclusions be identical homogeneous ellipsoids. Then, for the hypothesis H_1 and the polynomial conservation theorem (Sect. 3.3), the field $\varepsilon(x)$ inside each inclusion is constant, and the operator P in Eqs. (19.8), (19.9) and (19.11) is multiplication with constant tensor P

$$\left(P\varepsilon^*\right)(x) = P\varepsilon^*(x), \quad x \in V, \tag{19.12}$$

where

$$P = C^1(I + A(a)\, C^{1^{-1}}. \tag{19.13}$$

Here, the tensor $A(a)$ is defined in Eq. (4.89). Substituting Eq. (19.12) into Eq. (19.11), we yield the equation for the local external field $\varepsilon^*(x)$ in the composite medium

$$\varepsilon^*(x) = \varepsilon^0(x) - \int K\left(x - x'\right) P\varepsilon^*\left(x'\right) V\left(x'; x\right) dx', x \in V. \tag{19.14}$$

After averaging this equation by the condition that $x \in V$, we obtain the equation for the conditional mean $\varepsilon^{*(1)}(x) = \langle \varepsilon^*(x)|x\rangle$ in the form

$$\varepsilon^{*(1)}(x) = \varepsilon^0(x) - p \int K\left(x - x'\right) P \left\langle \varepsilon^*\left(x'\right) |x', x\right\rangle \Psi(x - x')dx'. \tag{19.15}$$

Here, $\left\langle \varepsilon^*\left(x'\right) |x', x\right\rangle$ is the mean of the effective field by the condition that points x and x' are in different inclusions. It is taken into account that according to hypothesis H_2, the following equation hold

$$\left\langle \varepsilon^*\left(x'\right) V\left(x'; x\right) |x\right\rangle = p\left\langle \varepsilon^*\left(x'\right) |x', x\right\rangle \Psi(x - x'), \tag{19.16}$$

$$\langle V(x)\rangle = p, \Psi(x - x') = \frac{\left\langle V\left(x'; x\right) |x\right\rangle}{\langle V(x)\rangle}, \tag{19.17}$$

where p is volume fraction of inclusions. Then, accepting the hypothesis of quasicrystalline approximation

$$\left\langle \varepsilon^*\left(x'\right) |x', x\right\rangle = \left\langle \varepsilon^*\left(x'\right) |x'\right\rangle = \varepsilon^{*(1)}\left(x'\right), \tag{19.18}$$

we obtain a closed equation for the mean $\varepsilon^{*(1)}(x)$

$$\varepsilon^{*(1)}(x) = \varepsilon^0(x) - p \int K\left(x - x'\right) P\varepsilon^{*(1)}\left(x'\right)\Psi(x - x')dx'. \tag{19.19}$$

In terms of the function $\varepsilon^{*(1)}(x)$, the means of the tensors $\varepsilon(x)$ and $\sigma(x)$ in Eqs. (19.8), (19.9) take the forms

$$\langle \varepsilon(x)\rangle = \varepsilon^0(x) - p \int K\left(x - x'\right) P\varepsilon^{*(1)}\left(x'\right)dx', \tag{19.20}$$

$$\langle \sigma(x)\rangle = \sigma^0(x) - p \int S\left(x - x'\right) B^0 P\varepsilon^{*(1)}\left(x'\right)dx'. \tag{19.21}$$

Because Eqs. (19.19), (19.20) and (19.21) are the equations in convolutions, after application of the Fourier transform operators to this system, we yield the linear algebraic system for the Fourier transforms of the functions $\varepsilon^{*(1)}(k)$, $\langle\varepsilon(k)\rangle$, $\langle\sigma(k)\rangle$

$$\varepsilon^{*(1)}(k) = \varepsilon^0(k) - pK^{\Psi}(k)\, P\varepsilon^{*(1)}(k), \tag{19.22}$$

$$\langle\varepsilon(k)\rangle = \varepsilon^0(k) - K(k)\, P\varepsilon_i^{*(1)}(k), \tag{19.23}$$

$$\langle\sigma(k)\rangle = \sigma^0(k) - S(k)\, B^0 P\varepsilon_i^{*(1)}(k). \tag{19.24}$$

Here, the functions with the argument k are the Fourier transforms of the corresponding original functions with argument x, and

$$K^{\Psi}(k) = \int K(x)\Psi(x)e^{ik\cdot x}dx. \tag{19.25}$$

Taking into account the equations

$$B^0 = \left(C^0\right)^{-1}, S(k) = C^0 K(k) C^0 - C^0, \sigma^0(k) = C^0\varepsilon^0(k), \tag{19.26}$$

the functions $\varepsilon^0(k)$, $\varepsilon^{*(1)}(k)$ can be excluded from Eq. (19.24), and the mean flux density $\langle\sigma(k)\rangle$ takes the form

$$\langle\sigma(k)\rangle = C^*(k)\,\langle\varepsilon(k)\rangle\,. \tag{19.27}$$

$$C^*(k) = C^0 + pP\left(1 - pA^*(k)P\right)^{-1}, \tag{19.28}$$

where

$$A^*(k) = \int K(x)\Phi(x)e^{ik\cdot x}dx, \quad \Phi(x) = 1 - \Psi(x). \tag{19.29}$$

Tensor $C^*(k)$ can be interpreted as the Fourier transform of the kernel $C^*(x)$ of the effective property operator in Eq. (19.2).

Keeping in Eq. (19.28) for $C^*(k)$ the terms of the order p^2 only

$$C^*(k) = C^0 + pP(1 + pA^*(k)P) + O(p^3), \tag{19.30}$$

and after transferring to the x-space, we yield

$$C^*(x) = \left(C^0 + pP\right)\delta(x) + p^2 PA^*(x)P + O(p^3). \tag{19.31}$$

Function $A^*(x)$ in this equation is product of the kernel $K(x)$ and the correlation function $\Phi(x)$

$$A^*(x) = K(x)\Phi(x). \tag{19.32}$$

The function $\Phi(x)$ has a finite support, and $\Phi(0) = 1$. As a result, the mean flux density $\langle\sigma(x)\rangle$ and field intensity $\langle\varepsilon(x)\rangle$ relate by the equation that follows from Eqs. (19.2) and (19.31)

$$\langle\sigma(x)\rangle = \left(C^0 + pP\right)\langle\varepsilon(x)\rangle + p^2 P \int A^*(x - x')\left\langle\varepsilon(x')\right\rangle dx' P. \tag{19.33}$$

Taking into account properties of the function $K(x)$ that is the second derivative of Green's function of the host medium, the integral in this equation is presented in the form (Appendix B.2)

$$\int A^*(x - x')\left\langle\varepsilon(x')\right\rangle dx' = A^0\langle\varepsilon(x)\rangle +$$
$$+vp \int K(x - x')\Phi(x - x')\left\langle\varepsilon(x')\right\rangle dx', \tag{19.34}$$

where $vp \int ..dx$ is the Cauchy principal value of the integral, A^0 is a constant tensor. Thus the mean flux density and field intensity in the composite relate by the equation

$$\langle\sigma(x)\rangle = \left(C^0 + pP(I + pA^0P)\right)\langle\varepsilon(x)\rangle +$$
$$+p^2 P vp \int K(x - x')\Phi(x - x)P\left\langle\varepsilon(x')\right\rangle dx'. \tag{19.35}$$

In this equation, the coefficient $C^0 + pP(I + pA^0P)$ is the local part of the effective stiffness tensor that coincides with the quasicrystalline approximation of C^* in Eq. (13.47) with accuracy to p^2. The integral term in Eq. (19.35) is the nonlocal part of the effective property operator. The kernel $K(x)\Phi(x)$ of the integral operator in Eq. (19.35) is concentrated in the region of the support of the function $\Phi(x)$, and the size of this region has the order of the correlation radius l of the random set of inclusions. This radius can be introduced by the equation

$$l^2 = \int_0^\infty r\Phi(r)dr. \tag{19.36}$$

Here, it is assumed that the inclusion set is isotropic in space, and therefore,

$$\Phi(x) = \Phi(|x|). \tag{19.37}$$

If the mean $\langle\varepsilon(x)\rangle$ changes slowly in space, the support of the function $\langle\varepsilon(k)\rangle$ (the Fourier transform of $\langle\varepsilon(x)\rangle$) is concentrated in the region of the k-space that is defined by the condition $|kl| << 1$. Thus for the calculation of the mean $\langle\sigma(x)\rangle$ from Eq. (19.35), it is possible to expand the symbol $C^*(k)$ of the operator C^* in Eq. (19.30) in the series with respect to $|kl|$ and to keep first terms in this expansion only. To reveal these terms we replace the function $e^{ik\cdot x}$ in (19.29) by the first three terms of its Taylor series

$$e^{\mathrm{i}(k\cdot x)} = 1 + \mathrm{i}\,(k\cdot x) - \frac{1}{2}\,(k\cdot x)^2\,. \tag{19.38}$$

As a result, the tensor $A^*(k)$ in Eq. (19.29) is approximated by the equation

$$A^*\,(k) = A^0 + \mathrm{i}A^{(1)}\cdot k - l^2 A^{(2)}\cdot(\mathrm{i}k\otimes \mathrm{i}k)\,, \tag{19.39}$$

$$A^0 = \int K(x)\,\Phi(|x|)dx\,,\; A^{(1)} = \int K(x)\otimes x\Phi(|x|)dx\,, \tag{19.40}$$

$$A^{(2)} = \frac{1}{2l^2}\int K(x)\otimes(x\otimes x)\,\Phi(|x|)dx =$$

$$= \frac{1}{2l^2}\int_0^\infty r\Phi(r)dr\int_{\Omega_1} K(n)\otimes(n\otimes n)\,d\Omega_n, \tag{19.41}$$

where Ω_1 is the surface of unit sphere in the x-space. For an isotropic conductive host medium, the function $K(x)$ is

$$K_{\alpha\beta}(x) = \frac{1}{4\pi c_0 r^3}\left(\delta_{\alpha\beta} - 3n_\alpha n_\beta\right),\, r = |x|,\; n_\alpha = \frac{x_\alpha}{r}, \tag{19.42}$$

and the tensors A^0, $A^{(1)}$, $A^{(2)}$ in Eq. (19.39) take forms

$$A^0 = \frac{1}{3c_0}I,\; A^{(1)} = 0,\, A^{(2)} = \frac{1}{15c_0}\left(E^2 - 3E^1\right). \tag{19.43}$$

From these equations follows the approximation of the symbol $C^*(k)$ in Eq. (19.30) with accuracy to $|kl|^3$

$$C^*\,(k) = C^{*0} - p^2 l^2 P A^{(2)} P\cdot(\mathrm{i}k\otimes \mathrm{i}k)\,, \tag{19.44}$$

$$C^{*0} = C^0 + pP\left(I + pA^0 P\right). \tag{19.45}$$

Because product with the vector $(-\mathrm{i}k)$ in the Fourier transform space corresponds to gradient operator ∇ in the x-space, we obtain the following approximation for the constitutive relation (19.35)

$$\langle\sigma(x)\rangle = C^{*0}\,\langle\varepsilon(x)\rangle - p^2 l^2 P A^{(2)} P\nabla\nabla\,\langle\varepsilon(x)\rangle\,. \tag{19.46}$$

Using the equilibrium equation for the flux density in the presence of external field sources $q(x)$

$$\operatorname{div}\,\langle\sigma(x)\rangle = -q(x) \tag{19.47}$$

and assuming existence of the potential $u(x)$ such as $\langle \varepsilon(x) \rangle = \nabla u(x)$, the governing equation for $u(x)$ takes the form

$$C^{*0}_{\alpha\beta} \nabla_\alpha \nabla_\beta u(x) - l^2 p^2 P_{\alpha\beta} A^{(2)}_{\beta\gamma\delta\mu} P_{\mu\nu} \nabla_\alpha \nabla_\gamma \nabla_\delta \nabla_\nu u(x) = -q(x) . \tag{19.48}$$

The analogy of this equation for elasticity is the governing equations of the moment theory for a medium with constraint rotations [2, 4]. The correlation radius l of the random set of inclusions plays the role of the parameter with the dimension of length that presents in the moment theories of elasticity.

Note that Eq. (19.48) contains two microstructural parameters only: the inclusion volume fraction p and the correlation radius l of the random set of inclusions. Other structural parameters, e.g., inclusion sizes, are not presented in this equation. This is a consequence of the assumption that the local exciting field $\varepsilon^*(x)$ is constant inside each inclusion. The dependence of the effective operator C^* on the inclusion sizes arises in the version of the EFM, where the local exciting fields inside inclusions are assumed to be nonconstant [5, 6].

19.1.1 Green's Function for the Effective Media with a Nonlocal Constitutive Equation

The governing equation for the potential $u(x)$ that follows from the constitutive equation (19.33) has the form

$$C^{*(0)} \nabla\nabla u(x) + C^{*(1)} \nabla \left(K^{\Phi} \nabla u \right)(x) = -q(x) \tag{19.49}$$

$$C^{*(0)} = C^0 + pP(I + pA^0 P), \; C^{*(1)} = p^2 P \otimes P, \tag{19.50}$$

where K^{Φ} is the integral operator with the kernel $K^{\Phi}(x) = K(x)\Phi(x)$

$$\left(K^{\Phi} \nabla u \right)(x) = vp \int K(x - x')\Phi(x - x')\nabla' u(x')dx. \tag{19.51}$$

Here, $\Phi(x) = 1 - \Psi(x)$, and $\Psi(x)$ is the correlation function of the random set of inhomogeneities in Eq. (19.17).

Green's function $g(x)$ of the operator in Eq. (19.65) is vanishing at infinity solution of Eq. (19.49) with the right-hand side $q(x) = \delta(x)$

$$C^{*(0)}_{\alpha\beta} \nabla_\alpha \nabla_\beta g(x) + C^{*(1)}_{\alpha\beta\gamma\delta} \nabla_\alpha \left(K^{\Phi}_{\beta\gamma} \nabla_\delta g \right)(x) = -\delta(x), \tag{19.52}$$

where $\delta(x)$ is 3D Dirac's delta-function. Further, we consider the structure of function $g(x)$ for the nonlocal constitutive equations (19.35) and (19.46).

For simplicity, we consider an isotropic conductive host medium with spherical homogeneous isotropic inclusions. In this case, the effective medium is isotropic and

$$C^0_{\alpha\beta} = c_0\delta_{\alpha\beta},\ \ C^{*(0)}_{\alpha\beta} = c^{(0)}_*\delta_{\alpha\beta},\ C^{*(1)}_{\alpha\beta\gamma\delta} = c^{(1)}_*\delta_{\alpha\beta}\delta_{\gamma\delta}. \tag{19.53}$$

Here, the coefficients $c^{(0)}_*$ and $c^{(1)}_*$ are reconstructed from Eq. (19.50) taking into account the equation

$$A^0_{\alpha\beta} = \frac{1}{3c_0}\delta_{\alpha\beta},\, P_{\alpha\beta} = \frac{3c_0(c-c_0)}{2c_0+c}\delta_{\alpha\beta}. \tag{19.54}$$

The function $K(x)$ in Eq. (19.51) is

$$K_{\alpha\beta}(x) = \frac{1}{4\pi c_0|x|}\left(\delta_{\alpha\beta} - n_\alpha n_\beta\right),\ n_\alpha = \frac{x_\alpha}{|x|}. \tag{19.55}$$

For homogeneous and isotropic distribution of inclusions in space, $\Phi(x) = \Phi(|x|)$. Application of the Fourier transform operator to Eq. (19.52) results in the equation for the Fourier transform $\widetilde{g}(k)$ of Green's function

$$\widetilde{g}(k) = \frac{1}{k^2(c^{(0)}_* + \overline{c}^{(1)}_*\Gamma(k))},\ \overline{c}^{(1)}_* = \frac{c^{(1)}_*}{c_0}, \tag{19.56}$$

$$\Gamma(k) = c_0 v p\int k_\alpha K_{\alpha\beta}(x)k_\beta\Phi(|x|)e^{\mathrm{i}(k\cdot x)}dx = 2\int_0^\infty j_2(kr)\Phi(r)\frac{dr}{r}. \tag{19.57}$$

Here, $j_2(z)$ is the spherical Bessel function. It follows from this equation that function $\Gamma(k)$ has the following asymptotic estimates at $k = 0$ and for $k \to \infty$

$$\Gamma(k) = O(k^2), k \to 0;\ \Gamma(k) = \frac{2}{3} + O(k^{-2}),\ k \to \infty. \tag{19.58}$$

Then, the function $\widetilde{g}(k)$ can be presented as the sum

$$\widetilde{g}(k) = \widetilde{g}_1(k) + \widetilde{g}_2(k), \tag{19.59}$$

where

$$\widetilde{g}_1(k) = \frac{1}{d_*k^2},\ \widetilde{g}_2(k) = \frac{\overline{c}^{(1)}_*\left(\frac{2}{3} - \Gamma(k)\right)}{d_*k^2(c^{(0)}_* + \overline{c}^{(1)}_*\Gamma(k))},\ d_* = c^{(0)}_* + \frac{2}{3}\overline{c}^{(1)}_*. \tag{19.60}$$

Here, the function $\widetilde{g}_2(k)$ decays as k^{-4} at infinity. The inverse Fourier transform of $\widetilde{g}(k)$ is

$$g(x) = g_1(x) + g_2(x), \tag{19.61}$$

$$g_1(x) = \frac{1}{4\pi d_* |x|}, \quad g_2(x) = \frac{1}{2\pi^2} \int_0^\infty \widetilde{g}_2(k) j_0(k|x|) k^2 dk. \tag{19.62}$$

Function $g_1(x)$ is classical Green's function of a homogeneous isotropic medium with the conductivity d_*, function $g_2(x)$ is finite at $|x| = 0$, and $g_2(x) = O(|x|^{-1})$ if $|x| \to \infty$.

For the effective medium with the governing equation (19.48), the Fourier transform of Green's function $\widetilde{g}(k)$ has the form

$$\widetilde{g}(k) = \frac{1}{k^2(c_*^{(0)} + l^2 c_*^{(2)} k^2)}, \tag{19.63}$$

where the positive coefficient $c_*^{(2)}$ is reconstructed from Eq. (19.48). The inverse Fourier transform of $\widetilde{g}(k)$ is

$$g(x) = \frac{1}{4\pi |x|} \left(1 - \exp\left(-\sqrt{\frac{c_*^{(0)}}{c_*^{(2)}}} \frac{|x|}{l} \right) \right). \tag{19.64}$$

This function is finite at $x = 0$, and $g(x)$ vanishes as $|x|^{-1}$ at infinity.

Note that application of the constitutive equation (19.46) for solution of the boundary value problems for the effective medium encounters difficulties in construction of a unique solution of the governing equation (19.48). For this is a partial differential equation of the fourth order, classical boundary conditions (Dirichlet, Neumann, or mixed ones) are not enough for the determination of all the functions in the general solution of this equation. Formulation of the additional boundary conditions is an open problem.

The governing equation (19.49) is an integro-differential equation that contains partial derivatives of the second order only. Therefore, classical boundary conditions allow constructing unique solutions of the corresponding boundary value problems.

19.2 The Homogenization Problem for Time-Varying Fields in Heterogeneous Conductive Media

In this Section, an infinite heterogeneous thermo-conductive medium subjected to some conditions at infinity and "zero" initial condition is considered. The time-varying temperature fields $T(x, t)$ in the medium satisfy the parabolic differential equation [7]

$$\nabla_\alpha C_{\alpha\beta}(x) \nabla_\beta T(x, t) - \eta(x) \frac{\partial T(x, t)}{\partial t} = 0. \tag{19.65}$$

Here, $C(x)$ is the tensor of thermo-conductivity coefficients, $\eta(x) = \rho(x)c(x)$, $\rho(x)$ is material density, $c(x)$ is specific heat (capacity), x is point of the medium, t is time. If $C(x)$ and $\eta(x)$ are homogeneous in space random functions, the temperature field $\langle T(x,t)\rangle$ averaged over ensemble realizations of these functions satisfies an equation of the same type as the original equation (19.65)

$$\nabla C^* \nabla \langle T(x,t)\rangle - \eta^* \frac{\partial}{\partial t} \langle T(x,t)\rangle = 0. \tag{19.66}$$

Here, C^* and η^* are parameters of an effective medium that is equivalent to the original heterogeneous medium. This statement is known as the hypothesis of equivalent homogeneity [8]. For media with periodic microstructures, this hypothesis is strictly proved in the asymptotic theory of homogenization [9, 10].

In the asymptotic theory of homogenization, a small parameter $\delta = l/L$ is introduced, where l is the period of the property oscillations and L is the characteristic scale of external field variations. The solution of the homogenization problem is looking for a series with respect to δ. It is shown in [10] that the principal terms of this series for heterogeneous media with fast oscillating property parameters $C(x)$ and $\eta(x)$ are as follows. The tensor of effective conductivity C^* coincides with such a tensor of the static problem when the time derivative in Eq. (19.65) is neglected. The effective parameter η_* is the average of the function $\eta(x)$ over ensemble realizations of the random heterogeneous medium

$$\eta^* = \langle \eta(x)\rangle, \tag{19.67}$$

and η^* is also constant. The static conductivity C^{*s} of composites with periodic microstructures have been extensively studied for more than three decades (see, e.g., the review [11] and references therein).

Because heterogeneous media are associated with space and time-dispersions, the coefficients $C^*$and η^* in Eq. (19.66) are not constant but they are nonlocal operators with respect to x and t-variables [1–3, 12]. If the characteristic scale L of changing of the averaged field $\langle T(x,t)\rangle$ is much larger that the characteristic spatial scale h of the microstructure ($\delta = h/L \ll 1$), the operators C^* and η^* can be approximated by constants with respect to x-variables, but they remain to be nonlocal operators with respect to time (memory effects). In the framework of the asymptotic theory of homogenization, an attempt to obtain nonlocal operators C^* for a layered medium was performed in [13, 14].

In contrast with the asymptotic homogenization of periodic composites with "controlled accuracy" parameter δ, for the composites with random microstructures, the homogenization problem can be solved by approximate methods or by numerical simulations. Reviews of approximate methods of solution of the steady-state (static) homogenization problems of thermo-conductivity for composites with random microstructures is presented, e.g., in [15, 16].

In this section, a homogeneous isotropic host medium containing random sets of spherical homogeneous isotropic inclusions is considered. For the solution of the non-stationary homogenization problem, the self-consistent effective medium method (EMM) is used. The steady-state effective conductivity coefficients c_* of the composites with spherical homogeneous inclusion were obtained by this method in [17]. It was shown that the effective conductivity c^* is a positive root of the equation

$$c_* = c_0 + 3p\frac{(c - c_0)c_*}{c + 2c_*}, \tag{19.68}$$

where c and c_0 are the conductivity coefficients of the inclusions and the host medium, p is the volume fraction of the inclusions. The solution of this equation is considered in Sect. 11.2. In this section, the EMM is extended to the case of non-stationary temperature fields to construct approximations for nonlocal effective operators C^* and η^* in Eq. (19.66).

19.2.1 The Governing Equation of the Conductivity Problem in the (x, ω)-Presentation

Laplace transform of Eq. (19.65) with respect to time variable t and "zero" initial condition has the form

$$\nabla_\alpha c(x)\nabla_\alpha T(x, \omega) - \omega\eta(x)T(x, \omega) = 0, \tag{19.69}$$

$$T(x, \omega) = \int_0^\infty T(x, t)e^{-\omega t}dt. \tag{19.70}$$

This equation is called the (x, ω)-presentation of Eq. (19.65). Further, we focus on the composite materials that consist of a homogeneous host medium (matrix) with the thermo-conductivity parameters c_0, η_0 and a homogeneous in space random set of identical spherical inclusions with the parameters c, η. In this case, the functions $c(x)$ and $\eta(x)$ in Eq. (19.69) are presented in the forms

$$c(x) = c_0 + c_1 V(x), \quad c_1 = c - c_0, \tag{19.71}$$

$$\eta(x) = \eta_0 + \eta_1 V(x), \quad \eta_1 = \eta - \eta_0, \tag{19.72}$$

where $V(x)$ is the characteristic function of the region V occupied by the inclusions: $V(x) = 1$, when $x \in V$, $V(x) = 0$, when $x \notin V$. For such a medium, Eq. (19.69) can be written in the form

$$(c_0\nabla_\alpha\nabla_\alpha - \omega\eta_0)T(x, \omega) = -\nabla_\alpha c_1 V(x)\nabla_\alpha T(x, \omega) + \omega\eta_1 V(x)T(x, \omega). \tag{19.73}$$

Application of the inverse operator with respect to the operator $L_0 = c_0\nabla_\alpha\nabla_\alpha - \omega\eta_0$ to this equation results in the integro-differential equation for the field $T(x,\omega)$

$$T(x,\omega) = T_0(x,\omega) + c_1\int \nabla_\alpha g(x-x',\omega)\tau_\alpha(x',\omega)V(x')dx' -$$
$$-\eta_1\omega\int g(x-x',\omega)T(x',\omega)V(x')dx', \quad \tau_\alpha(x,\omega) = \nabla_\alpha T(x,\omega). \tag{19.74}$$

Here, $T_0(x,\omega)$ is the external field that would have existed in the homogeneous host medium without inclusions, and the same conditions at infinity, $g(x,\omega)$ is the Green function of the operator L_0 that is a vanishing at infinity solution of the equation

$$c_0\nabla_\alpha\nabla_\alpha g(x,\omega) - \omega\eta_0 g(x,\omega) = -\delta(x). \tag{19.75}$$

Here, $\delta(x)$ is 3D Dirac's delta-function. The explicit solution of this equation is

$$g(x,\omega) = \frac{e^{-q_0 r}}{4\pi c_0 r}, \; q_0^2 = \omega\frac{\eta_0}{c_0}, \; r = |x|. \tag{19.76}$$

19.2.2 The Effective Medium Method

For homogeneous and isotropic in space random sets of inclusions, the macroscopic properties of the composite medium are also isotropic, $C^*_{\alpha\beta} = c_*\delta_{\alpha\beta}$, and it is reasonable to assume that the average $\langle T(x,t)\rangle$ of the temperature field over ensemble realizations of the random set of inclusions satisfies the equation similar to Eq. (19.66)

$$\nabla_\alpha c_*\nabla_\alpha\langle T(x,t)\rangle - \eta_*\frac{\partial}{\partial t}\langle T(x,t)\rangle = 0. \tag{19.77}$$

Because a heterogeneous medium is associated with space and time-dispersions, c_* and η_* in this equation are nonlocal operators with respect to x and t-variables. If the characteristic scale of changing of the field $\langle T(x,t)\rangle$ is much larger that the characteristic spatial scale h of the microstructure, the operators c_* and η_* can be approximated by constants with respect to x-variables, but they remain to be nonlocal operators with respect to time. It follows from homogeneity with respect to time that these operators are convolutions with time-dependent functions $c_*(t)$ and $\eta_*(t)$

$$(c_*\varphi)(x,t) = \int_0^t c_*(t-t')\varphi(x,t')dt', \tag{19.78}$$

$$(\eta_*\varphi)(x,t) = \int_0^t \eta_*(t-t')\varphi(x,t')dt'. \tag{19.79}$$

In the (x, ω)-presentation, Eq. (19.77) takes the form

$$c_*(\omega)\nabla_\alpha\nabla_\alpha \langle T(x,\omega)\rangle - \omega\eta_*(\omega)\langle T(x,\omega)\rangle = 0. \tag{19.80}$$

Here, $c_*(\omega)$ and $\eta_*(\omega)$ are Laplace transforms of the kernels $c_*(t)$ and $\eta_*(t)$ in Eqs. (19.78), (19.79). The objective of the homogenization problem is construction of the functions $c_*(t)$ and $\eta_*(t)$ for the given parameters of the heterogeneous medium.

Further, the self-consistent effective medium method (EMM) is used for solution of this problem. The method is based on the following hypotheses.

H_1. *Each inclusion in the composite can be considered as an isolated one in the homogeneous medium with effective thermo-conductivity parameters* c_* *and* η_*.

H_2. *The field that acts on each inclusion is the averaged field* $\langle T(x,t)\rangle$ *over the composite.*

These hypotheses reduce the problem of interaction between many inclusions in the composite to the one-particle problem. This is the problem for an isolated inclusion embedded into the infinite homogeneous medium with the effective parameters $c_*(\omega)$ and $\eta_*(\omega)$ and subjected to the external field $T_*(x,\omega) = \langle T(x,\omega)\rangle$. In the absence of heat sources, the field $T_*(x,\omega)$ satisfies the equation

$$c_*(\omega)\nabla_\alpha\nabla_\alpha T_*(x,\omega) - \omega\eta_*(\omega)T_*(x,\omega) = 0. \tag{19.81}$$

Further, $T_*(x,\omega)$ is taken in the form of 1D-field

$$T_*(x,\omega) = \exp\left(-\mathbf{q}^*\cdot\mathbf{x}\right),\ \mathbf{q}^* = q_*\mathbf{m},\ q_* = \sqrt{\omega\frac{\eta_*(\omega)}{c_*(\omega)}}, \tag{19.82}$$

where $\mathbf{m}$ is a unit vector of the external heat flux direction, $\mathbf{m}\cdot\mathbf{x} = m_\alpha x_\alpha$. The integral equation of the one-particle problem is similar to Eq. (19.74)

$$T(x,\omega) = T_*(x,\omega) + c_*^1(\omega)\int_v \nabla_\alpha g_*(x-x',\omega)\tau_\alpha(x',\omega)dx' -$$

$$-\ \omega\eta_*^1(\omega)\int_v g_*(x-x',\omega)T(x',\omega)dx'. \tag{19.83}$$

In this equation, v is the region occupied by the inclusion,

$$c_*^1(\omega) = c - c_*(\omega),\ \eta_*^1(\omega) = \eta - \eta_*(\omega), \tag{19.84}$$

$$\tau_\alpha(x,\omega) = \nabla_\alpha T(x,\omega), \tag{19.85}$$

and $g_*(x,\omega)$ is the Green function of the effective medium. This function coincides with the function $g(x,\omega)$ in Eq. (19.76), where the parameters c_0, η_0 should be changed with the ω-dependent effective parameters $c_*(\omega)$ and $\eta_*(\omega)$.

Let the solution of Eq. (19.83) be known, and the fields $T(x,\omega)$ and $\tau(x,\omega)$ inside the inclusion centered at the point $x = 0$ are presented in the form

$$T(x,\omega) = L(e^{-\mathbf{q}^*\cdot\mathbf{x}}),\ \tau_\alpha(x,\omega) = M_\alpha(e^{-\mathbf{q}^*\cdot\mathbf{x}}),\quad x \in v. \tag{19.86}$$

Here, L and $M_\alpha = \nabla_\alpha L$ are linear operators that are to be found from the solution of the one-particle problem. For an inclusion v_k centered at the point $x^k \neq 0$, the fields $T(x,\omega)$ and $\tau_\alpha(x,\omega)$ inside the inclusion are presented in the form

$$T(x,\omega) = L(e^{-\mathbf{q}^*\cdot\mathbf{x}}) = L(e^{-\mathbf{q}^*\cdot(\mathbf{x}-\mathbf{x}^k)})e^{-\mathbf{q}^*\cdot\mathbf{x}^k} =$$

$$= L(e^{-\mathbf{q}^*\cdot(\mathbf{x}-\mathbf{x}^k)})e^{\mathbf{q}^*\cdot(\mathbf{x}-\mathbf{x}^k)}T_*(x,\omega),\quad x \in v^k; \tag{19.87}$$

$$\tau_i(x,\omega) = \partial_i L(e^{-\mathbf{q}^*\cdot\mathbf{x}}) =$$

$$= \left(\partial_i A(e^{-\mathbf{q}^*\cdot(\mathbf{x}-\mathbf{x}^k)})\right) e^{\mathbf{q}^*\cdot(\mathbf{x}-\mathbf{x}^k)} \left(-\frac{1}{q_*}m_j\right) \partial_j T_*(x,\omega),\ x \in v^k. \tag{19.88}$$

It is taken into account that $e^{\mathbf{q}^*\cdot\mathbf{x}^k}$ is constant and $T^*(x,\omega) = e^{-\mathbf{q}^*\cdot\mathbf{x}}$. These equations can be written in the forms

$$T(x,\omega) = \lambda(x - x^k)T_*(x,\omega), \tag{19.89}$$

$$\tau_\alpha(x,\omega) = \Lambda_{\alpha\beta}(x - x^k)\tau^*_\beta(x,\omega),\ x \in v_k; \tag{19.90}$$

$$\lambda(y) = L(e^{-\mathbf{q}^*\cdot\mathbf{y}})e^{\mathbf{q}^*\cdot\mathbf{y}}, \tag{19.91}$$

$$\Lambda_{\alpha\beta}(y) = -\frac{1}{q_*}\left(\partial_\alpha L(e^{-\mathbf{q}^*\cdot\mathbf{y}})\right) e^{\mathbf{q}^*\cdot\mathbf{y}} m_\beta,\ \tau^*_\alpha(x,\omega) = \nabla_\alpha T_*(x,\omega). \tag{19.92}$$

Here, the functions $\lambda(x)$ and $\Lambda_{\alpha\beta}(x)$ are to be constructed from the solution of the one-particle problem for an inclusion centered at the point $x = 0$.

Let us introduce the stationary random functions $\overline{\lambda}(x)$ and $\overline{\Lambda}_{\alpha\beta}(x)$ in 3D-space. These functions coincide with $\lambda(x - x^k)$ and $\Lambda_{\alpha\beta}(x - x^k)$ inside the inclusion centered at the point x^k ($k = 1, 2, 3, \ldots$), and they are equal to zero in the host medium. Because $T(x,\omega)V(x) = \overline{\lambda}(x)T^*(x,\omega)V(x)$ and $\nabla_\alpha T(x,\omega)V(x) = \overline{\Lambda}_{\alpha\beta}(x)\tau^*_\beta(x,\omega)V(x)$, Eq. (19.83) take the form

$$T(x,\omega) = T_0(x,\omega) + c_1 \int \partial_i g(x-x')\overline{\Lambda}_{\alpha\beta}(x')\tau^*_\beta(x',\omega)V(x')dx' -$$

$$-\omega\eta_1 \int g(x-x')\overline{\lambda}(x')T^*(x',\omega)V(x')dx'. \tag{19.93}$$

After averaging this equation over ensemble realizations of the random set of inclusions and taking into account the equalities

$$T_*(x,\omega) = \langle T(x,\omega)\rangle\,,\ \tau^*_\alpha(x,\omega) = \nabla_\alpha \langle T(x,\omega)\rangle\,, \tag{19.94}$$

we obtain the integro-differential equation for the averaged field $\langle T(x,\omega)\rangle$

$$\langle T(x,\omega)\rangle = T_0(x,\omega) + pc_1 \int \nabla_\alpha g(x-x',\omega)H^c_{\alpha\beta}(\omega)\nabla_\beta \big\langle T(x',\omega)\big\rangle dx' -$$

$$-p\omega\eta_1 \int g(x-x',\omega)h_\eta(\omega)\big\langle T(x',\omega)\big\rangle dx'. \tag{19.95}$$

Here, $p = \langle V(x)\rangle$ is the volume fraction of the inclusions, $H^c_{\alpha\beta}$ and h_η are constant (with respect to coordinates) tensor and scalar that are defined by the equations

$$h_\eta = \frac{1}{p}\lim_{W\to\infty}\frac{1}{W}\int_W \overline{\lambda}(x)V(x)dx = \frac{1}{v}\int_v \lambda(x)dx, \tag{19.96}$$

$$H^c_{\alpha\beta} = \frac{1}{p}\lim_{W\to\infty}\frac{1}{W}\int_W \overline{\Lambda}_{\alpha\beta}(x)V(x)dx = \frac{1}{v}\int_v \Lambda_{\alpha\beta}(x)dx. \tag{19.97}$$

Here, W is the region of the composite that occupies the entire space in the limit $W \to \infty$. It is implied that the ergodic hypothesis serves for the homogeneous in space random functions $\overline{\lambda}(x)V(x)$ and $\overline{\Lambda}_{\alpha\beta}(x)V(x)$, and averaging over ensemble realizations of these functions coincides with spatial averaging over typical realizations of the random set of inclusions in the composite.

After application of the operator $L^0 = c_0\nabla_\alpha\nabla_\alpha - \omega\eta_0$ to Eq. (19.95) and taking into account that

$$L^0 T_0(x,\omega) = 0,\ L^0 g(x,\omega) = -\delta(x), \tag{19.98}$$

we obtain the equation of for the field $\langle T(x,\omega)\rangle$ in the form

$$\big(\nabla_\alpha C^*_{\alpha\beta}(\omega)\nabla_\beta - \omega\eta_*(\omega)\big)\langle T(x,\omega)\rangle = 0. \tag{19.99}$$

Here, the effective parameters $C^*_{\alpha\beta}(\omega)$ and $\eta_*(\omega)$ are defined by the equations

$$C^*_{\alpha\beta}(\omega) = c_0\delta_{\alpha\beta} + pc_1 H^c_{\alpha\beta}(\omega),\quad \eta_*(\omega) = \eta_0 + p\eta_1 h_\eta(\omega). \tag{19.100}$$

Because $\langle T(x,\omega)\rangle = e^{-q_*\mathbf{m}\cdot\mathbf{x}}$, the functions $C^*(\omega)$ and $\eta_*(\omega)$ relate by the equation

$$q_*^2 m_\alpha C^*_{\alpha\beta}(\omega) m_\beta - \omega\eta_*(\omega) = 0. \tag{19.101}$$

19.2.3 Solution of the One-Particle Problem

For solution of the one-particle problem of the EMM, we consider a spherical inclusion of the radius a in a homogeneous medium with parameters c_*, η_* subjected to the external field

$$T_*(x,\omega) = e^{-q_*\mathbf{m}\cdot\mathbf{x}}. \tag{19.102}$$

Let (r,φ,θ) be a spherical coordinate system with the origin at the inclusion center and the polar axis directed along the unit vector $\mathbf{m}$. In this system, the field $T_*(x,\omega)$ is presented in the form [18]

$$T_*(x,\omega) = e^{-q_* r\cos\theta} = \sum_{n=0}^{\infty}(-1)^n(2n+1)i_n(q_* r)P_n(\cos\theta). \tag{19.103}$$

$$i_n(z) = \sqrt{\frac{\pi}{2z}}I_{n+1/2}(z), \tag{19.104}$$

$I_n(z)$ is the modified Bessel function, $P_n(z)$ is the Legendre polynomials.

Integral equation (19.95) is equivalent to the following system of the differential equations

$$\nabla_\alpha\nabla_\alpha T(x,\omega) - q_*^2 T(x,\omega) = 0, q_*^2 = \omega\frac{\eta_*(\omega)}{c_*(\omega)}, r > a, \tag{19.105}$$

$$\nabla_\alpha\nabla_\alpha T(x,\omega) - q^2 T(x,\omega) = 0,\ q^2 = \omega\frac{\eta}{c}, r < a \tag{19.106}$$

with the boundary conditions on the inclusion surface

$$T^+(x,\omega)\Big|_{|x|=a} = T^-(x,\omega)\Big|_{|x|=a}, \tag{19.107}$$

$$c^*\left.\frac{\partial T^+(x,\omega)}{\partial n}\right|_{|x|=a} = c\left.\frac{\partial T^-(x,\omega)}{\partial n}\right|_{|x|=a}, \tag{19.108}$$

where $T^+(x,\omega)$ and $T^-(x,\omega)$ for $|x| = a$ are the limit values of the function $T(x,\omega)$ on the inclusion boundary Ω from the side of the external normal n to Ω and from the opposed side.

We introduce dimensionless coordinates $\varsigma_i = x_i/a$ $(i = 1, 2, 3)$ and change the derivatives with respect to x_i-variable with the derivatives with respect to ς_i. In the coordinates ς_i, $|\varsigma| = 1$ is the equation for the inclusion boundary, and the parameters q_* and q are changed with the dimensionless parameters q_*a and qa. Further, the multiplier a is omitted, and it is implied that q and q_* are the dimensionless parameters.

The solution of the system (19.105) and (19.108) can be found in the form of the series that automatically satisfy Eqs. (19.105) and (19.106) [18]

$$T(\varsigma, \omega) = \sum_{n=0}^{\infty} a_n i_n(q\varsigma) P_n(\cos\theta), \qquad |\varsigma| < 1 \tag{19.109}$$

$$T(\varsigma, \omega) = \sum_{n=0}^{\infty} (-1)^n (2n+1) i_n(q_*\varsigma) P_n(\cos\theta) +$$

$$+ \sum_{n=0}^{\infty} b_n k_n(q_*\varsigma) P_n(\cos\theta), \qquad |\varsigma| > 1 \tag{19.110}$$

In the last equation, $k_n(z)$ is the modified spherical Bessel function of the third kind

$$k_n(z) = \sqrt{\frac{\pi}{2z}} K_{n+1/2}(z), \tag{19.111}$$

$$K_{n+1/2}(z) = \frac{\pi}{2}(-1)^{n+1}\left(I_{n+1/2}(z) - I_{-n-1/2}(z)\right). \tag{19.112}$$

In Eqs. (19.109) and (19.110), a_n and b_n are functions of ω that have to be found from the boundary conditions (19.107) and (19.108) at $|\varsigma| = 1$. These conditions result in the following system of equations for the coefficients a_n and b_n

$$a_n i_n(q) - b_n k_n(q^*) = (-1)^n (2n+1) i_n(q^*), \tag{19.113}$$

$$c a_n q i_n'(q) - c_* b_n q_* k_n'(q^*) = (-1)^n (2n+1) c_* q_* i_n'(q^*), \tag{19.114}$$

where $i_n'(z) = di_n(z)/dz$, $k_n'(z) = dk_n(z)/dz$. After solution of this system, we obtain for the coefficients a_n that determine the temperature field inside the inclusion

$$a_n = \frac{\pi c_*}{2q^* \Delta(q, q_*)}(-1)^{n+1}(2n+1), n = 0, 1, 2, ..., \tag{19.115}$$

$$\Delta(q, q_*) = c_* q_* i_n(q) k_n'(q^*) - c q i_n'(q) k_n(q_*). \tag{19.116}$$

The expressions for the function $h_\eta(\omega) = h_\eta(q, q_*)$ in Eq. (19.96) follows from Eq. (19.109) in the form

$$h_\eta(q,q^*) = \frac{1}{v}\int\limits_v T(x,\omega)e^{\mathbf{q}^*\cdot\mathbf{x}}dx =$$

$$= \frac{3}{2}\sum_{n=0}^{\infty} a_n \int\limits_0^1 i_n(q\varsigma)\varsigma^2 d\varsigma \int\limits_0^{\pi} P_n(\cos\theta)e^{q_*\varsigma\cos\theta}\sin\theta d\theta. \tag{19.117}$$

Since

$$\int\limits_0^{\pi} P_n(\cos\theta)e^{q^*\varsigma\cos\theta}\sin\theta d\theta = 2i_n(q_*\varsigma), \tag{19.118}$$

after integrating with respect to ς, we obtain

$$h_\eta(q,q_*) = \sum_{n=0}^{\infty} a_n(q,q_*)g_n(q,q_*), \tag{19.119}$$

where

$$g_n(q,q_*) = \frac{3}{q^2 - {q_*}^2}\left(qi_n(q_*)i_{n+1}(q) - q_*i_n(q)i_{n+1}(q_*)\right). \tag{19.120}$$

For isotropic in space random set of inclusions, the tensor H_{ij}^c in Eq. (19.97) is isotropic

$$H_{\alpha\beta}^c = h_c\delta_{\alpha\beta}, \tag{19.121}$$

and from the definition H^c, we have

$$H_{\alpha\beta}^c\nabla_\beta T_*(x,\omega) = -h_c q_* m_i e^{-\mathbf{q}^*\cdot\mathbf{x}} = \frac{1}{v}\int\limits_v \partial_i T(x,\omega)e^{\mathbf{q}^*\cdot\mathbf{x}}dx e^{-\mathbf{q}^*\cdot\mathbf{x}} =$$

$$= \frac{1}{v}\left(\int\limits_S T(x,\omega)e^{q_* a\cos\theta}n_i dS - q_* m_i\int\limits_v T(x,\omega)e^{\mathbf{q}^*\cdot\mathbf{x}}dx\right)e^{-\mathbf{q}^*\cdot\mathbf{x}} =$$

$$= m_i\sum_{n=0}^{\infty} a_n\left(\frac{3}{q^*}i_n(q)i_n'(q_*) - g_n(q,q_*)\right)e^{-\mathbf{q}^*\cdot\mathbf{x}} \tag{19.122}$$

Here, S is the inclusion surface with the external normal $\mathbf{n}$, and the equation

$$\int\limits_0^{\pi} P_n(\cos\theta)e^{q^*\cos\theta}\cos\theta\sin\theta d\theta = 2i_n'(q^*), \tag{19.123}$$

is taken into account. The expression for the function $H_c(\omega)$ follows from Eq. (19.122) in the form

$$H_c(\omega) = H_c(q, q^*) = \sum_{n=0}^{\infty} a_n(q, q^*) g_n^1(q, q^*), \tag{19.124}$$

$$g_n^1(q, q_*) = -\frac{3}{q^*} i_n(q) i_n'(q_*) + g_n(q, q_*). \tag{19.125}$$

The system of equations for effective parameters $C^*(\omega)$ and $\eta_*(\omega)$ of the composite follows from Eqs. (19.100), (19.101) and takes the form

$$C^*_{\alpha\beta}(\omega) = c_*(\omega)\delta_{\alpha\beta}, \ c_*(\omega) = c_0 + pc_1 h_c(q, q_*), \tag{19.126}$$

$$\eta_*(\omega) = \eta_0 + p\eta_1 h_\eta(q, q_*), \tag{19.127}$$

$$c_*(\omega)\,(q_*)^2 - \omega\eta_*(\omega) = 0. \tag{19.128}$$

For each fixed Laplace parameter ω, the pair of unknowns (q_*, c_*) or (q_*, η_*) can be found from this system.

19.2.4 Asymptotic Expressions of Functions $c_*(\omega)$ and $\eta_*(\omega)$ for Small and Large Values of ω

Taylor series of the modified Bessel function $I_\nu(z)$ at point $z = 0$ has the form [18]

$$I_\nu(z) = \sum_{k=0}^{\infty} \frac{1}{k!\Gamma(\nu + k + 1)} \left(\frac{z}{2}\right)^{2k}. \tag{19.129}$$

As a result, for small values of ω, q and q_*, the principal terms in the series (19.119) for $h_\eta(q, q^*)$ follow from Eqs. (19.115) and (19.120) in the forms

$$a_0 g_0 = 1 + \frac{3}{5} q_*^2 - \frac{c_* - 5c}{15c_*} q^2 + O(\omega^3), \tag{19.130}$$

$$a_1 g_1 = -\frac{3c_*}{5(c + c_*)} q_*^2 + O(\omega^4), \tag{19.131}$$

$$a_n g_n = O(\omega^{n+1}), n = 2, 3, ... \tag{19.132}$$

The principal terms in the series (19.124) for $h_c(q, q_*)$ for small ω, q, and q_* are

$$a_0 g_0^1 = \frac{q^2}{15} + O(\omega^3), \ a_1 g_1 = -\frac{3c_*}{c + 2c_*} + O(\omega^2), \tag{19.133}$$

$$a_n g_n = O(\omega^{n+1}), n = 2, 3, ... \tag{19.134}$$

As a result, we obtain the equations for the values of the function $c_*(\omega)$ and $\eta_*(\omega)$ at $\omega = 0$

$$c_*(0) = c_0 + 3p\frac{c_1 c_*(0)}{c + c_*(0)}, \ \eta_*(0) = \eta_0 + p\eta_1. \tag{19.135}$$

Here, the first equation coincides with Bruggeman's formula (19.68) for the static effective conductivity of the composite $c_*^s = c_*(0)$, the second equation coincides with Eq. (19.67) (the rule of mixture), and we call it the static parameter $\eta_*^s = \eta_*(0)$.

For deriving the asymptotics of the function $c^*(\omega)$ and $\eta^*(\omega)$ when $\omega \to \infty$, the following presentations of the functions $i_n(z)$ and $k_n(z)$ can be used [18]

$$i_n(z) = \frac{1}{2z}\left(e^z \sum_{k=0}^{n} \frac{(-1)^k (n+k)!}{k!(n-k)!(2z)^k} + \right.$$

$$\left. +(-1)^{n+1} e^{-z} \sum_{k=0}^{n} \frac{(n+k)!}{k!(n-k)!(2z)^k}\right), \tag{19.136}$$

$$k_n(z) = \frac{\pi}{2z} e^{-z} \sum_{k=0}^{n} \frac{(n+k)!}{k!(n-k)!(2z)^k}. \tag{19.137}$$

Note that the principal terms of these series for large z do not depend on the index n

$$i_n(z) = \frac{e^z}{2z} + O\left(\frac{e^z}{z^2}\right), \ k_n(z) = \frac{\pi}{2z} e^{-z} + O\left(\frac{e^{-z}}{z^2}\right). \tag{19.138}$$

As a result, the principal terms of the coefficients $a_n(q, q_*)$ in Eqs. (19.115) for large ω take forms

$$a_n = \frac{2c_*(\omega) q}{cq + c_*(\omega) q^*} e^{-(q - q_*)} \left((-1)^n (1 + 2n) + O(\omega^{-2})\right), \tag{19.139}$$

and the temperature field inside the inclusion has the following asymptotic expression for large ω

$$T(\varsigma, \theta) \sim \frac{2c_*(\omega) q}{cq + c_*(\omega) q_*} e^{-(q - q_*)} \sum_{n=0}^{\infty} (-1)^n (1 + 2n) i_n(q\varsigma) P_n(\cos\theta) =$$

$$= \frac{2c_*(\omega) q}{cq + c_*(\omega) q^*} e^{-(q - q_*)} e^{-q\varsigma\cos\theta}, \ |\varsigma| < 1. \tag{19.140}$$

Finally, the principal terms of asymptotic expressions of the fields $T(\varsigma, \omega)$ and $\nabla T(\varsigma, \omega)$ for large ω are presented in the forms

$$T(\varsigma, \omega) \sim \frac{2c_*(\omega)q}{cq + c_*(\omega)q^*} e^{-(q-q^*)} e^{-q\mathbf{m}\cdot\varsigma}, \tag{19.141}$$

$$\nabla T(\varsigma, \omega) \sim -\frac{2c^*(\omega)q^2}{cq + c_*(\omega)q_*} e^{-(q-q_*)} e^{-q\mathbf{m}\cdot\varsigma} \mathbf{m}. \tag{19.142}$$

Then, the asymptotic expressions of the functions $h_\eta(q, q_*)$ and $H_c(q, q_*)$ can be assessed as follows

$$h_\eta(q, q^*) = \frac{1}{v} \int\limits_v T(\varsigma, \omega) e^{q^*\mathbf{m}\cdot\varsigma} d\varsigma \sim \frac{2c_*(\omega)q}{cq + c_*(\omega)q_*} e^{-(q-q_*)} F(q, q_*), \tag{19.143}$$

$$h_c(q, q_*) = \frac{1}{v} m_\alpha \int\limits_v \nabla_\alpha T(\varsigma, \omega) \frac{e^{q_*\mathbf{m}\cdot\varsigma}}{q_*} d\varsigma \sim$$

$$\sim -\frac{2c_*(\omega)q^2}{(cq + c_*(\omega)q_*)q_*} e^{-(q-q_*)} F(q, q_*), \tag{19.144}$$

$$F(q, q_*) = \frac{1}{v} \int\limits_v e^{-(q-q_*)\mathbf{m}\cdot\varsigma} d\varsigma =$$

$$= \frac{3}{(q - q_*)^3} \left((q - q_*) \cosh(q - q_*) - \sinh(q - q_*) \right). \tag{19.145}$$

We consider two situations.

1. $q > q_*$

In this case, the asymptotic expressions of $h_\eta(q, q_*)$ and $h_c(q, q_*)$ are simplified and they take the forms

$$h_\eta(q, q_*) \sim \frac{3c_*(\omega)q}{(q - q_*)^2 \, (cq + c_*(\omega)q_*)} = O\left(\frac{1}{\omega}\right), \tag{19.146}$$

$$h_c(q, q_*) \sim -\frac{3c_*(\omega)q^2}{(q - q_*)^2 (cq + c_*(\omega)q_*)q_*} = O\left(\frac{1}{\omega}\right). \tag{19.147}$$

Thus the functions $h_c(q, q_*)$ and $h_\eta(q, q_*)$ tend to zero when $\omega \to \infty$, and it follows from Eqs. (19.146), (19.147) that the asymptotic expressions of the functions $\eta_*(\omega)$ and $c_*(\omega)$ at infinity are

$$\eta_*(\omega) = \eta_0 + O\left(\frac{1}{\omega}\right), \quad c_*(\omega) = c_0 + O\left(\frac{1}{\omega}\right), \tag{19.148}$$

where c_0, η_0 are the conductivity parameters of the host medium.

2. $q < q_*$
In this case, the principal terms of the asymptotic expressions of the functions $h_\eta(q, q_*)$ and $H_c(q, q_*)$ at infinity are

$$h_\eta(q, q_*) \sim \frac{3c_*(\omega)q}{(cq + c_*(\omega)q^*)} \frac{e^{-2(q-q_*)}}{(q - q_*)^2}, \tag{19.149}$$

$$h_c(q, q^*) \sim -\frac{3c_*(\omega)q^2}{(cq + c_*(\omega)q_*)q_*} \frac{e^{-2(q-q_*)}}{(q - q_*)^2}. \tag{19.150}$$

Because $q - q_* < 0$, the both functions tend to infinity together with ω except the situation when $q_* \to q$ in the limit $\omega \to \infty$. Further, we consider equations for $\eta_*(\omega)$ and $c_*(\omega)$ in detail for two particular cases.

Let the coefficients of termo-conductivity of the inclusion and the host medium are the same $c = c_0 = 1$, and only the parameters η_0 and η of the phases are different. In this case, $c_1 = 0$, $c_* = c_0$, $q = \sqrt{\omega\eta}$, $q_* = \sqrt{\omega\eta_*}$, and the equation for the function $\eta_*(\omega)$ for large ω take the form ($\eta_*(\omega) > \eta$)

$$\eta_*(\omega) \sim \eta_0 + p\eta_1 \frac{3q \exp\left(2\sqrt{\omega}\left(\sqrt{\eta_*(\omega)} - \sqrt{\eta}\right)\right)}{(q + q_*)\left(\sqrt{\eta_*(\omega)} - \sqrt{\eta}\right)^2 \omega}. \tag{19.151}$$

This equation has a finite solution when $\omega \to \infty$ if the following equation holds

$$\eta_*(\omega) = \eta + O\left(\frac{1}{\sqrt{\omega}}\right). \tag{19.152}$$

In the case of equal η-parameters of the host medium and the inclusions ($\eta_0 = \eta = \eta_* = 1$) and different thermo-conductivity coefficients ($c_0 \neq c$, $q = \sqrt{\omega/c}$, $q_* = \sqrt{\omega/c_*}$), the asymptotic expression of effective conductivity $c_*(\omega)$ for large ω is defined from the equations ($c_*(\omega) < c$)

$$c^*(\omega) \sim c_0 -$$

$$- pc_1 \frac{3c_*(\omega)q^2 \exp\left(2\sqrt{\omega}\left(\sqrt{1/c_*(\omega)} - \sqrt{1/c}\right)\right)}{q_*(cq + c_*(\omega)q_*)\left(\sqrt{1/c_*(\omega)} - \sqrt{1/c}\right)^2 \omega}, \tag{19.153}$$

For large ω, this equation has a finite solution if the following equation holds

$$c_* (\omega) = c + O\left(\frac{1}{\sqrt{\omega}}\right). \tag{19.154}$$

In the general case when $c \neq c_0$, $\eta \neq \eta_0$, and $q < q_*$, the solutions of the system (19.100)–(19.101) for large ω are assessed as follows

$$c_* (\omega) = c + O\left(\frac{1}{\sqrt{\omega}}\right), \quad \eta_* (\omega) = \eta + O\left(\frac{1}{\sqrt{\omega}}\right). \tag{19.155}$$

Here, c, η are the thermo-conductivity parameters of the inclusions.

19.2.5 Deviation of the Parameters η in the Matrix and Inclusion Phases

Further, the dimensions of coefficients of thermo-conductivity are taken as $cal/(cm \cdot sec \cdot C^0)$ (calorie per centimeter, second, degree Celsius), the dimension of the parameter η is $cal/(cm^3 \cdot C^o)$. In these units, the parameters c and η of most metals have the order of 1 [19].

First, we consider the case of the composites with different η-parameters ($\eta \neq \eta_0$) and the same thermo-conductivity coefficients ($c = c_0 = 1$) of the inclusion and matrix phases. Let $\eta_0 = 1$, $\eta = 0.1$, and the volume fraction of the inclusions be $p = 0.2$. In this case, the effective static parameters of the composite are defined in Eqs. (19.135) and have the values

$$c_*^s = 1, \ \eta_*^s = 0.82. \tag{19.156}$$

The system (19.100), (19.101) is reduced to the equation for the effective parameter $q_* = q_*(\omega)$

$$q_*^2 = \omega\left(\eta_0 + p\eta_1 h_\eta(q, q_*)\right). \tag{19.157}$$

For the indicated values of η and η_0, $q_* > q$, and thus $\eta_*(\omega) \to \eta$ if $\omega \to \infty$. The function $\eta_*(\omega)$ can be presented in the form

$$\eta_*(\omega) = \eta + (\eta_*^s - \eta) f_\eta(\omega), \tag{19.158}$$

where $f_\eta(\omega)|_{\omega=0} = 1$, $f_\eta(\omega) = O\left(\omega^{-1/2}\right)$ when $\omega \to \infty$. The graphs of the function $f_\eta(\omega)$ obtained by numerical solution of Eq. (19.157) is shown in Fig. 19.1 (solid line) for $0 < \omega < 1000$. The dashed line in this figure is approximation of $f_\eta(\omega)$ by the function

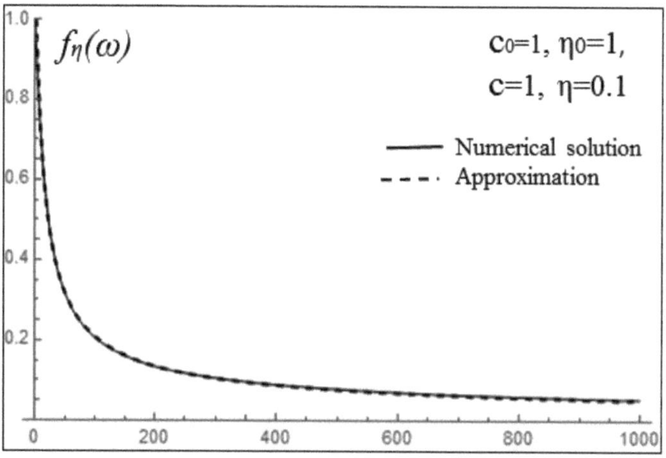

Fig. 19.1 The function $f_\eta(\omega)$ in Eq. (19.155) for the Laplace transform of the kernel of the operator η_*; solid line is the numerical solution of Eq. (19.154), dashed line is the approximation $\widetilde{f}_\eta(\omega)$ in Eq. (19.157)

$$\widetilde{f}_\eta(\omega) = \frac{1 + a\sqrt{\omega}}{1 + b\omega}, \; a = 0.12573, \; b = 0.09823, \tag{19.159}$$

and this function approximates the numerical solution with sufficient accuracy.

Inverse Laplace transform $f_\eta(t)$ of this function has the form

$$f_\eta(t) = \frac{1}{b}\left(e^{-t/b} + \frac{a}{\sqrt{\pi t}} - \frac{a}{\sqrt{b}} e^{-t/b} \operatorname{erfi}\left(\sqrt{\frac{t}{b}}\right)\right). \tag{19.160}$$

Here, $\operatorname{erfi}(z) = -i \operatorname{erf}(iz)$, and $\operatorname{erf}(z)$ is the error function. The asymptotic expressions of $f_\eta(t)$ at $t = 0$ and at infinity are

$$f_\eta(t) = O\left(\frac{1}{\sqrt{t}}\right), t \to 0; \;\; f_\eta(t) = O\left(\frac{1}{t^{3/2}}\right), \; t \to \infty. \tag{19.161}$$

The graph of the function $f_\eta(t)$ is shown in Fig. 19.2.

Thus symbol η^*in Eq. (19.79) is an integral operator that acts on an arbitrary time-depending function $\varphi(t)$ by the equation

$$\left(\eta_* f\right)(t) = \int_0^t \eta_*(t - t')\varphi(t')dt', \tag{19.162}$$

where

$$\eta_*(t) = \eta\delta(t) + (\eta_*^s - \eta) f_\eta(t) \tag{19.163}$$

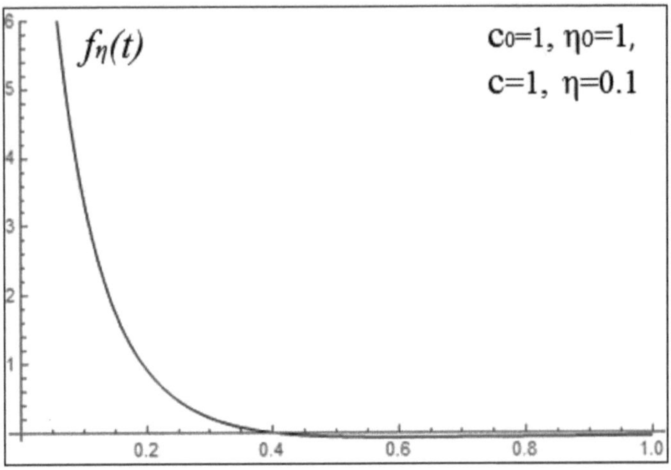

Fig. 19.2 Function $C(r)$ in Eq. (19.158)

and function $f_\eta(t)$ is defined in Eq. (19.160).

Let the parameters of the host medium and the inclusions be $\eta_0 = 1, c_0 = 1$ and $\eta = 10, c = 1, p = 0.2$. In this case, the static effective parameters of the composite are

$$c_*^s = 1, \ \eta_*^s = 2.8. \tag{19.164}$$

For the indicated values of η_0 and η, $q_* < q$, it follows from the previous analysis that $\eta_*(\omega) \to \eta_0$ when $\omega \to \infty$. The function $\eta^*(\omega)$ can be presented in the form

$$\eta_*(\omega) = \eta_0 + (\eta_*^s - \eta_0) f_\eta(\omega), \tag{19.165}$$

where $f_\eta(0) = 1, \ f_\eta(\omega) = O\left(\omega^{-1}\right)$ when $\omega \to \infty$. The graph of this function obtained by numerical solution of Eq. (19.157) is shown in Fig. 19.3 (solid line) for $0 < \omega < 200$. The dashed line in this figure is the function

$$\widetilde{f}_\eta(\omega) = \left(1 + 1.6111\omega^{1/2} + 0.3027\omega\right)^{-1}. \tag{19.166}$$

that approximates the numerical solution with sufficient accuracy. The inverse Laplace transform $f_\eta(t)$ of this function was calculated using Post's equation with $k = 12$ [20]

$$f_\eta(t) \simeq f_\eta(t, k) = \frac{(-1)^k}{k!} \omega^{(k+1)} \left. \frac{d^k \widetilde{f}_\eta(\omega)}{d\omega^k} \right|_{\omega = k/t}. \tag{19.167}$$

The graph of this function is shown in Fig. 19.4. For any k, this function is finite at $t = 0$ ($f_\eta(t)|_{t=0} = 3.304$), and it tends to zero at infinity as $t^{-3/2}$.

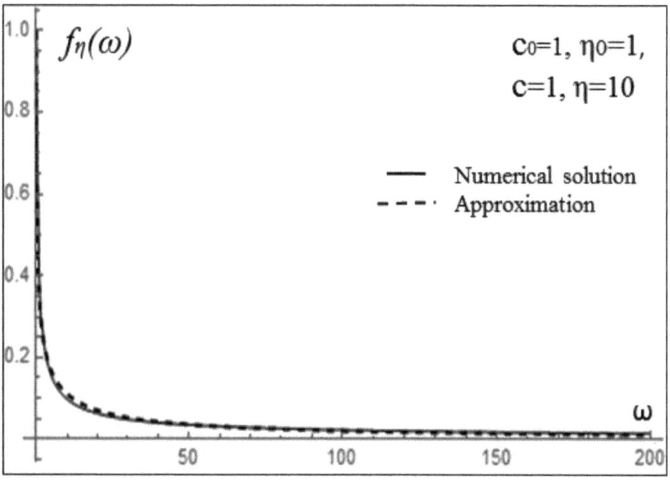

Fig. 19.3 The function $f_\eta(\omega)$ in Eq. (19.163) for the Laplace transform of the kernel of the operator $\mathbf{L}_1$; the solid line is the numerical solution of Eq. (19.155), dashed line is the approximation $\widetilde{f}_\eta(\omega)$ in Eq. (19.164)

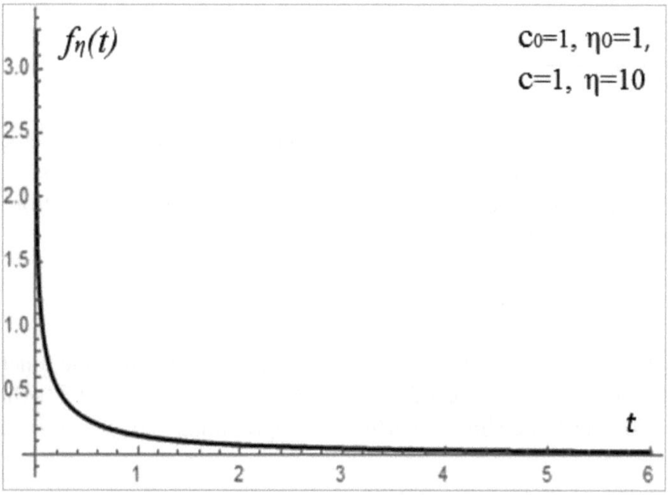

Fig. 19.4 Inverse Laplace transform of the function $\widetilde{f}_\eta(\omega)$ in Eq. (19.163)

19.2.6 Deviation of Thermo-Conductivity Coefficients c in the Matrix and Inclusion Phases

Let the η-parameters of the phases of the composite be the same ($\eta = \eta_0 = 1, \eta_* = 1$) but the coefficients c and c_0 be different. In this case, the system (19.126)–(19.128) is reduced to one equation for the effective parameter $q^* = q^*(\omega)$

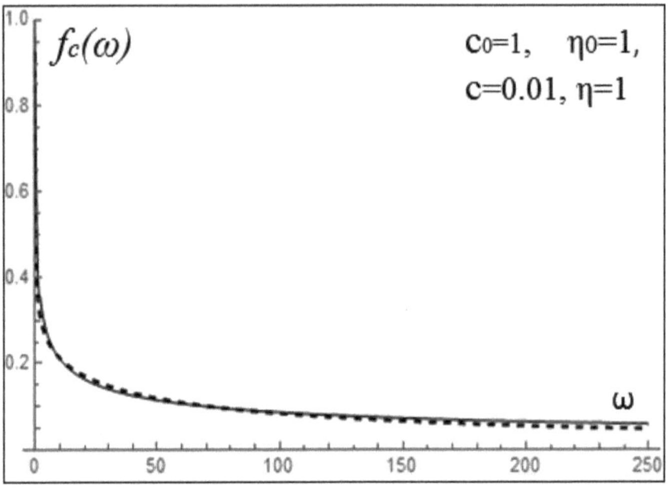

Fig. 19.5 The function $f_c(\omega)$ in Eq. (19.167) for the Laplace transform of the kernel of the operator c_*; solid line is the numerical solution of Eq. (19.166), dashed line is the approximation $\widetilde{f}_\eta(\omega)$ in Eq. (19.168)

$$q_*^2 = \frac{\omega}{c_0 + pc_1 h_c(q, q_*)}. \tag{19.168}$$

First, we consider the composite with the thermo-conductivity coefficients of the phases $c_0 = 1, c = 0.01,\ p = 0.2$. For these parameters, $c_*^s = 0.7051,\ \eta_*^s = 1$. In this case $q > q_*$, and therefore, $c_*(\omega) \to c_0$ when $\omega \to \infty$. The function $c_*(\omega)$ can be presented in the form

$$c_*(\omega) = c_0 + (c_*^s - c_0) f_c(\omega), \tag{19.169}$$

where $f_c(0) = 1,\ f_c(\omega) = O(\omega^{-1})$ when $\omega \to \infty$. The graphs of the function $f_c(\omega)$ obtained by numerical solution of Eq. (19.168) is shown in Fig. 19.5 (solid line) for $0 < \omega < 250$. The dashed line in this figure is approximation of $f_\eta(\omega)$ by the function

$$\widetilde{f}_c(\omega) = \left(1 + 1.7779\omega^{1/4} + 0.0549\omega\right)^{-1}. \tag{19.170}$$

The inverse Laplace transform $f_c(t)$ of this function is shown in Fig. 19.6.

For the values of thermo-conductivity coefficients of the phases $c_0 = 1, c = 100,$ $p = 0.2,\ (\eta = \eta_0 = 1)$, the static parameters are $c_*^s = 2.3134,\ \eta_*^s = 1$. In this case $q < q_*$, and $c_*(\omega) = c + O(\omega^{-1/2})$ when $\omega \to \infty$. The function $c_*(\omega)$ can be presented in the form

$$c^*(\omega) = c + (c_*^s - c) f_c(\omega), \tag{19.171}$$

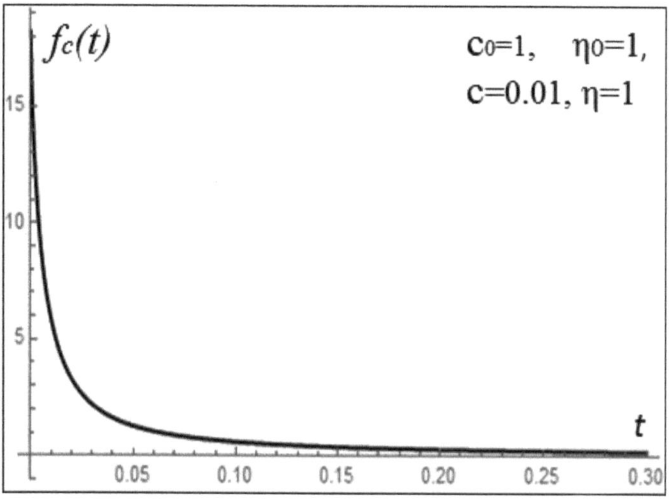

Fig. 19.6 Inverse Laplace transform of the function $f_c(\omega)$ in Eq. (19.168)

were $f_c(0) = 0$ and $f_c(\omega) = O(\omega^{-1/2})$ when $\omega \to \infty$. The graphs of the function $f_c(\omega)$ obtained by numerical solution of Eq. (19.168) is shown in Fig. 19.7 (solid line) for $0 < \omega < 1000$. The dashed line in this figure is an approximation of $f_\eta(\omega)$ by the function

$$\widetilde{f_c}(\omega) = \left(1 + 0.0101\omega^{1/8} + 0.0409\omega^{1/4} + 0.0292\omega^{1/2}\right)^{-1}. \tag{19.172}$$

The inverse Laplace transform of this function $f_c(t)$ is shown in Fig. 19.8.

19.2.7 Composites with Different Thermo-Conductive Parameters of the Phases

We consider the composite material with copper host medium ($c_0 = 0.93$, $\eta_0 = 0.8158$) containing glass inclusions with $c = 0.0028$ and $\eta = 0.44$, $p = 0.2$. The static effective parameters of the composite are

$$c_*^s = 0.6524, \eta_*^s = 0.7408. \tag{19.173}$$

In this case, $q > q_*$, and the Laplace transforms of the kernels of the operators c_* and η_* of the composite are presented in the forms

$$c_*(\omega) = c_0 + (c_*^s - c_0) f_c(\omega), \tag{19.174}$$

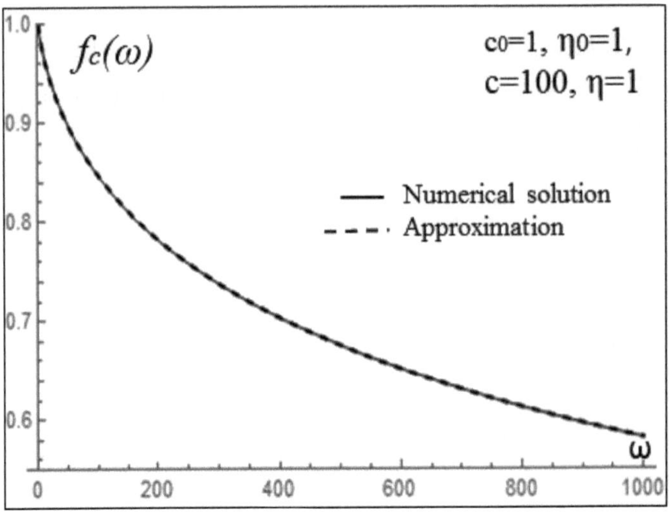

Fig. 19.7 The function $f_c(\omega)$ in Eq. (19.169) for the Laplace transform of the kernel of the operator c_*; the solid line is the numerical solution of Eq. (19.166), dashed line is the approximation $\widetilde{f_c}(\omega)$ in Eq. (19.170)

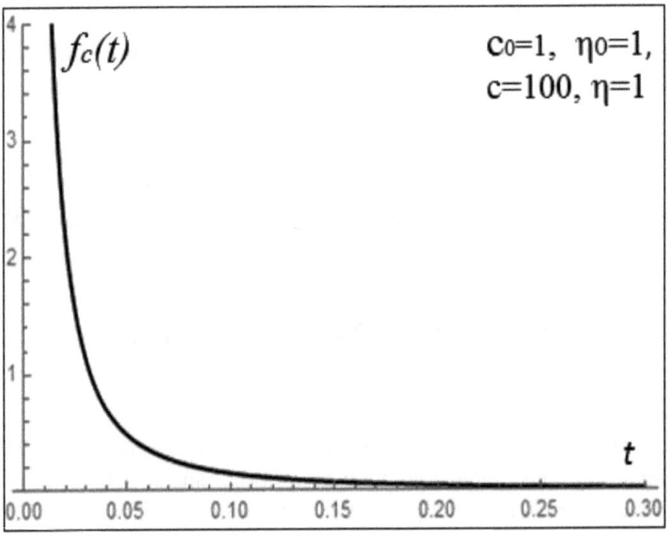

Fig. 19.8 The inverse Laplace transform of the function $\widetilde{f_c}(\omega)$ in Eq. (19.170)

$$\eta_*(\omega) = \eta_0 + (\eta_*^s - \eta_0) f_\eta(\omega). \tag{19.175}$$

Here, the functions $f_c(\omega)$ and $f_\eta(\omega)$ are equal to 1 for $\omega = 0$, and they tend to zero as ω^{-1} when $\omega \to \infty$. These functions are obtained by numerical solution of the

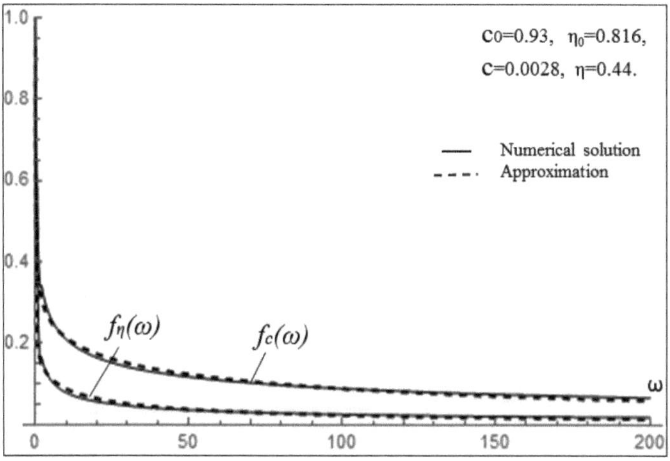

Fig. 19.9 The functions $\overline{K}_1(\mathbf{q})$ and $\overline{K}_2(\mathbf{q})$ in Eqs. (19.82), (19.83) for the Laplace transform of the kernels of the effective operators of the medium with spherical inclusions

system (19.126)–(19.128) and shown in Fig. 19.9 by solid lines. The dashed lines in this figure are the following approximations of these functions.

$$\widetilde{f}_c(\omega) = \left(1 + 1.8299\omega^{1/4} + 0.0457\omega\right)^{-1}, \tag{19.176}$$

$$\widetilde{f}_\eta(\omega) = \left(1 + 4.2597\omega^{1/4} + 0.2896\omega\right)^{-1}. \tag{19.177}$$

The inverse Laplace transforms of these functions $f_c(t)$ and $f_\eta(t)$ are presented in Fig. 19.10. These functions are finite at $t = 0$

$$f_c(0) = 21.88,\ f_\eta(0) = 3.45, \tag{19.178}$$

and their asymptotic expressions at infinity are

$$f_c(t) = O(t^{-5/4}),\ \ f_\eta(t) = O(t^{-5/4}),\ \ t \to \infty. \tag{19.179}$$

The kernels of the effective thermo-conductivity operators take the forms

$$c_*(t) = c_0\delta(t) + (c_*^s - c_0)\, f_c(t), \tag{19.180}$$

$$\eta_*(t) = \eta_0\delta(t) + (\eta_*^s - \eta_0)\, f_\eta(t). \tag{19.181}$$

For the composite material which phase inversion (the host medium is glass with $c_0 = 0.0028$ and $\eta_0 = 0.44$, and the inclusions are copper with $c_0 = 0.93$ and $\eta_0 = 0.8158,\ p = 0.2$), the static effective parameters are

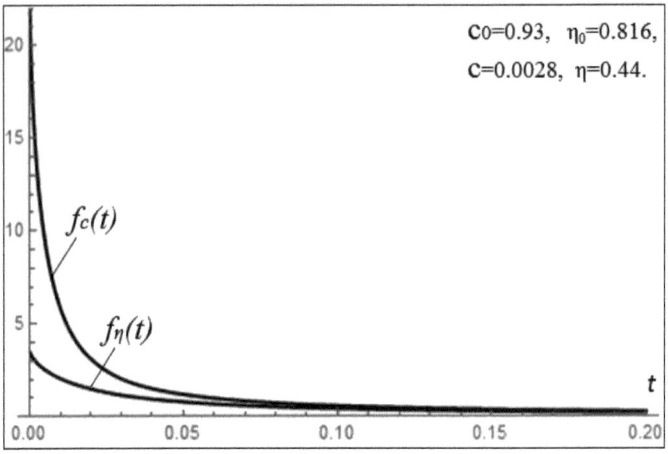

Fig. 19.10 Inverse Laplace transform of the function $\overline{K}_1(\mathbf{q})$ and $\overline{K}_2(\mathbf{q})$ in Eqs. (19.174), (19.175)

$$c_*^s = 0.00682, \eta_*^s = 0.5152. \tag{19.182}$$

In this case, $q < q^*$, and the Laplace transforms of the effective thermo-conductivity parameters of the composite are presented in the forms

$$c_*(\omega) = c + (c_*^s - c) f_c(\omega), \ \ \eta_*(\omega) = \eta + (\eta_*^s - \eta) f_\eta(\omega). \tag{19.183}$$

Here, the functions $f_c(\omega)$ and $f_\eta(\omega)$ are equal to 1 for $\omega = 0$ and tend to zero as $\omega^{-1/2}$ when $\omega \to \infty$. These functions are obtained by numerical solution of the system (19.126)–(19.128), and they are shown by solid lines in Fig. 19.11. The dashed lines in this figure are the approximations of these functions by the equations

$$\widetilde{f}_c(\omega) = \left(1 + a\sqrt{\omega}\right)^{-1}, \widetilde{f}_\eta(\omega) = \left(1 + b\sqrt{\omega}\right)^{-1}. \tag{19.184}$$

$$a = 0.17528, \ \ b = 0.22518. \tag{19.185}$$

Inverse Laplace transforms of these functions $f_c(t)$ and $f_\eta(t)$ have the following explicit forms

$$f_c(t) = \Phi(t, a), \ \ f_\eta(t) = \Phi(t, b), \tag{19.186}$$

$$\Phi(t, \alpha) = \frac{1}{\alpha\sqrt{\pi t}} - \frac{1}{\alpha^2} \exp\left(\frac{t}{\alpha^2}\right) \operatorname{erfc}\left(\frac{\sqrt{t}}{\alpha}\right), \tag{19.187}$$

$$\operatorname{erfc}(z) = 1 - \operatorname{erf}(z). \tag{19.188}$$

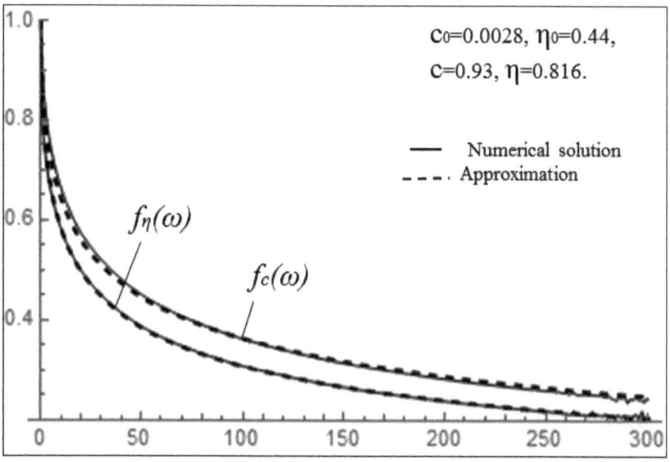

Fig. 19.11 The functions $\overline{K}_1(\mathbf{q})$ and $\overline{K}_2(\mathbf{q})$ in Eq. (19.190) for the Laplace transform of the kernel of the operators $\mathbf{A}_1$ and $\mathbf{A}_2$; solid lines are the numerical solution of the system (19.134)–(19.136), dashed lines are the approximations $\overline{K}_1(\mathbf{q})$ and $\overline{K}_2(\mathbf{q})$ in Eq. (19.191)

The asymptotic of $\Phi(t, \alpha)$ at $t = 0$ has the order of $t^{-1/2}$, and this function tends to zero as $t^{-3/2}$ at infinity.

The kernels of the effective thermo-conductivity operators K^* and η^* take the forms

$$c_*(t) = c\delta(t) + (c_*^s - c) f_c(t), \tag{19.189}$$

$$\eta_*(t) = \eta\delta(t) + (\eta_*^s - \eta) f_\eta(t). \tag{19.190}$$

Let $\varphi(t)$ be a slowly changing function of time t. In this case, actions of the operators c_* and η_* on this function are presented in the forms

$$(c_*\varphi)(t) \approx \int_0^t c_*(t - t')dt'\varphi(t), \; (\eta_*\varphi)(t) \approx \int_0^t \eta_*(t - t')dt'\varphi(t). \tag{19.191}$$

For large t, the upper limits in the integrals in these equations can be changed with infinity, and we have

$$\int_0^t c_*(t - t')dt' = \int_0^t c_*(t')dt' \approx \int_0^\infty c_*(t')dt' = c_*(\omega)|_{\omega=0} = c_*^s, \tag{19.192}$$

$$\int_0^t \eta_*(t-t')dt' = \int_0^t \eta_*(t')dt' \approx \int_0^\infty \eta_*(t')dt' = \eta_*(\omega)|_{\omega=0} = \eta_*^s. \tag{19.193}$$

Thus, for slowly changing fields and large time intervals, the effective operators c_* and η_* can be approximated by products with the static parameters c_*^s and η_*^s of the composite

$$\left(c^*\varphi\right)(t) \approx c_*^s\varphi(t),\ (\eta_*\varphi)(t) \approx \eta_*^s\varphi(t). \tag{19.194}$$

19.2.8 Green's Function of the Effective Medium with the Memory Effect

In the (ς, ω)-presentation, Green's function of the effective medium has the form

$$g_*(\varsigma, \omega) = \frac{1}{4\pi c_*(\omega)|\varsigma|}\exp\left(-q_*(\omega)|\varsigma|\right), \tag{19.195}$$

$$q_* = \sqrt{\omega\frac{\eta_*(\omega)}{c_*(\omega)}}. \tag{19.196}$$

For the static effective parameters $c_*(\omega) = c_*^s$, $\eta_*(\omega) = \eta_*^s$, the (ς, t)-presentation of Green's function is [7]

$$g_*^s(\varsigma, t) = \frac{(\eta_*^s)^{3/2}}{8(\pi c_*^s t)^{3/2}}\exp\left(-\frac{\eta_*^s|\varsigma|^2}{4c_*^s t}\right). \tag{19.197}$$

This function corresponds to the temperature field in a homogeneous conductive medium with the constant parameters c_*^s,η_*^s subjected to the external field source of the delta-type

$$q(\varsigma, t) = -\delta(\varsigma)\delta(t). \tag{19.198}$$

Here, $\delta(\varsigma)$ and $\delta(t)$ are 3D and 1D Dirac's delta-functions.

We consider an example of the composite with a homogeneous host medium and spherical inclusions, thermo-conductive parameters of the host medium are $c_0 = 0.0028$, $\eta_0 = 0.44$, and the same for the inclusions are $c = 0.93$, $\eta = 0.816$, the inclusion volume fraction is $p = 0.2$. For these values of the parameters, the effective static parameters are

$$c_*^s = 0.0068,\ \eta_*^s = 0.5152.$$

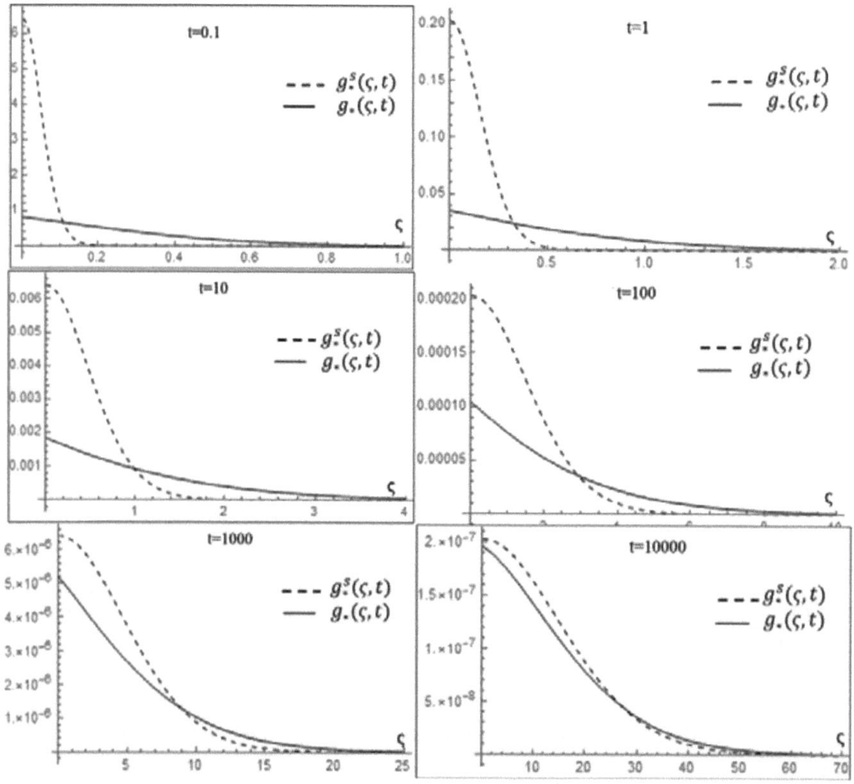

Fig. 19.12 Green's function $g_*^s(\varsigma, t)$ of the effective medium with static parameters η_*^s, c_*^s and the actual Green function $g_*(\varsigma, t)$ for the composite with spherical inclusions of volume fraction $p = 0.2$ and thermos conductivity parameters of the phase $c_0 = 0.028$, $\eta_0 = 0.44$, $c = 0.93$, $\eta = 0.816$

Comparison of the function $g_*^s(\varsigma, t)$ with the actual Green function $g_*(\varsigma, t)$ are shown in Fig. 19.12 for various time moments t. For construction of $g_*(\varsigma, t)$, Post's equation [20]

$$g_*(\varsigma, t) \simeq g_*(\varsigma, t, k) = \frac{(-1)^k}{k!}\omega^{(k+1)} \left. \frac{d^k g_*(\varsigma, \omega)}{d\omega^k} \right|_{\omega=k/t}, \quad k = 1, 2, \ldots \tag{19.199}$$

with $k = 12$ was used. Here, $g_*(\varsigma, \omega)$ is Laplace transform of the function $g_*(\varsigma, t)$ in Eq. (19.195). In the figures, dashed lines correspond the "static" Green function $g_*^s(\varsigma, t)$ are solid lines to the actual Green functions $g_*(\varsigma, t)$. It can be seen that $g_*(\varsigma, t)$ tends to $g_*^s(\varsigma, t)$ with time, but at small values of t, the difference between these functions is substantial.

For the composite with the phase inversion: $c_0 = 0.93$, $\eta_0 = 0.816$, $c = 0.0028$, $\eta = 0.44$, $p = 0.2$, the values of the effective static parameters are

$$c_*^s = 0.665, \ \eta_*^s = 0.7408.$$

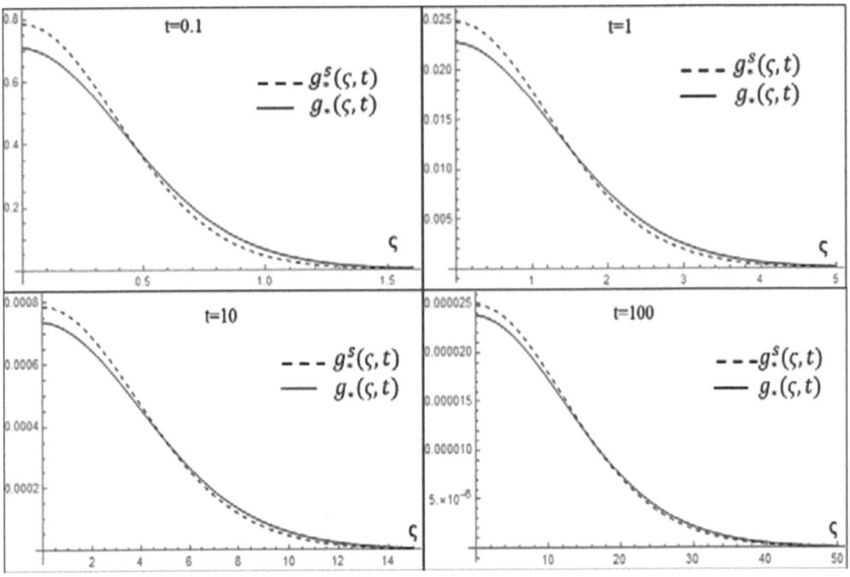

Fig. 19.13 The same as in Fig. 19.12 for the parameters $c_0 = 0.93$, $\eta_0 = 0.816$, $c = 0.0028$, $\eta = 0.44$, $p = 0.2$

The "static" and actual Green functions for different time moments are shown in Fig. 19.13.

The version of the EMM developed in this section for the revelation of nonlocal properties of composites with respect to time can be extended to the fields which Laplace transforms $T(x, \omega)$ has the form

$$T(x, \omega) = \psi(\omega) \exp\left(-\sqrt{\omega\eta_0/K_0 m} \cdot x\right), \tag{19.200}$$

where the function $\psi(\omega)$ growth not faster then $e^{a\omega}$ $(0 < a < \infty)$ at infinity. This is a class of 1D fields that can exist in an infinite homogeneous media without field sources. For such temperature fields, the proposed self-consistent method predicts time dispersion only.

19.3 Conclusion

In this chapter, the self-consistent methods are used to reveal nonlocal effective property operators of composite materials with respect to the space and time variables. Spatial nonlocality is pronounced for the fields which characteristic scale change is of the order of the correlation radius of the random set of inhomogeneities.

The time nonlocality can be observed for space and time localized impulses acting on the composite medium. The corresponding fields are substantially different from Green's functions of the medium with stationary effective parameters. The difference is notable at short time intervals from the moment of excitation, and it tends to diminish with time. The rate of diminishing depends on the contrast in the thermo-conductivity properties of the matrix and inclusion materials.

Nonlocal properties of composite materials with respect to spatial and time variables are notable in dynamic problems, e.g., wave propagation. Nonlocality results in a dispersion effect: dependence of phase velocity on frequency of propagating waves. These problems are considered in the second volume of the book.

19.4 Notes

Nonlocal properties of constitutive equations for heterogeneous media are discussed in [1–3, 12]. In the framework of the asymptotic theory of homogenization, an attempt to obtain nonlocal operators C^* for a layered medium was performed in [13, 14]. The authors assumed that the thermo-conductivity of one of two phases of the composite is small, such as the ratio of the conductivity coefficients of the phases has the order of a small parameter δ that characterizes spatial oscillation of the microstructure. Then, the nonlocal operator C^* with respect to time was obtained in the form of series of differential operators of increasing orders.

This chapter is based on the works [6, 21].

References

1. Furmanski P (1991) Effective macroscopic description for heat conduction in heterogeneous materials. Int J Heat Mass Transf 35(11):3047–3058
2. Kunin I (1982) The theory of elastic media with microstructure I. Springer, Berlin Heilderberg New York, p 291
3. Furmanski P (1997) Heat conduction in composites: homogenization and macroscopic behavior. Appl Mech Rev 50(6):327–356
4. Mindlin R, Tiersten H (1962) Effect of couple stresses in linear elasticity. Arch Rat Mech Anal 11:415–448
5. Kanaun S (1982) Effective field method in the linear static problems of composite media. Appl Math Mech (PMM) 46:655–665
6. Kanaun S, Levin V (1984) About the microstresses in composite materials in the regions of fast changing external fields. Mech Compos Mater 4:625–629
7. Tikhonov A, Samarskii A (1964) Equations of mathematical physics. Dover Publications, NY
8. Christensen R (1979) Mechanics of composite materials. Wiley-Interscience, New York, p 384
9. Sanches-Palencia E (1984) Heterogeneous media and oscillation theory. Mir, Moscow, p 472
10. Bakhvalov N, Panasenko G (1984) Homogenization of periodical media. Nauka, Moscow, p 352
11. Kalamkarov A, Andrianov I, Danishevs'kyy V (2009) Asymptotic homogenization of composite materials and structures. Appl Mech Rev 62. https://doi.org/10.1115/1.3090830

12. Kunin I (1983) The theory of elastic media with microstructure II. Springer, Berlin Heilderberg New York, p 272
13. Auriault J (1983) Effective macroscopic description for heat conduction in periodic composites. Int J Heat Mass Tranf 26(6):861–869
14. Arliault J, Royer P (1993) Double conductivity media: a comparison between phenomenological and homogenization approaches. Int J Heat Mass Tranf 36(10):2613–2621
15. Pietrak K, Wishniewski T (2005) A review of models for effective thermal conductivity of composite materials. J Power Technol 95(1):14–24
16. Burgen N, Laachachi A, Ferrios M, Lutz M, Toniazzo V, Ruch D (2016) Review of thermal conductivity in composites: mechanisms, parameters and theory. Prog Polym Sci 61:1–28
17. Bruggeman D (1937) Berechnung verschiedener physikalischer Konstanten von heterogenen Substanzen. III. Die Elastische Konstanten der Quasiisotropen Mischkoerper aus isotropen Substanzen. Ann Phys 29:160–178
18. Abramowitz M, Stigan I (1964) Handbook of mathematical functions. Nat Buro Stand, 1046
19. Carslow H, Jaeger J (1986) Conduction of heat in solids. Oxford Science Publication, London
20. Cohen A (2007) Numerical method for Laplace transform inversion. Springer, NY
21. Kanaun S, Levin V (2021) Effective thermal conductivity of particulate composites for time-varying temperature fields. Int J Eng Sci 164:103492

Appendix A
Special Tensor Bases of Fourth Rank Tensors

Elastic properties of solids are described by fourth rank tensors of special symmetry. For algebraic operations with these tensors, it is convenient to use special tensor bases. In this Appendix, the bases of fourth rank tensors that are widely used in this book are considered in detail. Explicit equations for the products and inversion operations with the tensors are presented.

A.1 E-Basis

The E-basis is constructed from a unit vector n and second rank unit tensor $\delta_{\alpha\beta}$.

$$E^1_{\alpha\beta\lambda\mu} = \delta_{\alpha\lambda}\delta_{\mu\beta}\,|_{(\lambda\mu)}, \quad E^2_{\alpha\beta\lambda\mu} = \delta_{\alpha\beta}\delta_{\lambda\mu}\,, \quad E^3_{\alpha\beta\lambda\mu} = \delta_{\alpha\beta}n_\lambda n_\mu\,,$$

$$E^4_{\alpha\beta\lambda\mu} = n_\alpha n_\beta \delta_{\lambda\mu}\,, \quad E^5_{\alpha\beta\lambda\mu} = n_\alpha \delta_{\beta\lambda} n_\mu|_{(\alpha\beta)(\lambda\mu)}, \tag{A.1}$$

$$E^6_{\alpha\beta\lambda\mu} = n_\alpha n_\beta n_\lambda n_\mu\,.$$

These tensors are symmetric with respect to the permutation of the indices of the first and second pairs, and form a closed algebra with respect to the multiplication operation defined by the equation (convolution over two indices)

$$\left(E^i E^j\right)_{\alpha\beta\lambda\mu} = E^i_{\alpha\beta\rho\tau}E^j_{\rho\tau\lambda\mu}\,. \tag{A.2}$$

The multiplication table of tensors E^i has the form

S. Kanaun, *Self-Consistent Methods for Composites*, Solid Mechanics and Its Applications 282, https://doi.org/10.1007/978-3-031-96739-9

	E^1	E^2	E^3	E^4	E^5	E^6
E^1	E^1	E^2	E^3	E^4	E^5	E^6
E^2	E^2	$3E^2$	$3E^3$	E^2	E^3	E^3
E^3	E^3	E^2	E^3	E^2	E^3	E^3
E^4	E^4	$3E^4$	$3E^6$	E^4	E^6	E^6
E^5	E^5	4	E^6	E^4	$(E^5 + E^6)/2$	E^6
E^6	E^6	E^4	E^6	E^4	E^6	E^6

(A.3)

It is seen that the E-basis is closed with respect to the product operation.

Elastic properties of isotropic materials are described by isotropic tensors symmetric with respect to the first and second pair of indices, as well as about the permutation of the index pairs. The tensors E^1 and E^2 can be used for the presentation of such tensors. In this case, it is convenient to introduce the basis that consists of two orthogonal tensors E^2 and $E^1 - \frac{1}{3}E^2$:

$$E^2\left(E^1 - \frac{1}{3}E^2\right) = \left(E^1 - \frac{1}{3}E^2\right)E^2 = 0. \tag{A.4}$$

$$E^2E^2 = 3E^2, \quad \left(E^1 - \frac{1}{3}E^2\right)\left(E^1 - \frac{1}{3}E^2\right) = \left(E^1 - \frac{1}{3}E^2\right). \tag{A.5}$$

If two isotropic tensors A and B are presented in the forms

$$A = a_1E^2 + a_2\left(E^1 - \frac{1}{3}E^2\right), \; B = b_1E^2 + b_2\left(E^1 - \frac{1}{3}E^2\right), \tag{A.6}$$

with scalar coefficients a_i, b_i, the product of these tensors is

$$AB = 3a_1b_1E^2 + a_2b_2\left(E^1 - \frac{1}{3}E^2\right) \tag{A.7}$$

From Eqs. (A.4), (A.5) follows that the inverse tensor $A^{-1}(AA^{-1} = A^{-1}A = E^1)$ has form

$$A^{-1} = \frac{1}{9a_1}E^2 + \frac{1}{a_2}\left(E^1 - \frac{1}{3}E^2\right). \tag{A.8}$$

A.2 *P*-Basis

This basis is constructed similar to the E-basis in Eq. (A.1) but second rank unit tensor $\delta_{\alpha\beta}$ is changed with the projector $\theta_{\alpha\beta}$ on the plane orthogonal to a unit vector n

$$\theta_{\alpha\beta} = \delta_{\alpha\beta} - n_\alpha n_\beta. \tag{A.9}$$

The P-basis consists of the following fourth rank tensors

$$P^1_{\alpha\beta\lambda\mu} = \theta_{\alpha\lambda}\theta_{\mu\beta}\,|_{(\lambda\mu)}, \quad P^2_{\alpha\beta\lambda\mu} = \theta_{\alpha\beta}\theta_{\lambda\mu}, \quad P^3_{\alpha\beta\lambda\mu} = \theta_{\alpha\beta}n_\lambda n_\mu,$$

$$P^4_{\alpha\beta\lambda\mu} = n_\alpha n_\beta\theta_{\lambda\mu}, \quad P^5_{\alpha\beta\lambda\mu} = n_\alpha\theta_{\beta\lambda}n_\mu|_{(\alpha\beta)(\lambda\mu)}, \tag{A.10}$$

$$P^6_{\alpha\beta\lambda\mu} = n_\alpha n_\beta n_\lambda n_\mu.$$

The elements of the P and E-bases (A.1) relate by the equations

$$P^j = \alpha^{ji}E^i, \quad E^j = \left(\alpha^{-1}\right)^{ji} P^i, \tag{A.11}$$

where the matrices α and α^{-1} have the forms

$$\alpha^{ij} = \begin{bmatrix} 1 & 0 & 0 & 0 & -2 & 1 \\ 0 & 1 & -1 & -1 & 0 & 1 \\ 0 & 0 & 1 & 0 & 0 & -1 \\ 0 & 0 & 0 & 1 & 0 & -1 \\ 0 & 0 & 0 & 0 & 1 & -1 \\ 0 & 0 & 0 & 0 & 0 & 1 \end{bmatrix}, \quad \left(\alpha^{-1}\right)^{ij} = \begin{bmatrix} 1&0&0&0&2&1 \\ 0&1&1&1&0&1 \\ 0&0&1&0&0&1 \\ 0&0&0&1&0&1 \\ 0&0&0&0&1&1 \\ 0&0&0&0&0&1 \end{bmatrix}. \tag{A.12}$$

Six tensors P^i form a closed algebra with respect to the multiplication operation (A.2). The multiplication table of these tensors has the form

	P^1	P^2	P^3	P^4	P^5	P^6
P^1	P^1	P^2	P^3	0	0	0
P^2	P^2	$2P^2$	$2P^3$	0	0	0
P^3	0	0	0	P^2	0	P^3
P^4	P^4	$2P^4$	$2P^6$	0	0	0
P^5	0	0	0	0	$P^5/2$	0
P^6	0	0	0	P^4	0	P^6

(A.13)

Note that tensor $P^1 - \frac{1}{2}P^2$ is orthogonal to all the tensors of the P-basis except P^1, i.e.,

$$P^1\left(P^1 - \frac{1}{2}P^2\right) = \left(P^1 - \frac{1}{2}P^2\right)P^1 = \left(P^1 - \frac{1}{2}P^2\right), \tag{A.14}$$

$$P^i\left(P^1 - \frac{1}{2}P^2\right) = \left(P^1 - \frac{1}{2}P^2\right)P^i = 0, \quad i = 2, 3, 4, 5, 6, \tag{A.15}$$

$$\left(P^1 - \frac{1}{2}P^2\right)\left(P^1 - \frac{1}{2}P^2\right) = \left(P^1 - \frac{1}{2}P^2\right). \tag{A.16}$$

The linear shells of the E and P-bases coincide.

Let the tensor A in the P-basis is presented in the form

$$A = a_1 P^2 + a_2 \left(P^1 - \frac{1}{2} P^2 \right) + a_3 P^3 + a_4 P^4 + a_5 P^5 + a_6 P^6, \tag{A.17}$$

where a_i $(i = 1, 2, ..., 6)$ are scalar coefficients. Then, the inverse tensor A^{-1} $(AA^{-1} = A^{-1}A = E^1)$ takes the form

$$A^{-1} = \frac{a_6}{2\Delta} P^2 + \frac{1}{a_2} \left(P^1 - \frac{1}{2} P^2 \right) - \frac{a_3}{\Delta} P^3 - \frac{a_4}{\Delta} P^4$$

$$+ \frac{4}{a_5} P^5 + \frac{2a_1}{\Delta} P^6, \quad \Delta = 2\left(a_1 a_6 - a_3 a_4\right). \tag{A.18}$$

Product of two tensors A and B presented in the form (A.17) with scalar coefficients a_i and b_i is

$$AB = \left(2a_1 b_1 + a_3 b_4\right) P^2 + a_2 b_2 \left(P^1 - \frac{1}{2} P^2 \right) + \left(2a_1 b_3 + a_3 b_6\right) P^3 +$$

$$+ \left(2a_4 b_1 + a_6 b_4\right) P^4 + \frac{1}{2} a_5 b_5 P^5 + \left(a_6 b_6 + 2a_4 b_3\right) b^6. \tag{A.19}$$

The P-basis can be used for presentation of the elastic stiffness tensor C of transverse isotropic materials. If n is unit vector of the material isotropy axis, then

$$C = c_1 P^2 + c_2 \left(P^1 - \frac{1}{2} P^2 \right) + c_3 \left(P^3 + P^4 \right) + c_5 P^5 + c_6 P^6, \tag{A.20}$$

where c_i are 5 independent elastic constant. The matrix presentation of a transverse isotropic tensor C in Foigt's notations is [1]

$$\left[c_{ij}\right] = \begin{bmatrix} c_{11} & c_{12} & c_{13} & 0 & 0 & 0 \\ c_{12} & c_{11} & c_{13} & 0 & 0 & 0 \\ c_{13} & c_{13} & c_{33} & 0 & 0 & 0 \\ 0 & 0 & 0 & c_{44} & 0 & 0 \\ 0 & 0 & 0 & 0 & c_{44} & 0 \\ 0 & 0 & 0 & 0 & 0 & c_{66} \end{bmatrix}. \tag{A.21}$$

The coefficients c_{ij} are expressed in terms of the coefficients c_i in Eq. (A.20) as follows

$$c_{11} = c_1 + \frac{c_2}{2}, \quad c_{12} = c_1 - \frac{c_2}{2}, \quad c_{13} = c_3, \tag{A.22}$$

$$c_{33} = c_6, \quad c_{44} = \frac{c_5}{4}, \quad c_{66} = \frac{c_2}{2}. \tag{A.23}$$

A.3 T-Basis

The T-basis consists of twelve linear independent fourth rank tensors symmetric with respect to permutation of the indices of the first and second index pairs

$$T^1_{\alpha\beta\gamma\delta} = n^1_\alpha n^1_\beta n^1_\gamma n^1_\delta,\ T^2_{\alpha\beta\gamma\delta} = n^2_\alpha n^2_\beta n^2_\gamma n^2_\delta,$$

$$T^3_{\alpha\beta\gamma\delta} = n^3_\alpha n^3_\beta n^3_\gamma n^3_\delta,\ T^4_{\alpha\beta\gamma\delta} = n^1_\alpha n^1_\beta n^2_\gamma n^2_\delta,$$

$$T^5_{\alpha\beta\gamma\delta} = n^2_\alpha n^2_\beta n^1_\gamma n^1_\delta,\ T^6_{\alpha\beta\gamma\delta} = n^1_\alpha n^1_\beta n^3_\gamma n^3_\delta, \tag{A.24}$$

$$T^7_{\alpha\beta\gamma\delta} = n^3_\alpha n^3_\beta n^1_\gamma n^1_\delta,\ T^8_{\alpha\beta\gamma\delta} = n^2_\alpha n^2_\beta n^3_\gamma n^3_\delta,$$

$$T^9_{\alpha\beta\gamma\delta} = n^3_\alpha n^3_\beta n^2_\gamma n^2_\delta,\ T^{10}_{\alpha\beta\gamma\delta} = 4n^1_\alpha n^2_\beta n^1_\gamma n^2_\delta|_{(\alpha\beta)(\gamma\delta)},$$

$$T^{11}_{\alpha\beta\gamma\delta} = 4n^1_\alpha n^3_\beta n^1_\gamma n^3_\delta|_{(\alpha\beta)(\gamma\delta)},\ T^{12}_{\alpha\beta\gamma\delta} = 4n^2_\alpha n^3_\beta n^2_\gamma n^3_\delta|_{(ij)(\gamma\delta)}.$$

Here, n^1, n^2, n^3 are three mutually orthogonal unit vectors. Twelve tensors T^i compose a closed algebra with respect to the product operation

$$(T^i T^j)_{\alpha\beta\gamma\delta} = T^i_{\alpha\beta\mu\nu} T^j_{\mu\nu\gamma\delta},\quad i, j = 1, 2, ..., 12. \tag{A.25}$$

Multiplication table of the T-tensors

	T^1	T^2	T^3	T^4	T^5	T^6	T^7	T^8	T^9	T^{10}	T^{11}	T^{12}
T^1	T^1	0	0	T^4	0	T^6	0	0	0	0	0	0
T^2	0	T^2	0	0	T^5	0	0	T^8	0	0	0	0
T^3	0	0	T^3	0	0	0	T^7	0	T^9	0	0	0
T^4	0	T^4	0	0	T^1	0	0	T^6	0	0	0	0
T^5	T^5	0	0	T^2	0	T^8	0	0	0	0	0	0
T^6	0	0	T^6	0	0	0	T^1	0	T^4	0	0	0
T^7	T^7	0	0	T^9	0	T^3	0	0	0	0	0	0
T^8	0	0	T^8	0	0	0	T^5	0	T^2	0	0	0
T^9	0	T^9	0	0	T^7	0	0	T^3	0	0	0	0
T^{10}	0	0	0	0	0	0	0	0	0	$2T^{10}$	0	0
T^{11}	0	0	0	0	0	0	0	0	0	0	$2T^{11}$	0
T^{12}	0	0	0	0	0	0	0	0	0	0	0	$2T^{12}$

The elements of the E-basis can be expressed in terms of the elements of the T-basis. For $n^3 = n$, the following equations for $E^i(n)$ hold

$$E^1 = T^1 + T^2 + T^3 + \frac{1}{2}\left(T^{10} + T^{11} + T^{12}\right),$$
$$E_2 = T^1 + T^2 + T^3 + T^4 + T^5 + T^6 + T^7 + T^8 + T^9, \tag{A.26}$$
$$E^3(n) = T^3 + T^6 + T^8,\ E^4(n) = T^3 + T^7 + T^9,$$
$$E^5(n) = \frac{1}{4}\left(T^{11} + T^{12}\right),\ E^6(n) = T^3.$$

The tensor E^1 plays the role of unit tensor I in the E, P and T-bases. If tensor A has the form

$$A = \sum_{i=1}^{12} a_i T^i, \tag{A.27}$$

the inverse tensor A^{-1} ($AA^{-1} = A^{-1}A = I$) is represented in the form

$$A^{-1} = \sum_{i=1}^{12} b_i T^i, \tag{A.28}$$

where the first nine scalar coefficient b_i in this equation are the solutions to the following system of linear algebraic equations

$$a_1b_1 + a_4b_5 + a_6b_7 = 1, \quad a_2b_2 + a_5b_4 + a_8b9 = 1,$$
$$a_3b_3 + a_7b_6 + a_9b_8 = 1, \quad a_4b_2 + a_1b_4 + a_6b_9 = 0,$$
$$a_5b_1 + a_2b_5 + a_8b_7 = 0, \quad a_6b_3 + a_1b_6 + a_4b_8 = 0, \tag{A.29}$$
$$a_7b_1 + a_9b_5 + a_3b_7 = 0, \quad a_8b_3 + a_5b_6 + a_2b_8 = 0,$$
$$a_9b_2 + a_7b_4 + a_3b_9 = 0.$$

Other three coefficient are:

$$b_{10} = \frac{1}{4a_{10}}, \quad b_{11} = \frac{1}{4a_{11}}, \quad b_{12} = \frac{1}{4a_{12}}. \tag{A.30}$$

The elastic stiffness tensor C of an orthorhombic crystal is presented in this basis T^i as follows

$$C = c_1 T^1 + c_2 T^2 + c_3 T^3 + c_4(T^4 + T^5) +$$

$$+ c_5(T^6 + T^7) + c_6(T^7 + T^8) + c_7 T^{10} + c_8 T^{11} + c_9 T^{12}. \quad \text{(A.31)}$$

Here, c_i are 9 independent elastic coefficients. In Voigt's notation, the stiffness tensor of the orthorhombic material has the form [1]

$$\left[c_{ij}\right] = \begin{bmatrix} c_{11} & c_{12} & c_{13} & 0 & 0 & 0 \\ c_{12} & c_{22} & c_{23} & 0 & 0 & 0 \\ c_{13} & c_{23} & c_{33} & 0 & 0 & 0 \\ 0 & 0 & 0 & c_{44} & 0 & 0 \\ 0 & 0 & 0 & 0 & c_{55} & 0 \\ 0 & 0 & 0 & 0 & 0 & c_{66} \end{bmatrix}. \quad \text{(A.32)}$$

Relations of the scalar coefficients c_i in Eq. (A.31) with the elastic constants c_{ij} are given by the equation

$$c_1 = c_{11}, \quad c_2 = c_{22}, \quad c_3 = c_{33}, \quad c_4 = c_{12}, \quad c_5 = c_{13},$$

$$c_6 = c_{23}, \quad c_7 = c_{55}, \quad c_8 = c_{66}, \quad c_9 = c_{44}. \quad \text{(A.33)}$$

For crystals of a higher symmetry, some of the nine constants in Eq. (A.31) coincide. For instance, for trigonal crystals only 6 constants are independent, and

$$c_1 = c_2, \; c_5 = c_6, \; c_8 = c_9. \quad \text{(A.34)}$$

For hexagonal crystals, there are 5 constants:

$$c_1 = c_2, \; c_5 = c_6, \; c_7 = \frac{1}{2}(c_1 - c_4), \; c_8 = c_9; \quad \text{(A.35)}$$

for cubic crystals 3 constants

$$c_1 = c_2 = c_3, \; c_4 = c_5 = c_6, \; c_7 = c_8 = c_9; \quad \text{(A.36)}$$

and for isotropic body 2 constants

$$c_1 = c_2 = c_3, \; c_4 = c_5 = c_6, \; c_7 = c_8 = c_9 = \frac{1}{2}(c_1 - c_4). \quad \text{(A.37)}$$

In the case of crystals of cubic symmetry, the elastic stiffness tensor C can be presented in terms of three basis tensors H^i

$$C = c_1 H^1 + c_2 H^2 + c_3 H^3. \quad \text{(A.38)}$$

In the T-basis, tensors H^i are presented as follows

$$H^1 = T^1 + T^2 + T^3,\ H^2 = T^4 + T^5 + T^6 + T^7 + T^8 + T^9, \tag{A.39}$$

$$H^3 = T^{10} + T^{11} + T^{12}. \tag{A.40}$$

The elements E^1, E^2 of the E-basis are expressed in terms of the tensors H^i

$$E^1 = H^1 + \frac{1}{2}H^3,\ \ E^2 = H^1 + H^2. \tag{A.41}$$

Relations of coefficients c_i in Eq. (A.38) with the components c_{ij} in the matrix form of Hooke's law for cubic crystals have the forms

$$c_1 = c_{11}, c_2 = c_{12},\ c_3 = c_{44}. \tag{A.42}$$

If tensors A and B belong to the shell of H-tensors

$$A = \sum_{i=1}^{3} a_i H^i, \quad B = \sum_{i=1}^{3} b_i H^i, \tag{A.43}$$

the product of these tensors is

$$AB = (a_1b_1 + 2a_2b_2)H^1 + (a_1b_2 + a_2b_1 + a_2b_2)H^2 + 2a_3b_3H^3. \tag{A.44}$$

The inverse tensor A^{-1} ($AA^{-1} = E^1$) is defined by the equation

$$A^{-1} = \frac{a_1 + a_2}{\Delta}H^1 - \frac{a_2}{\Delta}H^2 + \frac{1}{4a_3}H^3,\quad \Delta = (a_1 + a_2)(a_1 + 2a_2). \tag{A.45}$$

A.4 Averaging the Elements of the E, P and T-Bases

By solution of the homogenization problems, there appear necessity of averaging fourth rank tensors over orientations in 3D-space. In this Appendix, the equations for the elements of E, P, and T-bases averaged over the orientations are presented.

E-basis The elements of the E-basis averaged over orientation $\langle E^i(n)\rangle$ are the following integral

$$\langle E^i(n)\rangle = \frac{1}{4\pi}\int_{\Omega_1} E^i(n)\, d\Omega_n,\quad i = 1, 2, ..., 6. \tag{A.46}$$

Here, Ω_1 is the surface of the unit sphere in 3D-space, integration is performed over vector n on this sphere.

$$\langle E^1 \rangle = E^1, \ \langle E^2 \rangle = E^2, \ \langle E^3(n) \rangle = \langle E^4(n) \rangle = \frac{1}{3} E^2, \tag{A.47}$$

$$\langle E^5(n) \rangle = \frac{1}{3} E^1, \ \langle E^6(n) \rangle = \frac{1}{15} \left(2E^1 + E^2 \right), \tag{A.48}$$

P-basis. For the elements of the P-basis, we have

$$\langle P^i(n) \rangle = \frac{1}{4\pi} \int_{\Omega_1} P^i(n) \, d\Omega_m \, , \ i = 1, 2, ..., 6 \, , \tag{A.49}$$

$$\langle P^1(m) \rangle = \frac{1}{15} \left(E^2 + 7E^1 \right), \ \langle P^2(m) \rangle = \frac{2}{15} \left(E^1 + 3E^2 \right), \tag{A.50}$$

$$\langle P^3(m) \rangle = \langle P^4(m) \rangle = \frac{2}{15} \left(2E^2 - E^1 \right), \tag{A.51}$$

$$\langle P^5(m) \rangle = \frac{1}{5} \left(E^1 - \frac{1}{3} E^2 \right), \ \langle P^6(m) \rangle = \frac{1}{15} \left(2E^1 + E^2 \right) . \tag{A.52}$$

***T*-basis**. The elements of the T-basis are expressed in terms of three mutually orthogonal unit vectors n^1, n^2, n^3 in 3D-space. The averages $\langle T^i \rangle$ of these tensors over possible orientations of these three vectors have the forms

$$\langle T^i \rangle = \frac{1}{15} (E_2 - 2E_1), \quad i = 1, 2, 3; \tag{A.53}$$

$$\langle T^i \rangle = \frac{1}{15} (2E_2 - E_1), \quad i = 4, 5, 6, 7, 8, 9; \tag{A.54}$$

$$\langle T^i \rangle = \frac{2}{5} (E_1 - \frac{1}{3} E_2), \quad i = 10, 11, 12. \tag{A.55}$$

***H*-basis**

$$\langle H^1 \rangle = \frac{1}{5} (E^2 + 2E^1), \langle H^2 \rangle = \frac{2}{5} \left(2E^2 - E^1 \right),$$

$$\langle H^3 \rangle = \frac{2}{5} \left(3E^1 - E^2 \right). \tag{A.56}$$

A.5 Bases of the Fourth Rank Tensors in the 2D-Case

In the 2D-space, the E-basis consists of the tensors in Eq. (A.1), where Greek indices have the values 1, 2. In the 2D-case, not all the tensors $E^i(n)$ are linearly independent. Because of the equation

$$E^1 - E^2 + E^3 + E^4 - 2E^5 = 0, \tag{A.57}$$

five from six tensors $E^i(n)$ in Eq. (A.1) are independent only. It can be any four tensors from the first five and $E^6(n)$.

For the P-basis, linear dependence of the tensors in Eq. (A.10) is obvious because the tensors P^1 and P^2 coincide in the 2D-case. The multiplication table for the five linearly independent elements of the P-basis has the form

	P^1	P^3	P^4	P^5	P^6
P^1	P^1	P^3	0	0	0
P^3	0	0	P^1	0	P^3
P^4	P^4	P^6	0	0	0
P^5	0	0	0	$P^5/2$	0
P^6	0	0	P^4	0	P^6

. (A.58)

As a result, product of two elements (A and B) of the linear shell of the P-basis

$$A = a_1 P^1 + a_3 P^3 + a_4 P^4 + a_5 P^5 + a_6 P^6, \quad \text{(A.59)}$$

$$B = b_1 P^1 + b_3 P^3 + b_4 P^4 + b_5 P^5 + b_6 P^6, \quad \text{(A.60)}$$

where a_i, b_i are scalar coefficients, has the form

$$AB = (a_1 b_1 + a_3 b_4)\, P^1 + (a_1 b_3 + a_3 b_6)\, P^3 +$$

$$+ (a_4 b_1 + a_6 b_4)\, P^4 + \frac{1}{2} a_5 b_5 P^5 + (a_4 b_3 + a_6 b_6)\, P^6. \quad \text{(A.61)}$$

The tensor A^{-1} is the inverse tensor with respect to A if

$$A^{-1} A = A A^{-1} = E^1,$$

and A^{-1} is defined by the equation

$$A^{-1} = \frac{a_6}{\Delta} P^1 - \frac{a_3}{\Delta} P^3 - \frac{a_4}{\Delta} P^4 + \frac{4}{a_5} P^5 + \frac{a_1}{\Delta} P^6, \quad \text{(A.62)}$$

$$\Delta = a_1 a_6 - a_3 a_4 .$$

Relations between the elements of the E and P-basis in 2D-case are determined by the equations:

$$E^1 = P^1 + 2P^5 + P^6, \quad E^2 = P^1 + P^3 + P^4 + P^6, \quad \text{(A.63)}$$

$$E^3 = P^3 + P^6, \quad E^4 = P^4 + P^6, \quad \text{(A.64)}$$

$$E^5 = P^5 + P^6, \quad E^6 = P^6, \tag{A.65}$$

$$P^1 = E^1 - 2E^5 + E^6, \quad P^3 = E^3 - E^6, \tag{A.66}$$

$$P^4 = E^4 - E^6, \quad P^5 = E^5 - E^6, \quad P^6 = E^6. \tag{A.67}$$

Averaging the tensors of the E and P-basis in 2D-case have the forms

$$\left\langle E^i\left(n\right)\right\rangle = \frac{1}{2\pi}\int\limits_{l_1} E^i\left(n\right) dl_n\,, \quad i = 1, 2, ..., 6\,. \tag{A.68}$$

Here l_1 is a unit circle.

$$\left\langle E^1\right\rangle = E^1, \quad \left\langle E^2\right\rangle = E^2, \quad \left\langle E^3\left(n\right)\right\rangle = \left\langle E^4\left(n\right)\right\rangle = \frac{1}{2}E^2, \tag{A.69}$$

$$\left\langle E^5\left(n\right)\right\rangle = \frac{1}{2}E^1, \quad \left\langle E^6\left(n\right)\right\rangle = \frac{1}{8}\left(2E^1 + E^2\right), \tag{A.70}$$

$$\left\langle P^i\left(n\right)\right\rangle = \frac{1}{2\pi}\int\limits_{l_1} P^i\left(n\right) dl_n\,, \quad i = 1, 3, 4, 5, 6\,, \tag{A.71}$$

$$\left\langle P^1\left(n\right)\right\rangle = \frac{1}{8}\left(E^2 + 2E^1\right), \quad \left\langle P^3\left(n\right)\right\rangle = \left\langle P^4\left(n\right)\right\rangle = \frac{1}{8}\left(3E^2 - 2E^1\right), \tag{A.72}$$

$$\left\langle P^5\left(n\right)\right\rangle = \frac{1}{8}\left(2E^1 - E^2\right), \quad \left\langle P^6\left(n\right)\right\rangle = \frac{1}{8}\left(E^2 + 2E^1\right)\,. \tag{A.73}$$

A.6 Fourth Rank Tensors Non-symmetric with Respect to Permutation of Index Pairs

In equations of self-consistent methods for effective elastic stiffness tensors of composites, there appear fourth rank tensors that are symmetric with respect to the permutation of the indices of the first and second index pairs but not symmetric with respect to the pair permutations. Such tensors have 36 independent components that can be arranged in a non-symmetric 6×6 matrices. The matrix $\overline{T}_{pq}$ that corresponds to the tensor $T_{\alpha\beta\gamma\delta}$ is composed by the rule

$$\left[\overline{T}_{pq}\right] = \begin{bmatrix} \overline{T}^{11} & \overline{T}^{12} \\ \overline{T}^{21} & \overline{T}^{22} \end{bmatrix}, \quad p, q = 1, 2, 3, 4, 5, 6. \tag{A.74}$$

Here, $\overline{T}^{11}, \overline{T}^{12}, \overline{T}^{21}, \overline{T}^{22}$ are 3×3 block-matrices which elements are the components of the tensor $T_{\alpha\beta\gamma\delta}$

$$\overline{T}^{11} = \begin{bmatrix} T_{1111} & T_{1122} & T_{1133} \\ T_{2211} & T_{2222} & T_{2233} \\ T_{3311} & T_{3322} & T_{3333} \end{bmatrix}, \ \overline{T}^{12} = \begin{bmatrix} T_{1112} & T_{1113} & T_{1123} \\ T_{2212} & T_{2213} & T_{2223} \\ T_{3312} & T_{3313} & T_{3323} \end{bmatrix} \tag{A.75}$$

$$\overline{T}^{21} = \begin{bmatrix} T_{1211} & T_{1222} & T_{2333} \\ T_{1311} & T_{1322} & T_{1333} \\ T_{2311} & T_{2322} & T_{2333} \end{bmatrix}, \ \overline{T}^{22} = \begin{bmatrix} T_{1212} & T_{1213} & T_{1223} \\ T_{1312} & T_{1313} & T_{1323} \\ T_{2312} & T_{2313} & T_{2323} \end{bmatrix} \tag{A.76}$$

For such a definition, the sum and difference of two tensors correspond to the sum and difference of the matrices in the matrix representations of these tensors.

Let matrix $\overline{C}$ that corresponds to the product of two fourth rank tensors A and B

$$C_{\alpha\beta\gamma\delta} = A_{\alpha\beta\lambda\mu} B_{\lambda\mu\gamma\delta} \iff \overline{C} = \Pr(\overline{A}, \overline{B}). \tag{A.77}$$

The matrix $\overline{C}$ is calculated in terms of the matrix presentations $\overline{A}$ and $\overline{B}$ of the tensors A and B as follows

$$\overline{A} = \begin{bmatrix} \overline{A}^{11} & \overline{A}^{12} \\ \overline{A}^{21} & \overline{A}^{22} \end{bmatrix}, \overline{B} = \begin{bmatrix} \overline{B}^{11} & \overline{B}^{12} \\ \overline{B}^{21} & \overline{B}^{22} \end{bmatrix}, \overline{C} = \begin{bmatrix} \overline{C}^{11} & \overline{C}^{12} \\ \overline{C}^{21} & \overline{C}^{22} \end{bmatrix} \tag{A.78}$$

$$\overline{C}^{11} = \overline{A}^{11}\overline{B}^{11} + 2\overline{A}^{12}\overline{B}^{21}, \ \overline{C}^{12} = \overline{A}^{11}\overline{B}^{12} + 2\overline{A}^{12}\overline{B}^{22}, \tag{A.79}$$

$$\overline{C}^{21} = \overline{A}^{21}\overline{B}^{11} + 2\overline{A}^{22}\overline{B}^{21}, \ \overline{C}^{22} = \overline{A}^{21}\overline{B}^{12} + 2\overline{A}^{22}\overline{B}^{22}. \tag{A.80}$$

Here, $\overline{A}^{pq}\overline{B}^{rs}$ is conventional matrix product of the 3×3 bloc-matrices $\overline{A}^{pq}$ and $\overline{B}^{rs}$, $p, q, r, s = 1, 2$. Definition of the product operator Pr in Eq. (A.77) follows from Eqs. (A.78)–(A.80).

The inverse tensor A^{-1} with respect to fourth rank tensor A is defined by the equation

$$\left(A^{-1}\right)_{\alpha\beta\lambda\mu} A_{\lambda\mu\gamma\delta} = E^1_{\alpha\beta\gamma\delta}, \tag{A.81}$$

where unit fourth rank tensor E^1 is the element of the E-basis in Eq. (A.1). Matrix $\overline{A}^{-1}$ that corresponds to the tensor A^{-1} is the solution of the equation

$$\Pr(\overline{A}^{-1}, \overline{A}) = \overline{E}^1, \ \overline{E}^1 = \mathrm{diag}[1, 1, 1, \frac{1}{2}, \frac{1}{2}, \frac{1}{2}]. \tag{A.82}$$

Here, $\mathrm{diag}[a_1, a_2, a_3, a_4, a_5, a_6]$ is 6×6 diagonal matrix with the elements a_i in the diagonal. For a given matrix $\overline{A}$, Eq. (A.82) is an equation for 36 unknown components of the matrix $\overline{A}^{-1}$.

A.7 A Computer Code for Operations with Fourth Rank Tensors

For operations with tensors that belong to the shells of E, P or T-bases, commercial computational packages can be successfully used. In this Section, an example of the program for Wolfram "Mathematica" package for operation with the tensors of P-basis is presented.

```
(*TensP.nb*)
(*1. P-basis
A=a1*P2+a2*(P1-1/2*P2)+a3*P3+a4*P4+a5*P5+a6*P6
*)
(*
P2:=T[1,0,0,0,0,0]
P1-1/2*P2:=T[0,1,0,0,0,0]
P3:=T[0,0,1,0,0,0]
P4:=T[0,0,0,1,0,0]
P5:=T[0,0,0,0,1,0]
P6:=T[0,0,0,0,0,1]
*)
(*Unit tensor with multiplayer a*)
U[a_]:= T[a/2,a,0,0,2*a,a]
E1:=U[1]
(*2.Summ of two tensors*)
SU[T[a1_, a2_, a3_, a4_, a5_, a6_], T[b1_, b2_, b3_, b4_, b5_, b6_]]:=
 T[a1 + b1, a2 + b2, a3 + b3, a4 + b4, a5 + b5, a6 + b6]
(*3. Difference of two tensors*)
DI[T[a1_, a2_, a3_, a4_, a5_, a6_], T[b1_, b2_, b3_, b4_, b5_, b6_]] :=
 T[a1 - b1, a2 - b2, a3 - b3, a4 - b4, a5 - b5, a6 - b6]
(*4. Product of two tensors*)
PR[T[a1_, a2_, a3_, a4_, a5_, a6_], T[b1_, b2_, b3_, b4_, b5_, b6_]] :=
T[2*a1*b1+a3*b4,a2*b2,2*a1*b3+a3*b6,2*a4*b1+a6*b4,a5*b5/2,
a6*b6+2*a4*b3]
(*5. Inversion operation*)
IN[T[a1_, a2_, a3_, a4_, a5_, a6_]] :=
T[a6/(2*del[a1,a3,a4,a6]),1/a2,-a3/del[a1,a3,a4,a6],-a4/del[a1,a3,a4,a6],
4/a5,2*a1/del[a1,a3,a4,a6]]
del[a1_,a3_,a4_,a6_]:=2*(a1*a6-a3*a4)

(*Example:
Effective elastic stiffness tensor for a homogeneous isotropic host medium with identical
spheroidal inclusions*)
<<TensP.nb
(*Initial data*)
(*la0, mu0 - Lame parameters of the host medium,
```

```
la, mu - Lame parameters of the inclusions*)
k0= la0 + mu0; m0 = mu0;
l0 = la0; n0 = la0 + 2 mu0; kp = (la0 + mu0)/(la0 + 2 mu0);
(*C0 is elastic stiffness tensor of the host medium*)
C0 = T[la0 + mu0, 2 mu0, la0, la0, 4 mu0, la0 + 2 mu0];
la1:=la-la0; mu1=mu-mu0;
C1 = T[la1 + mu1, 2 mu1, la1, la1, 4 mu1, la1 + 2 mu1];
(*g is aspect ratio of spheroidal inclusions*)
A[g_] := T[A1[g], A2[g], A3[g], A3[g], A5[g], A6[g]];
A1[g_] := (b1[g] - 2 kp*b2[g])/mu0;
A2[g_] := -kp*b2[g]/mu0;
A3[g_] := -kp*b3[g]/mu0;
A5[g_] := (1 - 3 b1[g] - 4 kp*b3[g])/mu0;
A6[g_] := -kp*b4[g]/mu0;
b1[g_] := (g ^ 2/4)*NIntegrate[Sin[t] ^ 3/dell[g, t], {t, 0, Pi}]
b2[g_] := (g ^ 4/16)*NIntegrate[Sin[t] ^ 5/dell[g, t] ^ 2, {t, 0, Pi}]
b3[g_] := b1[g] - 5 b2[g];
b4[g_] := 1 - 10 b1[g] + 35 b2[g];
dell[g_, t] := g ^ 2*Sin[t] ^ 2 + Cos[t] ^ 2;
(*Effective elastic stiffness tensor*)
Ce[g_,p_]:= SU[C0, PR[U[p], PR[C1, IN[SU[E1, PR[U[1 - p], PR[A[g], C1]]]]]]];
```

Similar computer codes for operation with the tensors of the shell of the T-basis and forth rank tensors symmetric with respect to first and second pairs of indices are presented in Vol. II of the book (Appendix A).

Appendix B
Generalized Functions Associated with Green's Functions of Stationary Conductivity and Static Elasticity

Green's function $g(x)$ of a homogeneous conductive medium is the vanishing at infinity solution of the partial differential equation

$$\nabla_{\alpha} C^{0}_{\alpha\beta} \nabla_{\beta} g(x) = -\delta(x), \tag{B.1}$$

where $\delta(x)$ is 3D Dirac's delta-function, C^0 is the conductivity tensor of the medium.

Green's function $G(x)$ of static elasticity is the vanishing at infinity solution of the equation

$$\left(L^{0}_{\alpha\beta} G_{\beta\lambda}\right)(x) = \delta_{\alpha\lambda}\delta(x), \quad L^{0}_{\alpha\beta} = -\nabla_{\lambda} C^{0}_{\lambda\alpha\beta\mu} \nabla_{\mu}, \tag{B.2}$$

where $\delta_{\alpha\beta}$ is the Kronecker symbol and C^0 is the elastic stiffness tensor of the medium. In the book, this function is also called Green's function for displacements. Green's functions for the strain and stress tensors are expressed in terms of second derivatives of $G(x)$. Regularizations of integral operators which kernels are second derivatives of $g(x)$ and $G(x)$ are considered in this Appendix.

B.1 Green's Functions of Static Elasticity in the k-Presentation

The direct F and inverse F^{-1} Fourier transforms in the 3D-space are determined by the equations

$$(Ff)(k) = \widetilde{f}(k) = \int f(x) \exp(ik \cdot x)\, dx, \tag{B.3}$$

$$\left(F^{-1}\widetilde{f}\right)(x) = f(x) = (2\pi)^{-3} \int \widetilde{f}(k) \exp(-ik \cdot x)\, dk. \tag{B.4}$$

S. Kanaun, *Self-Consistent Methods for Composites*, Solid Mechanics and Its Applications 282, https://doi.org/10.1007/978-3-031-96739-9

For two integrable functions $f(x)$, $\varphi(x)$ in R^3, the following equation is called Parseval's formula

$$(f, \varphi) = (2\pi)^{-3} \left(\widetilde{f}, \widetilde{\varphi}\right), \tag{B.5}$$

$$(f, \varphi) = \int f(x)\,\bar{\varphi}(x)\,dx\,, \left(\widetilde{f}, \widetilde{\varphi}\right) = \int \widetilde{f}(k)\overline{\widetilde{\varphi}}(k)\,dx \tag{B.6}$$

Here, $\bar{\varphi}(x)$, $\overline{\widetilde{\varphi}}(k)$ are the complex conjugate function.

The Fourier transform $\widetilde{g}(k)$ and $\widetilde{G}(k)$ of Green's function follows from Eqs. (B.1), (B.2) in the form

$$\widetilde{g}(k) = \left(k_\alpha C^0_{\alpha\beta} k_\beta\right)^{-1}, \tag{B.7}$$

$$\widetilde{G}(k) = \left(L^0(k)\right)^{-1}, \quad L^0_{\alpha\beta}(k) = k_\lambda C^0_{\lambda\alpha\beta\mu} k_\mu \tag{B.8}$$

These functions are homogenous functions of k of the degree of -2.

For an isotropic conductive medium, $C^0_{\alpha\beta} = c_0\delta_{\alpha\beta}$, and

$$\widetilde{g}(k) = \frac{1}{c_0 k^2}, \; k^2 = k_\alpha k_\alpha. \tag{B.9}$$

For an isotropic elastic medium, $\widetilde{G}(k)$ is

$$\widetilde{G}_{\alpha\beta}(k) = \frac{1}{\mu_0 k^2}\left(\delta_{\alpha\beta} - \kappa_0 \frac{k_\alpha k_\beta}{k^2}\right), \quad \kappa_0 = \frac{\lambda_0 + \mu_0}{\lambda_0 + 2\mu_0}, \tag{B.10}$$

where λ_0, μ_0 are Lame's parameters of the medium.

Green's function for strains in homogeneous elastic media is defined in Eq. (3.15)

$$K_{\alpha\beta\lambda\mu}(x) = -\left.\nabla_\alpha \nabla_\lambda G_{\beta\mu}(x)\right|_{(\alpha\beta)(\lambda\mu)}. \tag{B.11}$$

The Fourier transform of this function has the form

$$\widetilde{K}_{\alpha\beta\lambda\mu}(k) = \left.k_\alpha k_\lambda \widetilde{G}_{\beta\mu}(k)\right|_{(\alpha\beta)(\lambda\mu)}, \tag{B.12}$$

For an isotropic medium, $\widetilde{K}(k)$ takes the form

$$\widetilde{K}(k) = \frac{1}{\mu_0}\left(E^5(m) - \kappa_0 E^6(m)\right), \quad m_\alpha = \frac{k_\alpha}{|k|}, \tag{B.13}$$

where $E^i(m)$ are the elements of the E-basis in Eq. (A.1).

Green's function $S(x)$ for stresses relates to the function $K(x)$ by Eq. (3.58)

$$S(x) = C^0 K(x) C^0 - C^0 \delta(x), \tag{B.14}$$

and the Fourier transform of $S(x)$ is

$$\widetilde{S}(k) = C^0 \widetilde{K}(k) C^0 - C^0. \tag{B.15}$$

For an isotropic medium, function $\widetilde{S}(k)$ is

$$\widetilde{S}(k) = -2\mu_0 \big(P^1(m) + (2\kappa_0 - 1) P^2(m)\big). \tag{B.16}$$

where $P^i(m)$ are the elements of the P-basis (A.10).

The function $\widetilde{K}(k)$ and $\widetilde{S}(k)$ are homogeneous functions of the zero degree, and therefore, the equations

$$\widetilde{K}(\lambda k) = \widetilde{K}(k), \quad \widetilde{S}(\lambda k) = \widetilde{S}(k) \tag{B.17}$$

hold for any scalar parameter λ.

B.2 Green's Functions of Static Elasticity in the x-Presentation

The x-presentations of Green's functions $G(x)$, $K(x)$ and $S(x)$ are the inversion Fourier transforms of the functions $\widetilde{G}(k)$, $\widetilde{K}(k)$ and $\widetilde{S}(k)$ in Eqs. (B.8), (B.11), (B.12). The table of the inverse Fourier transforms of generalized functions is presented in [2]; detailed derivations of these equations can be found in [3]. The inverse Fourier transforms of basic functions used for the construction of Green's functions in x-presentations are

$$F^{-1}\left(k^{-2}\right) = \frac{1}{4\pi|x|}, \quad F^{-1}\left(k^{-4}\right) = -\frac{|x|}{8\pi}, \tag{B.18}$$

$$F^{-1}\left(\frac{k_\alpha k_\beta}{k^2}\right) = -\frac{1}{4\pi}\nabla_\alpha \nabla_\beta \frac{1}{|x|} =$$

$$= \frac{1}{4\pi|x|^3}\left(\delta_{\alpha\beta} - 3n_\alpha n_\beta\right), \quad n = \frac{x}{|x|}, \tag{B.19}$$

$$F^{-1}\left(\frac{k_\alpha k_\beta}{k^4}\right) = \frac{1}{8\pi|x|}\left(\delta_{\alpha\beta} - n_\alpha n_\beta\right), \tag{B.20}$$

$$F^{-1}\left(\frac{k\otimes k\otimes k\otimes k\otimes k}{k^4}\right)=$$

$$=\frac{1}{8\pi|x|^3}\left(E^2+2E^1-3\left(E^3(n)+E^4(n)+4E^5(n)\right)+15E^6(n)\right). \tag{B.21}$$

Here, $E^i(n)$ are the elements of the E-basis (A.1). These equations and Eqs. (B.10), (B.13) and (B.16) result in the expressions for the x-presentation of Green's functions $G(x)$, $K(x)$ and $S(x)$ for isotropic media

$$G_{\alpha\beta}(x)=\frac{1}{8\pi\mu_0}\left(\frac{1}{|x|}\delta_{\alpha\beta}+\kappa_0\nabla_\alpha\nabla_\beta|x|\right)=$$

$$=\frac{1}{8\pi\mu_0|x|}\left((2-\kappa_0)\,\delta_{\alpha\beta}+\kappa_0 n_\alpha n_\beta\right), \tag{B.22}$$

$$K(x)=\frac{1}{4\pi\mu_0|x|^3}\Big(E^1-3E^5(n)-\frac{\kappa_0}{2}\big(E^2+2E^1-$$

$$-3\left(E^3(n)+E^4(n)+4E^5(n)\right)+15E^6(n)\big)\Big), \tag{B.23}$$

$$S(x)=$$

$$=\frac{\mu_0}{2\pi|x|^3}(1-\kappa_0)\left(2\left(E^1-E^2\right)+3\left(E^3(n)+E^4(n)-2E^5(n)\right)\right)+$$

$$+\kappa_0\left(E^2+6E^5(n)-15E^6(n)\right)\Big). \tag{B.24}$$

For anisotropic media, $G(x)$ is a homogeneous function of the degree (-1) integrable at $x=0$, $K(x)$ and $S(x)$ are homogeneous generalized functions of the degree (-3) with nonintegrated singularities of the order $|x|^{-3}$ at $x=0$. Actions of the generalized functions $K(x)$ and $S(x)$ on smooth functions $\varphi(x)$ with finite supports in R^3 are presented by formally divergent integrals. Regularization formulas for such functions are defined in [3]. For instance, the generalized function $K(x)$ acts on a finite smooth function $\varphi(x)$ by the rule

$$\int K(x)\varphi(x)\,dx=A(a)\,\varphi(0)+\lim_{v^\varepsilon\to 0}\int_{R^3\backslash v^\varepsilon}K(x)\varphi(x)\,dx, \tag{B.25}$$

$$A(a)=\lim_{v^\varepsilon\to 0}\int_{v^\varepsilon}K(x)\,dx. \tag{B.26}$$

Here, v^ε is an ellipsoidal region with the center at $x=0$, $R^3\backslash v^\varepsilon$ is complement of v^ε to the entire space. If a is the second rank tensor of linear transformation of x-space

($y = a^{-1}x$) that converts the ellipsoid v^{ε} in a unit ball, tensor $A\,(a)$ is expressed in terms of the Fourier transform $\widetilde{K}(k)$ of the function $K(x)$ (see Section 4.4)

$$A\,(a) = \frac{1}{4\pi}\int\limits_{\Omega_1} \widetilde{K}\left(a^{-1}k\right) d\Omega_k\,, \tag{B.27}$$

where Ω_1 is unit sphere in the k-space of the Fourier transforms. The integral in the right-hand side of Eq. (B.25) is expressed in terms of the Cauchy principal value (pv) of the integral

$$\lim_{v^{\varepsilon}\to 0}\int\limits_{R^3\backslash v^{\varepsilon}} K\,(x)\,\varphi\,(x)\,dx = (\det a)\;pv\int K\,(ay)\,\varphi\,(ay)\,dy\,, \tag{B.28}$$

$$pv\int f\,(y)\,dy = \lim_{\varepsilon\to 0}\int\limits_{|y|>\varepsilon} f\,(y)\,dy. \tag{B.29}$$

Similarly, the regularization formula of the generalized function $S(x)$ has the form

$$\int S\,(x)\,\varphi\,(x)\,dx = D\,(a)\,\varphi\,(0) + (\det a)\;pv\int S\,(ay)\,\varphi\,(ay)\,dy\,, \tag{B.30}$$

$$D\,(a) = \frac{1}{4\pi}\int\limits_{\Omega_1} \widetilde{S}\left(a^{-1}k\right) d\Omega_k\,. \tag{B.31}$$

Relation between $D(a)$ and $A(a)$ in Eq. (B.27) follows from Eq. (B.14) for $\widetilde{S}(k)$

$$D\,(a) = C^0 A\,(a)\,C^0 - C^0\,. \tag{B.32}$$

B.3 Green's Functions of Static Elasticity in 2D-Case

In the 2D-case, the Fourier transforms of Green's functions have the same forms as in Eqs. (B.8), (B.12) and (B.15) for the 3D-case, where Greek indices take the values 1, 2. For isotropic medium, the coefficient κ_0 in these equation is

$$\kappa_0 = \frac{1}{2\,(1-\nu_0)} \quad \text{(for plane strain state)}, \tag{B.33}$$

$$\kappa_0 = \frac{1+\nu_0}{2} \quad \text{(for plane stress)}. \tag{B.34}$$

In 2D-case, the inverse Fourier transform F^{-1} is defined by the equation

$$\left(F^{-1}\widetilde{f}\right)(x) = \frac{1}{(2\pi)^2}\int \widetilde{f}(k)\exp\left(-\mathrm{i}\,(k\cdot x)\right)dk, \tag{B.35}$$

In 2D-case, the inverse Fourier transforms in Eqs. (B.18)–(B.21) change for the following equations

$$F^{-1}\left(k^{-2}\right) = -\frac{1}{2\pi}\ln|x|\,,\quad F^{-1}\left(\frac{k_\alpha k_\beta}{k^2}\right) = \frac{1}{2\pi x^2}\left(\delta_{\alpha\beta} - 2n_\alpha n_\beta\right), \tag{B.36}$$

$$F^{-1}\left(k^{-4}\right) = \frac{x^2}{8\pi}\ln|x|\,,\quad F^{-1}\left(\frac{k_\alpha k_\beta}{k^4}\right) = -\frac{1}{4\pi}\left(\ln|x|\delta_{\alpha\beta} + n_\alpha n_\beta\right), \tag{B.37}$$

$$F^{-1}\left(\frac{k\otimes k\otimes k\otimes k}{k^4}\right) =$$

$$= \frac{1}{4\pi|x|^2}\left[3E^1 - \left(E^3(n) + E^4(n) + 10E^5(n)\right) + 8E^6(n)\right]. \tag{B.38}$$

Here, $E^i(n)$ are the elements of the E-basis in the 2D-case, $n = x/|x|$. Taking into account these equations, the expressions for Green's functions $G(x)$, $K(x)$, and $S(x)$ for 2D isotropic media are

$$G_{\alpha\beta}(x) = \frac{1}{4\pi\mu_0}\left((2-\kappa_0)\ln\frac{1}{|x|}\delta_{\alpha\beta} + \kappa_0 n_\alpha n_\beta\right), \tag{B.39}$$

$$K(x) = \frac{1}{2\pi\mu_0|x|^2}\left(E^1 - E^5(n) - \right.$$

$$\left. -\frac{\kappa_0}{2}\left(4E^1 - E^2 - 12E^5(n) + 8E^6(n)\right)\right), \tag{B.40}$$

$$S(x) = \frac{\mu_0\kappa_0}{\pi x^2}\left(E^2 + 4E^5(n) - 8E^6(n)\right). \tag{B.41}$$

Actions of the generalized functions $K(x)$ and $S(x)$ on finite functions in R^2 is defined in Eqs. (B.25), (B.30), where the tensors $A(a)$ and $D(a)$ are the integrals over a unit circle l_1 in the 2D k-space

$$A(a) = \frac{1}{2\pi}\int_{l_1}\widetilde{K}\left(a^{-1}k\right)dl_k\,,\quad D(a) = \frac{1}{2\pi}\int_{l_1}\widetilde{S}\left(a^{-1}k\right)dl_k\,. \tag{B.42}$$

Appendix C
Properties of Surface Potentials of Static Elasticity

C.1 Gauss' and Stokes' Integral Theorems

Let V be a bounded region in 3D-space with a piece-wise smooth surface Ω, and $n(x)$ is the external normal to Ω. If $A(x)$ is a tensor function with integrable first derivatives in R^3, the following relation holds [4]

$$\int_V \nabla_\alpha A_{\beta\lambda\ldots\mu}(x)\,dx = \int_\Omega n_\alpha(x)\,A_{\beta\lambda\ldots\mu}(x)\,d\Omega\,. \tag{C.1}$$

A particular form of this equation when $A(x)$ is a vector field is known as Gauss' theorem

$$\int \mathrm{div} A(x)\,dx = \int_\Omega n(x)\cdot A(x)\,d\Omega\,. \tag{C.2}$$

Let Ω be a smooth oriented surface (Lyapunov surface) bounded by a smooth closed contour Γ. The orientation of the surface is defined by a continuous normal vector $n(x)$ to Ω. The positive side of Ω is the side of the normal, the positive path-tracing of Γ is such as the positive side of Ω remains on the left-hand side; $\tau(x)$ is unit vector tangential to Γ and directed along the positive path-tracing along Γ. If $A(x)$ is a tensor field in R^3 with an integrable first derivative, Stokes' theorem for the surface is the following integral identity [4]

$$\int_\Omega \epsilon_{\alpha\beta\lambda}\nabla_\beta A_{\lambda\mu\ldots\nu}(x)\,n_\alpha(x)\,d\Omega = \int_\Gamma A_{\lambda\mu\ldots\nu}\tau_\lambda(x)\,d\Gamma\,. \tag{C.3}$$

where $\epsilon_{\alpha\beta\lambda}$ is the Levi-Civita tensor. The symbolic form of this identity is

S. Kanaun, *Self-Consistent Methods for Composites*, Solid Mechanics and Its Applications 282, https://doi.org/10.1007/978-3-031-96739-9

$$\int\limits_{\Omega} n(x) \cdot \mathrm{rot}\, A(x)\, d\Omega = \int\limits_{\Gamma} A(x) \cdot d, \quad d = \tau d\Gamma\,. \tag{C.4}$$

We introduce the projection operator $\theta(x)$ at each point x of the surface Ω

$$\theta_{\alpha\beta}(x) = \delta_{\alpha\beta} - n_{\alpha}(x)\, n_{\beta}(x)\,, x \in \Omega. \tag{C.5}$$

Vector field $a(x)$ belongs to Ω if it satisfies the equations

$$n_{\alpha}(x)\, a_{\alpha}(x) = 0, \quad \theta_{\alpha\beta}(x)\, a_{\beta}(x) = a_{\alpha}(x). \tag{C.6}$$

Similarly, tensor field $A(x)$ belongs to Ω if the following equation holds

$$\theta_{\alpha\beta}(x)\, \theta_{\lambda\mu}(x)....\theta_{\nu\rho}(x)\, A_{\beta\mu...\rho}(x) = A_{\alpha\lambda...\nu}(x). \tag{C.7}$$

The derivative ∂_{α} along the surface Ω is defined by the equation

$$\partial_{\alpha} = \nabla_{\alpha} - n_{\alpha}(x)\, n_{\beta}(x)\, \nabla_{\beta}, \tag{C.8}$$

where ∇ is the Nabla-operator in R^3. Then, the analogue of Gauss' theorem for surfaces can be formulated.

Let $n(x)$, $\tau(x)$, $e(x)$ be an orthogonal reference basis on the contour Γ—the boundary of a smooth surface Ω. Here, $n(x)$ is the limit value of the normal $n(x)$ to Ω on Γ, $\tau(x)$ is the positive path-tracing unit tangent vector to Γ, $e(x)$ is the unit normal vector to Γ in the plane tangent to Ω at point $x \in \Gamma$ and directed outside Γ. For an integrable with the first derivatives tensor field $A(x)$ that belongs to the surface Ω, the following analogue of Gauss' theorem (C.1) for surfaces holds [4, 5]:

$$\int\limits_{\Omega} \partial_{\alpha} A_{\alpha\beta\lambda...\mu}(x)\, d\Omega = \oint\limits_{\Gamma} e_{\alpha}(x)\, A_{\alpha\beta\lambda...\mu}(x)\, d\Gamma\,. \tag{C.9}$$

C.2 Derivatives of the Potential of the Simple Layer of Static Elasticity

Let Ω be a smooth oriented surface in a homogeneous elastic medium. The potential of the simple layer is the displacement vector $u(x)$ in the medium subjected to the force density $q(x)$ concentrated on Ω

$$u_{\alpha}(x) = \int\limits_{\Omega} G_{\alpha\beta}(x - x') q_{\beta}(x') d\Omega', \tag{C.10}$$

The potential $u(x)$ is continuous in the entire space for the integral in this equation has a weak singularity at $x' = x$ of the order $|x - x'|^{-1}$.

The strain tensor $\varepsilon(x)$ in the medium with the displacement field in Eq. (C.10) has the form

$$\varepsilon_{\alpha\beta}(x) = \int_{\Omega} \nabla_{(\alpha} G_{\beta)\gamma}(x - x') q_{\gamma}(x') d\Omega'. \tag{C.11}$$

Here, the integrand function has a non-integrable singularity of the order of $|x - x'|^{-2}$, and the potential $\varepsilon(x)$ is discontinuous on Ω. For revealing the jump of $\varepsilon(x)$, we present this potential as the sum

$$\varepsilon(x) = \int_{\Omega} \operatorname{def} G(x - x') \left(q(x') - q(x)\right) d\Omega' + \int_{\Omega} \operatorname{def} G(x - x') d\Omega' q(x) \tag{C.12}$$

and consider the limit of each term of this sum when $x \to \Omega$. Let $q(x)$ outside Ω be determined by an arbitrary smooth continuation. The first term in Eq. (C.12) is continuous on Ω as the integral with a weak singularity. The limiting value of the second integral

$$J_1(x) = \int_{\Omega} \operatorname{def} G(x - x') d\Omega' \tag{C.13}$$

is presented in the form

$$\lim_{x \to x_0 \in \Omega} J_1(x) = \int_{\Omega \setminus \Omega_\delta(x_0)} \operatorname{def} G(x - x') d\Omega' + \int_{\Omega_\delta(x_0)} \operatorname{def} G(x - x') d\Omega'. \tag{C.14}$$

Here, $\Omega_\delta(x_0)$ is a part of the surface Ω that is cut from Ω by a sphere of the radius δ centered at the point $x_0 \in \Omega$. Then, we introduce the local coordinate system y_1, y_2, y_3 with the origin at the point $x_0 \in \Omega$ and the y_3-axis directed along the normal $n(x_0)$ to Ω. In these coordinates, the second integral in Eq. (C.13) takes the form

$$\lim_{x \to x_0 \in \Omega} \int_{\Omega_\delta(x_0)} \operatorname{def} G(x - x') dx' = \lim_{y \to 0} \int_{\Omega_\delta(0)} \operatorname{def} G(y - y') d\Omega'. \tag{C.15}$$

In what follows, the parameter δ tends to zero. Therefore, $\Omega_\delta(0)$ can be considered as a planar disk $\Omega_\delta(0) = y_1^2 + y_2^2 < \delta^2, \;\; y_3 = 0$, and the second integral in Eq. (C.14) is

$$J^0(y) = \int_{\Omega_\delta(0)} \operatorname{def} G(y - y') d\Omega'. \tag{C.16}$$

Because def$G(y)$ is a homogeneous function of the order of -2, this integral does not depend on the radius δ of the disk $\Omega_\delta(0)$. It allows us to write

$$J^0(y) = \lim_{\delta\to\infty} \int\limits_{\Omega_\delta(0)} \mathrm{def}G(y-y')d\Omega' = \int \mathrm{def}G(y-y')\delta(y_3')dy', \tag{C.17}$$

where it is taken into account that $\Omega_\delta(0)$ in the limit $\delta \to \infty$ is transformed into the plane $y_3 = 0$; the last integral is calculated over the entire 3D-space. Using the convolution property and the expressions for the Fourier transforms of the integrand functions

$$F\left(\nabla_\alpha G_{\beta\gamma}(y)\right) = -\mathrm{i}k_\alpha \widetilde{G}_{\beta\lambda}(k), F(\delta(y_3)) = (2\pi)^2\delta(k_1)\delta(k_2), \tag{C.18}$$

the equation for $J^0(y)$ takes the form

$$J^0_{\alpha\beta\lambda}(y) = \frac{\mathrm{i}}{2\pi}\int k_{(\alpha}\widetilde{G}_{\beta)\lambda}(k)\delta(k_1)\delta(k_2)\exp(-\mathrm{i}\,(y\cdot k)\,)dk =$$

$$= \frac{1}{2}\delta_{3(\alpha}\widetilde{G}_{\beta)\lambda}(0,0,1)\ \mathrm{sign}\ (y_3)\ . \tag{C.19}$$

The invariant form of this equation is

$$J^0_{\alpha\beta\lambda}(y) = \frac{1}{2}n_{(\alpha}\widetilde{G}_{\beta)\gamma}(n)\ \mathrm{sign}\ (n\cdot y)\,, \tag{C.20}$$

where n is the normal to Ω at the point x_0.

Let δ turn to zero in Eq. (C.17). Taking into account Eqs. (C.17)–(C.20), we obtain the limits of the integral $J_1(x)$ (J_1^+ is the limit from the side of the external normal, J_1^- is such a limit from the opposed side) in the form

$$J_1^{\pm}(x_0) = pv\int_\Omega \mathrm{def}\ G(x_0-x')d\Omega' \pm \frac{1}{2}\Lambda(n^0)\,, \tag{C.21}$$

$$\Lambda_{\alpha\beta\gamma}(n^0) = n^0_\alpha \widetilde{G}_{\beta\gamma}(n^0)|_{(\alpha\beta)},\quad n^0 = n(x_0)\,. \tag{C.22}$$

Here, the Cauchy principal value of the integral exists because def $G(x)$ is an odd function of x. From this equation and Eq. (C.12) follows that the limit values of the potential (C.11) at point $x_0 \in \Omega$ are

$$\varepsilon^{\pm}(x_0) = pv\int\limits_\Omega \mathrm{def}G(x_0-x')q(x')d\Omega' \pm \frac{1}{2}\Lambda(n^0)q(x_0)\,. \tag{C.23}$$

The jump of this potential on Ω is determined by the equation

$$[\varepsilon_{\alpha\beta}(x_0)] = \Lambda_{(\alpha\beta)\lambda}(n^0)q_\lambda(x_0)\,. \tag{C.24}$$

Let us consider the potential of the double layer of static elasticity in Eqs. (3.42), (3.43)

$$u_\alpha(x) = -\int_\Omega \nabla_\lambda G_{\alpha\beta}(x - x')C_{\lambda\beta\mu\nu}n_\mu(x')b_\nu(x')d\Omega' . \tag{C.25}$$

The properties of this potential are similar to the properties of potential (C.11). The limit values of this potential on Ω are determined by the equation

$$u^\pm_\alpha(x_0) = -pv\int_\Omega \nabla_\lambda G_{\alpha\beta}(x_0 - x')C_{\beta\lambda\mu\nu}n_\mu(x')b_\nu(x')d\Omega' \pm \frac{1}{2}b_\alpha(x_0) . \tag{C.26}$$

Here, we take into account the equation

$$n_\lambda \widetilde{G}_{\alpha\beta}(n)C_{\beta\lambda\nu\mu}n_\mu = \delta_{\alpha\nu} , \tag{C.27}$$

that follows from the definition of $\widetilde{G}(k)$ (B.8).

Equations (C.23) and (C.26) hold for an arbitrary anisotropic medium. In particular, it follows from (C.26) that the jump of the potential of the double layer on the surface Ω is equal to the potential density $b(x)$.

C.3 The First Derivatives of the Double Layer Potential of Static Elasticity

Let Ω be a closed Lyapunov surface in R^3. We consider the potential

$$\varepsilon_{\alpha\beta}(x) = \int_\Omega K_{\alpha\beta\lambda\mu}(x - x')\, C^0_{\lambda\mu\nu\rho} n_\nu(x')\, b_\rho(x')\, d\Omega' , \tag{C.28}$$

$$K_{\alpha\beta\lambda\mu}(x) = -\nabla_\alpha\nabla_\lambda G_{\beta\mu}(x)\,|_{(\alpha\beta)(\lambda\mu)} , \tag{C.29}$$

where $G(x)$ is Green's function of static elasticity for a homogeneous medium with the elastic stiffness tensor C^0. The tensor $\varepsilon(x)$ in Eq. (C.28) is the symmetrized gradient of the vector function $u(x)$ that is the double layer potential of static elasticity in Eqs. (3.42), (3.43). For a continuous density $b(x)$, the fields $u(x)$ and $\varepsilon(x)$ are continuous everywhere except the surface Ω. The jump of the field $u(x)$ across Ω is defined in Eq. (C.26).

Consider integral (C.28) when $b = const$. Using Gauss' theorem for surfaces and taking into account that Green's function $G(x)$ satisfies the Eq. (B.2)

$$\nabla_\alpha C^0_{\alpha\beta\lambda\mu}\nabla_\lambda G_{\mu\nu}(x) = -\delta_{\beta\nu}\delta(x) , \tag{C.30}$$

we obtain

$$\varepsilon_{\alpha\beta}(x) = \int\limits_V \nabla'_\nu K_{\alpha\beta\lambda\mu}\left(x - x'\right) C^0_{\lambda\mu\nu\rho} dx' b_\rho =$$

$$= \int \nabla'_\alpha \delta\left(x - x'\right) V\left(x'\right) dx' b_\beta = n_\alpha(x)\, b_\beta \Omega(x)\ . \tag{C.31}$$

Here, $\Omega(x)$ is the delta-function concentrated on Ω; $n(x)$ is the external normal to Ω.

Then, let $b(x)$ in Eq. (C.30) be a linear function

$$b_\alpha(x) = D_{\alpha\beta}\left(x - x'\right)_\beta , \tag{C.32}$$

where D is a constant tensor. We consider the limiting values of the potential (C.28) with the density $b(x)$ in Eq. (C.32) when $x \to x_0 \in \Omega$. Let Ω_ρ be the part of the surface Ω located inside a sphere of a small radius ρ with the center at the point x_0, and $\overline{\Omega}_\rho$ be the complement of Ω_ρ to the whole surface Ω. Introducing the local Cartesian coordinate system y_1, y_2, y_3 with the origin at point x_0 and axis y_3 directed along the normal $n(x_0) = n_0$ to Ω, we present integral (C.28) in the form

$$\int\limits_\Omega K(y - y')\, C^0 n\left(y'\right) b\left(y'\right) d\Omega' =$$

$$= \int\limits_{\Omega_\rho} K(y - y')\, C^0 n\left(y'\right) D\left(y - y'\right) d\Omega' +$$

$$+ \int\limits_{\bar{\Omega}_\rho} K(y - y')\, C^0 n\left(y'\right) D\left(y - y'\right) d\Omega'. \tag{C.33}$$

Here, the integral over $\overline{\Omega}_\rho$ is a continuous function at point $y = 0$ $(x = x_0)$. For calculation of the limit of the integral over Ω_ρ when $y \to 0$, we take into account that in what follows $\rho \to 0$, and thus it is possible to consider Ω_ρ as a planar circular disk and $n(y') = n_0$. Introducing the variables $\xi_i = y_i/|y|$, we have

$$\lim_{\rho \to 0} \int\limits_{\Omega_\rho} K(y - y')\, C^0 n_0 D\left(y - y'\right) d\Omega' =$$

$$= \int K(\xi - \xi')\, C^0 n_0 D\left(\xi - \xi'\right) \delta\left(n_0 \cdot \xi'\right) d\xi , \tag{C.34}$$

where the integral in the right-hand side is calculated over the whole space R^3, and $K(x)$ is an even homogeneous function of degree -3. Then, the function $n_0 \delta(n_0 \cdot \xi')$ in Eq. (C.34) can be presented in the form

$$n_{0\alpha}\delta\left(n_0\cdot\xi\right)=\nabla_\alpha H\left(n_0\cdot\xi'\right),\quad H\left(t\right)=\begin{cases}1\,, & t>0\\ 0\,, & t<0\end{cases}, \tag{C.35}$$

and after integrating by parts in Eq. (C.34) we yield

$$\int K\left(\xi-\xi'\right)C^0 n_0 D\left(\xi-\xi'\right)\delta\left(n_0\cdot\xi'\right)d\xi'=$$

$$=-H\left(n_0\cdot\xi\right)D+\int K\left(\xi-\xi'\right)H\left(n_0\cdot\xi'\right)d\xi'^0 D\,. \tag{C.36}$$

Using the convolution properties of the integral on the right-hand side, we find

$$\int K\left(\xi-\xi'\right)H\left(n_0\cdot\xi'\right)d\xi'=(2\pi)^{-3}\int\widetilde{K}(k)\widetilde{H}(k)\,e^{-ik\cdot\xi}dk=$$

$$=\frac{1}{2}\big(\widetilde{K}(0)+\text{sign}\left(n_0\cdot\xi\right)\widetilde{K}(n)\big). \tag{C.37}$$

Here, it is taken into account that the Fourier transform of the function $H(n_0\cdot\xi)$ is

$$H^*\left(k\right)=H^*\left(k_1,k_2,k_3\right)=(2\pi)^2\,\delta\left(k_1\right)\delta\left(k_2\right)\left(\pi\delta\left(k_3\right)+ik_3^{-1}\right). \tag{C.38}$$

Let $J^+(n_0)$ be the limiting value of the potential in Eq. (C.34) when y tends to zero from the side of the normal n_0, and $J^-(n_0)$ be the same limit when $y\to 0$ from the opposed side. These limits are independent of the size of the region Ω_ρ and because of Eqs. (C.36) (C.37), they have the forms

$$J^+\left(n_0\right)=\frac{1}{2}\big(\widetilde{K}(0)+\widetilde{K}(n_0)\big)C^0 D-D\,, \tag{C.39}$$

$$J^-\left(n_0\right)=\frac{1}{2}\big(\widetilde{K}(0)-\widetilde{K}(n_0)\big)C^0 D\,. \tag{C.40}$$

While changing n_0 to $(-n_0)$, the integral (C.34) changes the sign, but its limiting values are defined by the same Eqs. (C.39), (C.40). As a result, the equation $J^+(n_0)=-J^-(n_0)$ holds, and it results in the equation $K^*(0)=(C^0)^{-1}$.

The limiting values of the potential (C.28) on Ω with the density in Eq. (C.32) follow from Eqs. (C.33), (C.39), (C.40). Letting ρ tend to zero in Eq. (C.33) $x\to\Omega$, we obtain the limiting values of the potential from the side of the normal $\left(\varepsilon^+\right)$ and from the opposed side $\left(\varepsilon^-\right)$ in the forms

$$\varepsilon^{\pm}_{\alpha\beta}\left(x\right)=pv\int\limits_{\Omega}K_{\alpha\beta\lambda\mu}(x-x')C^0_{\lambda\mu\nu\rho}n_\nu\left(x'\right)D_{\rho\delta}(x-x')_\delta d\Omega'\pm$$

$$\pm\frac{1}{2}\Lambda_{\alpha\beta\lambda\mu}\left(n\right)D_{\lambda\mu}\,,\quad \Lambda_{\alpha\beta\lambda\mu}\left(n\right)=\widetilde{K}_{\alpha\beta\nu\rho}\left(n\right)C^0_{\nu\rho\lambda\mu}-E^1_{\alpha\beta\lambda\mu}\,. \tag{C.41}$$

Here, $pv \int ...d\Omega'$ is the Cauchy principal value of the integral, and this integral exists because $K(x)$ is an even function.

Let $b(x)$ in Eq. (C.28) be a smooth function on Ω. The potential $\varepsilon(x)$ can be presented in the form

$$\varepsilon_{\alpha\beta}(x) =$$

$$= \int_{\Omega} K_{\alpha\beta\lambda\mu}(x - x') C^0_{\lambda\mu\nu\rho} n_\nu(x') (b_\rho(x') - b_\rho(x) - \partial_\tau b_\rho(x)(x' - x)_\tau) d\Omega' +$$

$$+ \int_{\Omega} K_{\alpha\beta\lambda\mu}(x - x') C^0_{\lambda\mu\nu\rho} n_\nu(x') d\Omega' b_\rho(x) +$$

$$+ \int_{\Omega} K_{\alpha\beta\lambda\mu}(x - x') C^0_{\lambda\mu\nu\rho} n_\nu(x') \partial_\tau b_\rho(x)(x' - x)_\tau d\Omega'. \tag{C.42}$$

Here, the first integral converges absolutely, and it is a continuous function in the entire space. When $x \in \Omega$, this integral coincides with its Cauchy principal value, and it can be presented in the form of the sum of two Cauchy's principal values of integrals with the densities $b(x') - b(x)$ an $\partial b(x)(x' - x)$, correspondingly. Each of these integrals exists in the mentioned sense. The second integral in Eq. (C.42) is the generalized function in Eq. (C.31), and the last integral is the potential in Eq. (C.33) with the density in Eq. (C.32). The jumps of this potential on Ω are defined in Eq. (C.41). It follows from these equations that the limit values of the potential $\varepsilon(x)$ in Eq. (C.28) on Ω have the forms ($x \in \Omega$)

$$\varepsilon^{\pm}_{\alpha\beta}(x) = pv \int_{\Omega} K_{\alpha\beta\lambda\mu}(x - x') C^0_{\lambda\mu\nu\rho} n_\nu(x') \times$$

$$\times (b_\rho(x') - b_\rho(x)) d\Omega \mp \frac{1}{2} \Lambda_{\alpha\beta\lambda\mu}(n) \partial_\lambda b_\mu(x), \tag{C.43}$$

where the tensor $\Lambda(n)$ is defined in Eq. (C.41).

Let us consider the potential $\sigma(x)$

$$\sigma_{\alpha\beta}(x) = \int_{\Omega} S_{\alpha\beta\lambda\mu}(x - x') n_\lambda(x') b_\mu(x') d\Omega', \tag{C.44}$$

$$S_{\alpha\beta\lambda\mu}(x) = C^0_{\alpha\beta\nu\rho} K_{\nu\rho\tau\delta}(x) C^0_{\tau\delta\lambda\mu} - C^0_{\alpha\beta\lambda\mu} \delta(x), \tag{C.45}$$

where Ω is a closed surface in R^3. The potentials $\varepsilon(x)$ in Eq. (C.28) and $\sigma(x)$ in Eq. (C.44) relate by the equation

$$\sigma_{\alpha\beta}(x) = C^0_{\alpha\beta\lambda\mu} \varepsilon_{\lambda\mu}(x) - C^0_{\alpha\beta\lambda\mu} n_\lambda(x) b_\mu(x) \Omega(x). \tag{C.46}$$

From Eq. (C.43) follow the limiting values of $\sigma_{\alpha\beta}(x)$ on Ω

$$\sigma^{\pm}_{\alpha\beta}(x) = pv\int_{\Omega} S_{\alpha\beta\lambda\mu}\left(x - x'\right) n_{\lambda}\left(x'\right)\left(b_{\mu}\left(x'\right) - b_{\mu}(x)\right) d\Omega' \mp$$

$$\mp \frac{1}{2}\widetilde{S}_{\alpha\beta\lambda\mu}(n)\,\partial_{\lambda} b_{\mu}(x), \quad x \in \Omega, \ . \tag{C.47}$$

$$\widetilde{S}(n) = C^{0}\widetilde{K}(n)\,C^{0} - C^{0}. \tag{C.48}$$

Because of Eq. (B.12) for $\widetilde{K}(n)$, $n_{\alpha}\widetilde{S}_{\alpha\beta\lambda\mu}(n) = 0$, and vector $n_{\alpha}\sigma_{\alpha\beta}(x)$ is continuous on the surface Ω. The equation for the vector $n_{\alpha}(x)\sigma_{\alpha\beta}(x)$ on Ω takes the form

$$n_{\alpha}(x)\,\sigma_{\alpha\beta}(x) = -pv\int_{\Omega} T_{\alpha\beta}\left(x, x'\right)\left(b_{\beta}\left(x'\right) - b_{\beta}(x)\right) d\Omega', \tag{C.49}$$

$$T_{\alpha\beta}\left(x, x'\right) = -n_{\lambda}(x)\,S_{\lambda\alpha\beta\mu}\left(x - x'\right) n_{\mu}\left(x'\right). \tag{C.50}$$

If the surface Ω is not closed, integration in Eqs. (C.43), (C.47) and (C.49) can be performed over any smooth closed surface that includes Ω by taking $b(x) = 0$ outside Ω. Let $\overline{\Omega}$ be the complement of Ω up to an arbitrary closed surface. Taking into account Eq. (3.53) for $S(x)$

$$S_{\alpha\beta\lambda\mu}(x) = \mathrm{rot}_{\nu\lambda}\mathrm{rot}_{\rho\mu} Z_{\alpha\beta\nu\rho}(x), \tag{C.51}$$

where $Z(x)$ is Green's tensor for internal stresses, and Stokes' theorem (C.9), we obtain the equation

$$\int_{\bar{\Omega}} S_{\alpha\beta\lambda\mu}\left(x - x'\right) n_{\mu}\left(x'\right) d\Omega' = \oint_{\Gamma} \mathrm{rot}_{\nu\lambda} Z_{\alpha\beta\nu\mu}\left(x - x'\right) d\Gamma'_{\mu}, \tag{C.52}$$

where $d\Gamma_{\mu}$ is the vector element on the border Γ of $\bar{\Omega}$. Hence, in the case of a non-closed surface, Eq. (C.47) can be written in the form

$$\sigma^{\pm}_{\alpha\beta}(x) = pv\int_{\Omega} S_{\alpha\beta\lambda\mu}\left(x - x'\right) n_{\lambda}\left(x'\right)\left(b_{\mu}\left(x'\right) - b_{\mu}(x)\right) d\Omega' +$$

$$+ \oint_{\Gamma} \mathrm{rot}_{\nu\lambda} Z_{\alpha\beta\nu\mu}\left(x - x'\right) d\Gamma'_{\lambda} b_{\mu}(x) \mp \frac{1}{2}\widetilde{S}_{\alpha\beta\lambda\mu}(n)\,\partial_{\lambda} b_{\mu}(x). \tag{C.53}$$

It follows from this equation and Eq. (C.43) that for a unclosed surface Ω, the limit values of potential (C.28) at point $x \in \Omega$ have the form

$$\varepsilon^{\pm}_{\alpha\beta}(x) = pv\int\limits_{\Omega} K_{\alpha\beta\lambda\mu}\left(x - x'\right) C^0_{\lambda\mu\nu\rho} n_\nu\left(x'\right)\left[b_\rho\left(x'\right) - b_\rho(x)\right] d\Omega' +$$

$$+\left(C^0\right)^{-1}_{\alpha\beta\lambda\mu}\oint\limits_{\Gamma} \mathrm{rot}_{\nu\delta} Z_{\lambda\mu\nu\tau}\left(x - x'\right) d\Gamma'_\delta b_\tau(x) \mp \frac{1}{2}\Lambda_{\alpha\beta\lambda\mu}(n)\,\partial_\lambda b_\mu(x)\,. \quad \text{(C.54)}$$

The equation for the vector field $n_\alpha(x)\sigma_{\alpha\beta}(x)$ on Ω, where $\sigma(x)$ is the potential in Eq. (C.44), has the form

$$n_\beta(x)\,\sigma_{\beta\alpha}(x) = -pv\int\limits_{\Omega} T_{\alpha\beta}\left(x, x'\right)\left(b_\beta\left(x'\right) - b_\beta(x)\right) d\Omega' +$$

$$+\oint\limits_{\Gamma} n_\beta(x)\,\mathrm{rot}_{\nu\lambda} Z_{\alpha\beta\nu\mu}\left(x - x'\right) d\Gamma'_\lambda b_\mu(x)\,. \quad \text{(C.55)}$$

C.4 Surface Potential Which Density Belongs to the Surface

We consider the potential

$$\varepsilon_{\alpha\beta}(x) = \int\limits_{\Omega} K_{\alpha\beta\lambda\mu}\left(x - x'\right) q_{\lambda\mu}\left(x'\right) d\Omega' \quad \text{(C.56)}$$

which density $q(x)$ belongs to surface Ω

$$n_\alpha(x) q_{\alpha\beta}(x) = 0\,,\quad \Theta_{\alpha\beta\lambda\mu}(x)\, q_{\lambda\mu}(x) = q_{\alpha\beta}(x)\,,\quad x \in \Omega\,. \quad \text{(C.57)}$$

Here, $n = n(x)$ is normal to Ω, and $\Theta(x)$ is the projector on the tangent plane to Ω

$$\Theta(x) = E^1 - 2E^5(n) + E^6(n) = P^1(n). \quad \text{(C.58)}$$

$E^i(n)$ are the elements of the E-basis in Eq. (A.1).

Using Gauss' theorem for surfaces (C.9) we obtain

$$\varepsilon_{\alpha\beta} = -\int\limits_{\Omega} \nabla'_\lambda \nabla'_\alpha G_{\beta\mu}\left(x - x'\right)\Theta_{\lambda\mu\nu\rho}\left(x'\right) q_{\nu\rho}\left(x'\right) d\Omega' =$$

$$= -\int\limits_{\Omega} \partial'_\lambda\left(\nabla'_\alpha G_{\beta\mu}\left(x - x'\right) q_{\lambda\mu}\left(x'\right)\right) d\Omega' +$$

$$+\int\limits_{\Omega} \nabla'_\alpha G_{\beta\mu}\left(x - x'\right)\partial'_\lambda q_{\lambda\mu}\left(x'\right) d\Omega' =$$

$$= \int_{\Omega} \nabla'_{\alpha} G_{\beta\mu}(x - x') \, \partial'_{\lambda} q_{\lambda\mu}(x') \, d\Omega' -$$

$$- \oint_{\Gamma} \nabla'_{\alpha} G_{\beta\mu}(x - x') \, q_{\lambda\mu}(x') \, e_{\lambda}(x') \, d\Gamma'. \tag{C.59}$$

Here, $e_{\lambda}(x)$ is the normal vector to Γ, and the identity

$$\Theta_{\alpha\beta\lambda\mu}(x)\nabla_{\lambda}A_{\mu\ldots\nu}(x) = \Theta_{\alpha\beta\lambda\mu}(x)\partial_{\lambda}A_{\mu\ldots\nu}(x), \tag{C.60}$$

is taking into account, where ∂_{α} is the gradient operator along the surface in Eq. (C.8). The first integral in Eq. (C.59) is the potential in Eq. (C.11) and its limiting values on Ω are defined in Eq. (C.23). Hence the limiting values of the potential (C.56) are

$$\varepsilon^{\pm}_{\alpha\beta}(x) = pv \int_{\Omega} \nabla'_{(\alpha} G_{\beta)\mu}(x - x') \, \partial'_{\lambda} q_{\lambda\mu}(x') \, d\Omega' -$$

$$- \oint_{\Gamma} \nabla'_{(\alpha} G_{\beta)\mu}(x - x') \, q_{\lambda\mu}(x') \, e_{\lambda}(x') \, d\Gamma' \pm \frac{1}{2} \Lambda^{0}_{(\alpha\beta)\mu}(n) \, \partial_{\lambda} q_{\lambda\mu}(x), \tag{C.61}$$

$$\Lambda^{0}_{\alpha\beta\mu}(n) = n_{\alpha} G^{*}_{\beta\mu}(n), \quad n = n(x), \quad x \in \Omega\,. \tag{C.62}$$

The principal value of the integral in this equation can be presented in the form

$$pv \int_{\Omega} \nabla'_{\alpha} G_{\beta\mu}(x - x') \, \partial'_{\lambda} q_{\lambda\mu}(x') \, d\Omega' =$$

$$= pv \int_{\Omega} \nabla'_{\alpha} G_{\beta\mu}(x - x') \, \partial'_{\lambda}(q_{\lambda\mu}(x') - q^{0}_{\lambda\mu}(x, x'))d\Omega', \tag{C.63}$$

where $q^{0}(x, x')$ is a tensor field on Ω that satisfies the equations

$$\partial'_{\lambda} q^{0}_{\lambda\mu}(x, x') = 0, \; q^{0}_{\lambda\mu}(x, x) = q_{\lambda\mu}(x), \quad x, x' \in \Omega - \tag{C.64}$$

Using Gauss' theorem for surfaces (C.9), the integral (C.63) is transformed as follows

$$pv \int_{\Omega} \nabla'_{\alpha} G_{\beta\mu}(x - x') \, \partial'_{\lambda}(q_{\lambda\mu}(x') - q^{0}_{\lambda\mu}(x, x'))d\Omega' =$$

$$= -pv \int_{\Omega} \partial'_{\lambda} (\nabla'_{\alpha} G_{\beta\mu}(x - x')) \, (q_{\lambda\mu}(x') - q^{0}_{\lambda\mu}(x, x'))d\Omega' +$$

$$+\oint_{\Gamma} e_{\lambda}\left(x'\right) \nabla'_{\alpha} G_{\beta\mu}\left(x-x'\right)\left(q_{\lambda\mu}\left(x'\right)-q^{0}_{\lambda\mu}\left(x, x'\right)\right) d\Omega', \tag{C.65}$$

where $e_{\lambda}(x)$ is the external normal to Γ in the tangent plane to Ω. The regular presentation of the limiting values of the potential (C.56) at points of surface Ω follows from this equation in the form

$$\varepsilon^{\pm}_{\alpha\beta}(x)=pv\int_{\Omega} K_{\alpha\beta\lambda\mu}\left(x-x'\right)\left(q_{\lambda\mu}\left(x'\right)-q^{0}_{\lambda\mu}\left(x, x'\right)\right) d\Omega' -$$

$$-\oint_{\Gamma} \nabla'_{(\alpha} G_{\beta)\mu}\left(x-x'\right) q^{0}_{\lambda\mu}\left(x, x'\right) e_{\lambda}\left(x'\right) d\Gamma' \pm \frac{1}{2} \Lambda^{0}_{\alpha\beta\mu}(n)\, \partial_{\lambda} q_{\lambda\mu}(x). \tag{C.66}$$

It follows from the definition of the tensor Λ^{0} in Eq. (C.62) that the "tangential" part $\Theta(x)\varepsilon(x)$ of the tensor field $\varepsilon(x)$ is continuous across Ω, and it is determined on Ω by the equation

$$\Theta_{\alpha\beta\lambda\mu}(x)\, \varepsilon_{\lambda\mu}(x)=pv\int_{\Omega} U_{\alpha\beta\lambda\mu}\left(x, x'\right)\left(q_{\lambda\mu}\left(x'\right)-q^{0}_{\lambda\mu}\left(x, x'\right)\right) d\Omega' -$$

$$-\Theta_{\alpha\beta\lambda\mu}(x) \oint_{\Gamma} \nabla_{\lambda} G_{\mu\nu}\left(x-x'\right) q^{0}_{\rho\nu}\left(x, x'\right) e_{\rho}\left(x'\right) d\Gamma, \tag{C.67}$$

$$U_{\alpha\beta\lambda\mu}\left(x, x'\right)=\Theta_{\alpha\beta\nu\rho}(x)\, K_{\nu\rho\tau\delta}\left(x-x'\right) \Theta_{\tau\delta\lambda\mu}\left(x'\right). \tag{C.68}$$

Appendix D
The Percus-Yevick Correlation Function of Non-penetrating Sets of Spheres

The Percus-Yevick correlation function $\psi(x)$ is the probability density of positions of the centers of non-overlapping spheres of radii a by the conditions that the origin ($x = 0$) is occupied by the center of such a sphere. In the 3D-space, this function is defined by the equations [6]

$$\psi(\zeta) = 1 + \frac{1}{12\pi p\zeta}\int_0^\infty \frac{w^2(s)}{1 - w(s)}\sin\left(\frac{s\zeta}{2}\right) s ds \ \text{if} \ \zeta \geq 2, \tag{D.1}$$

$$\psi(\zeta) = 0, \ \text{if} \ \ \zeta < 2, \qquad \zeta = \frac{|x|}{a}.$$

Here, $w(s)$ is the following integral

$$w(s) = \frac{24p}{s}\int_0^1 q(t)\sin(ts)t dt, \quad q(t) = -a_1(1 + p\frac{t^3}{2}) - a_2 t, \tag{D.2}$$

$$a_1 = \frac{(1+2p)^2}{(1-p)^4}, \ a_2 = \frac{-3p(2+p)^2}{2(1-p)^4}, \tag{D.3}$$

p is the volume fraction of spheres.

In the 2D-case, the Percus-Yevick correlation function is presented in the form

$$\psi(\zeta) = 1 + \frac{1}{8p}\int_0^\infty h(x)J_0\left(\frac{\zeta x}{2}\right) x dx, \ |\zeta| \geq 2; \tag{D.4}$$

$$\psi(\zeta) = 0, \ \ |\zeta| < 2.$$

S. Kanaun, *Self-Consistent Methods for Composites*, Solid Mechanics and Its Applications 282, https://doi.org/10.1007/978-3-031-96739-9

Here, $J_0(z)$ is the Bessel function of order of 0 and function $h(x)$ is

$$h(x) = \frac{s(x)^2}{1 - s(x)}. \tag{D.5}$$

The function $s(x)$ is the solution of the non-linear integral equation

$$s(x) = -8p\frac{J_1(x)}{x} - \frac{1}{2}\int_0^\infty \frac{s(t)^2}{1 - s(t)} \left(J_0^2(t) + J_1^2(t)\right) dt. \tag{D.6}$$

References

1. Nye J (1985) Physical properties of crystals, their representation by tensors and matrices. Clarendon Press, Oxford, p 329
2. Brychkov Y, Prudnikov A (1989) Integral transforms of generalized functions. CRC Press, 344
3. Gelfand I, Shilov G (1964) Generalized functions. Academic, New York, p 470
4. McConnell A (1957) Application of tensor analysis. Dover, New York, p 411
5. Vekua I (1978) Tensor analysis and the theory of covariants. Nauka, Moscow, p 296
6. Wertheim M (1963) Exact solution of the Perkus-Yevick integral equation for hard spheres. Phys Rev Lett 10:321–323

Index

S. Kanaun, *Self-Consistent Methods for Composites*, Solid Mechanics and Its Applications 282, https://doi.org/10.1007/978-3-031-96739-9

MIX
Papier aus verantwortungsvollen Quellen
Paper from responsible sources
FSC® C105338

If you have any concerns about our products,
you can contact us on
ProductSafety@springernature.com

In case Publisher is established outside the EU,
the EU authorized representative is:
Springer Nature Customer Service Center GmbH
Europaplatz 3, 69115 Heidelberg, Germany

Printed by Libri Plureos GmbH
in Hamburg, Germany